Kril


Michael A Jeffords

Dedication:

To my wife, Myra, who lit up my dark heart
with the bright flame of her love over four
decades. To Shawn and Craig, our children
becoming men. The three of them accepting
who I was, never complaining over the years we
were together. Supporting a daily raging,
broken-in-mind survivor, with unconditional love.

# Chapter 1

"The intuitive mind is a sacred gift, and the rational mind is a faithful servant. We have created a society that honors the servant and has forgotten the gift. It is entirely possible that behind the perception of our senses, worlds are hidden of which we are unaware."
Albert Einstein

## Omega Ending

I want to finally die this time. The ninety-five-year-old man was on his hospital bed; mind-floating in the dark. His morphine-drip causing in and out flashes of consciousness.

Flipping montages in his mind of his four children, his doctors, the attending nurses, and empty silent room events. Morphine numbed his pain, enhancing the only things he had left; his memories.

Having experienced multiple timelines, in his case life-ending was not final. How many others survived multiple-lifetimes? Not remembering those other existences, except at night recalling indecipherable dreams and nightmares.

The human race had elementary tunneling Bosons in their body's brain acquired from multitudes of long dead ancestors. Quantum jumps, spookiness, and tunneling waves containing information from those individuals existing long ago in the past. Interacting by 'spooky action'. A constant instantaneous synchronization with the universe.

Each succeeding timeline of life he experienced was uniquely different; yet similar. His

3

conscious mind forgetting past details; except in incomprehensible dreams during the night. His Boson-information, through quantum spookiness did not forget. His three living timelines became unreal. Each of his lives accumulated information.

His first life was painful and normal. While his second life was hazily physcotic. This third life was a combination of daily rages spiced with days of loving happiness.

The dying old man's mind was wandering. The Seech Field, starting his second life, was Oval Science devised. Inaccurate to name the Seech Field a Time Machine; it was not. To the Ovals, interdimensional Seeching was a fishing expedition, collecting the dead.

Never would or could explain to anyone my experiences. My second life began after two decades in my first lifetime. Field effects from Seech Fields and Osil-drives aren't unknown to me. The boson path is random, and I never knew the destination.

His Zabin-Kril brothers and sisters, who survived the Seech Field entering the Oval Universe, began another lifetime. With tens of trillions of Boson-informational memories.

The dying ninety-five-year-old, in his drifting-morphine coma, recalled his fellow Zabin-Kril's Earth existences.

The Roman in 9 CE, the exact date unknowable, Osnabruck County, Lower Saxony

The aqila incident caused Primus consternation when he pointed the bird out to his Centurion, Didius Flavus. An hour ago, Didius had observed the same eagle soaring with its mate over their unit. Both saw an arrow in the male's flank.

4

Then they lost the eagles in the clouds. Portending a bad omen of defeat; the wounded eagle flying away from their anticipated battleground. His friend and leader Didius had kept his silence. Primus could not be certain; omens of the gods were not within his skills.

They watched scudding clouds, touch the top of the trees, black and rolling from the advancing storm.

Ominously the clouds internally grumbling, the god Jupiter Optimus Maximus, throwing thunderbolts to Mars Cocidius. Primus hoped the low rumbling thunder sounds would aid the patrol in their mission. He was full of introspection during the pre-storm atmosphere. He witnessed the patrol lying on the damp dead-leaf-blanket upon the black earth watching the forest terrain in front of them. They were far beyond the Castrum unit encampment. But, better than stercus duty cleaning the latrines.

Far off crashing-thunder in the distance was ominous and increasing in fury and substance.

Gargilius Primus's gaze following the path of their superior behind them; stumbling through the forest. Normally, a Principale commissioned officer like Marcellus, would not be here. Primus mused. But the man could read, write, and his count abilities were valuable in assessing the strength of the enemy. This officer could send a runner with a written parchment of his assessment. Most legionnaires could not write. The officer Marcellus in his frustration constantly berated Primus and the other men in the patrol. Thanks to the gods, Marcellus reserved his abundant hate for Didius, Primus's friend.

This patrol from the Tenth Cohort of the Seventeenth Legion, was actually led by his friend

5

Didius; not Marcellus. Searching for Cherusci barbarians, a Teutonic tribe led by Arminius. Arminius served five years in the Roman Legions learning valuable lessons in warfare and earning Roman citizenship. A few men in this XVII Legion remembered serving in past campaigns with the barbarian leader.

"Is Marcellus our magnificent leader, still complaining?" Didius Flavus whispered to Primus, crouching and surveying the area in front of them. Flavus smiled with derision. "Marcellus left his family business, his wife and children, for adventure and glory." He mock-whimpered. Abruptly, his voice became sharp-edged. "Principale Marcellus is no better than a Mile, a common recruit, striving to be appointed Tribunus-laticlavius, and command our Calvary and Legion. Mark me. I will not assist his ambitions."

His friend Didius, was inflamed about Marcellus for good reason. Didius was three times champion in single combat with twenty-three scars, on the front of his body; not on his back. With over ten years of service in the Legions and forty battles, his friend held decorations of thirteen roman torques, eighteen phalerae, and thirty-five armillae.

He should have earned the highest award in Rome, the corona graminea. For being the first man over the ramparts at Gullacia. Only eight men earned graminea in the past, Augustus the emperor, being the last one. Didius held the rank of Centurion. For political reasons and family lineage, regrettably the prestigious device was not awarded to him.

Repeat flashes of lightening made them blink and hunch their shoulders anticipating the physical concussion waves and deep rumbles. It lasted minutes, first one crash then another.

Didius Flavus, before he was born, the servile wars had started in his home city of Capua, where the infamous slave revolt of the gladiator Spartacus started. Didius never spoke of Spartacus, avoiding any references to the slave revolt. Happening many years in Didius's past; the fact continued to embarrass him out of regional sensitivity and the Legion's long memory. He was a Roman Freeman he reasoned, not a slave. Much as every man in the Legion was; he was ambitious. Didius coveted higher rank than Centurion; but his family and fortune could not purchase him the rank of Tribune.

"Early upon this morning, Primus, I looked with disgust at the tears in his eyes when he wailed about his rent tunic."

Primus grinned. "More likely, less because of the damage, than the stoppage put on his pay to replace the garment."

They both softly laughed.

Didius Flavus said, "He whines mightily when the deduction from the fifteen AS's paid to us daily is made to the burial club." Didius sobered quickly. "Be still now, back our leader comes." Then Didius Flavus growled. "If he does not stop tripping over weeds and brushes, giving away our positions, I will drive a pilum through him and send him to Elysium."

A tremendous lightning bolt, followed immediately by a crash of thunder, annihilated a tree on top of a nearby hill.

Marcellus approached them, prudently whispering. "Retreat the patrol. You cannot capture the barbarians in the ravine. I have counted the size of their force, their dispositions, and will report to our new Camp Legate. They outnumber us. We should withdraw." He looked off into the distance. "The new Legate's name is part of these mists at the

7

moment. I hope I am able to get through the turves and the twelve-foot ditch when I go back to our camp."

"That seems wise." Didius Flavus said.

The irony in the voice of Didius was not lost on Marcellus. "I should have been immune from this duty. I am here because of the new Legate taking command. I know the numbers and am literate."

"I am commissioned by the Legate to keep you from harm." Didius said. "Do not you trust my expertise?"

Marcellus ignored him. "Only I count the barbarians and can explain to our Cohort leaders the situation. Only yesterday the Cornicularius, stated he will make me an equal to him with the title of libarius-legionis. I should be within our fortifications. The new Legate countermanded the Cornicularius. The Legate assured me I am with this patrol only to observe." He licked his lips nervously. "My duty is clear."

Flavus did not respond. He turned his back to Marcellus, in an obvious rebuff, moving closer to the top of the hill to observe the barbarians below them.

Primus Gargilius hand-signaled the rest of the patrol forward, while whispering in Marcellus's ear. "Flavus advises you do not leave until we spring the ambush. By the gods, you do not want to give alarm to the barbarians and lose the decorations you will surely earn."

That made much sense to Marcellus but he was tense; his legs quivering in anticipation. "My duty is to return immediately to report." He quietly squeaked. He groaned inside at the slowness of his fellow legionnaires as they moved into position.

Gargilius lie next to Didius, as his friend whispered in his ear. "Position your men along the top of this ridge. Stay hidden, no alarms or movement. Or by the gods, afterwards I will flail the offender." Flavus glanced over to Marcellus. "I thank the god Mars our brave scribe will not participate with neither counsel nor command. We will proceed in the attack thusly."

Drawing their plan of attack on a bare area of black dirt Didius made it appear simple, effective, and deadly, though they were outnumbered by the barbarians by three to one. Surprise was to be their ally.

Didius smiled, with resignation in his whisper. "My friend, once again we risk ourselves hoping the gods let us retain our arms and legs and heads. Breathe deep fellow, we have tiny minutes more of peace or an eternity of war with Mars."

His mood abruptly changed. "Use the slinger first. The slinger, a young man well practiced, flinging the hard missile to the mark with consistency. The young man was also practiced in throwing a one-pound rock by hand, with deadly force, and to accurately find the target. Both fearing the noise of the fluttering sling would prematurely alert a knowledgeable barbarian; the stone was to be loosed first, as Primus instructed him. The slinger was to act after Flavus pointed his pilum to begin the ambush.

"Loose the pilums immediately. Close in quickly with short sword. No hesitation. No slashing, use only kill thrusts. Mark our legionnaires though; I want prisoners. Tell them not to forget. The Legionnaire not in immediate battle must...'he must'...control the others locked in mortal struggle."

He grasped Gargilius by his forearm. "To the gods, glory, and friends, Primus.

Gargilius repeated. "Gods…glory…and friends." He was striving mightily to dampen down his rising fear.

"Now take your position." Didius motioned his head towards Marcellus. "Remember that not before the signal must anyone move triggering the ambush; only upon my command."

Forward of his Legionnaire's, Didius's hair was billowing and twisting in the rising wind. Thank the gods for the storm.

Primus silently set the men in attack formation above the barbarians. The ticking minutes felt like an hour in their stealth; not a buckle trapped a bush, or helmet came off clanging to the ground. If a mistake happened, he knew it would occur in those brief moments during this storm when the wind ceased, when the thunder was silent, and the forest held its breath. Even veteran men, he knew made stupid errors at such times.

Marcellus was watched closely by Primus during this period of time; his anxiety palpable. Their eyes locked. Stay, fellow, do not force me to widow your good wife. Marcellus accurately interpreted the silent message.

Moving silently off the crest of the hill and grabbing Marcellus's shoulder, in a whisper Didius Flavus said. "If you make noise or bolt prematurely, a sword will cut your throat. Your silent screams will bubble away with the mists. Remain silent and immobile. Stay fellow, until we attack. Then you may run through Hades reporting to your Legate."

Marcellus glanced down to the earth. He looked to retch as if in a public bathhouse.

Flavus moved back to his commanding position.

10

Above them, over the treetops the black rolling clouds were flickering with lantern fire from the gods. With approval, Primus witnessed his men moving or adjusting their positions only when the thunder crashed around them. The fierceness of the storm accelerated moment by moment.

Below them, the Cherusci barbarians were cooking antelopes and assorted small game over wooden spits, ignoring the storm gods. Dressed in animal skins and heavily bearded; they were a fierce-looking lot. Primus Gargilius credited them their fighting skills. Underestimating any enemy was an excellent way to move into the underworld. Any veteran of even one vicious hand to hand battle with them concurred. Only non-combatants made sport of a barbarian's prowess; not the men who faced them.

These barbarians lived as creatures of the forest. Squatting anywhere to soil their camp and pissing indiscriminately. Swilling like sows in the same area and sleeping with no order. Walking away from their camp in the morn, to build another camp as haphazard as the first; or none if the mood struck them.

Primus knew his enemy and their women. Ho, what a disgusting thought. It was well documented; the skinny barbarian women were ugly as a hunting bat in the night. Though a good rape never bothered a needy legionnaire.

Primus Gargilius watched Didius Flavus place his battle helmet on his head securely and raise his pilum to get the attention of the slinger, Gargilius, and the Legionnaires. Expectation was making their hearts throb. Didius kept his weapon high. Moments felt ominously like long minutes.

What is he doing? Why is he grinning? Primus thought.

Gargilius spotted Marcellus and knew Flavus was toying with the frightened man. In anger, Primus fiercely gestured to have done with it! Command the ambush!

On the adjoining hill a lightning bolt and blast were as one, matching Didius' demeanor. His friend was grinning; maliciously squeezing Marcellus's nerves from afar.

Two things occurred at once, surprising Primus.

First, Marcellus bolted from the ambush screaming loudly out. "I go to report! I go to report! I…the legate must know…happening …at…how many…! Parts of his sentences drowned out by multiple thunder.

The barbarians below bolted, scattering into the forest.

A second after Marcellus ran off, the battle tense legionnaires felt the hairs on their bodies from their head to their scrotum rise stiffly as if the gods touched them. Followed by a brain-numbing-white-flash-crash of lightening obliterating his friend Didius Flavus.

Lightning plasma hit the tip of Flavus's pilum engulfing him, rising to the heavens and created in an instant a whirling pool of spinning light around him. The bolt instantaneously changed from an intense white, to a rainbow of colors, then to a black spot within its center. Immediately a blast of wind was enveloping and knocking down some of the men preparing for ambush.

When the blast-wave receded, abrupt silence dropped upon them. Raindrops became a waterfall. The omen of the eagle and its mate returned to him with vengeance. To have the gods rip a man from the earth without a trace terrified him. They had not been in a battle; yet the eagle-caused-lightening forebode ominous tidings for his Cohort, his Legion, the accompanying Legions, and even Rome itself. Their Roman gods were abandoning them to the barbarian's gods.

The barbarians, having seen Didius obliterated by the lightning strike, hastened their retreat. Cursing, Primus ran back along the trail in pursuit of his own frightened patrol anxious to get off this cursed ground.

In Primus's future he would die witnessing his Legion losing their Imperial Standard in the barbarian massacre of them. Disgraced, those doomed Legions would never again be formed by Rome.

Tiny wisps of smoke, dying in the deluge of rain, rising from charred ground; a curious circle from the Seech Field Vortex branded the spot where Didius had once stood.

Before he left, Primus had seen the glowing marker embedded in a rock. The symbol of the Oval's Seech Field. On planet Arna, the dead human Didius Flavus, in the Oval Universe would become KrutEk.

An American Patrol in1966 CE near the Sông Yên River, Quang Nam Province, Danang, Vietnam.

As a United States Marine Lieutenant, Steven was wobbly, trying to clear his head after the explosion. "God, that was close."

Twenty feet behind him, an insignificant gray pall of smoke innocently drifted into the blue sky, smelling of scorched jungle, rising from a small crater. Scuz's boots were charred and smoking close to the hole. Churning, unbroken legs attached to the boots, spasmodically tried to run. Pfc Jones, named Scuz by his friends, legs flopped, stopped, and then started again punctuated by gagging groans from the wounded man. "Uhhhh...Uh...Uhh...Me hurt...oh, it hurts...can't breathe."

In the field the platoon was ordered to use Lieutenant Steven's first name; not his last. <u>My Polish surname would be corrupted into 'Ski'. I don't want familiarity.</u>

The Lieutenant assessed the explosion was a booby-trap. Now they were being bracketed by mortars. He could see other mortar rounds growing grey dandelions around his platoon. Lieutenant Steven was shaking his head attempting to clear out black and white blinking fireflies hampering his vision.

He lay for a minute retrieving his senses to feel for wounds and blood on his body. Rolling onto his back to get his face out of the sewer of the jungle floor, his left ear was ringing incessantly. He distantly heard "Corpsman Up! CORPSMAN!"

Steven heard Corporal Ord, nick-named Snake, over the hand-held radio yelling, "Don't bunch up! Move away from the mortars! Don't bunch up!"

Stunned, he analyzed the situation. From the stink he guessed they were ambushed with a homemade mine wire-hand-detonated in the middle of the squad where he and the radioman Jones were walking. They had not tripped a wire. The mortar barrage came afterward.

Watching the Corpsman work, crashing sounds from up the trail pulled Steven's eyes away from Scuz. An arm and a swinging rifle barely missed Steven's head.

Sergeant Huggy, as the platoon called him, was running past, yelling furiously, "Incoming...INCOMING, SIR!" Steven tried to trip him and cursed himself when he missed.

Steven was fuming; losing track of the Staff Sergeant.

The crackling corn-popping sound from American M-14's firing, and the buzz-like AK-47's returning fire were followed by adjusted incoming mortars. Those mortar rounds receiving his immediate attention. We've stumbled into a shit blizzard, he thought.

Sa-spblam! Sa-spblam! Sa-spblam!

Steven's mind raced with his heart in the confusion, feeling completely alone and isolated. His watch said only five minutes elapsed.

Sa-spblam!

On the ridge across from them and through the gray overhanging smoke he could see North Vietnamese regulars with putty-brown uniforms against the ridgeline disappear into the jungle. They were scrambling; bunched up in places, hand over hand where the slope pitched, firing back towards this ridge. In the jungle they could barely be seen, only an occasional blue/green tracer streaked towards Steven's men. The "Whoompf" of their mortar tubes could be heard during the shortening intervals of no incoming rounds. The NVA mortar crews were on the hidden opposite side of the ridgeline.

Rock fragments and mud pelted his face as AK-47 rounds cracked and scorched past his body.

Someone heaved into place next to him. Frantic at first, he calmed down recognizing Corporal Ord. The squad leader was scooping mud with his face.

Lieutenant Steven yelled over the noise. "You digging a basement, Corporal?"

Corporal Ord or Snake had extended his first tour from his former battalion to join this one. Ord had been in I Corps for a long time and knew this particular terrain, rivers, and rice paddies as well as the VC.

Lieutenant Steven trusted the Corporal to do his job.

With his face and helmet scrunched tightly to the ground, flinching when the incoming shells landed, Ord yelled, "I think you need this…"

"We have no time for excavating, son." Son hell, Corporal Ord is only a year younger than me. He stared incredulous as Ord handed him a black handset. "You got the radio?" Steven asked.

Ord must have run through the mortar fire to snatch the radio from the wounded radioman Scuz. He gave Steven the map he had dropped near Scuz after the booby-trap went off.

You have balls, Corporal. He thought. Steve grabbed the PRC-10 and terrain map from him. "Outstanding!"

"Sir...I already called for supporting fire. The Captain wants you to confirm the co-ordinates." Ord looked scared and angry. As an afterthought, he said. "Sergeant Huggy was zapped."

Steven considered Ord's remark. Their Captain had indicated Snake was to be promoted to meritorious Sergeant when they got back to the area.

Snake jumped to his feet. "I'm taking my squad attacking the NVA's pulling back." His eyebrows raised, waiting for Steven's okay. "Be nice if the rest of the squads got out of the kill-zone and supported us."

16

No time to issue a frag order. Steven was about to order the same tactic. "Move your ass, Ord. When the platoon gets on line I'll order 2$^{nd}$ squad to be base-of-fire and 1$^{st}$ squad to flank 'em. Get it done!"

Sa-spblam...spblam...blam!!!

The incoming fire was intensifying. Snake took off running back to his squad.

The lieutenant collapsed into a firing position, hung his head and trembled spasmodically; fear and adrenaline battling in his bloodstream. Steven mumbled out loud to himself. "I have to get higher up." Steven ran disjointedly and in bounds taking cover when he could.

Within a minute he reached a small knob on the hill where a tiny Buddhist shrine, tangled with green strangling vines, crumbled in the jungle. A burst of automatic fire rained punctured shrine brick-dust down on him.

"Bright Crown...this Track Three Actual." Hissing background noise answered him.

"Track Three Actual...to Bright Crown...come in... come in Bright Crown!" More hissing and static. "Sweet Jesus!" he cursed.

Then an electronic disembodied voice crackled. "...what situation...be...mov...upport...ere...ver." Static washed out of the receiver again.

"Bright Crown, this is Track Three Actual, you're weak and garbled. Say again...over!" Steven was bellowing against the crashing noises around him. His platoon near the bottom of the ridge. He had to stop this barrage. "Bright Crown, say again...say again, this is Track Three Six Actual...over!"

17

The Company's reply became a humming racket. Raised and lowered in pitch, undecipherable, drowned out by the hissing static. As if the handset on the Company end was being keyed then released.

"Can...at...lly...hundred...efore...ordinates correct? Over."

Steven maintained his cool. Beforehand he had palmed his compass to get some idea of direction; knowing the trail to the village of Bin Bao and the Catholic Church in the boonies was South of the Song Yen river. The ridge the gooks were on he confirmed on his map. "Request fire mission on the co-ordinates that were sent...! Do you read, Bright Crown?"

Come on. Come on, he inwardly begged. He knew by experience the Captain wouldn't pull this kind of crap; radios were notorious for going out when they were most needed.

The radio squawked loudly. "Track Three Actual...this is Hardcase...Whiskey Papa on the way."

The lieutenant was relieved; the Captain had handed off control to his FO with the company. He knew Hardcase was the Captain's Forward Observer.

Steven blinked. A large white phosphorus shell exploded behind the ridge but off mark. Seconds later he heard the flat 'whoomp' of an explosion. The radio squawked, "Spotter should be impacting."

Steven's heart chugged when he heard. He saw the white hydra smoke-cloud from the spotter impact appear and he was calculating time and distance in his head.

"Hardcase, this is Track Three Actual. Adjust right Two Hundred...up One-Fifty. Request HE's and for effect. Out!" He jumped down into the muck of

18

the jungle hoping they heard him and waited.  Time dragged.

Buzzing AK-47's increased their fire answered by the popping M-14's and sweet-sounding M-60 machine guns chugging.  The NVA mortars had petered out, before the expected American artillery counter-fire arrived.

Steven glanced down the hill at Corporal Ord and watched him fire, methodically aiming, then turn and scream at the portion of the platoon he could see, urging them to charge up the ridgeline.

Corporal Ord was soon to be Sergeant.  Excellent Marine, good instincts under fire.  Sweet Jesus, I hope we all do.

GODDDAMMMNN!  Steven thought in awe.

A huge red-yellow-black explosion erupted near Ord enveloping the Corporal in an instant.  The jungle around him went from lush green to grey and pulsating white, followed by a strange rainbow of color spinning in a circular pattern.  Changing to a deadly flat black, after which the jungle turned blurry green again.  One second he had been there; the next sad moment Ord disappeared.

Lieutenant Steven winced from the tremendous concussion that followed, Steven thought simply, Well, Ord's dead...  Followed by another thought.  Shit!

Slight wobbling hisses turned immediately into cloud ripping screeches, then seconds later into mind-numbing, back-bone aching grinding as if huge steel fingers were scraped viciously down a black board.  Eight inch self-propelled guns!  Steven identified.

He was glad the bastards were dying.  Not having compassion or sympathy for the enemy was a

19

part of him. He lost his humanity months before in this war.

"Hardcase...this s Track Three Actual...cease fire! I say again...cease fire! We're advancing on the objective; stand by. Cease fire, my lead squad is closing on your concentrations."

Steven wanted to cry for Snake; but Steven couldn't. War and misery and time in country, had taken emotion out of him.

Ord's correct co-ordinates brought the artillery in on the NVA. By his actions, Ord had been killed. Steven guessed one of the rounds from the Eight-Incher's had fallen short. Figuring out why or what had happened was not his immediate concern.

Steven analyzed in his mind. <u>An infantryman's nightmare. Ord had lived and died in it. Wonder if he knew he had killed himself? Damn! A short round.</u>

Knowing he was about to endure a lot of paperwork when Steven came off this patrol did not help his disposition. He would face grilling, as the officer in charge during the incident. And second guessing about the co-ordinates. Miscalculations happened sometimes and there would be hell to pay at first, while an investigation occurred. But soon forgotten after more pressing crises came to the forefront. The Corps investigates, makes their decisions, and adjusts. War moves forward; even after disaster.

On the ground after the action, no one saw it, but in the jungle hillside, embedded in a rock was the Seech Field Vortex symbol. The sign marked the place where future Zabin-Krils for the Ovals on Arna were Seeched. On Planet Arna, Corporal Ord as Snake, became KrutChan.

20

## The Mongol on the Manchurian Steppe 1139 AD.

The Mongol reconnaissance party had been patrolling long and silent in boredom. They were used to monotonous riding. Each of them kept their thoughts inward not wanting to distract the other. That was their job, to observe and report accurately back to their command. They were assigned to be intelligence gatherers; experts ahead of the famed Mongol Horde. They had been scouting for days; the only individuals on this part of the steppes.

When the dragon appeared without warning and fearsome force; they were frightened. Nogaikan was signaling with a burning torch the day's information back to their unit when engulfed by a Dragon's flaming fireball.

The god-sign was ominous. Lightning in winter? In a fleeting moment their friend disappeared into the blue-white crackling dragon-world. Nogaikan had been swallowed. The dragon's tail of lightening, as the Monster fell from the sky, expelled fire and smoke from its terrible mouth. They would never speak of the dragon sighting, fearing it's wrath. From their detailed descriptions their Mongol leaders would understand. Dragons were rich. Lethal towards any man who approached their gold and treasures according to Mongol legends.

White luminescent clouds accented the night sky while the remaining pair of horsemen circled the smoke and heat from the scene in front of them. They were outriders from the Mongolian Army destined to write world history as the Golden Horde.

They were dressed in Mongol kulets: brown tunics identifying light cavalry. They wore gray trousers lined with fur as were their quilted winter tunics. Laced thick leather boots completed their basic uniform. A mandatory silk shirt underneath their uniform, an ingenious remedy against an arrow strike. A Mongol surgeon skillfully pulled the edges of the silk shirt rerouting the arrow out the same hole with the same twist.

One of the riders wore a dog skin cap; the other man's helmet was made of goat skin: neither were officers. Each carried a leather covered wicker shield and two bows; one long and another short. A full quiver of fifty to seventy arrows hung on their sides with a dagger strapped to their opposite forearm. The saddle held a lasso, a small sword, two javelins, and hide saddlebags with their personal provisions.

They were the eyes of their Irbin, scouting for their Toomen division size force. What they witnessed had stopped them in the cold biting wind. Old ghosts and fears drifted into their minds.

They were puzzled and frightened, but also curious about the event trying to assess what had happened in order to report to their Khan. When their spotty vision cleared after minutes, they felt relief when they saw the body. Death was something they could understand, as were corpses. Disappearances by dragons made them uneasy. They rode slowly and silently to the deceased, dismounting and illuminating it with a torch.

The body was charred and smoldering, wisps of smoke traveling to the gods. The creature was unrecognizable as a human. The Dragon vomited another person down to replace Nogaikan? The entire episode was outside their experience.

Preparing a report to their commander was not taken lightly by any Mongol. Each man soaked up

22

the scene, observing in their professional way the entire area of the episode, missing nothing. Outriders had been executed for incorrect or exaggerated claims. By training neither talked to the other so as not to pollute the impressions of the other.

When they observed an alien patch on the cadaver's chest burned on the edges; they would not have known what they were looking at; nor care. The small blue line tattooed on the right cheek visible through the burnt flesh hardly drew their attention. Satisfied the cadaver was not their Mongol friend; but a dragon demon.

They stared at the smoldering body for only a moment longer before remounting their short stocky pony's and galloped furiously back to their commander to report on the day's events. They would record their observations of the velocities, positions of moments and instants in time, which they had observed that day. It was the Timeline Arrow of Reality which had been observed.

After the Mongols galloped off, on the ground still smoldering, glowing red and embedded in the dirt was the Seech Field Vortex symbol. On Arna, the Mongol's dead friend Nogaikan, taken by the gods, would be renamed PacalMo by the Ovals.


A Gurkha unit near Delhi, India.

It is the year of our Lord 1857 Anno Domino, Duncan thought. Not the bloody dark ages. One would think the Sepoy Mutiny in May at Meerut would have begun for proper reasons instead of over pig and cow fat. It is outrageous! Mutiny could

23

hardly be tolerated in this enlightened day and age, rather.

He recalled what inspired the Mutiny; the American Armorer Samuel Colt, sold his design of the Enfield rifle to the Parliamentary Committee on Small Arms in London. The Enfield cartridges were greased and had to be bitten prior to inserting in the musket.

That started an awful row. The mutineers said the cartridges were coated with pig's or cow's fat; a religious taboo repugnant to Muslim and Hindu alike.

*Yet my Gurkhas fellows, most of whom were Hindu, request using the greased cartridges. I might say, they dislike the connection with the kala log black folk, as they call the Sepoys.*

Blistering heat beat onto his head through his field helmet. Lieutenant Duncan Shore squinted against the glare of the sun. He was a Queen's Guard Officer of a company of hundred-twenty Gurkhas, one of eight companies in the Sirmoor Battalion.

From atop his horse he looked down at his Havildar Battis. The Gurkha non-commissioned officers like Battis did not advise the officer directly, but through example, to his fellow Gurkhas.

Their behavior did not go unnoticed. Duncan's Gurkhas were fierce, intensely loyal and resolute in battle. It was observed by him many times that their knives, kukris they called them, and their shouts of 'Ayo Gurkhali!' meaning simply 'Gurkhas are upon you!' were more feared by the enemy in battle than shot and shell.

Gurkhas did not molest enemy women they captured; a trait unheard of on the Asian Continent, or any war on any continent.

*Damn shame the East India Company had formed them.* Duncan had reported, like many of his

24

fellow British officers, requesting this matter be resolved, by bringing the Gurkhas under British Arms in the King's Service. They certainly earned it.

His Havildar's name was Dilbhakta Thapa (Loyal in Heart) nicknamed Battis (number thirty-two). Using his serial number because many names of the Gurkhas were the same.

Battis was born a Hindu on a farm in Nepal, which raised Jute. He joined the British Army as a Mercenary to earn money for his family. He now served in the Second King Edward VII's own Gurkha Rifles.

Presently, the Siege of Delhi, as it was laughingly called, was stalled. The British force was too small to cover six miles of the city's circumference. To restrict supplies and arms going into the city. From these ridges the British force's days consisted either of painfully watching the commerce of the city go on or launching attacks against the Sepoy Mutineers on the ramparts. It had taken them two months to get near the city.

Reliable rumors stated British High Command distrusted the loyalty of the Gurkhas and issued secret orders for the artillery to fire on them if there was the slightest hint of the Gurkhas joining in the Indian Sepoy Mutiny against the Crown.

Lieutenant Shore did not believe his wogs were disloyal for a moment. His little fellows were splendid fighters. The East India Company profited exceedingly after the dying was done, he knew.

Duncan looked down at Battis. "We require supplies from the bunker before tomorrow; must have ourselves at the ready by nightfall."

Shore said to Battis. "Be a good lad, take this list and move at the quick ascertaining what supplies are available. Report to the Havildar-Major upon

25

completion and he will form a party for requisitioning."

"Yes Lieutenant, we will kill many Sepoy." Grinning, he was off; but not before jiggling the testicles of Lieutenant Shore's horse. The animal whinnied and stood on its hind legs almost throwing the lieutenant to the powdery dust.

<u>Damn his eyes!</u>

The Gurkhas sense of humor, while liberally on the macabre side, possessed a streak of practical joke definitely on the lewd side. Their humor could not be understood by anyone but another Gurkha.

Lieutenant Shore rode perimeter the next hour, oblivious to the occasional musket attempting to wound him or his mount. He spent a few minutes discussing tactics for the early morning attack with his Havildar-Major.

Looking into the distance at the bunker, Shore said good day to his subordinate. He witnessed the lightning strike, as wide as the artillery bunker exploding into the air. Curiously, rising from the ground to the heavens.

Shore fought to control his horse when a solid hot wind and explosive rumble reached him. The circular blast and concussion wave was petering out as he watched. A dervish wind was spinning in multi-colors throughout bunker rubble. Then abruptly shifted to grey-black with a yellow roiling cloud speckled with debris-raisins throughout.

After his initial shock subsided, he galloped towards the inferno, yelling for the Havildar-Major to take charge of the Gurkha unit. He rode to the middle of his position expecting an attack from the Bastion within moments. He neither heard nor saw cannon fire from the mutineer ramparts. Shore was thinking perhaps sappers had gotten through the pickets.

Half-hour later, coming out of the dust cloud stirred by shifting wind blowing over the Gurkha emplacements, a British runner from the 60th regiment reported to Lieutenant Shore. He identified himself and advised Shore the Regimental Commander had dispatched him to Duncan.

The Lieutenant was absorbing the news, nodding his head, the report lasted minutes and he was reciting it back to ascertain he heard it correctly. "The elephants had been off-loaded then left. The entire bunker of black power, cartridges and mortar bombs all exploded simultaneously."

Shore then asked. "How many men were underground?"

"Yes sir exploding as one bomb. Only one man was in the bunker, sir. No sappers were observed. The colonel states there is no evidence of earthwork tunneling. Though it would be frightfully premature to come to a definite conclusion, colonel said sir."

"Very good lance corporal. You can be off. My compliments to the Colonel."

He read the young man's anxious face and asked. "Is that all? There is more? Speak up, lad."

"The gentleman should know, sir. It was witnessed your Gurkha Battis was seen going into the bunker just prior to the incident."

Duncan's eyes glazed over and hooded his disappointment and guilt. He had forgotten Battis. Regretting he had sent the little fellow.

Shore knew what high command thought of the Gurkhas during this time of mutiny. Surely, they could not believe this wog would blow himself up. The last thing a Gurkha was capable of was sneaking about." Rubbish.

"The Colonel suspects Battis 32?  I sent him there."

"Oh no sir, never would the Colonel say that! I only report the disappearance of the Gurkha, your lordship."

"Is there anything else?"

The man saluted to leave, thought a moment, and then made a decision to speak.  "Witnesses saw queer things.  A humming sound coming from the bunker area prior to detonation.  Even the elephants were prancing about, shaking their heads and anxious. The witnesses heard other sounds, similar to a flight of bees, no, similar to locusts.  Loud, and at the same time distant; humming vibrations.  Then occurred the detonation.   Very queer, if you ask me, your lordship...sorry sir, I mean Lieutenant."

Duncan    thought.    <u>Sounds    more    like superstitious wog rot.</u>  "Not terribly queer, lad.  The sun beating on us the entire afternoon was palpable. Heat can produce strange thoughts in a wandering mind.  Mark it as a mirage with sights and sounds."

He sat up straighter in his saddle and slowly exhaled.  It was disturbing he was the only one to see the lightning strike.

<u>No sense in me adding to the poppycock.</u> Perhaps he had simply seen the explosion itself at the precise moment of detonation?

"Report back to your unit, lance corporal." Shore smiled with reassurance.  "Remember dancing elephants, locusts or bees cannot fire Enfield muskets or cannon, or destroy bunkers."

The man smiled in relief at the explanation. "Of course not, that would be impossible."  He ran off immediately before the Lieutenant upbraided him for familiarity.

Duncan mumbled to himself. "The heat could have jolly well ignited the ammunition."

Sitting alone on his horse in the blistering sun, watching the smoke trail from the bunker still rising into the sky; Duncan swore silently to himself.

Blast the luck! Bloody infernal sun! Bloody Sepoys! Bloody damn country! Bloody dust! Bloody dead Gurkha's!

He jerked the reins of his horse in frustration and spurred his mount to gallop in the direction of headquarters at Hindu Rao's House. Bloody fate! He would never ride to the obliterated bunker out of respect to his dead man.

Unnoticed, near the bunker, embedded in the debris field was the Seech Field Vortex symbol. On Arna, the dead man, Dilbhakta Thapa (Battis-32), would became known by the Ovals as TzenalAh.

Chapter 2


Twilight on Arna, home planet of the Oval Confederation Empire. The Seeching process had finished a half-KinUt two weeks ago. The terrible Five Uayebs would occur within KinUt months, in the NOW future.

On a parapet high above the churning crackling Seech Field, the Queen Xmucane of the female Ovals, watched with narrowed eyes and an indifferent attitude. She glanced down at the masses of scorched dead human flesh, vegetation, and other beasts that had spewed from the Seech field, into indistinguishable mounds.

Searching Quantum Many Worlds dimensions for fighting males to be controlled by the Oval species. Through trial and error and Osil Quantum possibilities, they eventually found the human species. The Seeched humans evolved from a basic reptilian brain. Compatible to the Willows and Ovals.

Xmucane was the current OvalChanHalach. Queen of the Oval Confederation Empire. She reigned over a thousand Galaxies. Galactic Empires arose and went extinct. The Oval Empire was one of those continuing to thrive.

The universe so huge, existing Galactic Empires concerned themselves with their own similar type-specie's administrations and ignored other empires for various biological and scientific reasons. Why would a slime-species want to make contact with a rock-species? Both with different logics, incompatible religious beliefs, sexual drives, and intelligence.

After the Ovals had exiled and gained control of their male Willow Dictatorship; a Science Oval located a hidden, non-light emitting distant nebula, near the Dark Matter threads. That Oval scientist, using unknown means, assimilated a Quantum planet. This Science Oval hid the location of that planet from the Oval Empire.

Ovals named the hidden planet StelaBalaam; collating timeline experiences of all the sentient beings in the Oval Confederation of Galaxies. The quantum planet repaired itself, administered by the Oval who created it. It controlled nothing in the Oval universe, simply storing individual timeline information.

Handpicked solitary Oval science teams belonging to no Nests, in effect secret exiles, helped the present Science Oval.

The Ovals and their Confederation Empire could access information, but the StelaBalaam's information could not be erased. Information obtained at a sub atomic level could only be updated. Xmucane and her predecessors, ruled those thousands of Galaxies aided by the StelaBalaam quantum planet.

Xmucane thoughtfully observed the lumps of matter engorging from the Seech Field. Those mounds would become, if surviving, reconstituted humans entering the Maluayeb Arena. Other plant and animal species surviving their Seechings would be seeded elsewhere on Planet Arna.

KinUt months ago, Xmucane while absorbed in her nightly time-pause; experienced a flash from her NOW future timeline. Forecasting a newly Seeched Zabin-Kril becoming KrilChan of the Kril Legions after her KrutEk met TOTL by the new Kril's hands. Her vision-pause indicated a timeline possibility she did not favor occurring. Frightening though it was, she was aware Timeline branching could not be stopped. However, branching could be astutely adjusted by a Queen. She not would endure infinite end-point timeline possibilities.

Upon awakening, Xmucane initiated her Top-secret Queen's plot, in case her premonition came true. No guarantees future visions were totally accurate. In her vision there occurred no names, physical descriptions, or identity hints about the individual in the future timeline branching. Whether the indicated Kril was a new or an older individual in the Kril Legions.

At present her extreme interest in the latest masses vomited out by the Seech Field was hidden deeply within her Queen's memory; never to be divulged to any of her subjects. Queens ruled and could adroitly function only if their secrets remained concealed.

Approaching much too soon for her tastes were the Five Unholy Uayeb Days. Her duties these coming pre-Uayeb Kin days was to attend the de-activation of the Seech Field reconstitution process for this Tun year. She would not remain in attendance near the Seech Field for the entire Tun year. As OvalChanHalach, she constantly was being pulled away. She had an Oval Confederacy Empire to rule. Later she would see the surviving reconstituted human masses of beings, at the Kril's Maluayeb Arena in her NOW future.

Xmucane now ordered Seech Field hibernation. The Queen, nodded to her entourage behind her, melding with them feeling relief, going to address more mundane matters in her vast Empire.

Planet Arna revolved around its Om in a three hundred-sixty-five-day cycle. Two KinUt months later, on the last Kin day before the coming Uayeb, Queen Xmucane was alone. Looking into a three-dimensional wall portal, down onto the Maluayeb Arena waiting for the ceremony of the Kril inductions to begin.

Xmucane was OvalChanHalach of the Ovals; the present Queen in a long continuing line of Oval Leaders. She was beautiful, by earth standards. Her skin was clear, with a yellow Asian tint to it. The tops of her ears were highlighted by faint scales, denoting Oval reptilian evolution over Gryles of Time.

This new Kin day preceding the start of the five 'unholy days' of Uayebs, was not a time of celebration on planet Arna.

Historically, the Uayebs were the time of Tun year in the Oval's past when their Willow males were conquered. Not in battle, but by Oval science superiority and banished by the female Ovals many Tuns ago.

Xmucane was waiting for her entourage to arrive. She was fearful of these coming Five Kins, the Uayebs. During her reign her anxiety grew each time the unlucky days approached; her inner mood dark.

The Oval Queen spent the other three hundred sixty Kin days much happier. Though the underlying plots, counterplots, secrets, and political considerations normally took up much of her time. She, as did most Ovals, did not think about Krils, except during those unlucky Kins.

The Ovals were a matriarchal society affirming life and its continuance. Ovals were forbidden violent behavior, to directly cause TOTL,

33

or to behave in any way like their banished male Dinarchy Willows. When Oval dominancy was threatened by the aggressive threatening Dinarchy race; Ovals unleashed their surrogate Krils.

Their Willows had forced the Ovals into an impossible life and death situation. By scientific error during an aborted Invasion attempt, sterilizing Willow males. Oval females reacted to keep their species from extinction.

After exiling the male Willows using their Oval Osil drives and by refining their Oval Seech Field technology, the Ovals found a way of continuing their bloodline. Not an easy task, nor inexpensive.

Importing human males from another universe-dimension. They would control and use the imported Earth-males to defend themselves against their male Dinarchy-Willow enemies. The status quo remained for Tun years in the Oval Confederation Empire.

Past OvalChanHalach Queens organized the Krils to fight the banished Dinarchy. The Seeched human Zabin-Kril's mentality resembled their Willow males before they were exiled. With arrogant attitudes and war-power dominant behavior.

The Kril Command hierarchy were allowed to perform their Maluayeb Arena ceremony a Kin day before the first Uayeb, to induct their Seeched brethren. The OvalChanHalach and the previous Oval queens controlled the Arena event so the Kril hierarchy did not become outrageous. By imitating the blood rites of their cursed Willows.

The Maluayeb Arena ceremony was an orgy of excessive Kril cruelty; abhorred by the ruling Ovals.

A minority of Ovals, who missed their
34

Willow-male ways, condoned the induction process. The Arna Ovals of the Confederation would be present in the Arena. Except for the Oval minority, none liked the ceremony. For them it was an annual spectacle to be endured. The Maluayeb Arena ceremony was Kril tradition. Oval coping with the spectacle fulfilled their tradition.

Jaguar, Xmucane's Intelligence Oval, melded into Xmucane's area. "You are honored my OvalChanHalach. As you bid me; I am here."

Xmucane smiled and gestured for Jaguar to be comfortable, offering her what she was drinking. Oval Osil-Wine was a narcotic their species consumed in their past for reasons long forgotten.

Keeping their body's young, reproductive organs fruitful, and their intelligence and self-awareness at a high level. Under their surface even-tempered-demeanor constantly lurking was reptile aggressiveness. Controlled calm the side effects of the Osil-Wine.

An undercurrent of highly charged emotions created a sense of mistrust of other's intentions, producing constant scheming to get what they wanted. Their plots involved amassing metallic treasures, an all-consuming driving ambition. Questing for a higher social position above their present timeline.

Xmucane and Jaguar both sat on invisible furniture in the cool air of the room, floating across from each other. Oval vison could perceive the entire electromagnetic spectrum, choosing the spectrum they wished to see using their moveable four nictitating membranes.

Human Krils could only see the small visual spectrum; considering the Oval abilities 'Oval Magic'.

"You are honored by me, Jaguar. What is happening in the Confederation Empire this Kin?"

Many times, in conversations, Ovals spoke in the third person. It was a rhetorical question, more of protocol then substance, hiding Xmucane's suspicious nature.

"Dirva is her usual self." Jaguar said.

Xmucane spread her hands in agreement. "Dirva is Dirva. I thank the Cosmic Egg she will never change."

"She is spreading discontent among the Ovals in the Tribunal. She is again intimating you not do support the Kril Legions these Uayebs."

"I need information from you of plots I not am aware of..." Xmucane raised her forefinger. "KrutEk, my own KrilChan, constantly berates me for the same reasons."

"She expounds in the Tribunal you are weak."

"I would be leery if she not did object to me." Xmucane brushed her hand through her hair. "It is well known my distaste for TOTL and Willow-Dinarchy ways, especially in my Oval Confederacy Empire."

"A Confederacy of Ovals at odds with itself." Jaguar said.

Xmucane reminded Jaguar. "A Confederacy I administer which includes the Empires of the Cunacks, Dagots, and the Zars."

Quietly Xmucane was staring at the end of her glowing Oval controlling glove-wand on her hand, as if seeing it for the first time. "Our Oval political system not is perfect, but is the best of the other Empires in our universe. Our other sentient alien

36

leaders in their Empires not do have a problem with my Oval feelings. Their subjects who listen to Dirva are their problem...not mine."

"You are Queen. It pains Dirva and her minions you advocate power should reside in the Oval females seated in the Tribunal. Dirva abhors your policy."

"An Oval who not does covet their own policy is an imbecile." Xmucane took a deep drink of her Osil wine. "The original OvalChanHalach set up our Republic form of government. I carry on the tradition." Xmucane said.

"Which is what Dirva resents. She honors you but desires to again be Queen."

"Dirva has too much of our Willows old totalitarian ways in her heart. She believes in Willow-Dinarchy absolute power."

Jaguar reminded her Queen. "Dirva believes the Ovals need a strong Monarchy led by a strong female. What allows her and her sisters to survive is your wishes to let other Ovals with your mindset govern unhindered in thousands of galaxies under your Tribunal guidance. Dirva's minions use your inaction against controlling them to plot against you." Jaguar said.

"Precisely why I must guard against her minions." Xmucane said. "Dirva's political ambitions are dangerous. Her and her sisters, dream daily of their perfect control of Oval society; yet they cannot convince the majority of Ovals to think their way. They are patriotic but strive to abolish the Tribunal."

Jaguar nodded. "Your enemies curry your favors to allow them continued existence. Dirva and her ilk need our female Monarchy, assisted by the Tribunal, for when they take control."

37

"Are you aware of any female Oval politician who not did seek power?" Xmucane asked. "Dirva is a 'known' anti-tribunal Oval; that is as I wish."

Jaguar advanced another thought. "No one understands the political logic of our non-Zabin-Kril males and the Dagots, Zars, and Cunacks."

"Keep our male ArnaMals and Germs ambitions and plots where they belong. Dirva's minions are more controllable. Willow-Krils are alike in their DNA nature."

"Yes, my OvalChanHalach. Dirva treats her Soothed Krils in her Nest much as Dinarchy Willows."

"That is her right under our Oval missives and laws! Changing our political society, I never will allow. Dirva's reign as OvalChanHalach was short and disastrous. The Ovals have come too far. They will resist her should she attempt to become my successor."

"I have no illusions or desire to succeed you."

"Not do take that tact with me." Xmucane smiled. "There is no Oval in our Confederation who not would become Queen, if offered…even you. Not do speak foolishness. I respect you because your enemies are my enemies."

"If I were Queen I would…" Jaguar stopped. It was dangerous to be presuming, quickly adding. "Perhaps you should nullify Dirva…or exile her?"

"Every Tun year before the unholy Uayebs when we meet, you continuously educate me of plots I am well aware. Of that I tire." Xmucane frowned. "You and I have discussed Dirva many times in our THEN. The Tribunal would never allow me to nullify Dirva without cause."

"Expelling her would make your crown less burdensome."

"A known threat is easier to deal with than an unknown one." Xmucane shook her head. "Why would I encourage her minions to revolt? You must find reasons for me to cancel her ambitions out. I not am a Dinarchy believer in the Willow Creed of governing. I believe in the Cosmic Egg creed." Xmucane said.

The Cosmic Egg was an evolving scientific-based-religion of the Ovals. Ovals were too old a species to maintain belief in religious conundrums.

"My timeline is limited. Uayebs drain on me." Xmucane said. "I wanted to speak with you alone. I honor your loyalty but not your redundancy in these matters. Many former OvalChanHalachs succumbed to plots against them."

Jaguar was thinking. <u>Which one of us is the more devious with their flowery praises.</u> "I am honored by your faith in me."

"My other Oval sisters, who are in the line of succession, not do have loyalty to me. That is how it should be; every Oval out of power always wants to be Queen, thinking they can govern better. That is the bane of my Royal position and I accept it as one of my burdens of power. I not can always do as I wish in these matters which are what my detractors seize upon as a flaw."

"You are correct, as always, my Queen." Jaguar said. "There is another concern; along those lines of plots against you. Your KrilChan, KrutEk, not is to be trusted. He plots to overthrow you with his Legions and become a male dictator."

"I will control KrutEk's actions." Xmucane said. "His former empire on Earth feeds his ambitions. Is it called what?"

Jaguar smiled. Without the StelaBalaam no creature could remember every bit of trivia. "He was in the Roman Empire in his THEN past."

"Yes, of course, the Romans." Xmucane smiled back. "He has that quaint male attitude of loyalty to his former leaders. He will protect me in any coup he hatches. However, I will never allow his coup to occur. I will subvert him with his own loyalty, by misdirecting his efforts."

Jaguar respectfully nodded. "May the Cosmic Egg assist you."

"We must hurry." Xmucane was a stressed Queen. "I summoned Dirva to attend me before we go to the Maluayeb arena. Have you set your plan in motion regarding the new creature-Kril?"

"I have done as you instructed." Jaguar said. I inspected the reconstituted specimen from earth. She thought. Introducing Red AnticArna potion into his brain; not the usual Blue. He will be uncontrollable but serviceable for the Queen's purposes.

Their plan was initiated a half Tun year ago in their THEN by the OvalChanHalach. Secretly manipulated by Jaguar. Jaguar could not foresee the consequences. Timeline branching determined NOW futures; not individuals. By command of the Queen, sworn to secrecy, they both would never speak of the details.

Xmucane nodded her understanding.

Jaguar continued. "I must caution you, the selected new creature may be too unmanageable induced by my AnticArna red-potion. A plan of this kind has never been attempted before in the Gryle of the Ovals. Your risk is great."

"Risk-taking is demanded of a Queen." Xmucane said. "We shall observe the result of 'your'

40

work in the Maluayeb Arena. Timelines cannot be controlled, only nudged. I keep secret my other options if 'your' plot fails."

Jaguar knew the blame of failure would fall on her; not her shrewd Queen.

Xmucane grew impatient. "Enough of these speculations and plots and subplots; let us speak more of what is 'right' in our Confederation Empire. We not will have time to do so during the Unlucky Uayebs." Xmucane sighed.

Jaguar spoke for only an Obet minute, before she melded out. As the former OvalChanHalach Dirva melded in to receive her instructions before the Maluayeb Arena ceremony began.

Corporal Ord couldn't recall being helicopter Dust-off evacuated. At the time he was thinking, "I'm wasted." The next instant he was unceremoniously colliding with the ground. Trying to plow a furrow into rock and not succeeding. After which, he again lost consciousness.

Now Ord was walking within a long line exiting a jungle terrain of immense trees. Not Vietnam foliage; different. Ord saw a huge stadium-like building someone was guiding his line towards. Reminding him of a field route march.

He reveled in the flower-smells wafting through the atmosphere of his surroundings. Snake's senses perked, a vague memory of a tender moment, a clean soft-woman essence calmed him for an instant. Then the precious perfumed smell abruptly disappeared.

Then a stronger stinking-man-smell assaulted his nose; the odors from barracks, the urinals, and the pungent-reek of troops upwind of him.

There were lines of soldiers to his right and left; while to his immediate front four columns were leading into four different huge doors.

The sound of clubs viciously hitting bodies, red flashes of red, followed by zipper-like sounds discharging, had him on alert. Moans, the screams, and the intense praying of victims had been going on for over an hour.

Strange, he thought. What happened to me?

None of the men in the line glanced at each other, avoiding direct looks. The prisoners were coping, stoic and feeling each other's pain.

Snake assessed his own condition. Dragging his feet, beaten-down-numb; he tried to count the aching wounds he felt over every inch of his body. Even his ass and nuts were painful. Ord was bleeding from his torn skin, slowly leaking. Raw contusions were seeping along with scars from the artillery blast and his half-ripped off ear. He was dragging his untied ragged boots, shuffling clouds of tan dust from off the trail. He saw a blue-white lighted area coming ahead of them. The light muted.

The rest of the prisoners in line were shackled like him with green-colored vines. What's going on? He fought to control himself.

Ord shrugged off the thought. Forget it. The others were in the same shape he was, though they looked worse off to him. All of them were old. Not threatening. Those dudes couldn't rip the wrapper off a condom. He assumed he was just as old. The other's eyes easily read by him because of their mutual misery.

The beatings and killing continued, on a guard's whim, and random.

Snake followed his line of prisoners into a room constructed of huge blocks with a yellow

42

candle-like flickering glow on the walls. No fires were evident. Reminding him of a primal fortress and smelling like one.

Snake saw other marines-army-navy-air force, VC's, NVA's, and other nationality's from other eras throughout earth history. <u>Man, this is weird shit.</u>

Other races, other times in history; he noticed in his befuddlement. He recognized Civil war vets, WWI soldiers, veterans of the Boer War, Romans, black, red, white, and yellow men. Japanese Samurai, Cossacks, and WWII vets. And ancient warriors he couldn't begin to identify by their uniforms. Fighting men from his own past, as bewildered as he was staggering in those lines. Anonymous individuals herded by clubs and red blasts.

Snake saw quite a few uniforms he should have, but didn't recognize. Some of the troops wore German helmets wearing unfamiliar utilities with American flags on their shoulders. Which was really a shock to him; causing his mind to stutter-step. <u>Where the hell were those guys from?</u>

Other prisoners dressed in unrecognizable uniforms looking foreign and some others futuristic. The only similarity was their abject misery. Indiscriminately being killed as they moved forward.

He was part of three hundred men, similar in size of his own line company.

The conversations, the screams, the cursing, and intermingled voices in a multitude of languages filled the area with noise. His ears ringing incessantly, drowning out the details.

Tall Guards in purple hooded-robes were pacing between the lines. Silently watching the lines, bored by the detail; ready for any break in the ranks. The guards carried arm-long-solid thick staffs, stubbed on the end making lethal clubs. Goading the men in the formation along.

A line from the 23$^{rd}$ psalm ran through his head. 'I will fear no evil.' He wasn't comforted. He was angry. The hell with that noise; I'm scared shitless.

A small NCO-looking man, his uniform unrecognizable to Snake, appeared from a high door on an altar-type dais and started right off with the rules. Snake assumed were for all of them. The little sergeant recited in a boring rote tome, speaking in an unknown language.

Snake was amazed. The NCO's lips were not in synch with the English translation in Snake's mind.

"All of you have been granted another lifetime. Absolutely no talking. You have about one hour to live. You must obey absolutely and briskly. No questions."

This reminds me of receiving barracks at boot camp, but without the yellow footprints. Snake had never heard of anyone murdered in marine boot camp. Then he wondered how he was receiving the translation?

One man, from an ancient tribe, began loudly complaining in a foreign tongue, at least it sounded like a complaint. One of the Guards pointed his staff at the man and a round red plasma ball struck the victim in the chest, exploding in a shower of blood and gristle killing him; drenching the bystanders near him. Many of the other prisoners grumbled in unison, which began the bloodbath that followed. The complainers died similarly. Methodically, the guards were eliminating prisoners.

Snake hit the deck with his shoulder and protectively hugged the cold stone floor, staring at a crack. The hairs on his body bristled when a guard's staff passed over him. He tensed in expectation of a blow.

He remembered back to MCRD San Diego during his first week there. The DI's had been screaming and punching guys for what seemed no reason at all; calling the recruits 'pukes'. In Dago the recruits could do nothing right that first week.

This place was reality-of-death-incarnate with mercy an out-of-date concept. The guard slowly raised his weapon off Ord and moved on.

They were down to less than one hundred prisoners before those zipper weapons and the blood bath stopped.

"Are you beginning to understand?" The NCO guy continued, in that foreign translated voice. "This place is not heaven-not-hell-not-earth." He let that sink in, and unceasingly went on. "You are in another plane of existence, another universe. Concentrate. If you do not respond correctly immediately, you will die!"

Another man in English called the little Noncom a son of a bitch. The nearest Guard instantly killed the man with a red-plasma burst.

Far in the distance a large feminine crowd could be heard by Snake, yelling and murmuring in a prayer-like-séance. Who the hell are they; perpetrators or spectators?

Ord was confused. Can I escape? No helmet, no rifle; he had his cartridge belt, plus all the attached items including a bayonet and a K-Bar. Other men were carrying similar weapons available to overcome guards, but the tight bounds around their arms hampered resistance.

Some struggling prisoners able to rid themselves of those green vines, died instantly

45

attempting to incite a riot. Or the guard just didn't like their looks. The guards made the death decisions. Snake heard screams of fear; real pain, real agony the closer they got to the door.

In the background he heard women screaming. Women were cheering, whooping or snarling. They were watching something cruel.

Ord stood when the rest of the men did. His mind raced, trying to decide what he was to do and when? He never had time; there was too much happening. Snake focused on staying alive by paying attention.

The prisoners with Snake, moved through an arch into a massive arena. Wary men were being pushed out of the line, especially if they protested verbally or by their actions singled out. Snake noticed snarling tough guys were selected first, while another NCO watched from high up on another stone parapet. Those unruly men were blasted by zipper weapons, cleaved with their own knives, garroted, and murdered by the guards in blue cloaks and hidden faces. It was random systematic and lethal with no logical reason behind it.

Snake was thinking. If the guards wanted to scare the crap out of us prisoners; they were succeeding.

Swallowing his intense fear, Snake was in awe, taking in the sight of the massive arena he was entering. Though he had never been in Rome; he had seen pictures. This arena made that Coliseum look infinitesimal by comparison. The sheer size of the arena shrunk everything he remembered in his limited experience. A floating clear golden dome above the arena added to his puny-size feelings. The arena caused him to wonder if he was one of the players or one of the doomed Christians. He definitely was bait for the lions.

He noticed women his mind-translator said were female Ovals, arranged inside the arena, in ascending steps. Oval females wandered through the floating seated Ovals. The invisible-effect looked strange. Crudely, he was thinking. <u>With those tits and asses; those are definitely females, buddy- boy. And turn off your sex fantasies, if ya wanna survive.</u>

Watching the females, he forgot where he was and paid for his distraction. Snake was struck by a blue bolt from a zipper weapon. He stumbled kneeling to the deck. He was thankful it was not a red bolt and he was not dead, like the others. They weren't trying to kill him; just getting his attention.

Snake shut his mouth, working hard at reducing his ragged breathing. He didn't give a shit anymore. <u>Do your best, assholes! Do what you want; screw you!</u> He did not speak and did not look at the guard, who eventually moved on looking for other victims. Then he arose as if nothing had happened, resolutely reminding himself to keep his focus.

Ord endured a half hour of waiting. He witnessed other men terminated during that time. His prisoner ranks were down since entering the arch. The guards were moving prisoners out of one line into another nearby. Guys were added to his line. Those that did not move fast enough were killed. This culling-out was senseless and irrational, serving no purpose.

There were only twenty men left in his surviving batch and were stopped at the top of a raised ramp. <u>Why twenty guys?</u> His mind questioned.

Someone or something answered him in his mind. Twenty is a traditional Oval number. It is not your concern.

Snake surreptitiously noticed numerous other ramps, each with twenty prisoners he estimated. Around and above the huge floor of the arena, feeding a central sand area. The arena was so immense the end was lost in perspective. There must be thousands of prisoners being processed.

Ord observed covertly, not wanting to draw the attention of any of the guards. He looked into an amphitheater watching men like him being led, one at a time, to granite-looking pillars at the bottom of the huge arena.

He listened to foreign languages in his immediate area, but he heard—English, strange-sounding English, coming directly into his head. Loudspeakers, of some kind, were whispering in the distance to his ears, not brassy or squawking. And he realized those voices were speaking in myriads of different languages the crowd's conversations. English was being spoken in his mind.

Remain where you are now! The Oval-Servicers will call you individually and you will move up to your assigned pillar!

That order came from another voice in his head, not from the little Noncom.

Each leading man of the group was called and was directed towards a column beyond the bottom of the ramp. One man clockwise moved forward from each of the ramps nearby and in the distance. They were killed instantly if they hesitated. Their bodies thrown down into a black pit at the side of the ramps.

With their hair rising on the backs of their heads, the prisoners watched other guards in different

48

robe-uniforms moved from left to right down the line of prisoners.

The watching prisoners on the ramps cringed, expecting a flash of red, killing the men on the pillar stones. The prisoners on the columns, in addition to the green vines they arrived with, were hung by black vines controlling their movement.

Each prisoner attached to a pillar was struck with a smaller device. There were intermittent flashes of deep purple-bluish charges discharging from the devices of the guards all over the arena. The prisoners on the pillars were fully dressed and many shouting in agony. Must be an internal pain, like electricity, he surmised.

Screams from the ones hanging on the granite-pillars were raising hell with the waiting prisoner's resolve. A smell of ozone in the air was gagging them as they waited their turn. Some screamers died from the immense charge they received; taking one blow when their heart exploded. Their arms were cut by the black vines and their bodies were abruptly released and thrown into another deep black pit.

Most of the ones not dying were freed and they were singularly herded towards circles of females in a different perimeter area of the Arena. The ones who fainted were carried off through another door. One man, after enduring a blow from that vicious mechanism, Snake noticed, was chosen by a female and went through another door under the parapet. The man had not fainted, or died when struck, Ord noticed.

He had a flash of insight. Take the pain and live.

Snake's mind flashed back, remembering the instant after the artillery had exploded in Vietnam. There had been a kaleidoscope of colors, swallowing him into a magnetic vortex. <u>If I survived dying; I can handle this shit!</u>

Xmucane was with Jaguar on a platform dais above the Arena. Dirva met them. "I have instructed your Princess Reela regarding her duties. As usual she is confused."

"Were you forthright in your explanations?" Jaguar asked.

Dirva bristled. "Reela is weak by nature. She was commanded by the Queen to control the new Ovals choosing appropriate Two-Strike Krils surviving the Pillars."

"Clear to me. Reela not is strong." Xmucane said.

"This is her first time having Soothing duty over new young Ovals." Jaguar said.

"She not needs your sympathy, Jaguar." Xmucane said. "I selected her for the duty. Failure is shameful to Ovals, but she must have her chance."

"The new Ovals are honored not to have to pick a One-Striker ignoramus." Dirva insisted.

Jaguar laughed. "Ovals argue constantly whether a Kril taking more than one strike is more insane."

"Enough!" Xmucane was irritated. "We witness a Kril tradition. Ovals control the Kril Legions excess, not do enjoy the spectacle."

In the Arena, Ord's wandering mind cleared. Snake was unconsciously listening to a whisper from someone else...not his translator. Not able to tell if the inner voice was female or male. The voice steady. Unclear to him as to who it was; telling him to be strong and not fail. He made a decision.

*I've had enough of this bullshit! They'll probably kill me, but I'll die my way, in my time, under my control.*

From the original twenty men, Ord was left with five others. The prisoner in front of the group panicked. He was thrown down the ramp, grabbed by guards, was blasted to death before being thrown into the black pit.

*That does it!* Convincing Ord to make his move. Another prisoner was pushed to the front. Snake butted that prisoner aside with a shoulder-shove and moved forward into the other man's place.

Before the guard could stop him, hiding his tremendous fear, Snake walked down into the arena unescorted. He coldly stared at the pillar Guard, daring the creature to fire a red bolt into his chest. Genuinely surprised he was not killed for jumping the line.

Snake hoped other men, under similar circumstances, did as he had done. Wanting to get it over with. There was no bravery involved. In his unconscious mind he was being coerced by that inner voice.

Snake heard translated words from the pillar guard about his inferior racial past. The column guard's voice interpreted in his head with a time-lag. The voice from the guard whispering a preamble to him.

*This creature-human is Alien. As a Kril, should he survive, he will be an instrument of his Oval. The Ovals choose Krils to service them. Humans are useless for anything except perpetuating the Oval species, fighting, and meeting TOTL.*

In front of his assigned salmon-colored-pillar he noticed his guard. She was an Oval female according to the voice in his head. Replacing the

51

male guard from the ramp. She did not have her hood up; her brown hair framed her face. Purposefully, she walked up to him with one of those smaller devices glowing purple and held it close to his face.

Snake was nervous. His skin puckering with overwhelming fear but he kept a stoic expression on his face while he licked his dry lips. <u>This is gonna hurt</u>.

The woman guard was asking him in a quiet whisper, while the translation occurred in his brain. "Do you wish to be released without pain, wanting castration and life as a Germ?"

<u>Castration? Germ? I don't think so, sister!</u>

"Or do you accept the purple strike of Oval service to become a Kril? You must decide. Speak your Choice! Life? Or Strike and Service?"

Snake had a long history of training himself to accept pain, hiding his long forgotten stutter. Not wanting to hear the stuttering hiss of his S's when he said the words 'strike or service'; he refused to speak.

She was giving him a way out to avoid the pain. At seven years old in the county home, a teenager inmate had raped him explaining grownups made love that way. When she found out what had happened, the matron had used corporal punishment on the teenager.

Immediately afterward in her office, the strict elderly matron paddled Ord's bare bottom. She never stopped until he was crying. Her beating taught him much at a young age. He never cried again. She had taught him pain could be endured. Her lesson hardened his hatred for authority for the rest of his life.

As he listened to the female guard, he realized all the prisoners had been asked the same question.

Thinking quietly in his mind. <u>Thump away, you bitch!</u>

The female guard glanced away as if looking for advice on how to handle his non-compliance. She asked again. "Answer me!"

Not speaking, he grunted in answer to her question.

Knitting her eyebrows in confusion, the female guard looked away again at someone.

Giving him time; his apathy and deadness-of-heart focused him. It was not courage. Rather his acceptance and not caring. It was a technique he used in his past during 'thump call' in boot camp. He shut out every sound and sight while staring at a small dark spot on the pillar. He focused on that spot, his mind moving out of his immediate time.

One strike!

The crowd was distantly yelling something in happiness as other men around the arena in his position were being struck at the same time.

Snake was drooling, unable to stop. <u>Son of a bitch…that smarts!</u> How the female guard could find a clean spot, not damaged by old wounds from the explosion that had killed him, was beyond his comprehension.

Ten minutes passed as the ceremony moved on to other prisoners on pillars. During the interval, dead prisoners were removed, along with prisoners quitting not wanting the pain.

After taking the first blow, the majority of prisoners begged to be released. The remaining one-strike victims, like Snake, were struggling to compose themselves.

Eventually the process continued.

53

Snake was listening to his tormentor's muffled female voice, sounding far off in the distance. She asked him again in that whisper. <u>Choose Service.</u> <u>Rather than death. Or submit to life by accepting no</u> <u>more strikes. You are an insignificant creature.</u>

Watching the other prisoners hit with the purple device after him, he noticed the staffs strangely left no whip-mark. That observation was no comfort to him in his pain. He was concentrating on wiping spittle from his mouth on his upper arm.

A second strike!

That blow was harder than the last and Snake was in agony. <u>Jesus Christ!</u> He could not take more of this. His unconscious mind, listening to an alien voice, urged him to accept the strikes. His knees were trembling. His bladder release urgent.

The yelling from the female Oval crowd was slackening. Their voices sounding more surprised with murmuring expectation.

Without fanfare, most of the Krils on the other pillars gave up. The unwilling ones wanting no more strikes. Dead prisoners were taken away.

Another slow ten minutes ticked away.

There were prisoners left on other pillars with Snake, hanging as wobbly as he was. Prisoners who had absorbed the second strike; refusing to quit.

Through his throbbing pain Snake glanced at his Oval female guard who had used the purple weapon on him. She was licking her lips, not in triumph, but acting as if she really did not want to be doing this to him.

Snake's left shoulder radiated a burning excruciating pain. Feeling his pulse pounding in his neck. The bile was bitter rising in his throat; forcing himself to swallow. That mouthful allowed him to gasp through clenched teeth, focusing back onto the small dark spot; the only reality he owned.

The murmuring crowd was quiet during the pause while the arena was cleared of passive two-striked Krils. Indicating they had taken enough.

High above the Maluayeb Arena, Xmucane was laughing while Dirva giggled about some OvalChanHalach-in-joke Dirva had said. The two Queens, a former one and the current one, were observing protocol, portraying indifference to the Kril ritual in the Maluayeb Arena. Neither one of the Queens could personally stand the other one, but amenities had to be preserved.

Xmucane frowned when Jaguar sheepishly glided up to them. "What is it? You not should interrupt two queens in conversation."

"Respectfully, I honor both of my queens, but I wish to advise you..."

Dirva snapped at Jaguar. "Why is your spy so impertinent to interfere with our discussion of Confederation business?"

Jaguar finished her sentence. "...that the two-strike Krils have been dealt with..."

Dirva's disposition remained irritated. "Not is that the purpose of the Maluayeb Arena? Why bother us about such trivial matters?"

Ignoring Dirva, Xmucane calmly said. "Finish what you were saying, Jaguar."

Jaguar bowed to her queen. "There are only eleven creatures left. We have never witnessed so many to be serviced to the Ovals for their third strike."

Xmucane glanced away and gently took Dirva's arm, moving closer to the Arena. "I not did realize the ceremony had progressed this far. Thank you for bringing it to our attention, Jaguar. Dirva, we are shirking our duty. Let us pay more attention to this abominable event."

"I also yearn to see this historical event." Jaguar said, as she followed behind them.

"These earth males will be the highlight of the Maluayeb Arena ceremony." Dirva happily said. "I 'am' curious as to how many, if any, will survive the third service strike."

Xmucane locked eyes with Jaguar; their thinking was in synch with the possibilities of time-branches that could occur. The two co-conspirators anxiously awaited the results of their secret plot.

Do not falter…you must continue…. A quiet voice was coming through Snake's painful fog taunting him.

That Guard hitting me doesn't like her job, but she's good at it. He thought.

The female guard was distressed, looking at him to ascertain if he remained conscious. Should he faint; her job was done.

Snake could feel her perfumed breath on his cheek. "You have honorably absorbed service twice. Do you wish to quit Service to your Oval? Or do you wish to endure another strike? Speak!"

The female guard was looking around for someone to help her. "Simply say no more strikes and you will be Soothed." Her voice hesitated. "Be warned. You may possibly TOTL if you do bear another strike."

It was impossible for him not to stutter saying 'strikes'. The voice in his head translated. TOTL means to die. The pain came in intervals and waves; rehabilitating his resolve. He rebelled at answering her.

She delivered the third strike!

Holy shit! The higher-setting blow from her staff applied more concentrated force than the other two. Landing near his kidneys and sent charged waves throughout his entire being. That hit caused mind-blasting bright balls of light. His snot exploded in a thick line leaving his nostrils. He slumped in pain, cursing. His pain was overwhelming. He was ready to quit.

Minutes passed in the Arena while other prisoners who had died or quit after surviving the third strike were being pulled off their X-pillars.

After his appalling strike, Snake's arms were still pinned to the pillar, so he concentrated on crudely blowing his mucus-dripping nose towards his feet. Attempting to clear his sinuses and drag his mind off his pain. He was uncontrollably shaking.

He was about to plead for mercy, when Snake noticed one other prisoner a few pillars away, was still hanging as he was. The guy looked to be in rough shape. Kneeling with his arms pulled down, about to dislocate, barely holding his position. Snake was glad he was not alone on these columns. He wondered if they were the only ones left.

The Oval women in the arena were nervously laughing or muttering in awe at the both of them.

In his peripheral vision, four or five hooded faces were near him, out of focus. Each taking turns coming close to stare at him. He had bitten his lip and coughed involuntarily. His bloody spittle painted the pillar. He hoped the guards did not assume he was being defiant. In truth, he was broken.

He forced a grim smile, in reality a smirk, to form on his face. It wasn't heroism. It was a willingness to take whatever the guard could dish out.

He did not want to die. He was remembering some of the KIA's in his past. <u>Death makes the pain go away. When I die…I win.</u>

Snake was concentrating on hyperventilating, continuing to stare at his dot on the pillar's face. He gathered himself, waiting for the next excruciating blow to be delivered to his body.

Inside his mind in a constant refrain, that goddam insinuating voice was spurring him to not quit. His conscious and unconscious mind mightily holding his fear in check; not easy. Sweat and a blinding-white pain filled his vision. Watery snot was dribbling from his nose now that his sinuses were clearing.

The voices of the hooded male guards around him were angrily consulting each other in confusion as to what to do with him.

His Guard female was whispering to Snake again. "Do you wish to TOTL? Are you ignorant? You not do have to continue. I can stop your pain at your request. Ask me to cease your service strikes."

<u>You don't know it, lady, but I can't say 'stop' either.</u> He glanced at the prisoner further down the line of pillars, still held up by the black vines; crumpled in agony. The other prisoner was trying to pull himself up.

Forcing the issue, Snake was deciding for his guard. By sheer inner will or stupidity, with a tremendous effort he slowly straightened. He stood stiffly erect. Inward, he was slowly coming apart, dissolving into his personal time of panic and acceptance. His intense personal fear was all he felt as he was preparing himself to die.

Snake heard the same question addressed to the only man still hung on his personal column, like him.

The other guy answered. Was untied and led off. <u>He must have said he had enough.</u> Snake thought. If he would have been able, Snake would have saluted the prisoner for having more brains than Snake did. That foreign manipulating voice in his head was still stupidly advising him to persevere.

The entire arena was hushed in expectation. The silence causing a pause in the ceremony. The spectators were holding their breath, watching the sole prisoner left on a pillar in the Maluayeb Arena.

Next to Xmucane, Jaguar heard her Queen mutter. "This obscenity must stop…"

Dirva was in awe. "He is the only one left. He must surely TOTL. He deserves what is about to happen." She barely contained her excitement. "He has much stubborn Willow in him. I will choose to Sooth him against his will should you cease this ceremony. I would cherish such a Kril in my Nest."

Xmucane's personal Kril, the Roman KrilChan KrutEk snarled. "You mistake defiance for courage. He is mocking all of you."

With a worried voice Jaguar asked. "My Queens…what if he survives four strikes? What then?" She shared conspirator glances with her Queen.

Xmucane studied Jaguar. The OvalChanHalach seemed to be in internal pain and mumbled out loud. "Yes, you are correct." The Queen thinly smiled. "What then?"

Dirva added. "He is about to meet TOTL; he will not survive the next strike. Perhaps you 'should' halt this?" Dirva had no concern for the man on the pillar; wanting Xmucane to violate the duties of an OvalChanHalach. Dirva would gain much if her advice was followed.

"Should I let this continue?" The Queen rhetorically asked her KrilChan.

KrutEk, the Commander of the Kril Legions, was immediately next to Xmucane. "My Queen...my Oval...we must continue the service of this creature. Our Kril traditions demand he be struck again. You must not discontinue his servicing to an Oval!"

Jaguar shrugged at her Queen. "Servicing is the way of the Krils. The Ovals not do condone but control timelines." She knew how much the majority of the Ovals detested the Servicing ritual, especially Xmucane. "As Queen you can command they stop..." Seeing Xmucane's angry glare, Jaguar continued. "...or continue this abuse."

KrutEk snapped. "Let the process complete. I do not have use for a new Kril who is incomplete."

"My mind would be more at ease if you not did get pleasure out of torturing these creatures." Xmucane said in resignation. "Do as you wish, KrutEk."

Snake heard a command bellow over the confusion by the guards. It was a man's voice, in another translated foreign tongue.

"Finish the Rites of Maluayeb Arena Service!" The man was shouting, sounding very much like he had never been disobeyed in his life. Snake's mind-hate focused on whoever he was.

Far off in the distance, Snake heard another question coming from the woman guard addressing him. "Do you accept another touch-strike of Oval service? Or do you wish me to stop, allowing you to continue your timeline? Not do be foolish." The female Oval voice from his guard was weary and beseeching. "Plead with me to stop servicing you!"

60

Her sympathy for him was, in her way, trying to be helpful; begging him to quit.

Snake shrugged with indifference and stared at the black dot, accepting the reality of that tiny dot.

An immense fourth strike collided on his neck meeting his shoulders!

Vicious pain exploded, blowing furiously into his brain! <u>The bastards increased the voltage of that damn thing!</u> He experienced agony and had no more thoughts. Snake's body was squirming in uncontrollable spasms; his knees were buckling. Tears were flowing down his cheeks. His greying tunnel-vision was blanking out his peripheral sight. Snake struggled to keep conscious for some insane reason.

He fought his urge to uselessly hang not wanting to drop to his knees. It would be so easy to succumb and pass out. He forced his squinting concentration onto the black spot at the center of his assigned column far above his crouching body. Curiously to him, the spot began glowing gold back at Snake.

Ord heard the Oval females in the Maluayeb Arena. Snake was the only victim left. They were chanting in some kind of religious ecstasy. A wall of sound causing the atmosphere to pulsate with their incantations.

Snake slowly forced his knees to lock by sheer will and determination. <u>Come on...come on...come on...stand up; make your legs work! If ya die trying to stand, so fuckin what!</u>

He felt the black ropes binding his body and arms begin to slightly loosen as he struggled to rise. He urinated. Embarrassed, he turned his head, crudely rubbing and purposely scraping his cheek against the pillar's material. Causing an abrasive

61

blood-speckled burn to form on his face. Snake grunted in an effort to remain upright. His inner will against his conscious overwhelming agony.

"He is standing." The female guard whispered in awe. Knowing the inevitable fifth strike was about to happen, the female guard was pleading directly to Xmucane. "He is insane! I...cannot strike him...again!" She yelled.

KrutEk, the KrilChan next to the queen, hand signaled.

Another male guard was assigned by KrutEk taking the female guard's place.

Snake was miraculously vertical after a minute; in a tunnel-vision-daze. His rage was overpowering, but he felt his fear would finally force him to submit. His failure was imminent. He wiped his tears on his upper sleeve, his lungs gasping. In absolute desperation; he refused to collapse. Inside, without the translator and the urging voice to focus, he felt more like himself.

The Maluayeb arena erupted with translated cries of happiness, encouraging shouts, and some vindictive-hate chants for him to meet TOTL. The din sounding like a caged lion, roaring in indignation for raw meat offered just out of reach.

The shouts in the Arena were joyous and ugly at the same time; some calling for more torturing of the bound victim.

"As leader of the Kril Legions, KrutEk commands his guard to complete his Servicing!" KrutEk was furiously shouting at the Arena guard who had replaced the hesitating female.

The OvalChanHalach pointed at the male guard. "You not will obey KrutEk's command!" Xmucane shouted. "Your Queen rules here!"

All eyes in the Maluayeb Arena were on Xmucane and KrutEk.

Turning his attention away from the golden spot, Snake looked at the female Oval leader on her parapet. The inner voice in his head informed him, in translation. She is the OvalChanHalach, the Queen here named Xmucane. She not can save you. Your life is in your own hands.

Mentally Snake taunted the Queen. Kill me, like the dumbass bitch you are! He concealed his thoughts; his face an angry mask.

The translator in Snake's head was advising Xmucane what his thoughts were towards her. His pain was palpable to her; their translator-minds both connected to each other.

When Xmucane remained silent, his pain was screaming for release and his thoughts came out in a crude curse. "Kill me cunt!" In his past life no woman would take that kind of disrespect from a man. He would finally die and get some relief.

His words were translated to all present.

With one royal hand-gesture the Oval Queen silenced the shocked and mewing Ovals in the Maluayeb Arena. She had taken control of the proceedings. Then she spoke directly to Snake. "Do you wish the other Seeched creatures to be in Oval service?"

The OvalChanHalach was pointing at the waiting prisoners on the other ramps transfixed on him.

Xmucane and Snake were the center of attention. The prisoners on the other ramps waited. Everyone was watching Snake and their Queen in this test of wills.

"Yes? No?" She asked.

*Crazy place…she'll do just the opposite of what I ask anyway. Those other guys on the ramps are on their own.* He chose to shrug indifferently, wearing a mocking angry frown.

His translator's voice came into his head again. *You are a very stupid male to taunt the OvalChanHalach. You are assuredly insane.*

After receiving a translation of his thoughts, the Queen smiled at him. "Very good. With your silent answer I will assume you mean yes."

She addressed the other ramps. "He has by his silence, made them all Krils. The other prisoners must accept at least one strike applied to them. Not more than three. As is traditional. Let them rest, then offer them the choice. If they survive the Strikes, Sooth them to an Oval immediately."

A tremendous roar of approval vibrated the arena. The intensity of their reaction overwhelmed the atmosphere in the arena; the climax of the day.

Silencing the spectators again, the OvalChanHalach quietly said. "This Maluayeb Arena ceremony is coming to an end."

Snake heard that hated man's voice again. "KrutEk commands no! This is a Kril ceremony! Strike him again! He is not ready to be Soothed! He resists the Kril Legions and the Queen! Strike him again! He will never survive!"

"You not will obey KrutEk!" The OvalChanHalach Xmucane shouted to the guard. "I command here!" Pausing a moment in thought, and then coldly addressed the male Guards around Snake. "KrutEk shames the Ovals! This Kril Strike-ceremony is concluded." She looked directly at the angry KrutEk in irritation. He was sputtering and complaining.

"Release the Four-Striker." Xmucane turned to lecture Snake. "You will be chosen. The young Oval's circle of choice must be made now."

Snake was untied from his column, dragged by his green-bound arms and led to the floor of the arena. Snake's stumbling walk over the sand was the longest walk he had ever experienced. Including the forced marches in the FMF and lengthy runs as a Raider.

Releasing an inaudible sigh of relief, Snake's body was swaying and stumbling to maintain his balance as he was pushed along. He was viciously dragged towards a contingent of Oval females.

They were stunning and young; spectacular in negligee-thin red smocks. They slowly formed a circle around him.

He noticed another Oval outside their circle. His inner manipulating voice, not the translator's, in his reeling unconscious mind compelled him to look just at her. That Oval was obviously in charge of the other Ovals, wearing a different solid deep-blue smock, with shiny-ebony hair. As beautiful as the rest of them. She led the circle of women to him and was gently pushing individual females to approach him one at a time.

Reela's mind was full of conflicting emotions. This Kril is a Four-Striker. Her heart was palpitating with anticipation and fear. She had never been present at a Maluayeb Ceremony in her past to observe any human accepting Service Strikes of that magnitude.

He is near exhaustion. He would make any Oval full of lust because of what he has achieved. I wish I could choose him. Her sister Xmucane possessed the only Four-Striker in the Oval Empire, KrutEk.

She was fighting her emotions wanting to reach out for him. Her choosing could never occur, she knew. By Oval traditions she could not choose him; only a new Oval in the circle could.

Snake was wobbling on his feet; struggling to remain conscious. He made the females in the circle wait. They're scared of me!

He stopped walking a couple of times to catch his breath. His old and new wounds were congealing. Probably by that goddam Strike weapon. He thought. His body felt hot-plate blistered, cooling after disconnected from the pillar.

Within the circle of Ovals, he stared at the black haired Oval circle-leader as she sent one of her choices at him.

The young female was sexually undulating, beckoning for him, to get his attention. He fixated on the leader of the Oval group as he was instructed by that prodding inner voice.

They look like strippers. He coldly thought. In his painful condition they were wasting their time. That fucking Strike-weapon is as good as saltpeter. He thought.

He noticed as he staggered, the green bindings of the other serviced men who had been led off, were lying on the ground outside the circle. Glancing down, his own binding-strips, which had tightly held his arms, were gone.

Reela was shouting at him. "Let one of my new Ovals choose you!"

Snake's mind was adrenaline-hyped. He hooded his eyes and pretended to be semi-conscious and groggy. He ignored the young Oval females, in

66

turn. He was blatantly shoving each of them away and not very gently.

"They must choose you!" Reela hissed at him with exasperation in her voice; the translation clear. She changed her position to the center of the circle, closer to the women and was gently prodding a young Oval forward every time he rejected another one.

Reela was distraught and in a panic because the OvalChanHalach and all the Ovals were witnessing her failure. Each time one of the women reached out for him he slapped their hands away. Reela yelled. "These Ovals are trying to honor you! Let one of them choose you!"

A beautiful new Oval with a strong will and no fear ran to grab Snake. He pivoted is hip and sent her crashing with embarrassment to the sand of the arena at his feet. Snake turned his back on her and prepared to meet another Oval.

Reela was beside herself with shame. "You are injured and in pain. Do as I command you!"

In answer, his pretense of grogginess was dropped and he was invigorated to fulfill his mission. His bonds had fallen to the floor of the arena. Snake roughly grabbed the Oval circle-leader Reela by the wrist with his left hand.

The Maluayeb Arena exploded in outraged sound when he held Reela captive. The new Ovals ran away in panic and disappointment from Snake and Reela.

Snake was instantly confronted by a blue-hooded guard. By reflex and training, his K-bar was in his right hand. He was now in control, but only momentarily.

He crouched, keeping the guard in front of him; facing his attacker. As he always did as a boy, when he got into a knife fight in the Milwaukee ghetto; his emotions went flat and his fear left him. Quelling his imagination, knowing he was going to be stabbed focused him then. He coldly concentrated on the task at hand; a lethal-no-holds-barred-fight for survival.

Wearing a blue robe and hood his opponent looked much larger than he was; the creature proficient with the Zipper-Lance-weapon he carried. The tip of the lance was glowing red charging up. His only chance was to get under the lance before his opponent terminated Snake.

He guessed the guard was not firing because Snake was holding onto this Oval female. Snake might throw her into the path of the bolt weapon. Snake was not in the mood to be chivalrous. His Oval prisoner might die by mistake at the same time Snake died. The blue-robed-guy was being extremely cautious not to hurt Snake's Prisoner-Oval. It was a logical assumption. Snake was counting on hesitation by the guard as a small advantage in this fight.

The blue-robed-guards in another outer circle surrounded him but were holding back. Snake assumed they were ordered by the Queen to not interfere. They could have ended this standoff with one lethal thrust into his back. The trick was, could they kill him without killing the Oval he held so tightly? Whatever the reason they were abstaining. Snake ignored them for now.

The only thing he could count on was his former street-knife-fighting, his hand-to-hand training in the Corps, and his battle experience. He never underestimated his opponents. Neither he nor his opponent were amateurs and both of them circled each other warily, waiting for an opening.

A macabre dance of two professionals gauging their opponent for a minor weakness. The big blue-robed guard swung the lance in a probing arc once, twice, then three times, trying to make Snake let go of the Oval; waiting for Snake to drop his guard.

Time slowed to an intense crawl; each combatant not wanting to make an error.

Then the guard lost his footing slightly in the sand, on the fourth swing of his lance. It was a small error, a minor adjustment for the guard. That is all the marine had waited for to happen. Instinctively Snake moved in seeing his opening under the spear-like weapon.

Without loosening his grip on his Oval prisoner and viciously dragging her with him; he expertly stabbed in the area of the Guard's jugular with his K-bar fighting knife. The thrust went through the blue robe of the tall guard until only Snake's hand holding the hilt of the knife was seen. Snake twisted the knife, before yanking the K-Bar out, so the wound would not close in a clean slice. The thrust and withdrawal took only seconds.

Snake backed up in a crouch in case he missed the jugular, waiting for any further resistance from the Blue-Robe guy.

The guard, mortally wounded, slumped over without a sound, like a witch in a bad movie and died, with a death rattle, in a final spasm. Their combat ended immediately.

Snake knew it wasn't over with the other guards and he turned to face them, preparing to die.

The entire populace of Oval spectators was hushed in bewilderment, as if sucking in all the air in the arena. A universal groan erupted from them.

The slow-motion fight had now speeded up in Snake's mind to normal time as the entire crowd in the arena let out a sigh of indignation and resignation. Witnessing a violation of Oval and Kril protocol.

Snake was spinning in a circle with his captive, anticipating an attack from the outer guards. Murder was not condoned, even on this killing field; that option was reserved for the Blue-Robed guards.

The Kril leader, the Roman guy, standing next to the OvalChanHalach leader on her platform screamed apoplectic, "TOTL HIM!" The Roman finished that shout with an oath of indignant fury.

"The guards not will obey!" Xmucane was yelling as loud as KrutEk. "The guards will hold and stand!" She was in command.

Snake put the point of his K-bar on his own neck daring the hooded guards to kill him before he killed himself.

Xmucane was yelling again this time to Snake. "I said hold! You will cease this shameful killing!"

She spun around to her KrilChan. "You will be silent, KrutEk!"

Grinning cruelly, displaying a calm indifference he did not feel, Snake waited. Holding his captive tightly in his grip; the Oval was frozen with fear. He thought to himself. Not bad for a creaky old man, huh?

He felt his body, with its twenty-four-year-old reflexes, were viable under his older veneer. That's a plus jarhead; gotta remember that fact. I may be old. I may be physically torn up. I may be worn out and about to die, but I ain't beaten...yet!

Snake brought his aching old-young-body to a position of attention facing the dais of the female Oval leader. He nodded, his stare boring into the guy in the Roman armor.

The Roman KrutEk was in a protecting stance in front of the OvalChanHalach, in case the barbarian launched his knife at her. KrutEk spoke, over his shoulder, directly to Xmucane, "He was not chosen by any of the Ovals in the circle. He did not have the right to choose! I will deal with this barbarian in the Kril Legion way."

Snake recognized the Roman armor on the man, inwardly amused the man's translated voice did not have a British accent, as all the actors did in the movies. When playing the role of Romans, who were Italians, the actor's spoke the King's English.

Someone's doing a hell of a job translating this guy's Italian for me. This KrutEk dude is some piece of work. He sounds like a General who just had his dress shoes pissed on. Could he be my fucking leader?

A one-word answer from his translator came into his head. YES!

The OvalChanHalach Xmucane was speaking to KrutEk. "He has chosen my sister. She was not his to choose. I agree. But I cannot change his timeline. The creature has absorbed Four Strikes; by my command his choice is final."

"He is a criminal!" KrutEk was shrieking. "You should have allowed five strikes on him."

"Five strikes usually bring TOTL to a Kril." Xmucane said. "Twenty Katun's, 'four hundred of your years', has time-branched since a Kril took Five Strikes. The Ovals are historically aware of that traitorous Five-Strike Kril. This creature has earned being Soothed; whether KrutEk approves or not do."

BalamEk, a disinterested third party and the keeper of the StelaBalaam, quietly spoke. "There is precedent in the Gryle for a Four Striker to choose. You are correct, my Queen."

71

That pronouncement was heard, in their Stels, by everyone in the Maluayeb Arena.

With BalamEk's declaration, Xmucane snapped a command. "Let Reela and her new Kril, proceed to her Soothing room. His Service to her will commence immediately!"

Anticlimactically, as Snake was prepared by the guards to be led out of the arena with his captive Reela, the Maluayeb Arena hushed.

Snake's captured Oval Reela walked beside a shuffling but still dangerous Snake out an arch while he held her wrist handcuffed by his fist.

In the tunnel under the Maluayeb Arena, Reela was whispering to Snake. "Our Queen has saved you. You are safe. Not do precipitate any rash actions. You have won much with your bravery."

Snake scoffed at her reasoning and whispered back. "I ain't no hero and I ain't won shit, honey."

They were surrounded by blue hooded guards. The guard's hostile attitude towards Snake was highly evident. He did not trust the guards and grasped his unsheathed K-bar tightly in his hand, letting them know he knew they were not 'his' protectors.

KrutEk turned in disapproval to the OvalChanHalach, "You are the one making him a Kril. He would have TOTLed, like the barbarian he is. I command the Krils."

"What if he 'not' had met TOTL, my KrilChan? What if he survived Five Strikes?" Xmucane asked. "Of course my sister not does deserve him. Of course she not does have the right to him. But he has chosen; which was his decision by absorbing Four Strikes. A minor tradition stored in Gryle records; part of the StelaBalaam."

72

"The OvalChanHalach is correct." BalamEk said.

Xmucane informed KrutEk. "If he survived absorbing a Fifth Strike, 'he' would have immediately been pronounced the Kril leader of the Legions. Is that what you wish for?"

"He would never have survived five strikes." The Roman was adamant.

Xmucane was softly speaking to KrutEk with a confidential air. "That creature you despise would have become my KrilChan if he absorbed five strikes. Sometimes you think as a Dinarchy fool. You would have been summarily demoted on command of the Tribunal, if he survived. Krils are stupid in political ways. Why would you want to chance him replacing you? Do you have that much hate in you?" She snarled.

"He would never have survived five strikes." He repeated.

Ignoring KrutEk, Xmucane said, knowing the entire crowd in the Arena and her Empire was listening. "I could...but I not would have allowed that new Kril to absorb a Fifth Strike going into the five unlucky Kin of Uayeb! A timeline possibility abhorrent to the Ovals and to me personally."

She angrily looked past KrutEk at the older past OvalChanHalach Dirva, standing next to her. Dirva was ordered by the Queen to instruct Reela in the anteroom what was expected of her. "Why would Reela lose control and allow this to happen?" She rhetorically asked Dirva.

The older Oval Dirva held her hands at her sides with open palms and cocked her head, as if to say, I tried, my queen.

Xmucane asked. "What is your analysis Jaguar? Is this a plot against me?"

73

Their intense shared glance was stronger in meaning than Jaguar's answer. "Timelines branch; plots do not. I perceive no plot against you. The Cosmic Egg has decided."

KrutEk shrieked. "This has nothing to do with your gods!"

"My Cosmic god is stronger than your puny herd of earth immortals." Xmucane said.

Put in his place by the Queen, he was diligently toning his attitude down. "Apologies." KrutEk was fuming. "He will not survive another Kin."

Xmucane consoled KrutEk. "You will remain at my side. You will retain your power over your Legions of Kril. Not is that power enough for you?"

The Roman said. "He is mine and my alone now. I will deal with him in my way; the Kril way."

"Not do incur my Queen's wrath; should you unlawfully TOTL him."

To alleviate any appearance of disobedience to his Queen, he added. "The Tribunal must decide!"

Xmucane's eyes were slits. "Not do test 'my' powers, KrutEk. I rule this Confederation Empire. Not do you know our history; what occurred before you were Seeched and reconstituted. The StelaBalaam reminds the Empire each Tun year, of the last cretin who took Five Strikes. I choose he lives. I will be obeyed by you."

His lowered eyes, his submission to her authority was obvious to all in the Maluayeb Arena.

KrutEk continued pleading with his Oval leader. "My OvalChanHalach, he defamed your traditions and laws and my Kril protocol and laws. He has killed one of my Krils. He is an uncontrollable barbarian and he must answer for this.

The Tribunal should be convened to determine where he is to TOTL and when."

"Be careful, my KrilChan, you try my patience." The Queen said. "Not now KrutEk. It is not the time!" Xmucane was hurrying along. "NOT THIS OB! I will deal with this angry new Kril and my sister. You will obey me, if you value continued command of your Legions."

KrutEk bowed in submission to his Queen.

The OvalChanHalach abruptly pushed past him. He followed obediently behind her. She was still talking to him. "The Tribunal session you yearn for, at my command, is assembling presently. We must hurry; there is much to do when we are so close to the first Uayeb."

Walking away, Xmucane's eyes bored into Jaguar's. The Queen's plot was hatched. As an afterthought, she said to Jaguar. "Ovals dread these unlucky Uayeb days for a good reason."

The Roman hurried to get protectively in front of her. KrutEk thought grimly. Should the Tribunal fail in its duty. This new uncivilized Kril will die my way, at my command, before another Kin day passes. That barbarian will never survive to go on his first Uayeb Invasion.

Chapter 3

The wall of KrutEk's tactical cubicle was replaying the Maluayeb Arena Ceremony in three dimensions; as if occurring in the next room. His room decorated for human, not Oval vision; having sparse utilitarian furniture. When Snake killed the guard, KrutEk waved his hand, the scene stopped, went grey, then black, and the yellow marble-motif wall reformed.

"Who is this barbarian?" He asked. "Why was he not decimated by our Kril guards before arriving in the Maluayeb Arena?" As the KrilChan leader of all the Krils, he wanted answers.

The Krils of his High Command standing before him were listening intently. Not wanting to bear the brunt of his anger, knowing better than to make uninformed excuses. The senior Krils looked at each other meaningfully. A rhetorical question they adroitly decided not to answer.

Two were former Legionnaires of ancient Rome. One Roman his executive officer KrutSeet, Assistant Commander of the Kril Legions. As his subordinate, KrutSeet was there for consultation and advice as second in command.

AinAcbal, a human Viking, was a hulking presence; nothing fazed him. Considered by KrutEk best of his field Commanders in the Kril Legions. The Viking's flaw was his need for action in battle. Queens or women of any kind, he considered weak. He was not of the ilk for becoming KrilChan; which made him dependable.

Another ancient Roman was ZacNaab. He commanded the half-Malkril cycle the new Zabin-

Krils would be assigned to in the coming five unlucky Uayebs.

ZacNaab responded first. "Sir, the barbarian will be in my Malkril cycle as Reela's Kril. I will dispose of him before this Kin day is complete."

KrutEk waved a dismissive hand. "Our new barbarian-Kril has to serve a full Tun year of training before going on a Uayeb invasion. Speaking before you analyze grates me. What do 'you' think, AinAcbal?" KrutEk asked.

AinAcbal's Viking blood was showing. "He took Four Strikes. He is brave and thinks on his feet. He could become a valued Kril." Quickly seeing that was not the answer his KrilChan wanted, he added. "He is also a threat to you and all of us."

KrutEk pointed to his senior advisor. "And what do you say, KrutSeet?"

"I was stunned by his murderous action." KrutSeet said.

"How refreshing. You were stunned." KrutEk repeated. He rubbed his forehead mocking KrutSeet's answer. "This abortion, out of the Seech Field from Earth insulted me, he insulted the Kril Legions, and he insulted our code of honor. While you are stunned. What is occurring here, KrutSeet?"

"I think we should be discussing this situation...alone...between each other."

ZacNaab spoke quickly. "You have much to do before the invasion, KrutSeet. I will assassinate this minor character; this new Kril."

KrutSeet snapped back. "The same way you did during the last Uayeb? That was extreme even for you, ZacNaab..."

"My leadership qualities are well known to our KrilChan..."

KrutSeet shouted. "Be still! I'm your commanding officer, and…"

KrutEk stood, silencing them both. "I believe, I command the Kril Legions, in this WHEN."

The Viking AinAcbal looked amused.

Silence followed KrutEk's words before he went on. "I want to resolve this problem with this aberration among us…immediately. Thus the purpose of this meeting."

KrutSeet took the opening given to him. "He should be sent to the lowest ArnaMal Legion unit we have, and never be heard from again."

"Interesting. It is your advice I send a Zabin-Kril, who has taken four strikes in the arena, into obscurity. Assigned to a common lower ArnaMal unit. A Kril, in the ArnaMal ranks, who has four strike marks? You choose this course of action in order to hide him? That is a brilliant solution. You 'are' aware…are you not…concealing him is now impossible?"

ZacNaab added his thought. "Hopefully, this abomination will die as soon as he mingles with the Kril Legions."

"Oh yes…we can only hope." KrutEk said with disdain at what he considered a stupid remark. "Now there is a strategic thought from ZacNaab. He 'hopes' the Legions will assume our command responsibilities." He looked to the ceiling. "If that were the case, why do my Legions need officers?"

AinAcbal scoffed. "That is ZacNaab's officer way of leading…from his rear."

KrutEk paced the room in thought and then turned to them. "And conversely, if we follow KrutSeet's suggestion which smells of Dirva's political policies. We exile our problem barbarian as if he had never existed. We should act as if he never

78

absorbed four strikes, never took an Oval by force, and never killed one of our Krils."

AinAcbal was not hiding his Viking scorn. "Banishing him is a woman's way of problem solving. Dirva is an Oval alien female."

KrutSeet said. "My Oval Dirva commanded me. Suggesting to KrutEk he banish the barbarian into her Nest before sending him to the ArnaMal Legion. This was not my choice of action."

KrutEk shouted. "You disagree with your Oval, yet you vomit out her words! What is your own solution; that is what I asked."

KrutSeet tried to stay on neutral ground. "He is a threat to all of us, particularly you, my KrilChan. He should TOTL as soon as possible." He felt agreeing with his leader was prudent.

"ZacNaab and KrutSeet are both wrong!" KrutEk stared at AinAcbal.

"We all honor you...and your wisdom." ZacNaab said.

"Be still, you spineless Two-Striker." AinAcbal said, displaying his Three-Striker pride.

KrutEk was speaking softer. "Attending the meeting of the Oval Tribunal, I will advocate 'they' TOTL this abomination."

"What if the Tribunal pardons him?" ZacNaab quietly asked.

"If the Tribunal shirks their duty; he becomes ours to deal with." KrutEk walked to his subordinate ZacNaab. "If the Tribunal equivocates and does not TOTL him. ZacNaab will supervise his training. Not you yourself, but through others you designate. As he has demonstrated by his actions in the arena; pain, scourging, or beating him will only harden his resolve. If he does not change his attitude under your

79

guidance…" KrutEk let his threat hang in the air briefly.

"Should that occur, what then are my options, my KrilChan?" ZacNaab was hoping KrutEk would command him to kill the barbarian. "What if I cannot control…?"

KrutEk responded. "Then I will have 'you' fifty-count scourged for failure and find myself another Malkril Commander. As an example of 'my' control. Do you realize what 'you' have to lose?"

They heard the man gulp. "Yes, my KrilChan. I honor and trust you." ZacNaab's fearful expression said it all.

KrutEk dismissed him with a flick of his hand allowing a visibly nervous ZacNaab to meld.

With disdain, KrutEk said. "Ovals use the word honor as a whimpering salutation." Pacing again, KrutEk said. "ZacNaab suckles at my cock to gain favor. He is incapable of understanding we need those anonymous Legion Krils to fulfill our primary mission."

"ZacNaab trusts you!" AinAcbal laughed.

"Trust is a whore's word to gain monetary compensation." KrutEk smiled at the others. "I am alone. I trust 'none' of you."

AinAcbal was chuckling. "Spoken as a great Konungr, our Viking king."

"Overtly, and I stress overtly." KrutEk said. "Assassinating this new barbarian Zabin-Kril will not sit well in our Legionnaire's bellies, no matter how much ZacNaab yearns. The Kril Legions must obey us and follow our lead or we will fail in this war."

"Failing is not your way…" AinAcbal agreed.

KrutEk smiled. "We must enable fairness in our leadership. Our Kril Legions fully understand cruelty, but will never condone insincerity. The common legionnaires I have observed, abhor that attitude. If we treat this barbarian badly outside our Code, they will see our action as irrevocable Mingo-pissing on their battle respect."

"The OvalChanHalach ordered you not to kill him. She will not be pleased." KrutSeet said in warning. He was anticipating KrutEk's thoughts.

Waving his hand in dismissal, KrutEk said. "Xmucane, as Queen, cares nothing for this insignificant Zabin-Kril." KrutEk's opinion of the Ovals was well known. "Her position and ambitions rest on a higher plane. She is concerned with her timeline and her legacy in her galaxies."

"Xmucane is well aware of your ambitions." AinAcbal said. "By her very nature the Queen plots against you"

His leader dismissed that idea. "As a Queen, I expect no less of her. She possesses more balls than many Krils I know." KrutEk said. "I command the Krils by right of performance. The Queen commands by right of succession. In loyalty, I obey her in principle."

AinAcbal and KrutSeet were nodding agreement to the realities.

"As a Zabin-Kril I war on my principles, not hers. I will not soil my sword on this minor creature. I will not outright kill another Kril, who has survived four strikes. That is against the Kril code and my personal honor. I will not betray our Kril Code as he has done. I am better than he. I am the KrilChan. He will TOTL, but not by my hand."

AinAcbal was grinning in his deadly Viking way. "My personal honor will not allow me to kill him for you. He is your problem; not mine." His eyes narrowed. "Inferring a solution is the same as an

81

order. There need be no direct command to kill this Zabin-Kril."

"As your loyal subordinate I must counsel you to be careful." KrutSeet said, trying to reason with his commander. "As KrilChan you are walking on quicksand, even if you are not directly ordering his death. You may be falling into a political trap set by Xmucane. This barbarian will not be a memory a Tun year from now and will be no threat to you. He sealed his NOW future with his hasty act of aggression."

KrutEk was deep in thought. "You sincerely believe this barbarian is part of a plot by Xmucane?"

"I suggest you be careful. If she chose this plot to negate you, it appears to have failed because of what happened in the Maluayeb Arena."

"In other words, you disagree with my decisions. Are you plotting to be KrilChan?"

"I diverge from your solution to the problem should the Tribunal not TOTL him. The barbarian has killed himself. The decision is yours, sir. As long as the solution does not come back to haunt you." KrutSeet said.

Just before melding, with AinAcbal in tow, KrutEk whispered to KrutSeet. "Never again disagree with my decisions or you will be crucified into oblivion lasting Kins days, KrutSeet. Along with that barbarian."

AinAcbal, who heard the threat was laughing loudly as he melded together with KrutEk.

KrutSeet was sweating, not from the heat, but from his own precarious position. He had been warned and not in a subtle way.

AinAcbal, that Viking, is more barbarian than this new Zabin-Kril. May that new Zabin-Kril meet TOTL, cross over into hades, and soon.

His own Oval, Dirva, had a plot developing for KrutSeet to kill KrutEk in the next Uayebs, making her OvalChanHalach again. It was clear to him Dirva had no compunctions using him as a pawn to gain power and could not protect him from KrutEk.

Oval law decreed only a Three Strike or more Kril could advance a Queen, when the OvalChanHalach's personal KrilChan was eliminated in battle. The Tribunal voted the succession of any old or new OvalChanHalach. A Queen, not perpetrating malfeasance during her reign, ruled supreme. Voting down the successor if it did not have the support of the majority.

Kril Code decreed if another Kril unlawfully murdered a KrilChan, in or out of battle for personal gain; the responsible Kril met TOTL for an illegal killing. The Oval Tribunal voted into office a new Oval Queen with a new KrilChan she chose. Or the old OvalChanHalach's choice to replace her murdered KrilChan.

Prior history drove the Confederation Empire's traditions. Their violent Willow males were a tyranny-driven patriarchal powerful military society. The pre-Oval females known as Vessels, could not own property. Vessels were not allowed higher stations, or control of Soothing-family groups. Willow males retained the wealth and inheritances; never a Vessel. Disciplining their bottom-naked Vessels with their Willow-whips. Willow society controlled all aspect of life in their Empire; particularly on home planet Arna.

83

Such was the status-quo, until the Vessels revolted on the last five Kin of their Tun year. Occurring after the dominant ruling Willows caused the near extinction of their species. The Vessels renamed themselves Ovals after their insurrection.

Ovals memorialized 'the unlucky Kins' of their Five Uayebs. Uayebs were deeply imbedded in cultural memories, recalling the Oval Revolt against their Willow males. In revulsion, their Oval uprising was resurrected in their personal Stels from the StelaBalaam records before beginning the Uayebs at the end of their Tun year.

After Oval Independence, the Ovals decreed Willow traditions and disciplinary rules cease. After Seeching became nominal, the Ovals allowed the Kril-controlled Maluayeb Arena ceremony become Kril mandatory. At the insistence of the Zabin-Kril leaders inducting the Seeched humans; they required pain-acceptance to cull out the unfit recruits.

As a precaution, Ovals decreed the Kril Service-strikes should never produce blood, physical

84

marks, or wounds of any kind. Behavior too prevalent in the Tuns when their Willow males ruled Arna and the galaxies.

Instead of banning the ceremony, the first OvalChanHalach decreed not more than five strikes, signifying the Five Uayebs, be applied upon willing prospective Krils; to eliminate Kril exuberance. Oval distaste for violence mandated them to qualify the Maluayeb Ceremony in the Arena be administered under tight Oval oversight.

Four-Striker KrutEk became KrilChan when Xmucane became Queen. At the time of his Seeching, he believed in Willow discipline. Scourging was used in his Roman Empire on earth, to encourage compliance with Roman Legion discipline. To his dismay Oval decrees were in effect after he was Seeched.

On the selection ramp, watching others in the Maluayeb Arena; he thought the Maluayeb Ceremony was useless punishment. When he experienced the pain in the Arena himself, he changed his mind; deciding the bloodless ceremony was as effective as Roman scourging.

Since Oval independence, only once had a Seeched Kril taken Five Strikes. That single Five-

Strike Kril immediately became the KrilChan of the sitting OvalChanHalach. Against her wishes, the presiding Queen was forced to Sooth him.

During his KrilChan Command, being an ambitious and mentally-defective male, he conspired to bring down the Oval Confederacy Empire. That KrilChan was destroyed, by his own Krils loyal to the Queen, following his failed coup d'état. The Ovals rejoiced.

The ruling OvalChanHalach immediately Soothed a more controllable Kril. Because of the previous disastrous KrilChan; subsequent OvalChanHalachs tightened their control.

By superstition, negative history, and bad experiences, five strikes were considered unlucky by the Ovals in the Confederation. Only the ruling OvalChanHalach could authorize a fifth strike. Five was an unlucky number.

So prevailing was their Oval fear, another previous reigning OvalChanHalach decreed no child-bearing beings in the Oval Confederacy Empire could issue a fifth hatchling from any one Soothed Kril.

One and two Strikes were common occurrence in the Arena. In the history of the Maluayeb Arena Ceremony only four Kril individuals accepted more than three strikes. Accepting the intensity of Four Strikes became a tiny minority and considered insane by the Ovals.

Of those four Krils, one Kril was a cretin, quickly disposed of by attrition. Three others, including KrutEk, became outstanding military leaders.

Oval society throughout the Empire now buzzed with the news of a fifth Four-Striker. Five a bad omen for the Cosmic Egg theology and prior

86

Oval history.    Snake survived due to prior Oval tradition, believing a Five-strike Kril a challenge to the OvalChanHalach's reign.    The laws of the Confederacy Empire protected the Queen.

The present KrilChan, KrutEk, was the last Four Striker in Oval-Kril memory.    A new Four Striker, especially being the fifth one to survive; Snake was not welcomed by the Krils or the Ovals. Rather seen as a usurper.    The Baktun Malkril Commanders in the Legions saw this new Four Strike survivor as threatening their ambitions to become the KrilChan, upon the dismissal or death of the Kril leader.

A Kril had his accepted strikes marked on his forehead with purple lightening marks; a sign of his status in the Kril hierarchy.    As a citation of rank for identification.    A Kril after absorbing mandatory strikes would have to earn his future positions in the Legions during Uayeb-battle.

Upon being strike-serviced, the Krils were

chosen by handpicked Ovals to be Soothed.  During

Soothing the Oval bred him to obtain his essence.

Once the Kril was integrated by Soothing to the Oval;

he became a part of her Nest family as her mate.

Traditionally, Soothing was an appropriate necessity

for continuing the Oval species bloodlines.

Snake staggered into Reela's Soothing Room, grabbing a wall for support. His chest heaving from exertion and pain. Panting, he coldly stared at his Oval and another female.

His Oval captive spoke. "Not do be afraid. We are here to Sooth you. My name is Reela. Though you have committed a grievous error, my Queen has commanded me to be your Oval. After I perform the Soothing ceremony, you become part of my Nest family."

The translations he received in the Arena were still in effect.

Sweat, from his hair and forehead was cascading on his face, stinging many of his facial lacerations. "Soothing-wise means you'll be healing me?"

"Yes, you are correct in your way." Reela was an Oval twenty Tuns of age, chronologically much older in earth time. Because of her age she was shamed when this Earth-Seeched Zabin-Kril had chosen her. Mortified by not controlling his actions; allowing him to dishonor her sisters and her Queen.

She lost two hatchlings children from prior two-strike Krils she had Soothed into her Nest. Losing progeny made her an outcast to some scheming Ovals. The future existence of the Ovals demanded healthy children. It was her former attending ArnaVal's fault according to her Oval logic. The ArnaVals, her multi-racial humans were always responsible in these matters, according to Oval folklore.

To add to her woes, both of those other two strike Krils she had previously Soothed were killed early in their Uayeb Invasions; not bringing her much honor or fortune. Her princess status did not rate higher strike-Krils.

She upheld the traditions of the Ovals by living every Bot of her existence dedicated to her Queen and the Oval Sisterhood. The Arena Service ritual had been explained to her beforehand. But she had failed again. Her fault lay in being chosen by the Kril before her. She felt she would never survive her shame.

Thanks to Xmucane, with this new Four Strike Kril in her Nest, her timeline possibilities were enhanced. She would behave as a Princess of the Ovals. Advancement and controlling Krils was a time honored Oval trait.

Still dazed, Snake heard Reela say, pointing at the other woman in the room. "Hortim is my Oval-trainee, a being from your hostile race's future timeline." The way she pronounced the word 'race' indicated she would rather pet a rattlesnake.

Hortim was a Hispanic diminutive black haired woman, obviously human, with the full lips and the hips of a Spanish ancestor. She had helped Reela drag Snake further into Reela's Soothing room. At her Oval's silent command, the small woman backed into a corner to observe a tradition she was aware existed for all Seeched male humans.

"How ya doing, Whore-Time."

"Watch your mouth. My name is pronounced W-h-o-o-Team! Show respect or shut up."

Hortim wore a United States Naval uniform with the rank of Lieutenant Commander and pilot's wings. A mystery to Snake. He had attended Escape and Evasion schools in the Marines. He wondered if Hortim was a Soviet Intelligence officer. One of their prisoner tricks was wearing American uniforms to pretend they were a friend.

After she was Seeched, Reela re-named her Hortim. Assigned to Reela for training. Known on earth in her circle of friends as a damn good pilot; a consummate professional. Selina was the eldest daughter in her family, with two younger brothers.

A woman fighter pilot trained by the United States Navy. She wanted to fly in space. To achieve her goal, she had to fly jets. Luckily, the Armed forces at the time, were being pushed to get women into war billets.

Despite Higher Command old-boy pettiness, outright sexism and discrimination she got her Navy wings. She earned her spurs in the first gulf war and led a squadron in the longer second war. Between the wars she earned her master's degree continuing to work toward her doctoral degree in astrophysics. Her plan was to leave the Navy after her enlistment; becoming an Astronaut in NASA.

Seeched during a night explosion on the ground, with two others. It rankled her she was the only one surviving. Her only Oval perk was she never had to absorb the Zabin-Kril pain in the Maluayeb Arena. Females reconstituted from the Seech Field were not required to endure that crucible.

The Ovals never allowed a Seeched female to be degraded by the Krils; reminding them of Willow traditions. Hortim hated that particular Oval decree. In her former life she competed against males and earned her billets in the Navy. She felt, in her heart, she could have handled the pain of the Maluayeb arena.

Embraced half-heartedly by the Ovals as one of their own; she acclimatized and adjusted into this matriarchal universe. She was not yet an Oval; more an auxiliary sister of the Confederation Empire.

Upset to find out she was expected to breed with a human male. Keeping Zabin-Krils subservient to the Oval cause. Hortim disliked being told what her future was to be without her consent. In this case being told who to screw. She was determined to adapt to this universe and its rules even if the selective breeding part was distasteful to her.

Though she seethed inwardly at his insubordination; she waited as directed, beyond sight of the bedraggled male specimen before them.

As Snake wobbled, he was aware of the ever-present translator in his head. He could understand Reela, though her voice was distant.

"Focus your mind! You must pay attention, Kril!" Knowing he was in a pain-stupor, Reela was curtly yelling at him. "Your name is given to you by me."

"I know my name, lady."

Leaning forward, she said. "You are now KrutChan meaning, 'Snake of the Kril'. A name of one of our constellations; part of our Divine Cosmic Egg tradition. Never will you use another. You are to be Soothed into my Nest. We will clean you, removing signs from the Seech Field and Maluayeb

94

Arena."

"Where the hell am I?" Not caring, he was slowly acclimatizing himself; his powerful fear swept away by rising anger at his treatment. "How did I get here?"

Reela indicated her Oval distaste for pain and suffering. "The Seech Field is an abomination to sentient creatures, as is the Maluayeb Arena Ceremony."

"That doesn't answer my question."

She appeared to empathize with him. "You must comply with my decisions. Never again will you choose your timeline branching." She swung her arm. "All you experienced will be explained to you during your Tun year of training."

"That's clear as mud." Snake said. Bile was burning his throat as his fear of capture was returning. "Where am I...like in what location?"

The translator voice in his head told him he did not need that information at this time.

Ignoring the voice, he asked Hortim. "Well...where am I?"

Reela hid a biological glove; an Oval Control-wand on her hand. Her loud voice quavered with indecision he did not notice. "We reside in the Oval city of Tikumyax, the capital of the planet Arna. You must be patient for more information."

"Why should I wait? Am I a prisoner?"

"Not any more..." Hortim said. "You're about to be family. Think of it as a shotgun wedding. You ever been married? You can't still be afraid?"

"Fear? You gotta be kidding, Commander. I died miserably. Afterward, that fucking female guard stomped me pretty good."

"I wasn't there in the Maluayeb Arena, but I saw you in my Stel recording." Hortim said.

"You've been trained to hide your fear. Still shook up?"

"Not anymore. Those goddam guards thumped that feeling out of me." He wasn't about to ask what she meant by a Stel.

"That female guard saved your life. In the old Kin days, a male Kril guard would have scourged you to death."

"Wow, am I lucky." KrutChan said. "The guard was compassionately beating the hell outta me?"

Silencing Hortim with a gesture, Reela changed the conversation. "How do you feel?"

He said. "Like I'm daydreaming. Everything is out of focus…even you guys"

Reela signaled Hortim again to remain silent. "You will obtain more references as your timeline proceeds. Be still and absorb your new life."

"Prisoners are kept in the dark so they can't plan escape." He grinned, realizing all those escape and evasion schools the marines sent him to were not in vain. He had been taught his chances of success were better closer to the time of his capture. He was already contemplating breaking out.

"A time-pause will teach you much."

Snake-KrutChan tried another question. "I want to know where the hell Ticom-…wherever-the-hell…is located? Is it a country, a state, a city, a town, a village…or what?"

"Pay attention old man." Lieutenant Commander Hortim frowned. "Her EkSeet advises Reela your translator says you're planning to escape." Hortim said. "You and I aren't captured. If we were prisoners, I would tell you."

"Tell her Eck-shit guy to stay out of my head."

96

"That voice you hear is the Arena-Translator."
Reela took control. "You must cease analyzing such matters. You have no frame of locus to compare our planet Arna to…stop questioning!"

His heart skipped a beat. "This's a bummer. When's this bad scene over?"

"Reela's EkSeet is having trouble translating." Hortim said. "Your confusion is normal. Your pain during the Servicing ceremony in the Arena should've convinced you this isn't a dream."

Reela was silent, in a conversation with her EkSeet.

He touched his forehead and winched, feeling the raised swelling, from being branded. He felt four small parallel lines shaped like lightening symbols.

Reela looked at him. "Those marks were imbedded by the Krils before you were released from your pillar in the Arena." She did not want to be accused by him as the perpetrator.

Hortim said. "You're on the planet Arna, old guy. There's more pleasantness to come."

He glanced around. The glowing yellow room was empty as a padded cell. "I'm on another planet called Arna?" He said for verification.

Hortim was laughing. "You're further away. You're out of our old universe."

KrutChan's mind was reeling, trying to absorb the consequences. If little green men abducted me in their flying saucer; I would remember.

"You must listen to me!" Reela was waving a driving-gloved hand in his face. "Not Hortim's earth drivel."

KrutChan was used to having irritated people yell at him. His nickname on Earth was Snake. As a child, he stuttered when pronouncing his S's. The bullies in school called stammering 'dummy' talk.

97

The Oval Reela could not know the paradox of his Oval name. Snake of the Kril was as good as my old nickname, I guess.

"Relax grandpa." Hortim said.

At the moment he didn't care. Pain filled his mind. All he wanted to do was pass out. "Take me to my cell with the other prisoners. I have rights under the Geneva Convention."

Reela's questioning face calmed when Hortim explained to her in a whisper what he was saying. "Not do you understand me? Remain silent, Hortim." She said. "You must heed my words, my Kril."

KrutChan thought Reela had a lousy translator or he did. Why were her words screwed up; mixing her precedence of verbs for Chrissake! Then again, she was a translated alien! "Okay...okay..." He wearily raised his hand and waved it. "...okay, you win. I'm drafted."

Dragging him into a corner of the yellow lighted room, Reela stripped him with Hortim's help.

Being naked in front of them bothered him. At his enlistment, he'd been naked in a crowd. During his FMF Fleet Marine Force tour he never was alone at any time, except on guard duty. In the Marine Corps, even heads, showers, or four holers he was never by himself for long.

His crusted wounds from Vietnam started bleeding again; his clothes sticking to his body, making the process excruciating. He moaned. His green jungle boots were a minor problem; but disposed of along with his heavy-green wool socks, with a brief swing of Reela's left hand containing her glowing orange control wand.

KrutChan staggered against the walls when Reela placed him into an upright container appearing

98

in the yellow wall.  Once he was in it, the shower-like container began to hum quietly.  His body firefly-glowed for a brief minute, and then he was covered in a brownish-white ash.

Reela swiped the glowing orange hand-wand over him again.  He was cleansed.  His open wounds cauterized and his seeping lacerations instantly healed.  Pain was a constant reminder he was not completely healed.  His body steaming.

The females, not gently, dragged him out of the enclosure.  KrutChan looked into a mirror on a wall appearing in front of him, that wasn't a mirror.  It was a 3-D projected hologram.

He expected the swollen face looking back at him with cuts, blue-black abrasions, and a fat lip.  A thick scar parted his hair on his left side and meandered down his face ending under his right chin.  Except for the brown at the top of his head, his hair side-walls were silver in color.  Moon cratered pits marked his face, pebbles of growths in uninjured areas accentuated the swelling, and his mouth slightly drooped on one of the corners.

"Calling you ugly would be a compliment, grandpa." Hortim said.

Reela explained, after conferring with her EkSeet.  "Your TOTL on Earth and the Seech Field not was kind to your appearance.  We Ovals not do expect you to be handsome, only functional."

"My Oval Reela means you'll be scaring the hell out lots of kids on Arna.

KrutChan's appearance meant nothing to him; he had seen worse battlefield corpses.  His anxiety under the scars, was caused by his recognition of his scowling father looking back at him from the mirror projection.  Dad indicating, he had screwed up again.  He half-joked.  "How old am I for Christ's sake?"

Hortim answered him. "You're in your early eighties. Your aging will be explained later."

She lowered her eyes in submission when Reela glared at Hortim for continually speaking to him without her permission.

"You must relax." Reela said.

"If I get any more relaxed I'll fall flat on my face."

When the Oval Reela took his elbow, he jerked his arm away, keeping his focus on the image in front of him. After a moment, he was glad she had touched him. The face in the mirror-thing looked less like his dad. "I can't even Cheshire grin right."

Reela asked her female EkSeet in her mind. Translate please...what did he mean?

The sentence is gibberish. He is hallucinating. Her EkSeet answered in her mind.

Hortim assisted Reela in directing him to the middle of the glowing blue floor, pushing him backwards.

"What the hell you guys doing?"

"I wish you to recline." Reela said.

KrutChan looking anxiously over both his shoulders. "On what? There ain't nuthin there!"

Reela gestured to Hortim giving her permission to speak.

"Quit fighting Reela. Trust us not to let you fall." Hortim said. "You'll survive." Her bedside manner was cool and detached.

Both females lowered him. KrutChan falling without resistance until he was floating flat on his back. A black aura under him supporting his body. He couldn't see the cot or bed or whatever he was on but he felt it under his body.

Reela passed her glowing wand over him, projecting a cone of pink light, bathing him in its radiance. His eyes were wide open in apprehension, but he was exhausted; not caring.

Sensing Reela when she tenderly touched his crotch, Snake moaned. "Forget it, honey. You said I'm eighty years old. Turn off your engine. You're beating a dead horse." He loudly laughed.

"Translate please; what is bringing TOTL to an equine have to do with my touch?" Her EkSeet translated his words and Reela admonished KrutChan. "Your organs are functional for my purpose. This traditional ceremony must be done."

Hortim added to the discussion. "Your Nest-Oval is telling you your body's alive; needing a stimulus. What she's doing is normal here on Arna. She'll take care of the mechanics."

Snake did not like Reela's probing. "Goddammit, tell her to leave me alone!"

Hortim watched in amusement, reacting with her woman's observation. A typical male response. She knew, from her own mild Soothing service to the Ovals, his compliance was imminent. Though Hortim's human eggs had not been gathered.

Reela would collect his 'essence' during his Soothing. Huge numbers of spermatozoa would suffice for her purposes. Reela would own them. They would be stored, part of her wealth.

There were no Oval males. They had sterilized themselves out of the Oval species. A peripheral reason for the acquisition of human males as a species necessity.

Reela would have to process his 'Essence'. Which had undergone faster than light tunneling in the Seech field; to be viable to her alien genes in this Arna universe.

Snake was brought to Arna to become a Kril. A human male used to acquire racial continuity,

101

wealth, fame, and elitism for his Oval Reela. Expected to die protecting her race.

Passing out upon the invisible bed inside the pink aura from the Oval device's effect; he was immediately dreaming. KrutChan saw a woman coming out of a blue-sparkling glacier lake towards him. He didn't recognize her. Her flowing cherry-blond hair covering most of her face. If he had ever met her, he would definitely have remembered her young body. The mystery woman seemed pleased to see him. Her manner projected her passion of the moment. KrutChan was astonished by the vision of a woman he 'should' know.

The woman wore a white floor-length negligee that stuck to her from the water. He felt his arousal lewdly anticipating her as she approached. She didn't say anything. She raised her gown over her head, paused a moment for effect, and then mounted him. The wet dream he experienced went from blissful aching and raunchy, to full blown crazy-lust as his body responded. His mind blotting out; his senses went into overload. The woman in the dream was awesome.

He startled awake. KrutChan painfully groaned.

Reela collected the essence she wanted into a small container in her left palm. After she obtained his 'essence', she transferred the container to a hip opening in her blue robe.

She turned off the controlling wand on her right hand speaking softly to compose him. Reela was whispering a prayer-like intonation.

"You are my Kril I call KrutChan. You will, in time-pause, meet the rest of your Kril cycle. Have knowledge imported into your unconscious system as to your duties, functions, and my expectations." She was whispering close to his face. "You are one with me as part of my Nest family. Earn much and learn to protect my honor and being."

KrutChan wasn't in the mood to whisper. Unexpectedly he grabbed Reela's wand hand and snarled at her, "You ain't even human! What in hell are you doing Alien?"

Getting Reela's permission Hortim told him. "Understand why we women on earth were so pissed at guys treating us like objects? How does it feel not to have a choice? The U S Military were the worst. The Ovals aren't any different with males in matters of sex, then those men in my past were towards women."

"Don't gimme that propaganda shit!" KrutChan yelled.

"Not all women back then accepted their submissive role in Earth society." Hortim grinned. "And not all Ovals are as gentle as Reela."

"Oh yeah? Whores are gentler or they don't get paid." He tightened his grip on Reela's hand. "Back off lady!"

Speaking calmly, as if she was used to Kril anger. She was holding a pewter colored disc in her hand. "This implant biological Stel I am attaching onto your chest, contains your personal EkSeet. Your Stel is blocked to your interference."

Changing the subject, KrutChan asked. "Who's the woman in my dream? She looked familiar. I never saw her before. I don't think we ever met. That your work?"

Reela felt his forehead as she listened to her female EkSeet. "Timelines have three observing positions; the past, the present, and the future, as your earth called them. If you not do remember the woman, she not is from your past. Obviously, she not is from your present. It is extremely rare, but timelines contain possibilities. She may be remembered by you, from your future."

Hortim explained. "It's more likely the woman is a figment of your imagination due to Reela's device. Before you woke up...that is..."

"You mean before this Alien raped me."

"Stop saying that! It was a collection! Your ugliness goes deep, I see." Hortim said.

"When does this sexual crap end?"

"You're not home free...yet."

KrutChan sighed in resignation. If Lieutenant Commander Hortim leads out a sheep on a rope, I'm really going to be pissed off!

Reela was explaining. "On Uayeb Invasions you will be facing, your life experience will be stored in your Stel and transferred to the StelaBalaam."

She adjusted the pewter-colored Stel device, with no chain, to his breast-bone. The biological Stel adhering to his body, becoming a part of his skin. "You now possess an EkSeet of your own, part of your Stel, to advise you of Oval laws, ceremonies, and traditions. Your Arena translator not is present."

"That translator idiot could've at least said good-by." His joke was disregarded by the females. He ran his hand over his craggy face. "What the hell is an Eeek-Shit, for Christ's sake?" His anger overriding his fear.

The correct form of the word is E-C-K-S-E-E-T...pronounced Ek-Seat. A voice in his head similar to his Id spoke. I am here to educate you and to assist you in your timeline.

"You're late." KrutChan said aloud.

"Your EkSeet is specifically translating my

104

words to you as of this Ob moment." Reela said.

For example, moment is your word translated by her EkSeet; Ob is an Oval word meaning approximately the same.

"When you are sent on an Invasion Uayeb, you will receive more information." Reela said. "When and if you return from the Uayebs, more information will come to you that will help you to survive here and on a Uayeb."

Hortim grabbed KrutChan's shoulders. "Come on. Consider this experience to making out in the back seat of your car without the foreplay. Don't be a baby."

"You wouldn't be so tough if she had mauled you, Lieutenant Commander. Tell her to back off!"

"I've absolutely no control over her. The Ovals consider males ignorant. As you'll hear a million times; Kril means Knowledge-Retained-Intelligence-Limited. You're a perfect example." Hortim said.

Reela's device lit up green forcing him to reluctantly relax his hold on her hand. His arms became locked again with green cable-like-vines to his sides.

Hortim was thinking. Typical Marine male, arrogance always on the offensive; not an appropriate response for a serviced Kril. He'll learn, when I have anything to say about it.

Reela reacted as if she expected his anger. "We will now complete your Soothing cycle." Reela calmly said.

Hortim recognized the female hidden anxiety in Reela's voice and face.

Reela's wand changed color to purple.

KrutChan passed out immediately; a purple aura bubble-like blanketing his body. The green vines dissipating.

A voluptuous younger black woman walked into the Soothing Room. Her short hemmed yellow skirt flowed around enticing hips. Reela's ArnaVal Akna.

"Is he the Four-Striker given to you?" Akna asked, after closely bending over him to focus on his face. "I remember this Kril from the reconstition area."

"ArnaVals not are Oval; too many human genes." Reela said to a nodding Hortim. "Akna has Beekav love for many Krils."

Speaking to Akna, Hortim asked. "You remember this one?"

"This Kril's mind was shattered by AnticArna potion." Akna was frightened she would be blamed. "I appear before Reela. Is he prepared to be Soothed by me?"

Reela was tiring of the questions. "Be still. You are interfering and needed not at this Ob moment. Return to your cubicle. I will summon you when his further essence collecting is near."

"I not am interfering." Akna blushed. "The Queen's ArnaVal informed me Xmucane summons you. Our Queen not did want to interrupt your Soothing tradition."

Pointing her finger towards the door at Akna with her Oval wand hand, Reela was not happy.

Akna melded in a rush.

Reela took a deep breath, approaching Hortim. "I must leave to meet the OvalChanHalach and discuss the cycling of the other Soothed Krils by their new Ovals. Stay with him and make no mistakes."

"Yes, my Oval."

"Finish Soothing KrutChan to my Nest family

106

and collect his essence. When I return I will summon Akna. Observe him until his Soothing-cycle by you is over. Then move him into his barrack cubicle. Be vigilant of your responsibilities; appropriate for him and his WHEN present."

Hortim's EkSeet translated the order. She nodded her understanding.

Reela melded from the Soothing room.

KrutChan lay sweating, cursing and tossing about in a nightmare.

Hortim was wondering if his experience would be different from her own. Knowing he was a candidate for meeting TOTL; she instinctively knew this guy would not die peacefully. KrutChan mumbled during his cycling. Hortim stood over him. Waiting for him to regain consciousness when she would teach him Oval discipline.

A long, long montage of fear induced images were cascading through his mind, seemingly for hours. He was shuttled from one scene to other scenes, one grotesque image after another with the same theme.

Surrounded by screaming faceless enemies, unable to break the claustrophobic terror he couldn't escape them. Fighting down a gagging-vomit reflex. Near the end of his nightmares he experienced a clear monstrous feeling of dread, comingled with realization he was in another multi-universe. There was no escape. That terror-thought exploded bringing him instantly awake, startled and frightened.

Reela melded into the Tribunal ceremonial room, advancing quickly towards her OvalChanHalach. Xmucane did not look happy;

107

aloof to Reela. Her displeasure ill-disguised. Reela stood where she was directed behind Xmucane.

The room full of higher caste Ovals of the court, with their Krils behind them, silent and watchful. A large room, yet claustrophobic at the same time, because of the low ceiling, totally different from the adjacent huge Tribunal Chamber.

"Late and last as usual, my younger sister arrives!" The OvalChanHalach Xmucane said. The other Ovals present mingling on a separate immense veranda were laughing and tittering in deference at their Queen's joke.

To start the Uayeb ceremony of the five unlucky Kin's cycle, Xmucane gazed silently at the crowd as a hologram-like recitation overflowed an invisible dome structure forming over them, immersing the crowd in Oval history.

It was a memory they recalled every Tun year, before the Uayebs, boring history to some of them too jaded by life. The Oval majority were exhilarating in the recall, reminding them of who they were, where they came from, and what was at stake.

In Arna's distant past the Willow males of the female Vessels cruelly ruled their society. Their species evolved from reptiles, becoming bipedal. Their reptile instincts by evolution, were intense. Of a violent nature, they enforced punishments on each other by scourging, maiming, or killing the non-compliant disobedient to Willow edicts.

Not wanting to eliminate mating by treating their weaker females as harshly as the males, they punished their defiant Vessels with whippings on bare bottoms; forcing submission. Killing the stronger untamable females.

Using their scientific female's-inspired crude Osil drives for invasions, the Willow males spread

108

their violent nature across a thousand galaxies in continuous warfare, ruthlessly dominating the other species they encountered.

The Willows used weapons of mass destruction to achieve their goals. Extremely capable in the sciences, the Vessels had helped the males to amass their weapons.

The Willows were chagrinned, over countless battles, that not all species they encountered submitted weakly to their order. The conquered other species, stole, reengineered and used their modified horrific weapons on the Willow invaders.

Misusing the Seech Field, developed by their Scientific Vessels, generated a one-time Boson-Osil wave throughout the known universe. Sterilizing the majority of the Willow males. Nearly causing female Vessel extinction.

The Vessels renamed Ovals had had enough.

With their newly developed Controlling Wands, the female OvalChanHalach, when the Willow males became impotent, seized power. The Ovals destroyed all the Willow mechanical devices and mass-destruction weapons; forever prohibiting Oval use. Forcibly banishing their Willow-males into their Invasion modified crude Osil-drives. Sent beyond the home galaxy, never to be allowed to return. Retaining a tiny minority of loyal potent Willows able to be controlled.

After they were exiled, some Willows were still capable of reproduction, had brought their loyal Vessels along. Using the Vessels banished with them for breeding stock to bring back their population of potent males. The banished Willow males, far out in the recesses of the universe, conquered two existing Empires, forming anew a patriarchal dictatorial Dinarchy Empire.

The Willows spread virus-like, colonizing foreign Empire's galaxies, planets, systems by systems. Gaining strength, planning on destroying the Oval Confederation Empire.

Unknown to the Willows, the Oval scientific females re-engineered Osil dark-tunneling biologicals; more advanced than the Willow's Osil-drives.

Though galaxies and immense distances separated each species, the Oval females determined they needed a military force to confront the Dinarchy males. After the Willow catastrophe, in defense of their culture, clashing with the Dinarchy threat, the Ovals enhanced the ability of the Seech Field. Importing life forms from another multi-dimension-universe to fight the Dinarchy.

Most Seeched individuals were unsuited for the Oval plans; either useless for breeding or had no expertise in fighting. The Ovals strived to find a suitable sentient species.

The males from the Earth Universe became the optimum lifeform for Oval purposes. As a precautionary measure, the female Ovals refused to allow Seeched earth-male life forms to take the place of their Willows.

Those human males would never be accepted as equals in Oval society. Imported Zabin-Kril's were to stop the Dinarchy and would be controlled by the Ovals until meeting TOTL. Krils would not be allowed to gain significant power or wealth or substance within Oval society. The humans were destined to become Krils; the Ovals their Controllers.

Complications arose. There was no battle-incentive for the Zabin-Kril warriors. Being paid for their services by the Ovals solved one dilemma. A system the Seeched warriors were used to in their

110

Earth universe. The Ovals installed a controllable leader of the Krils, solving another dilemma. A KrilChan the warriors would obey and fight for under the Ovals. That leader could amass a great fortune but never as much as the least of the rich Ovals. The Oval plan worked well.

The Oval government system slowly became extremely proficient after Seeched Zabin-Krils from the Earth-Universe-Empires took control of the Krils.

Through trial and error, the KrilChans advanced the abilities of the Kril forces. The Ovals mistrusted those KrilChan leaders and their Krils, making provisions for their replacement. By attrition, outright Oval-Tribunal scheming to replace them, or more commonly by meeting TOTL.

If the KrilChans tried to gain influence beyond the duties of the Kril-mandate the Ovals ordained; the Ovals removed them. Through Tribunal decree the Ovals decreed the leaders of the Krils, were to always be in the most exposed frontlines of any battle. In this way the Ovals culled any cult of allegiance to any one KrilChan.

KrutEk, the present Roman Kril, in a line of thousands of preceding Zabin-Krils from numerous Earth races, proved to be adept and skillful, taking his leadership to an optimum level. This fostered huge resentment of him by other races. Whose leaders in the past had held his position. He succeeded where they did not. Those past KrilChans lost too many Uayeb campaigns.

Under the leadership of KrutEk the Legions fought the Dinarchy; halting their advances over Tuns, a quarter century. The confrontation between the Oval and Dinarchy genders for power remained murderous and vicious.

111

Oval-Kril history being projected on the dome came to an end. The remembering time was over.

The Uayebs became, for both the Ovals and Krils, an anniversary of contention. A way for the Ovals to try to move into peace with more prosperity, while attempting to control the surviving Kril's murderous tendencies.

The Queen melded to the Tribunal High podium. "Begin the proceedings of the Confederation Empire." Xmucane said waving her arm, causing a holographic projection, 3-D dimensionally for all attending, showing their galaxies. "We will collectively agree this Kin upon our goals to defeat the Dinarchy during these five Uayebs."

Reela watched Xmucane leaving the podium, drift away to another wall portal, looking bored. Confederation policy required updating, deleting, and ratifying the existing Joint Declaration at the end of the last Tun year. The Empire needed War Objectives at the beginning of the coming unlucky Five Uayebs.

Xmucane had reviewed the Declaration a KinUt ago, as she did every year. She now busied herself with other mundane tasks. Reela sympathized with her sister's responsibilities. In the background she heard the Tribunal speeches droning on.

The Stels reverberated the final vote. "As approved, this Long Count Confederation Declaration is stored in the StelaBalaam as the most current War Missive and intentions of the Confederacy this date until obtaining the unconditional surrender of the Dinarchy and its allied minions."

With that pronouncement, Xmucane walked back to the High podium. "Now!" With that single command Xmucane immediately took control of the Confederation meeting as the War declaration projections disappeared. "This Tribunal's Confederation Declaration is fact and has passed with no further objections."

Their mission fulfilled, many ambassadors left for their other specific species duties.

After a major bit of reshuffling of hovering seats, Xmucane continued. "Uayeb 13 will commence as decreed. Upon the end of this Kin day, the beginning of Uayeb One will commence on our zero timeline when the first of our unholy Kins occurs."

BalamEk went to the Queen's side and said. "I rise calling for the Oval Tribunal to conclave."

Xmucane took BalamEk's hand. "This ends this official discussion. The remaining Confederation delegations are adjourned." She held up her palm. "My Oval Tribunal members will remain."

Though her demeanor indicated her distaste of such things, Xmucane immediately started the discussion. "We move to the matter of the new Four-Strike Kril. The Tribunal will discuss his fate. My KrilChan KrutEk, has already advised many members on the Tribunal who not did witness what occurred in the Maluayeb Arena this Kin."

"You are honored..." BalamEk said.

Xmucane was in a hurry, admonishing her Tribunal. "Timelines are short." Xmucane said. Regally adjusted the sleeve of her robe, pointing to her Kril KrutEk. "Let us quickly dispense with this minor creature. Afterward, we will finish more important actions and matters."

KrutEk spoke quietly with uncharacteristic deference to the Tribunal. With his usual Roman aloofness of politics, he was in his element. War was his authority, his total focus, his core.

The Ovals and other species delegates did not confront him. They despised him and his subordinates. They had no intention of ever being like him.

"My Kril Legions in the other Galaxies are on Five-Kin Standby. The Arna Kril forces have been alerted. Arna's Invasion Krils will leave before sunrise on the First Uayeb. I won't divulge where the attack will occur for security reasons. My Kril legions will not...and I will not...fail."

KrutEk pointed his white baton to the walls projecting a flashback to the Maluayeb Arena. "It is obvious the barbarian has defied the Ovals and broken his Kril Code."

"But was he a Kril at the time?" An older Oval Tribunal member constantly involved in small details, asked.

"Cease the Tribunal's numerous questions. Many of you were present in the Arena." Speaking to her KrilChan, Xmucane softly said. "I require KrutEk to be brief."

Reela shut off her mind to KrutEk's sputtering.

Various Oval factions on the Tribunal were debating pro and con, for a few minutes.

Abruptly, Xmucane held up her device-hand emitting a red aura over the entire Tribunal to stop the debate. "I have heard enough. The five unlucky Kins will start in the next nine Ki. We not do have time to endlessly repeat the infractions. What is your Tribunal decision regarding this matter?"

Dirva shouted. "Vote to TOTL this creature!"

Xmucane calmly said. "This is a trivial contention to be wasting our timeline upon. This minor Kril has disrupted our meeting. We must move on to more pressing matters. We have to prepare for the first Uayeb."

KrutEk supported his Oval. "My Queen Xmucane is correct. I will handle the details. Vote for me to rid this Kril from our ranks. He was improperly serviced."

Xmucane was angry at KrutEk's breach of protocol. "My human KrilChan forgets I am the OvalChanHalach. I am his Queen; not his Nest mate here. I alone, not he, will handle the details, however you vote. Kril ways are Dinarchy ways. The Ovals command here. What are your wishes as a Tribunal?"

The majority on the Tribunal then voted into their Stels.

Dirva, with much pleasure, as titular head of the Tribunal, immediately announced their decree; through their Stels. "The majority of the Tribunal has decided this 'KrutChan' is to meet TOTL for treason before the first Uayeb begins."

The recorded decree was announced by the StelaBalaam. Xmucane accepted it.

KrutEk was overjoyed, but kept silent.

The Tribunal moved on to other things, mainly the strategy of the coming Uayeb battle.

KrutEk began his briefing.

Xmucane took Reela aside, back into the small alcove, as her KrilChan continued to speak to the Tribunal.

"Have you Serviced your new Kril?"

"Somewhat. I have his essence. I came as soon as you summoned me."

"Do you always have to be half done in all your endeavors?" Xmucane said with exasperation and shook Reela's arm. "Your command to your Nest Oval-trainee Hortim to finish your Kril's Soothing was not obeyed by Hortim. She has let her human emotions overrule your authority. She has decided to discipline your Kril by using our Pavilion chastisement traditions. Does she rule your Nest or do you?"

Reela bristled inside. "It is my Nest and I command all in it." Reela said, hurt by her sister's words.

"Command your Nest to follow your instructions without modifications. My EkSeet, through your Stel, will give you the facts about the error your Nest Oval-trainee is planning for your Kril before you leave here. Must I always be admonishing your performance, Reela?"

"I obeyed your command to Sooth him. Dirva shamed me again."

Xmucane brushed Reela's hair from her eyes. "Dirva obeys my commands by twisting them to her advantage. She has taken you under her wing to guide you. I want you in her inner circle to assist me."

"May I ask how you would have controlled him?"

Xmucane hissed in annoyance. "My duties as Queen not do include your Oval training, or controlling the mind of a minor Kril. Complete his Soothing. Later, bring your Kril immediately to the Memorial gathering."

"I not do understand." Reela said.

"What is causing your misperception?" Xmucane was sighing.

"You wish me to bring…my new Kril…to the Ovut ceremony." Reela stammered. "The Tribunal has voted to TOTL him. Do you wish his remains to be brought to the ceremony?"

Xmucane grimly smiled. "Your EkSeet's ability to become so immersed in details; your mind constantly wanders. Coupled with your inability to ascertain your perspective of the whole. Why do you believe I stopped your new Kril from receiving a fifth strike?"

Reela's was confident in her answer. "You followed Oval Law. You not did want my new Kril to replace your KrilChan if he survived the Fifth Strike."

"You are correct, however…." Xmucane took away the compliment with her next sentence. "…any Oval, ArnaVal, or female Germ understood what I did. Your thoughts not are clear as usual."

It seemed to Reela she could not be accurate at any time as far as her sister the Queen was concerned.

Xmucane walked over to an ornate golden table that appeared at her approach. She poured herself sweet Oval-Osil wine, offering none to Reela. She sipped slowly, savoring the liquid while she was in thought. "My reasons are never divulged to anyone. Why do 'you' think I ordered you to Sooth him?"

"I was astonished…and grateful to you."

Xmucane's eyes narrowed. "Precisely the effect to infer on you…and Dirva. As a former OvalChanHalach, Dirva had the right to take him from you. She salivated at the chance to choose him to correct his error and Sooth a new Four-Strike Kril into her Nest."

117

Reela was smiling in understanding. "You are a great Queen. I obey you, not Dirva."

Correcting Reela to understand the reality, Xmucane said. "You are a princess of the Ovals with a Four-Striker in her Nest."

Xmucane turned and glanced into the Tribunal to see at what point KrutEk was in his briefing.

"I am ashamed. I not do deserve a Four Striker." Reela said.

Xmucane said. "To correct my oversight, I am changing your position as Oval princess to OvalChanHalach-Consort to me. Putting you on equal terms with Dirva and any other possible OvalChanHalach-Consort plotting to succeed me."

"I am overwhelmed, my queen." Reela was stunned. "I have no glimmer of understanding of what you are giving me."

"Oval Princesses not are given anything! They serve the Queen." Xmucane straightened her back. "It will be difficult for you as a Consort. This new Kril of yours has been condemned to TOTL by the Tribunal. You must use your new position to control him until then. Ovals command and he has to obey. If you not do put aside your princess attitudes, you will fail again. Dirva will find a way to banish you to the level of a Germ."

"I honor you and understand."

Xmucane put down her empty cup. "I shall dispense the Tribunal ruling against your Kril at the Ovut ceremony, after our Memorial tradition. I not can overrule their decision. You must control your new Kril until that time."

Whatever does she mean? How can I control a Kril who will meet TOTL? Her indecision was welling up in Reela. "But, what if I not am capable of controlling KrutChan?"

118

The Queen glanced to the ceiling. "Then your incompetence will doom you to obscurity." Xmucane turned her back to Reela. "I not will fail. See that you not do. If you disappoint me, as you usually do, I have an alternate plot. No matter what you do, I will survive. Not do fail me or yourself."

It was Reela's turn to stand tall with newly instilled confidence and mental-will. "I honor you, my Queen. I will make you proud of me."

"Becoming a respected OvalChanHalach-Consort will make you proud of yourself. If you believe in yourself and who you are; strength will follow."

Reela remained transfixed, anticipating the worst.

"Retire to your Soothing room immediately and correct your trainee Hortim." Xmucane said.

The Queen did not indicate her plans. Politics were not Reela's forte.

Chapter 4

KrutChan stumbled through the entrance he expected to be his barrack. When his head stopped spinning, KrutChan realized the room wasn't like any barracks he remembered.

A man approached him. "I am your EkSeet. Your translator from your Stel Reela placed on your chest. This is the training area for new Krils."

KrutChan mind was somewhere else. "What's your name and where did you come from?"

An angry man wearing Marine herringbone dungarees from World War II thought KrutChan was asking who he was. "The Ovals renamed me CheChun. Pronounce it Chee-Choon, boot."

He was pointing at KrutChan. "Nobody uses their past name here." This KrutChan is a pussy. CheChun glowered at KrutChan, sanding toe to toe to KrutChan. "None-a yer biz-ness where we're from. This whole bunch of Krils don't care where the fuck you came from, boot."

KrutChan bristled at being called a 'boot'. He had over two and a half years in the FMF, Fleet Marine Force, as a grunt. He wasn't impressed by this guy and threw out his chin in defiance. "You the fucking leader here, old Corps?"

CheChun drew his thumb across his neck. "If I was, sonny boy, my K-bar would slit yer fat neck."

Just what I need; some salty old bastard getting on my case. Tough guy. Welcome to the Kril Legions. KrutChan was thinking. Grinning at the World War II guy, KrutChan asked. "You planning on tightening me up, hardcase? Ain't no anchor holding your ass down." Never back down. KrutChan was thinking.

"Be silent CheChun, this time-pause is his; not yours." The EkSeet interrupted. "KrutChan is a Four Striker. After his Tun year of training he may become your leader."

"I don't know shit about leading Krils." KrutChan said.

CheChun countered. "Yawl don't know nuthin, boot."

EkSeet said. "The Kril Legion's Uayeb Invasions across the universe on other worlds will educate you. If you survive."

"Invasions?" KrutChan hid his feelings of terror. "I've never been on another planet."

"Like ah said, so what?" CheChun added.

"We've never even been to the moon." KrutChan said. "Much less ridden in a space ship."

"Makes no difference." CheChun said. "In my timeline on earth, Buck Rogers was fighting Emperor Ming in the serial flicks. Besides numbnuts, a Burseeosil ain't a spaceship. It's like a troop transport."

"You are confusing him." EkSeet said. "Stop interfering."

Shaking his head, KrutChan was absorbing his present reality. Troop-transports, Mike boats, Armored Personnel carriers and helicopters were more familiar. These guys were speaking as if spaceship invasions were standard operating procedure. He was mentally in trouble and he knew it. He decided to shut up.

"You will be indoctrinated during the coming Tun year." His EkSeet said. "You have learned too much already." He turned to the old corps marine. "This is my class not yours, CheChun."

"A couple mugs over the years took four strikes; this guy ain't no different." CheChun was chuckling. "Four Strikers don't live long. You ain't goin' ta be another MacArthur, boot. You'll be in charge of a Kril-Bot starting out. The guys in your Bot-unit will be the unlucky bastard's yawl'll lead while you're learning."

"CheChun is throwing a spanner, filling your mind with inaccuracies." EkSeet said.

A Negro marine, from the look of his utilities, was Vietnam Seeched; moved in and faced KrutChan. "Name's XibEk...pronounced Z-I-B E-C-K." No smile. No handshake. The man was distant; for some reason angry towards KrutChan, though they had never met.

"That lifer-overseer CheChun is fukin' wid ya. He likes to swell up his balls like he knows everything."

KrutChan tried to link with his past. "You're a marine from 'Nam. What unit were you with?"

"None-a your mutherfukin business." XibEk said.

So much for 'once a marine; always a marine'. KrutChan thought. In Nam, off patrol, the Negroes kept to themselves.

He stared at KrutChan. "I been on as many Uayeb planets as this poor white trash." XibEk said. "That cracker ain't shit. Your Kril Bot-unit'll be the size of a platoon."

"Never commanded forty men." KrutChan said.

"I have..." A slight built future US Army soldier said, wearing a Wehrmacht-style helmet.

XibEk groaned. "Yer a doggie butter-bar lieutenant." XibEk snarled. "You don't know where your mulatto asshole is. Both ayou pig-chucks is lifers." That insult defined XibEk. Calling them

122

'lifers' said all he would ever say about any of his designated leaders, past or present.

Since he was born, KrutCheebel's mulatto past had the white folks considering him black and negroes disliking his whiteness. His year of Kril training had recently ended; never changing attitudes. He had never been on an invasion.

CheChun added, speaking to KrutChan. "If that uppity doggie did it; you outa be able to handle it. On an Invasion Burseeosil yah'll have somebody in charge over you, telling you where to go and what to do. What's the difference between your past THEN and your future NOW?"

KrutChan was shaking his head, not understanding the Kril Legion terms being used. He was living a nightmare. Though feeling like real time.

His first meeting with Kril Legion forces was under a green aura dome looking out onto a plain of crumbling broken pyramids and mounds of ancient destroyed habitats. Miles upon miles of debris, broken and weathered by time, graced the landscape.

His EkSeet instructed him. <u>What you see is a former Willow containment area. The Kril Legions were confined here by the Ovals for training after their Willows were banished.</u>

A billowing dust devil was winding around itself, off in the distance, approaching them at meandering speed. He watched its white-yellow-brown snake-like twisting shape. Nearby minor wisps of wind buffeted against the aura. The ruins did not look destroyed by war or violent natural events. Passing of time seemed to be the reagent. There was bright sunlight interrupted by clouds of all shapes; moodily adding to the desolation.

"Tikumyax is ultra-modern by comparison." One of the awestruck new Krils whispered.

This training area contained hundreds of groups of other new Zabin-Krils. Each group surrounding one EkSeet. From the expressions on the other foreign Krils, they were receiving the same information KrutChan was getting.

"Though Kril idioms hardly qualify as King's English, EkSeets translate for a modicum of understanding. All of my brethren you see here are named EkSeet meaning Star Wisdom. The Ovals do not name us individually. I am your teacher-translator. Questions and answers will be imparted."

He turned to the group. "This newly arrived Four-Striker Kril is KrutChan. Where does KrutChan think we should start?"

"This place is a bummer." KrutChan said. "After my psychedelic trip I was brought to a yellow-glowing empty room with invisible furniture by a winking Alien female."

"That ain't a question. Your female is an Oval." An older veteran Kril said.

"You will be informed." EkSeet said.

"The Oval chick waved some doo-hickey on her wrist, to knock my ass out." KrutChan said. "The machinery used in this puzzle-palace blows my mind."

Another seasoned Kril said. "We call the impossible offal, 'Oval Magic'.

"My Oval disappeared through the wall when she left. Is that a magic trick?" Another new Kril said.

"Not magic; it's called melding here." EkSeet said. "You observed one of their most intriguing abilities…electron splitting, or fractionalization."

124

KrutChan snorted. "Glad you cleared that up."

"You will see many strange things during your timeline here; do not dwell on them." KrutCheebel said.

EkSeet continued. "Let me give you an overview." Speaking slowly, EkSeet explained. "Through the Seech Field you have arrived in another Quantum Many Worlds universe, or Multiverse if you prefer, instantly exceeding the speed of light."

"Bullshit!" A new Kril shouted. "Einstein's Theory states nothing can exceed the speed of light."

"Total destruction does not occur in either Universe." EkSeet sighed. "Mass and Energy only changes form. FTL is not in your Standard Model; incorrect in our Model."

"Cut the science crap. Answer my questions. What about the chairs and tables I couldn't see?" KrutChan asked.

"The invisible furniture you referred to is biological. Not able to be seen by our human visible-light wavelengths. You will collide with the invisible furniture; so walk slowly."

"How come the Oval ain't crashing into them things?" Another new Kril asked.

"Ovals visualize in the entire electromagnetic spectrums by adjusting their four multi-nictitating membranes. That is why his Oval appeared to be winking at KrutChan."

A voice shouted out in the group. "Yeah, bitch wasn't winking. Ovals hate our guts!"

EkSeet frowned. "Her vision ability is not Oval Magic. The powerful creature on your Oval's wrist is also biological. It is not a mechanical device or wizardry."

"She'll tune you up when she activates that

125

bracelet of hers; getting your attention."

Another veteran Kril said. "She dropped you into this time-pause."

KrutChan asked. "What kinda pause?"

"Oval Science manipulates Time."

"Their machinery is awesome." A new Kril said.

"Ovals don't build or use machinery." The veteran Kril said. "Everything they have is biological. And before you ask another stupid question, does an ant need machinery to heavy lift?"

A new Kril, like KrutChan, asked. "Gotcha. What's a time-pause?"

The crowd of Krils simultaneously shouted for the new guys to shut up.

Stupid or not, KrutChan asked his EkSeet anyway. "We're all asleep?"

"It does not matter." EkSeet said.

The good news for KrutChan, none of these Krils were acting like those crazy superior females.

The threesome of XibEk, CheChun, and the new Kril KrutCheebel were arguing with each other. And then, for a few minutes, the two Kril veterans were picking on and razzing all the new Krils.

"Enough horseplay!" EkSeet shouted and went directly to the point. "These two nutter's advice should be disregarded. As I am speaking, the new Zabin-Krils are being subconsciously fed information allowing them to intelligently survive in this Oval universe."

"Are you the commander of these Kril Legions?" Someone else asked.

"An ancient Roman KrilChan rules the Legions. This Four Strike Kril KrutChan, will be your leader after his yearlong training is complete." EkSeet paused. "That is to say, if he does not soon

126

meet death-TOTL. His future NOW, is currently being decided by the Tribunal."

"If the Oval Tribunal decides to kill em, why's he in his time-pause?" XibEk asked.

"Sheee-it..." CheChun hissed. "I be go to hell. Dis lashup is getting more fucked-up by the minute."

"Your pathetic use of English hurts my ears." EkSeet snapped at CheChun. "You have existed here the longest of these Bot-unit Krils. Speak coherently without your irrelevant discrimination."

"Are you an Oval or a Kril?" Another ancient soldier from Israel asked EkSeet. The man wore Judean armor leather with a leather skullcap.

"I am not Kril." EkSeet said. "I am more valuable to the Ovals as a teacher. Nor am I an Oval. Obviously I am not female."

That remark brought on coarse snickers.

Another new ancient Zabin-Kril from Egypt, with braided cloth and bronze plates, asked. "Moments ago I was alone in my barrack."

"Slowly chaps...let us begin slowly." The EkSeet admonished them. "I will answer one question at a time. You will not understand many of the rudimentary answers I give you."

"Glad you cleared that up." One of the new troops, with an American flag on his shoulder and wearing a Wehrmacht-style helmet, was asking EkSeet. "Are you trying to mind-fuck us?"

"I warn the new Krils to restrain themselves from using profanity. It will be misinterpreted by other cultures and species. KrutChan's time-pause will advance more quickly."

"You're the one grossing me out." KrutChan said.

"Let's clear something up right now for the

127

FNG's…fuckin new guys." XibEk said. "This EkSeet dude's an Oval translator of our thoughts and our words. EkSeets fuk up everythin' a dude says. Don't trust 'em. He's an Oval spy, a Lifer. Get my drift?"

CheChun agreed. "Hell yeah. We called guys like EkSeets 'stool pigeons' back in my time."

"I do not see the relevancy. This is my forum." EkSeet sounded peevish.

A Kril veteran WWII British soldier said to KrutChan. "This EkSeet has no peerage. He speaks like the lord of the manor, but he's a commoner. His own idioms are gibberish. A cockney would swing him on a gibbet; the upper class would cashier him for incompetence."

KrutChan saw from XibEk's reactions, he and CheChun may have been on the same invasions in their past; but their hate for each other was obvious. Not my problem. He thought.

The shouting, shoving, and confrontations between the races were escalating among the crowd of new Krils. Tempers rising.

"Silence!" EkSeet shouted.

After they calmed down to a simmering manageable hate, EkSeet continued. "Now that the idiots from lower first form are done, we can prioritize. Enough babbling. You are not required to like each other. You must fight alongside your past enemies; not against them."

A new future American soldier asked. "Ok, who the Christ…. Where did you come from…?"

EkSeet's WWI field uniform was distinctive in this mass of Krils. " I was born in the year 1885 in Coventry, England. A physicist by training. I was serving as part of the British Army in the Allied forces during the Great War of 1914; many of you called it World War I."

"Why are you not a Kril, old Sot?"

EkSeet ignored him. "On the opening day of the battle of Albert on the Somme, 1 July of 1916, I died. That battle was the beginning of our offensive some of you know as the battle of the Somme. The Somme was the goriest day of the British Army when a bloody nineteen thousand two hundred forty men perished. I disintegrated in a bloody 'Big Bertha' shell. My Seeching comparable to yours. As many in the current United Kingdom say, 'I was shagged'."

"Bloody hell, you say." A British soldier snickered. "Bullocks! You just told us not to fucking curse."

"I monitor and translate hundreds of Krils." EkSeet said. "By osmosis while educating them, I have developed Kril bad habits. I refuse correction by you or anyone else here!"

"Sensitive, huh?" KrutChan asked.

EkSeet facilitated the discussion asking them if they had any questions pertinent to the individual new Krils present.

"How did I get here? In fact, where the hell is here?" An anonymous Kril in the back of the group asked.

EkSeet looked pained. "It is inconceivable to me any of you are confused about your arrival. You tunneled through the Seech Field re-appearing on planet Arna. Seech means in the Oval language: To fish. This planet has one massive continent in the middle of one blue ocean. Arna translates as blue."

Not caring, KrutChan asked, with a furrowed brow. "What about those Invasions? Going to another planet in my solar system was hundreds of years away in my time."

129

XibEk snarled. "Deep into dis universe. Not ta Mars. Da Kril Vets here came off our last Invasion yesterday."

"Of course." EkSeet tried another approach. "Information will be imparted to you during time-pause unconsciousness. You will learn."

KrutChan was getting an intense headache. "I vanished from the face of the earth?"

"Vanishing in respect to an observer." EkSeet spoke to KrutChan and the new Krils heard him simultaneously. "Matter does not disappear. Changing into energy, as in a nuclear explosion. On a subatomic scale, a particle-wave can communicate with its twin through any obstacle, over huge distances, from one end of the universe to the other end, in the same instant. You never moved within the Seech Field at the end of your timeline on Earth. Your duality-wave occurred in this universe."

"That's as clear as a shit window, professor." KrutChan said.

EkSeet was intensely waving for silence. "My explanation is vastly oversimplified. I am quite unable to explain Osil scientific theory to any of you."

A new Kril asked. "Sounds like you're hiding the truth EkSeet."

"Bloody hell! I do not comprehend Seeching engineering, melding abilities, Baryon-Osils, or Osil-drive principals! Quantum Mechanics was a theoretical fact; but no one 'understood' its reality."

KrutChan would not be put off. "Seeching wasn't a gas, dude. What you're telling me is; I tunneled through nothing, into another nothing, and reappeared as nothing, in another nothing, someplace else. All from nothing." KrutChan rolled his eyes. "What a pile of horse shit."

130

Speaking to the new Krils, other EkSeets in the groups continued, trying to help one of their own. "When the Ovals use the Egress Seech Field, the field instantaneously inputs an equal amount of mass into the field on this side. Simply speaking, your Seeched lump-mass appeared here as such, before the Ovals reconstituted you. The Ovals simply re-localized your memory Bosons during reconstitution."

"Simple hell!" KrutChan's headache was growing exponentially.

"The Ovals made a mistake in their reconstruction template, don't you think?" KrutCheebel said. "A lot of us were in our teens and twenties when we died. They should correct their data."

EkSeet replied. "We said reconstituted, not reconstructed. Your Virtual Bosons arrived here as quantum faster than light. You Zabin-Krils aged forward. Your age-regressing backwards over your time here. Seech Field effect, nothing more."

"More crapola..." KrutChan said. "...I got a Lieutenant Commander I saw who was Seeched and she wasn't as old as me."

EkSeet was shaking his head. "That woman officer was as old as you after Seeching. She is a female. The Ovals adjusted her metabolism, regressing her backward during reconstitution; regaining her youth."

"Her former rank doesn't mean nuthin. Ignore her lifer ass." XibEk said.

CheChun added. "Legion ranking is by number of Arena Strikes. Kril under-the-chin two-fisted salutes are what is used here. We don't do any 'yes sir' or 'yes ma'am' ass-kissing."

131

"Right on." KrutChan said. He asked EkSeet. "Then how come those guys CheChun and XibEk are young?"

"Cause you ain't earned the right, Hoss." CheChun said.

Speaking perfect diction, XibEk said. "New Krils haven't survived five Uayeb Invasions yet. Later, when we survived the Uayebs, our Ovals adjusted our Timelines."

None of the new Zabin-Krils comprehended. The shouts from the ancient Krils were unanimous on one point. "We do not speak the language of the gods!"

KrutCheebel said. "Gods…Magic…Spells…and Unidentified Objects are simply unrecognized science."

"Your science background will assist you well in this universe." EkSeet said.

A couple of modern Krils formed a circle with their thumb and forefinger around their lips and made a loud kissing sound.

KrutCheebel said to the naysayers, smiling with superiority. "Occam's Razor confuses ignorant people."

EkSeet continued. "The majority of ancient Krils do not understand translated scientific explanations. They receive a translation from we EkSeets of magic and demons and gods; for their understanding references."

One of the new Krils from KrutChan's future said. "If I understand you, we're zombies?"

An ancient Kril asked if the future Kril was referring to dragon's teeth being sowed to create warriors, as in the ancient fables.

EkSeet explained. "A Zombie is inaccurate, as are fables like dragon's teeth; and is not germane

132

to this discussion. We will move on now."

Raising his palm to interrupt, KrutCheebel asked. "Is it possible the Ovals have cloned us?"

EkSeet explained briefly what cloning was to the entire group. "You are not clones. The Ovals initially tried cloning their male Willows, but forbade the process once they saw the cretins they created. Your information-memories are reconstituted from particle-waves of duality."

"More science from an ass-kissing-professor." The new Kril from the future said. "Move on. We don't need to understand nothing plus nothing equals something...somewhere else."

"Amen. This is a waste of time." KrutChan was moaning and mumbling. "Are we done? At this rate, this day is never gonna end."

CheChun was laughing. "Yer the dumbass, boot. This training meeting was put into your brain during reconstitution. Don't bother asking how. How do ya like them apples, boot?"

KrutChan did not respond. He was looking beyond the aura walls, seeing the huge dust-devil, now of tornado size, sliding closer to their area, then suck upwards, disappearing into the sky. The event mystified KrutChan.

"Explain why Ovals have no machinery, no robots, no drones, or no smart weapons to fight for us." A future Kril named NoKoch said. Noticeably, he had six fingers on each hand; each thumb with a can opener extra digit.

"You bring up a valid point, thank you." EkSeet said. "Oval decrees forbid constructing machines of any kind. Their willows possessed androids or robots, smart weapons, and AI computers.

133

Silicon-based creatures cannot reproduce biologically, therefore are useless to the Ovals. Silicon creations are distrusted, banned from inventing or use by Cosmic Egg strictures."

This future Kril NoKoch knew the answer, being a longtime veteran Kril, but asked the question to help the new Krils. "Are the Osil-drives, TunToobs, and Burseeosils, machines?"

EkSeet rolled his upper body in a circle to loosen his stiff back, holding his impatience in check. "You have been here a long time as a Kril; you know they are biological creatures. Including the pyramids and other buildings. I do 'appreciate' you having me explain the concepts."

"That's some weird shit." KrutChan said. "In other words, Earth Science Laws don't apply here?"

"Wrong. Gravitational force, electromagnetic force, strong interaction force, and the weak force were well known in our old universe. However, this is another universe with minor differing degrees of the same forces." He paused a moment for effect. "With an added Baryon-Osil force utilizing Graviton-waves."

KrutChan hung his head. "Education-wise, this time-pause is fucking up my brain housing."

EkSeet pursed his lips in thought.

The new Krils were muttering and raising their voices creating more confusion.

Finally, a veteran Kril stated. "Just forget about it."

XibEk was saying. "Get on with yer new life before ya TOTL."

EkSeet smiled. "TOTL infers death to the Ovals, but more correctly means extinction."

"When I died; it wasn't inferred to me." KrutChan snapped in irritation.

134

EkSeet was somber and controlled. "We are approaching the end of this time-pause."

"Thank you." KrutChan said.

Another new Kril asked. "How come all these women look beautiful? That's groovy, but something's fishy."

Many of the new Krils shouted their agreement.

EkSeet rubbed his face, eager to move on. "Ovals transmit an invisible aura around you after the Maluayeb Arena. Enabling you to see what they wish you to see. The Ovals use Auras to project in your mind their attractiveness, enhancing your lust."

At the other end of the training area, crashing sounds of plates, metal pans, chairs thumping, and tables overturning, announced a vigorous altercation between old enemy factions. New ancient warriors, Israeli's and Canaanites the EkSeets said. Past grievances were anchored deep in their races. A common occurrence among other newly Seeched races. The fighting was intense, gore-filled, holding nothing back.

Until cooler EkSeet heads prevailed, untangling and warning the combatants to cease before a Kril met TOTL. The killer would be crucified or beheaded; his remains thrown into the Seech Field.

KrutChan was happy to see normal friction and hate between men in the Legions was standard here. These guys weren't lovable brothers in a common cause. What the hell, in my past units, we weren't much better at getting along all the time.

"Breeding Oval future generations is a better analogy." EkSeet stated. "They need your seed

135

collected now, in case you TOTL. There is nothing sinister about their methods. Sex is one of the driving forces of life, wherever you came from in your past."

"I didn't like their Soothing. The process was a rape." KrutChan said.

An Afghanistan veteran Kril on the side of the group stated. "Our Four-Striker don't like being fucked; that don't compute! The new guy's gay."

More laughter ensued from the other Krils.

EkSeet quieted them down. "I see this conversation is degrading in principal. After your first Uayeb the Ovals have no further need to use their aura to get you in the mood. Soothing is similar to being married to her."

He moved to the center of the circle again. "Concentrate on surviving and not facing TOTL."

"Oh yeah. Since you brought it up…how long do we live before we die?" KrutChan said.

CheChun said. "Don't chew worry Four-Striker; you're at the top of the list."

The group was laughing with CheChun, even XibEk, who hated him. Evidently racism is forgotten with humor; as long as somebody else bore the brunt. KrutChan thought.

"KrutChan's fate should be of no concern to any of you!" EkSeet reminded them.

"Dying concerns me a lot." KrutChan said.

After he spoke, the laughter was uneasy. The only respect he would receive from the Kril Legions.

Winds of hurricane force were buckling the bubble around the training area. KrutChan saw the constraining Aura was holding; he felt better. He could not see the distant crumbling pyramids through the sheets of monsoon weather. The outside violence matched the inner mood.

136

Failing to comprehend their EkSeet's explanations, the new Krils exploded in anger, astonishment, rage, prayers, and curses.

"SILENCE!" EkSeet waited for the tumult to subside. "You were brought here; not to save the universe, not to enjoy your existence, but to serve your Ovals. In your pasts, you took oaths to protect your families, tribes, or countries. In the Kril Legions, you must band together and protect 'her' first, your brothers second, and yourselves last."

"Brothers my aching ass..." XibEk said.

EkSeet replied. "Failing to follow those Kril Legion rules will cause you to be crucified or beheaded, and thrown into the Seech Field, without mercy, without thought."

"That is one rule I will not ever forget." KrutCheebel said.

As EkSeet began speaking again, he was looking directly at KrutChan. "This rule has been violated already by one of you...." EkSeet left his words hanging for effect.

The Krils focused on KrutChan.

He was staring sullenly at EkSeet. He formed a circle with his hand and pumped it in and out from his crotch.

Laughter exploded again from the assembled Krils. After getting Kril support, KrutChan was in nonchalant indifference.

"How crude." Sighing in frustration, EkSeet continued. "Refusal to reproduce is considered selfish and unpardonable in this Oval universe. But not by force. Rape is never condoned."

Outside their protective Aura, the weather was changing for the worst, a haze obscuring details.

137

Hurricane force-winds buffeted the protective shield. While inside the Aura the atmosphere was calm and comfortable; sealed against the foreboding weather.

KrutChan whined. "Move on for Christ's sake."

"I say end his time-pause." NoKoch said.

EkSeet was assembling his thoughts. "On earth, from the first caveman evolving, males controlled their societies by brute-force. Willow males were the same."

That brought on catcalls and cheers from the group, depending on individual attitudes.

"Let me finish! On earth, wife and daughter beatings occurred with frequency; males controlling their females. Even killing females who would not submit to their male rules. Whippings, strappings, spankings, punching, and physical or mental abuse of women were prevalent in your era's; including mine. Is there anyone here can say chastising females never happened?"

"I say…" KrutChan snidely announced. "I ain't interested in former history."

"Your past lives count for nothing!" CheChun shouted. "In this Oval universe all your personal memories are ignored. You wanta survive? Concern yourself with the coming Uayebs. Forget about the goddamn Oval Pavilion ass-whoopings."

What in the hell does that mean? KrutChan was thinking.

Outside the Aura, KrutChan was distracted by the heat-shimmering effect after the storm passed. Smaller dust devils were sticking fingers of wind into the wet sand. How is that possible? Dry desert terrain formed those things. In the distance a yellow-grey front was starting to clear the weathered and age-corrupted pyramids.

EkSeet sighed. "Back on Earth, strong women were berated, and insulted. Unable to ascend to positions of power, political or economic, and vilified. The women who did were ostracized."

"Now can we stop?" KrutChan asked. "Jesus Christ! Get me the hell out of here!"

"The majority of 'normal men' protected their females." A Wehrmacht soldier interrupted.

"You boys are so full of crap!" A veteran future Kril-Oval shouted. A human female Captain, wearing American desert camouflage. Her beauty and femininity was obvious, even in her field uniform. "Protection? You guys are in heat all the time."

The Kril group of males exploded with catcalls and laughter.

"You couldn't shake hands with a woman without fantasizing!" She said. "Always adjusting and scratching your genitals. Give me a break! You guys continuously contemplated penetration; not protection."

EkSeet was laughing out loud. "Excellent Kelel! These cretins needed a woman's point of view!"

Kelel looked irritated. "For the new Kril's benefit, K-e-l-e-l is pronounced 'Key-Lee', the last 'L' is silent."

CheChun grunted. "Who cares?"

"Don't mean nuthin, girl." XibEk snarled.

"What's the difference?" KrutCheebel asked.

"Works for me." Kelel said. "I'll call CheChun 'Grand Wizard', XibEk 'Littledick', and KrutCheebel 'Chirp'. You like those names, boys?"

139

"Fair enough." KrutChan said. "You're on the rag. We got your message, lady."

Before Kelel exploded, EkSeet jumped in. "By Oval decree, Key-Lee must endure the Uayebs. She will become an Oval when they decide. As a Kril-Oval she has seen as much combat as any of the male Krils here. Kelel has an expertise as lethal."

KrutCheebel said. "I agree with EkSeet and the lady. In my earth war, I deployed with women."

Kelel pointed at the front of KrutCheebel's trousers. "Don't patronize me. Get your hormones under control."

"She's pussy-whipping you, boy!" A new Kril shouted.

She glared at a smirking male. "Make another wise crack, sonny. And I'll so add your tiny pathetic weenie to my necklace of 'dead dicks'."

"Yes ma'am." He said.

The catcalls, the pursing lips, and the insults aimed at the new Kril ran on for a couple of minutes.

KrutChan would not be happy having a woman with him in battle. They don't belong there.

"Enough gender bias!" EkSeet shook his head at their ignorance.

CheChun groused. "How 'bout ending this kindergarten bullshit, EkSeet?"

"Right on. Enough of this 'birds and bees' class." KrutChan mumbled.

"Correct!" EkSeet was visibly trying to keep his composure. "This is not a proper discussion for this group."

Kelel smirked. "Explain to these new macho-men about the Pavilion."

EkSeet nodded. "Some of you have experienced Oval Corporal Punishment already and some will witness or be subjected to the Pavilion

140

ceremony." EkSeet said. "Ovals believe a few One-Strike Krils need maternal correction. The majority of you should dismiss any thought of it."

KrutChan whined. "I'm falling asleep."

EkSeet glanced to one side. "Yes, KrutChan, be patient."

"I was taught perverted Oval chastisement is allowed to exist." KrutCheebel said.

"Pavilion correction is not perverted. Ovals consider only one non-act perversion. In the Oval Confederacy, hedonistic self-centered beings who only think of themselves and their own pleasure, are considered perverted. Wasting the essence of a species by not reproducing is considered an aberration. That includes oneism."

"He means playing with Joyce and her five daughters." CheChun said.

XibEk laughed. "Meaning you white folks are jackoffs."

EkSeet nodded. "Crudely stated. Many creatures met TOTL before the Arena ceremony because of their sexual leanings. The Ovals will condone obtaining your own essence, but do not discard it, if you wish to remain alive."

KrutChan muttered. "Lame-ass rule."

EkSeet added. "Simply stated, the Ovals will not permit to exist those who will not reproduce. Those with differing sexual leaning survive as long as they reproduce with the opposite gender."

KrutCheebel said. "In other words, the hedonists, the Don Juan's, the lonely guys and girls, homosexuals of any kind, the bestiality lovers, and the sodomites had better watch out."

"Not quite true. Some of the non-reproducers survive the culling before the Maluayeb Arena Ceremony." EkSeet said. "However, they usually

141

cannot control their urges. A few are found out and eliminated. Others of them willing to comply with Oval mating decrees continue to live."

CheChun asked no one in particular. "What'sa hedonist?"

XibEk pointed at CheChun. "It'sa plantation overseer with a busted bullwhip who penetrates sheep wid his needle dick."

KrutChan chimed in. "Use a condom for your 'essence '. If you paid attention during Soothing, the Alien Ovals save what they collect."

EkSeet's patience was at the breaking point. "Gentlemen, this argument has gone on for far too long. The males ruled in our universe." He glanced around the group. "Ovals are as strict as the controlling males were in our earth universe. Females rule this Empire."

Growing impatient, KrutChan threw the chair he was sitting on, over the heads of ducking Krils, against the green Aura. "I'm sick of this shit. Let's move the fuck on."

"That's the first decent idea you've come up with, boot. For a sonofabitch." CheChun said.

KrutChan huffed. "My mother wasn't a dog, old corps."

Quietly EkSeet told them. "Ending this time-pause is an excellent idea. Both universes are the same. Only the ones in control have changed."

EkSeet was regaining his equilibrium, waiting for them to absorb information. The veteran Krils were shouting for quiet. The scientific mumble-jumble was irritating.

"Let us now move on with your THEN past, and your NOW future, in your languages, and explain more of your Kril existence. Timelines progress forward. Ovals define the past as THEN, the present as WHEN, and the future as NOW."

"Cut this crap session!" XibEk yelled. Growing tied of Oval science lessons he already knew; he threw a verbal jab at KrutChan. "FNG told you he needs his beauty sleep."

Placing his locked fingers behind his head and pulling forward, EkSeet was relieving a crick in his neck.

He pointed out the alien species in the group one by one. "One last point to make. There are five Kril Species: Zar, Cunack, Dagot, ArnaMal, and Earth Seeched Zabin-humans."

"Jesus Christ, end this shit!" KrutChan yelled.

"Part of my Tun year of training taught us the Dinarchy enemy turned some of our forces or captured them." KrutCheebel said. "We will be fighting those humans and Aliens."

KrutChan could not resist looking at the aliens in the group.

"New-Krils consider those Aliens you see here as different from them; but are considered Krils. They do not possess human-like heads. The Zar are a transparent gorilla sized creature's adept at excavating, mole-like. Their planet is constantly bombarded by asteroids. For protection and mining purposes, they tunnel. They are unable to pronounce much of human language. Enemy Zars brains are in their lower right chest. Aim your weapons correctly.

The Cunack are fish-like and beetle shaped in appearance, extremely intelligent to a snobbish degree. One of the oldest empires in this galaxy. With an octopi's appendages, communicating in the entire electromagnetic spectrum, from numerous flashing cells on their body. They also do not have a head. Shoot for their shoulders to hit their brains.

Dagots are amphibians. They have skin similar to eels and communicate by electromagnetic pulses of visible light, also in ultraviolet and infrared. They have been known to perceive gamma rays and ultra-high radio wave frequencies. Dagots are nearly invulnerable to TOTL except by suffocation. Shoot prodigiously, as their brains are in their lower back. When hurt they dehydrate to ten percent functionality. With enough holes in them Dinarchy Dagots expire; unable to keep skin integrity. They are not violent around unthreatening Krils.

The ArnaMal Krils, not the Zabin-Krils from the Seech Field, were procreated by Zabin-Kril essence, Ovals, and human bred ArnaVals. Dinarchy ArnaMals are human. Kill them with a head shot."

KrutChan yawned, then re-focused on EkSeet, who was staring at him.

"These species, like you, have unique fighting abilities. But are as lethal as you in combat. My fellow EkSeets and I will be interested to see how you will all fare on your upcoming invasions."

XibEk snarled. "The new dudes see why this Dinarchy war has lasted hundreds of years. The EkSeets bore the shit outta ever-body."

EkSeet took center of the group in front of KrutChan. "Meld to your barrack rooms. Back to your WHEN present on Arna."

"About fucking time!" CheChun said.

Wanting to conclude this Tun yearly trial, EkSeet said. "In summary, The Oval word Kril means: Knowledge Reconstituted Intelligence Limited. You do not trust the EkSeets. Female Germs, ArnaVals, and countless females do not trust you. Ovals do not trust you. In fact, you remind

144

them every Bot of their former Willow males.

"That's why you birds are expected to die before the Ovals begin to like you." CheChun said.

EkSeet rubbed the back of his neck to ease his tension. "Your former professions made you aware, so TOTL should not be confusing, of course. Your Oval took your essence in your WHEN while it is still viable. Though she must modify her collection to create future ArnaMal Krils."

"My mother baked cookies in the oven; not Aliens." A new Kril said.

EkSeet groaned. "Never mind. This time-pause has gone on long enough."

KrutChan was the lone Kril loudly clapping.

"I am tired and you are in a stupor." EkSeet spun on his heels. "Luck to you all or should I say, may your possibilities of life catch up with you before your TLEM Time Line Event Mode, of course. KrutChan, may you live forever and meet me again in another future time-pause."

"Not hardly, if I can help it, dude." KrutChan said.

EkSeet dissolved before KrutChan's eyes, followed in bunches by other Krils.

KrutChan was angry at EkSeet's use of the phrase, "Of course", like they were children. Maybe they were newbies, or 'boots' in this universe like that CheChun guy said, but KrutChan still didn't like it. This meeting, briefing, time-pause, or whatever, was muddier than a swamp full of bullshit. He thought. I ain't gonna like this Drill Instructor EkSeet very much.

His last thought. What now?

Chapter 5


Waking in a standing position felt weird to KrutChan. The Kril training he was about to start irked him. He had to admit, the training he received in the Marine Corps changed him from a civilian to a Marine. I'll be lost without Kril training. EkSeet said time was suspended for the new Krils. Looking around, evidently KrutChan's timeline pause was still in effect. His fear occupied his mind. A year of this stuff? He could handle that.

The EkSeets and the other Krils were gone. He was in what he assumed was his barracks. He investigated the room, walking around with no purpose. Arranged like his barracks in Las Pulgas at Pendleton, minus the other troops, making him feel comfortable and at home. A single rack to sleep on instead of the usual multiple bunk beds. He was happy the barrack held none of that spooky Oval floating furniture.

He became aware of a presence. She was standing behind a clear-glass door. Unconsciously he recognized she was an ArnaVal. A lower class, the ArnaVals were the earth-bred women of Arna. His mind somehow knew the information about Akna; from the on-the-fly EkSeet time-pause?

ArnaVals usually bred with a Nest Kril. Akna's name meant "moon." The daughter of Taban, a Zabin-Kril of Sudanese descent and of Grace, daughter of a Viking; another ArnaVal bred from humans and Ovals.

Akna had flowing black hair, with large brown eyes, and high cheekbones. She was a small-framed black woman; comfortable with her body.

146

But had reservations about parts of herself she considered outsized, undersized, or too much of or too little of the whole. A beautiful woman, self-conscious of her image; part of her nature.

She looked radiant to KrutChan. Akna was a negro, making KrutChan uncomfortable. Black women and men, when he grew up in the ghetto, were self-assured and distant to him. During his time in the ghetto, he learned they had the same wishes, wants, drives, vices, race-consciousness, and distant demeanor as he had.

Her entrance faced into his cubicle. "May I enter?" she politely asked him.

KrutChan beckoned her with his hand. As she entered, he had a feeling something was off-kilter with her. He walked away from her effect on him saying. "Still hanging around the cave, Akna?"

Akna lectured him. "Leather winged creatures you speak of suspend themselves to cavern rocks during the day. I not do hang around during the day or night of any Kin day."

I think I'd better be more clear in what I say to her.

The voice in his head said. That would be advisable if you want conversation.

Shut up, EkSeet!

"Your EkSeet speaks the truth. He is trying to help you."

"He tells you everything?"

"Not all. Only when I need clarification."

KrutChan nervously realized his thoughts were shared. He waved to her. "You're in. Whadda ya want?"

Her gait was ooze-like, quick-silvery, and attention-getting in a room full of Spartan non-elegance.

147

She was definitely noticed by him. Her presence was softer, not like Reela and Hortim. Akna wearing an appealing see-through Roman-toga flowing off her shoulder. She caused the stirring he felt at her sight.

"What is wrong?" She asked as she neared him.

"You're looking off to the side; not at me. Something wrong with you?" KrutChan asked, not very diplomatically.

She answered matter-of-factly. "I nearly am blind. I see your form, as an out-of-focus blurry figure from a distance."

KrutChan felt a moment of embarrassment. "Can't the almighty Ovals fix your condition with their 'Oval magic'…like with glasses…or perform some other medical procedure?"

"Oval society not does require perfection."

In his mind, KrutChan scoffed at her remark. Maybe ArnaVals couldn't use the Oval Auras to achieve body perfection.

"You will meet many beings on Arna with 'conditions'; including loss of limbs, no appendages. What is other frailties, EkSeet?" She waited. "Deafness and maimed faces and bodies. The Ovals and the Cosmic Egg not do interfere or adjust timelines. Creatures remain as they are hatched. The Ovals believe coping with 'imperfections' makes a creature stronger. The creature develops other abilities. I was blessed by the Cosmic Egg I not am totally sightless."

KrutChan didn't agree but he accepted the fact. "Don't take this wrong, but if you are half-blind, hanging around…er…lurking must be a lousy way to live."

148

"You use a terrible word, when I am doing what is required by my Oval." Akna calmed her emotions and her eyebrows arched in a question. "How do you recollect my name?" Perhaps he remembers me from his reconstitution?

"I don't know." KrutChan shrugged. "It just came into my mind." Could it be recall from my time-pausing. Who the hell knows? Maybe it was that vestal virgin...er...EkSeet guy. "Your face is familiar. Your body's tense; maybe that's what I recall about you. I don't remember. I feel like we have we met before? Before now, I mean."

"Being anxious is my nature." Akna came closer. "Your face is horribly disfigured. Our Oval not will correct your...ugliness. I wonder if my condition is exaggerating your features." She blushed at revealing her thoughts.

"What you see is what you get." KrutChan was cursing himself. Making that kind of crack to people with sight-problems is not cool. "Sweetie, from seeing myself in that spooky mirror you guys have; you're right on. You're the first person who's spoke to me like a man, not some turd in the road. I appreciate that. But your face is vague and cloudy to me."

"Your memories will be coming back. It happens slowly after the violence of the Seech Field and the Maluayeb Arena. My forgotten memories not are always pleasing to me."

He grinned, but it didn't help his countenance; looking more like a sneer. "I usually remember happy times. They keep me mellow." He joked.

Akna frowned. "Not do lie to me." She glanced down to the floor in thought. "Your memories will be fleeting and confusing. The Ovals

149

are accurate in providing familiar objects from your THEN past timeline?"

Remembering the religious artifacts and the McNamara clothing he intended on throwing, he bluntly said. "The Oval cookie-cutters aren't entirely accurate."

"You supplied the information during reconstitution."

He tersely snarled. "Reconstituted hell; you mean when I resurrected?" His moment of meanness passed quickly. KrutChan attempted to soften his tone to a less critical level. "My 'other' thoughts are confusing and unprintable."

"Begin to cope with those thoughts. Have you remembered anything else?"

KrutChan had to admit her EkSeet had her speaking fluent American. Oval-flipping some words in her sentences. "I told you, you look familiar, but you must hear that all the time from the males you hang around...er...play with...uh...meet. In my time, it was a pickup line and not a very good one."

"I not do understand 'pickup'; my EkSeet translated it to mean 'lifting a weight' but that is out of context, I think. What is the perspective of 'pickup'?"

"You wanta handle that question, EkSeet?" KrutChan asked, his eyes on the ceiling.

Within a second, his EkSeet spoke through KrutChan's mouth. "In your ArnaVal's frame of reference, the word means a prelude to essence collection."

'I not do mate with prohibited Krils." Akna was shocked. "Unless authorized by my Nest Oval, she forbids me to collect the essence of another Kril not from her Nest or an ArnaMal male." Instead of

150

blushing like a human female, Akna seemed irritated. "You must learn to speak without idioms."

"Fuck it. I speak like me, think like me, react as me. I'm me and nobody else."

She put her hands on her hips in exasperation. "Your EkSeet's admonishments about cursing during your time-pause not did adhere with you I see."

He couldn't resist needling her. "You mean during that time-suspension training meeting? My former marine training taught me to resist manipulation."

"You should gain understanding; not disobedience."

"I don't remember." KrutChan was angry as he listened to his words because they were manufactured by his EkSeet. He didn't sound to himself like himself.

Akna and KrutChan paced slowly around each other, like opponents in a ring, before she began speaking less patronizing, trying to force him to face reality. "I was assigned to Reela by the Oval Confederation's Queen a Tun year before I helped reconstitute you from the Seech field. Perhaps that is where you remember me?"

"If you say so…"

"My former Kril I Soothed met TOTL during the last invasions."

"Sorry about him dying…I mean your loss." He bit his tongue at his inane sympathy for her dead Kril.

"Your attitude reminds me of my Kril who met TOTL." Her demeanor signaled, let us forget the past. "I was present at the time of your awakening. I medicated you. I successfully salvaged your consciousness information. During the Arna months of your timeline before they were to select you for

151

Maluayeb Arena Service to a new Oval. I attended others of your species, helping them also. Your time here on Arna is short."

Holy shit, months have been cut out of my life; rather my second life. "Right on. And why is that; my time being short, am I dying again?" He already knew the answer, waiting for her chapter and verse explanation.

"Was TOTL death different in your universe?" Akna asked. "Not did you begin dying an Ob after your birth? Life always ends."

His anger flashed. "You know goddam well what I mean!"

Akna moved around the room before answering him, picking her words carefully; many Zabin-Kril became abnormal once understanding came upon them. A schizophrenic episode was not unusual. The timeline-loss and frame of relevance overwhelmed many Krils. A case of, in a MishMell dream or out of a dream. The ones with mental defects could not distinguish the difference.

"In earth words, you will live for the five days of Uayeb; the 'unlucky days'. Ageing backwards after each day of Uayeb, until the five days are over. If you survive those days, you can choose to stay on Arna. Transfer to another galaxy in the Confederation Empire, volunteer to rejoin the Legions. Or you will be dead at what is for you a younger age.

"The bottom line is I got five days to live?"

"During your stay on Arna for the next Tun year; your timeline will be on hold-pause. "Is hold the correct word, EkSeet?"

"Forget the explanations." KrutChan loudly groaned. "Let's not start wallowing in science again."

152

"After your training Tun year, on each Uayeb Kin you will be sent to a warzone planet. That Uayeb Invasion will last, in your past timeline vernacular, a day, a week, a month, a year of your timeline. When you return to Arna; you will be twenty years younger."

"Might as well throw away my calendar."

Akna did not get his humor. "While your timeline exists you will learn as you experience, from your time-pauses and EkSeets." She said.

"Help me out..." He was scratching an itch on the palm of his hand; looking at the ceiling, talking to EkSeet. "...nobody's explained what the difference is between a Kril and a Zabin-Kril."

She smiled, knowing he was berating his EkSeet. "ArnaMal Krils in the Legions were bred from Ovals and ArnaVals with Zabin-Kril essence, trained to fight for the Oval Empire. They are the bulk of the Legions."

He was not paying attention.

"Zabin-Krils were fished from Seeched Earth masses. An entirely different class. Though most Zabin-Krils are arrogant, Ovals are pleased you live again."

His mind was swimming, wishing he had never asked. "If the Ovals love life that much; I'm all for it. You earned your position, I earned mine. This Stel of mine, is it for real?"

"I am sure EkSeet explained it to you."

"I don't remember."

Akna paused, listening to her translation. She then stamped her foot. "My EkSeet says you are lying, you do remember. The Stel records your thoughts and experiences as they happen." She was looking away from him. "Did Hortim complete Soothing you?"

"How the hell would I know?" KrutChan's voice was weary. "This is too much information, and most of its bullshit to me. Get to the point."

"We must breed before you go to the Memorial of the Ovals. In my carelessness, I obtained your essence and have your germ in me."

A bomb went off in KrutChan's mind. "Whoa! Let me understand you." KrutChan was flabbergasted, staring wide-eyed in disbelief at her. "You already got my essence? But you need to breed with me again? You're saying you're slightly pregnant?"

"No one knows but Reela. We must Sooth again to seem normal."

"And what about my goddamn EkSeet blabbermouth?"

His EkSeet spoke in his mind. EkSeets must maintain secrets. We do not blindly report everything. You and I were not aware of what she has done. I keep your secret and hers for mutual defense.

"Your EkSeet has explained to you?"

"Yeah, in his usual superior way."

"Please not do make an issue of this! To the Oval Empire, after we breed again, this egg is ours."

KrutChan began pacing. "Shit! Back up a minute. Trust me; screwing is not something I would forget." KrutChan felt bad because she was tearing up.

"Please not do speak of your former moral code. My last Kril met TOTL on the last Uayeb. ArnaVals are bred and commanded by Oval degree not to remain barren. Incorrectly and selfishly I mated with you after reconstituting. That was wrong. Reela knows and has forgiven my shameful act. You fail to understand our traditions."

"Bet your ass…."

"She instructed me to say you impregnated me when we Soothed. She is a good Oval. She could have had me thrown into the Seech Field. Our Germ saved me. Ovals hate destroying life. After we mate again in her Nest; I will be safe."

"What about when and if someone asks 'me' how the hell I…made love…to you when I don't remember?"

"Stop being difficult! When Reela Soothed you, you became part of our Nest…family. No one will question you. Acceptance of life is an Oval trait."

KrutChan shook his head in defeat. "Every universe has bastards, I guess. In my earth past we would get married to preserve your honor."

She continued. "You must go now KrutChan. As they say in your universe, good luck."

"Just for the record, where am I going now?"

"You begin another time-pause. You will awake in Reela's Soothing room."

She reentered her area off his personal quarters before he could explode. With a wave of her hand she indicated he leave now for Reel's Soothing room.

Irritated, KrutChan stayed and sat on the folding marine issue chair, next to the four-foot square table in his area. EkSeet was humming in his ear and he ignored him.

Akna touched his back softly and he instantly spun around and grabbed her throat, his eyes wide, flat with rage. She was immobile and petrified.

KrutChan released Akna when he recognized her. He said ominously. "Don't sneak up on me like that again!"

155

<u>I attempted to warn you.</u> His EkSeet said,

"I only wanted to…" Her voice trailed off; she was swallowing, clearing her windpipe when he released her.

"Don't ever get that close to me again without making me aware. Sneeze or cough or do something to warn me."

"I was mistaken." She did not look repentant, angrier than sorry. "I will hopefully remember in your NOW future." She went to her cubicle to exit again, her resentment displayed with her thrown-back shoulders.

<u>How can I apologize for a startle reaction?</u> "What did you want?" KrutChan asked as he calmed down.

"Like you…I not do remember." She was throwing his attitude back at him. "In your earth words, should I belch, or clap my hands, or grunt before I leave?" Her annoyance indicating, she had her pride too.

"Leaving is not the problem; sneaking in without warning is; don't act stupid. You're too intelligent."

He was talking to himself. He was alone.

Chapter 6


Watching KrutChan during his time-pause, his body ghostly flickering under the aura, Hortim's patience was waning. She dressed him in a knee length blue-shift suitable for the Oval Pavilion; without his boots.

Waiting for KrutChan to gain consciousness and angry at his insubordinate attitude, Hortim had made her decision to apply corrective measures. Reela's indecisions irked Hortim. Her Oval was too concerned about what the Oval leadership thought of her. Reela is far too lenient with KrutChan than she should be. Now that Reela was gone, Hortim decided to act using her mother's rules.

Hortim remembered the StelaBalaam recording events. Activating the Stels of the Confederation Empire Ovals when errors were perpetrated against Oval traditions; exposing her actions.

To protect her secret, Hortim diligently removed her Stel. Too many prying eyes and ears around this place.

Hortim slid the Soothing room invisible platform, carrying an unconscious KrutChan, into his barrack. Upon arriving she hovered over him. KrutChan looked as other Kril males she had corrected in her past timeline on Arna.

Waiting for him to wake up, Hortim went to her cubicle and retrieved a former Willow whip. She fondly wished she had her mother's hairbrush. The Willow device had three control buttons on the handle. Hortim tested each one on her thigh; going from a crackling electrified whip, another button morphing into a paddle, and another setting creating a

157

long leather quirt.  She had a choice of all three settings.

She returned to KrutChan's barrack.

A few minutes passed.  While drinking her Oval Osil-wine, she recalled the Oval Pavilion.  After certain Uayebs, when their Krils became insubordinate and disobedient, the Ovals brought them near the beach for a ceremony called 'The Nesting Rituals of Renewal'.

KrutChan definitely is an unruly candidate for that Oval tradition.

She was going to give him a taste of what happened in the open-air Pavilion.  But without the ogling Oval Stel witnesses.  The Ovals were high on not keeping anything secret between themselves and their misbehaving Krils.  My mother's sessions were private; not for non-family.

Showing him the Willow device set on whip-mode when he regained consciousness, she snap-cracked it in front of his face.

Using the controlling wand on her wrist, she took a moment hanging him on his tip-toes in front of her.  She raised the hem of his blue smock to his shoulders.  Stepping back to admire her handiwork.

KrutChan was grinning.  "You planning on siphoning?  Remember, don't spit when you hit pay dirt."

His mocking laughter gave her second thoughts about her effectiveness.  Hortim never had that happen before.  His reaction was off-putting to her intentions.  She glared at him.  "My mother never stood for unruly behavior.  You're no different.  My brothers were as excited as you before I started on them."

158

"Did your daddy spank his little girl too much?" KrutChan grinned.

Hortim's eyes were slits. "My 'mother' wore the pants in our family. My two brothers spent a lot of bare-ass time over her knees. Even in college. When mom was gone I was designated her substitute." She said. "My wimpy father spent long disciplinary sessions in her bedroom."

"Well, that takes care of penis envy." KrutChan laughed again. "Mommy really spaced out your mind, didn't she?"

Hortim cracked the whip. "You'll soon find out I'm not joking."

Remembering EkSeet's briefing during his time-pause, KrutChan was getting the drift of what she intended to do. "Don't start something you'll regret."

"Your own mother should have bared you to teach you manners." Hortim approached him, waving the electrified whip. "After I warm your bottom with this, I'll finish your correction over my lap using the paddle."

"Go for it whipper." He was amused. "Don't dribble on my floor in excitement."

His old butt has a young man's bubble shape. A beautiful target for my purposes. Like my brothers, he won't be so smug after I finish with him.

Akna's EkSeet and her Stel were broadcasting to Reela what Akna was witnessing.

Reela responded immediately. "I am aware! My melding is near."

"Hortim is insane!" Akna shouted. Blurting out what Hortim was doing to KrutChan. "I have witnessed her actions with other of your Krils. She is relentless!"

159

"What is KrutChan's reaction?" Reela was worried about his AnticArna drug overdose.

Akna was gulping air. "KrutChan is laughing at her efforts! He is mocking her exertions!"

"His reactions not will stop Hortim." Reela said. "Use any means! Calm yourself. Stop her!"

Reela slowed when approaching his barrack. Wanting Hortim to shame herself with her out-of-control actions. She intended to act as a Consort, gaining control of her Nest as Xmucane had commanded.

KrutChan heard Akna knocking over a metal folding chair coming out of her cubicle into his barrack and glanced over his shoulder.

Hortim had paused at Akna's stumbling entrance, breathing heavily; her Willow-whip sparkling. KrutChan's continuous laughter spurring on Hortim's anger. She faced Akna, purposely advancing towards her.

Akna's mood was calm and calculating. A ju-jitsu participant preparing to engage her opponent. "Reela commands! Hortim will cease this activity immediately!"

Hortim wore a smirk on her face. "Don't try my patience, Akna. You'll get more than you bargain for from me. My military training's superior to your ArnaVal protection classes."

While they talked, KrutChan struggled to release the green vines restraining him. Worried about the tiny half-blind ArnaVal putting herself in jeopardy.

Reela went unnoticed, standing in the barrack doorway, assessing the situation. Her Oval wand charged. Reela was detached.

Hortim's eyes went wide in astonishment at Reela's arrival, but her formidable will would not allow her to stop. Having Reela as a spectator to KrutChan's embarrassment satisfied Hortim.

Akna viciously slapped and then kicked-slammed Hortim hard again against the wall. Reaching down, Akna ripped the Oval controlling wand off Hortim's wrist, discarding it. It skittered away across the room, coming to rest under KrutChan's locker.

Her free hand steadying herself against the barrack wall, the Willow-whip Hortim held in her other hand was flailing uselessly, as a warning.

Akna judo-tripped Hortim to the floor with the ArnaVal on top. She did not require sight to fulfill her mission.

With little success, Hortim was attempting to force Akna off of her. She needed that Oval wand. She lashed out at Akna in order to get away from her tight grasp.

"Remove her weapon, Akna." Reela commanded. "Not should this nonsense continue."

Grabbing Hortim's wrist holding the Willow implement, Akna twisted Hortim's hand making her drop the whip. She forced Hortim to stand and slammed her against the wall.

Reela remained a silent spectator with her arms folded. Akna had the upper hand. More than the two humans fighting, KrutChan's drug overdose concerned her. His eyes were red-fired, not blue from AnticArna.

KrutChan snarled. "Break her goddamn neck!" And then seeing Reela's indifferent attitude, KrutChan forced himself to calm down. Changing his mind. Too many people have been hurt today.

Got to admit:   Near-blind little Akna is one strong broad!

Another head butt from Akna's forehead put Hortim close to unconsciousness.   Akna feeling tremendous joy; getting revenge for the last Tun year of having to suffer under Hortim's verbal abuse.

Reela shouted.   "This animosity will cease! Akna that is enough…enough…release her!"   Reela's command left both the combatants breathing heavily, staring intensely at each other.   Hortim struggling to regain her composure.

When Akna backed off, Reela grabbed Hortim's arm and picked up the Willow whip from the floor.   Reela spun Hortim away from Akna.   "I require you to control yourself!"   Reela barked at Hortim.

"He asked for it!"   Hortim shouted.

Reela sternly silenced Hortim, with a look.

KrutChan saw Reela's expression of disgust; thinking he was the culprit.   After Reela released him, he backed away from Hortim's weapon in Reela's hand.   His chest was heaving, his adrenaline pumping his system; in persistent rage from his AnticArna overdose.

Reela clucked her tongue in empathy at the angry welts on his bottom.   She adjusted her controlling wand.   Green confining vines enveloped him, pinning him to his locker.

"You the second string?"   KrutChan stared sullenly at Reela.

She approached him jerking his chin to force his attention.   "My EkSeet not can translate your words."   She walked over and picked up the Hortim's discarded Oval glove-wand Reela had given to Hortim.   "Calm your Kril anger.   I have need to reason with you."

162

"A reasonable Alien? You're bent enough to be part of this bad scene? Don't make me laugh."

Reela tightly grasped KrutChan's face forcing him to focus. "My EkSeet not can translate your words literally. I promise you after I explain my feelings to her; Hortim will understand." Reela was calm and in control. <u>Xmucane would be proud of me.</u>

KrutChan was dissipating his rage. <u>This fucking Oval universe, with their female rules is pissing me off more and more.</u> His adrenaline began filtering out, now that Reela backed off.

Moving in from behind Reela, Akna approached him and briefly touched his chest with her small hand. "I am contrite I not did act sooner to protect you."

His chest was heaving as he stared daggers at Reela. "Never be sorry, Akna. You helped me plenty." KrutChan focused on Akna's black face, with his eyebrows arched. "Remind me never to get in a fight with you."

Reela abruptly dismissed Akna to her cubicle.

After Akna left, Reela again tenderly caressed KrutChan's face, pinned his vines to his locker, and then pulled Hortim aside.

Reela was frowning at Hortim. "You not are appropriately dressed for Soothing." Reela waved Hortim's Willow whip in her trainee's face. "This Dinarchy Willow tool is not part of Soothing!"

Seeing Reela changing the Whip to paddle mode, Hortim feared Reela was about to revenge KrutChan.

Reela purposely held Hortim's Oval controlling wand. "This is a very useful tool, is it not?"

"You instructed me." Hortim formally answered.

KrutChan was looking at the ceiling.

"Evidently my instructions not were concise or I not was clear in my instructions to you." Reela's face was inches away from Hortim's face.

Sensing the danger of responding incorrectly to Reela's question Hortim became defensive. "I respect your authority."

"I not will discuss how you used this in his barrack room." Shaking her head Reela said. "I am disgusted with you. You are within Kins of becoming an Oval." Reela's eyes narrowed. "Why do you continue to ignore me?"

"I don't get what you mean?" Hortim lied understanding too well.

"With your constant EkSeet's guidance can you have a memory lapse? Was your EkSeet silent after I departed my Soothing room?" Reela was tilting her head, her own EkSeet speaking to her. "Oh, I see. You wear the controller I gave you, yet took off your Stel again."

The Soothing room felt warm to Hortim. Looking at the floor, Hortim realized her error. "I understood you perfectly." Hortim said.

Reela's husky voice dropped. "You and KrutChan will be changed...by me." Reela raised her Oval wrist wand and charged it. "My Four Strike Kril was fished here to do battle for the Ovals and to protect me. Brought here to fight during the Uayeb invasions to conserve our Confederacy Empire. KrutChan is destined to TOTL in my service. You have diverted him from his required service to the Ovals and to me. I will see you regret your errors."

"I stand corrected." Hortim sounded sincere.

The orange aura from Reela's device immediately half-stripped Hortim naked; leaving only her Navy blouse on her upper body.

164

Hortim's eyes widened in anticipation.

Reela walked up to Hortim, asking quietly. "Is this my Nest? Not do I command here?"

Answering quickly, Hortim said. "Yes you do my Oval."

Reela waved her glowing wrist. "You and he will be controlled."

"I totally understand."

Informing Hortim, Reela said. "You will go on the next Uayeb under BalamEk's tutelage. Uayebs not are as pleasant as Soothing." Reela backed up and her device expanded a green aura around Hortim; vines instantly bound Hortim's arms and legs.

Hortim started to shake; fighting to still her quivering knees.

"I come now to your greatest omission." Reela was assessing Hortim's attitude. Then Reela was in EkSeet conference. "Why not did you breed KrutChan? When I melded, I commanded you to finish his Soothing. All females in my Nest must collect his essence."

Hortim was paying attention; not wanting a swat from the Willow paddle.

"You refuse to answer?" Reela's eyes were slits. "Do you think KrutChan learned his obedience lesson from you?"

"I didn't have the time." Hortim's heart sank; she noticed Reela's questioning lifted eyebrow.

Her voice ominous, Reela said. "What you have taught him strengthens his beliefs the Ovals are perverted."

"He's unruly. He wants to kill me. Krils cannot murder Ovals."

KrutChan wore a tiny smile.

"You are an imbecile." Reela was shaking her head, remembering back to the Maluayeb Arena.

165

"You are speaking Earth-female drivel. KrutChan kills when his survival demands. You not do frighten or threaten his existence."

"He's insubordinate!"

Reela turned her back and beckoned a gloomy Hortim. She strolled towards KrutChan, with Hortim crying and wriggling her feet to gain forward motion behind her Oval.

Reela recalled her own Oval-trainee lessons. Sympathy was not a strong point in an Oval's lessons. Ovals considered pity useless, especially this Oval princess; who never had sympathy shown to her. Reela had her own problems with her sister, the Queen.

KrutChan had a different perspective of who Reela was and her problems. KrutChan was naked from his hips down, standing bound. He faced Reela and Hortim. He had recovered and was looking at Hortim's semi-naked distress.

Reela walked up to KrutChan, holding the Willow whip; towing Hortim by her green vines.

KrutChan couldn't resist the temptation. "I see you brought her feather duster along."

Reela yelled. "Our Willows were extremely capable of inflicting the same disgusting punishments on Ovals! We not do want our Krils to be Willows...never ever Willows! This Willow device was hated by many Ovals enduring their repulsive actions." Reela said. She forced Hortim to turn her naked bottom to KrutChan.

"This lady hasn't gotten the Oval message." KrutChan said. "Maybe she needs further instructions about the Willows."

Hortim groaned in fear.

EkSeet silently instructed him and KrutChan tapped his head against Reela's forehead in the Kril sign of acceptance.

Reela was observing his body's reaction to Hortim's nakedness. "Listening to me seems to suit you, KrutChan." She said, glancing at Hortim. "You never will again experience this degrading violation on your person by an Oval."

KrutChan did not look convinced. "Are you sure your Queen agrees?"

"My sister Xmucane will be pleased at my success." Reela said.

Approaching Hortim, Reela forced her Trainee-Oval face down on the cot, raising the crying female to her knees, then pushed her head back down. Hortim's shame flooded onto her red cheekbones. She was helpless to resist. Hortim saw KrutChan grinning, through her tousled hair. She closed her eyelids not wanting to look at his detested male face.

Reela approached KrutChan and held the Willow-whip she had changed to a paddle for his inspection. "Do you wish revenge?"

"Hell yes, but I can't with these on." He twisted his body, indicating his own vines were restricting him. Waving her Oval wand towards KrutChan; his green bindings disappeared.

Hortim moaned.

"Oval intelligence sources say some human males never relinquish the chance to apply this terrible thing…" Her opinion of the Willow implement spoke volumes. She held it out for him to use.

KrutChan glowered. "No way Jose. Don't believe in that sh…stuff." KrutChan shook his head. "She's the crazy one."

"Can you be that forgiving?"

167

"Sure I wanted to break her neck at the time; but I don't abuse women."

From her exposed position on the bunk Hortim let out a loud and huge sigh of relief.

Reela was pleased. "The answer I long to hear from a Kril in my Nest!" Waving her purple-glowing wand at KrutChan enhancing potency. "Look at her, my Kril. Study her positioning." Reela had a knowing smile on her face.

This Oval has one devious Alien mind. KrutChan did not have to be instructed to look at Hortim.

Reela solemnly slid the Willow-device into her smock. "You are here to Sooth; not to harm her."

"If you don't trust me." He snapped. "If you think I'll hurt her. Call off this Soothing crap?"

"Soon you will have to do battle for the Ovals. I demand you save your human violence for the coming Uayeb Invasions in your NOW."

NOW means your future. His EkSeet whispered to him.

Turn off your damn dictionary. I know what the word means. Shut up. I'm busy...

Addressing Hortim, Reela said. "You ignored my command to Sooth him correctly. You failed me other times in your THEN, by not following my orders for you to breed. As Xmucane has admonished me, time after time, you still remained barren."

Reela brushed Hortim's hair to one side. "Your condition is a glaring error on my part. Your infertility is in defiance of Oval decree and dishonors me personally. I intend assisting both of you."

She nodded at KrutChan.

Akna watched from her quarters in fascination, staying near her glass door.

168

"Please Reela...I'm begging you." Hortim said.

Reela again touched her purple-glowing wand directly on KrutChan. Staring intently at Hortim, KrutChan felt his twenty-something-year-old hormones kicking into gear.

Hortim was moaning trying to find a way to get out of her predicament. Reela touched her purple-glowing device to Hortim's squirming body. A purple aura enveloped Hortim. "I heighten your acceptance of him, Hortim."

"Please don't do this Reela." Hortim pleaded.

The Oval, showing no compassion, ignored her trainee's pleas. "You shall complete his Soothing."

Not seen by the others, Akna was peering at the sight, her breath growing shallow, delighting in Hortim's humiliation.

Reela said, in an indifferent tone, to KrutChan. "I order you to give her your essence; but not by rape." She held up her forearm. "My controller not will TOTL you with a high setting. But I not do wish an incompetent drooling Kril in my Nest."

KrutChan's ornery streak surfaced. "Aw, give me a break. I ain't gonna hurt your 'precious' Oval."

Reela spun around. "Akna, depart immediately!"

Grumbling, Akna pulled her head into her cubicle. She stayed out of sight, listening until Reela turned off Akna's Stel.

Reela pushed KrutChan at Hortim.

KrutChan did not need encouragement. Hortim was moaning, her hormones heightened by Reela's purple aura; flooding over any objections she had.

Having second thoughts, Reela was not satisfied with KrutChan's Soothing positioning of Hortim. She corrected her mistake by laying KrutChan on his back. Reela placed Hortim superior on KrutChan.

Reela was thinking. Dirva and her minions Sooth their Krils with their Krils behaving as Willows in control. Reela decided teaching Hortim such Willow ways were in defiance of the Oval Soothing rituals.

KrutChan began. Grunting after only a minute.

Hortim was relieved he was done. She expected to be released. "He's finished. Please let me go, Reela." She was emotionally drained of resistance.

Reela shook her head at her trainee. "No, it is my wish he proceeds. He remains potent. Your exposure excited him. The time you spent punishing him lasted much longer. You will Sooth him two more times in my presence."

KrutChan lay physically worn out for his age. His wheezing breath telegraphing his weariness. His mind telling himself. I'll never understand these goddam Alien Ovals. I was raped when I arrived here hours ago and now I've stood stud. When is this shit going to end?

His EkSeet whispered in KrutChan's ear. Stop cursing and face your reality. Do not Cock-up. You reside in another universe.

Very funny pun, EkSeet.

Two more times, after Reela heightening hormones in both of them, over KrutChan's objections the Soothing procedure continued.

170

Hortim's complaints were forgotten in her mind after she was bathed in Reela's aura. She actually enjoyed the session.

Their double Soothing ended.

Reela released Hortim's bindings; with a wave of her controller wand, replacing the bottom of Hortim's clothing. Hortim adjusted her uniform in silence. Speaking to both of them. "KrutChan has shared his essence. These incidents are to be forgiven by you both, but never forgotten." Reela commanded them. "Both of you cannot rest. We not do have timelines to waste."

She pointed at her trainee. "Hortim, converse through your Stel with BalamEk to further your training for the Uayeb. All in my Nest will go to the Ovut ceremony with me. KrutChan, I have summoned Akna to receive your essence."

KrutChan was not happy. "Even a whore gets break-time."

She was assessing his condition. "My EkSeet prompts me. Are you okay?"

KrutChan hung his head. "Come on lady...come on...come on...enough of this Soothing stuff. I sure ain't a Hereford bull."

"You will follow me, Hortim." Reela said, departing for the Soothing Room.

KrutChan had his hands on his knees, breathing deeply, looking at Hortim. "I'm feeling old and lousy. This is dehumanizing."

In their Stels they heard Reela yelling. "Hortim, you are wasting timeline!"

Hortim stopped and turned at the exit, telling KrutChan. "You've got a lot of training to learn over the next Tun year. After I come back from Uayeb One's invasion I'm due to get my own Nest. Paying

171

Reela for your services, I'll enjoy finishing what I started."

"You'll die trying, lady." KrutChan's cold stare met hers. "Whipper, I already forgot you." He obscenely grinned. "You got a nice ass, commander."

Hortim held her fury inside. "I'll never forget." Hastily, Hortim left, responding quickly to Reela's angry shout.

Blushing, Akna strolled from her cubicle, where she had been sequestered from the Soothing ritual with Hortim and KrutChan. Ovals were not the only ones affected by Soothing.

They were interrupted.

In an afterthought, a flustered Reela reentered briefly and sat a reluctant KrutChan upon his barrack's chair. Reela touched him with her purple-glowing wand. Akna obediently straddled him; not needing Reela's hormone help.

"I swear you're going to break it in half..." KrutChan said, tired and sore but ready because of her controller. "Get it over with..." He weakly said, speaking to Akna.

"Please forgive my secret." Akna whispered."

Reela was back in her commanding mode, speaking to KrutChan. "When Akna has completed her task, clean yourself and dress appropriately in your fashion. Regain your strength in a time-pause."

These Alien females are fucking unbelievable! He thought.

Reela would force KrutChan and Hortim to forgive each other. She instinctively knew they would never forget. KrutChan had been Soothed properly.

172

Cringing as she left, Reela was aware she had broken another Oval tradition-rule. Some anonymous Oval would review the records in the StelaBalaam and report to the Queen. Reela had allowed KrutChan to be Soothed five times. Counting her own essence collection.

It not does matter. She thought. When the OvalChanHalach enforced the Tribunal's condemnation decree at the Ovut ceremony to have KrutChan meet TOTL; his essence would remain under Reela's control.

Chapter 7


KrutChan and Hortim were outside of his barrack. After their intimacy, neither could meet each other's eyes. They kept their own counsel. Akna had stayed in her cubicle-area not wanting to get between them should their tempers flare.

When Reela entered with a short muscular Roman Auxiliary preceding her, she quickly sized up the situation.

"Unhappiness still abounds in this Soothing room I see." She appraised them. "You are both ceremonially dressed...excellent."

Reela introduced the black warrior with her. "This is CuXiu, pronounced Coo-Icks-You; one of my Two-Strike Krils. After the last Uayebs, a Tun year ago, he was transferred into my Nest. One of my Krils met TOTL during the last Uayeb. The KrilChan calls him a Beneficiarius, a special soldier in the Roman Legions."

KrutChan's EkSeet translated. <u>This man is from the ancient kingdom of Kush. A Beneficiarius Auxiliary Legionnaire who bore special duties. Normally the Auxiliary units preceded the regular Legionnaires in battle</u>

CuXiu wore a similar helmet of a Roman legionnaire, with a chain-scale shirt, leather trousers, and a shield. He held a spear in his right hand and a short sword on his belt. This guy had bronze rectangular balls, small towel-sized for protection; hanging in front of his crotch area.

<u>Be careful of this guy...</u> KrutChan thought.

CuXiu was appraising KrutChan in a similar vein. "Three other Krils with me in the hall will lead

174

the way." CuXiu pronounced as he pointed at KrutChan. "Position yourself in front of Reela. Any creature behind us while we are moving must be dealt with..."

Nodding his understanding, KrutChan said. "Got it."

Hortim reiterated CuXiu. "Your rule of engagement is to eliminate anyone posing a threat of any kind to Reela."

CuXiu had a tight grin on his face. "KrutChan is adept at his craft. Reela should fear the mob."

Feeling the hair on the back of his neck rise; KrutChan didn't like the way CuXiu said 'mob'. "That's interesting." KrutChan sneered. "I gathered from the Maluayeb ceremony the Ovals don't want me wasting anyone. Rules of Engagement change again?"

Reela yelled at KrutChan. "The Arena guard not was my enemy! That guard was part of my Nest; a Kril of good standing!"

KrutChan thought. Oops.

"You shamed my heritage and me."

He made a lousy curtsy movement, as a joke. "Yes, ma'am, oh magnificent one."

Out of the corner of his eye, KrutChan saw CuXiu lower his shaking head.

A burst instantly came from Reela's wand; exploding a red flash into his face. Slamming him against the wall with tremendous force. Whatever hit him felt remarkably like he was eating the grill of a Mack truck moving at sixty miles an hour.

It took a full minute to clear the wooziness, the twinkling fireflies in his vision, and the nauseous feeling in his stomach as he arose from the floor.

CuXiu grinned in empathy, helping KrutChan. "A reminder of who rules here, Kril."

175

"Whoo…holy shit…that cleared my sinuses." KrutChan said. He wrinkled his bleeding nose to see if it was broken. It was not; but numb from the blunt trauma of the blow.

Speaking with deliberation, Reela asked. "Not-do-you-understand, KrutChan?"

KrutChan adjusted his demeanor, dipping his pisscutter forward on his head for semi-protection. "Loud and clear. I got the word…my Oval."

Hortim nodded at Reela, who was helping with her translation of his words; indicating his answer was more respectful. In her mind Hortim thought. Way to go, sister, show him your powerful side.

Reela was pleased with herself. "I happily observe you have addressed me correctly as your Oval. That is a good beginning." Reela beamed to hide her nervousness. "I feel wonderful."

The ancient warrior from Kush told KrutChan. "Maintain silence. Mobs fear us. When we arrive at the Memorial Ceremony we do not speak."

Reela glared at him. "CuXiu, remain vigilant. KrutChan needs not your counselling. KrutChan's EkSeet advises him."

"I'll protect; not shame you." KrutChan dutifully said; or rather his EkSeet said it.

"Remember the lesson of my anger you just received." Reela waved her Oval wand-glove in front of him. She did not refer to the Tribunal Decree for his immediate TOTL. She told herself. He not does need to know of the Tribunal's sentence until later.

KrutChan was rendering the salute of a Kril; doing it unconsciously. EkSeet and time-pause training are in my mind after all; la-dee-fuggin-da.

"You will follow CuXiu." Reela commanded.

176

Waving to Hortim. "Let's go Commander..." KrutChan said.

There was a loud 'crack' from KrutChan's swinging palm lifting Hortim on her tiptoes. "Lead me to the lions..."

Hortim hid her irritation.

KrutChan was happy inside, finally doing something useful.

CuXiu whispered to KrutChan as they melded. "The mists of your past life are forever gone. You are entering the Arna world of the Ovals in Tikumyax."

KrutChan did not know he was condemned to die.

Chapter 8


Exiting their Oval Step-pyramid; Reela, KrutChan, Hortim, and Akna, following CuXiu's detachment, were immediately engulfed in a claustrophobic mass of people. Reela's coterie synched into a column with the new Oval entourages behind them. As they moved forward, other Oval detachments joined behind them.

So much for guarding her..., KrutChan thought, ...this is gonna be hairy.

The crowd formed a wiggling motion-mass, going somewhere and everywhere; a pointless mob under any circumstances and self-absorbed. The crush of people, jostling and elbowing each other for position; keeping a respectful rifle and spear length from the Krils escorting their Ovals.

The mob consisted of Tikumyax residents. Sprinkled with ArnaMals, Germs, huge translucent Zars, grey-slug-like Dagots and beetle-like Cunacks intermingled with Ovals, ArnaVals, and their Krils protecting them. He heard the majority of the residents in the crowd speaking different tongues, laughing, singing, having some kind of a party. Drinking something, yelling across lanes to other people; a festival atmosphere.

The killings and beatings of the mob by other protecting Krils lasted an eternity for KrutChan. Weaving through a claustrophobic conduit of frenzied people. Though coming close a few times; Reela was protected.


After a tense half hour, they finally cleared out of the crowd. They were in an open plaza area. Reela's retinue paused for the other Oval entourages

178

to catch up, moving further into the plaza to provide more room. Milling around Reela's group, the other new Ovals kept Reela's coterie at the beginning of the procession.

KrutChan was bent over, his left hand pushing on his knee, while his right hand held his bayonet-studded rifle on his hip. He was sucking air. Reela's trembling hand on his shoulder attempted to calm him down. She yelled to CuXiu she wanted to stay for a moment. KrutChan guessed to help alleviate his adrenalin overdose.

Though they had done this before, KrutChan noticed CuXiu and his bunch were puffing and sweating profusely too. He was thinking of the mob. Evidently, not everyone loves the Ovals.

Kril guard-sentries in their blue robes were body-blocking the entrance to the plaza from the crowd. They kept everyone in the mob out with pointed spears, Zipper-weapons, shields, Dagot blasters, and brute strength; admitting only the new Ovals and their associated Krils.

It took about two Bots, when all the new Krils with their new Ovals assembled behind Reela. Converging into formations by class, they exited the plaza, walking in the opposite direction from the mob.

Reela led her entourage and KrutChan. CuXiu and his crew had dropped away. She was leading all the new Ovals with their new Krils, between two parallel ancient walls, longer than American football fields.

Seen in perspective view, the end point converged at an apex far in the distance. On each side of the parallel walls there was a median line. A mark pulsating in black, indicating activation. Reela led the procession into the distance for a long time and stopped.

179

KrutChan's mind was triggered by his EkSeet whispering. This is called the Memorial of the Oval Protectors. Your past will be revealed to you.

Reela pointed; directing KrutChan to the throbbing black line on the side of the ancient biological wall. Peripherally he noticed all of the new Krils left their Oval's side the same time as he did. They cautiously moved forward towards the wall, not knowing what to expect.

When KrutChan got within arm's reach, the black pulsating line immediately expanded to immense size, blotting out the entire barrier. The wall revealing a tunnel. He looked back, for a moment noticing Reela, the new Ovals, and their entourages were indifferent to the transformation.

EkSeet explained to KrutChan. They cannot see what the Krils observe. They are not Krils. Only the Krils can see their THEN past."

KrutChan entered further into the opening and moved forward. When he broke the black threshold, he was engulfed by the darkness. The other new Krils unseen.

Gradually, as he stood still for a moment, his eyes could make out columns and rows and rows of three dimensional golden figures. As far as he could see to his front and left and right. The figures were in formation, like facing-mirrors projecting the same image to infinity. Statues in front of him were moving silently towards him.

His mind internally shouted. Damn! It's a graveyard!

Full of golden statues that went on for acres. He could not see the other Krils he had entered with; they were invisible to him. The statues kept coming at him as he moved through their ranks.

180

Standing in front of one figure, KrutChan thought he should recognize the figure, but he couldn't see the full face, only the eyes inside a golden hood. Whoever, or whatever it was, the guy was obviously dead. Something about the figure's eyes strained his brain. KrutChan was uncomfortable; yet didn't know why.

He walked on, not knowing who the figure was representing. The grim reaper maybe?

KrutChan wandered through the moving ranks of golden mirages. He stopped in horror when he recognized guys from his battalions who had been killed in Vietnam. Their gold hoods dropping off their heads. How the hell could the Ovals know of those KIA's?

If you recall, the Ovals rebuilt your mind after you were Seeched; your memories held that information. Was EkSeet's offered explanation.

He didn't speak; mentally distancing himself from them. Hairs bristled on his neck when a few of his friends moved in front of him. He realized they were aligned by the order of their deaths; from the first man killed to the last casualty.

Straining to recall their names and angry he couldn't remember. Friends and acquaintances from his platoons; in his former two Companies.

He fought down his nausea. Snake had buried their names deep into his heart, right after their deaths. These Oval alien females are assholes for dredging up these guys.

EkSeet broke into his thoughts. Your experience is real; yet your ghosts are not here. They are from your memory.

Sweat was dribbling down his face and his soaked shirt front, as he recalled each of those statue's moments of deaths. Remembering afterward

181

their poncho-wrapped bodies or pieces loaded onto dust-off choppers. None of the figures spoke or tried to interact with him. He was overjoyed they didn't accuse him of responsibility for causing their deaths.

KrutChan's details of his memory were gone. He decided this mausoleum he was visiting was a time-pause incident again.

Abruptly, KrutChan was pulled out of the wall. Reela was leading him back to her entourage.

He looked over his shoulder. The black line was closing on the marble orchard. His mind was distancing him from the golden statues. It was a relief to be out of there.

As he followed Reela, he wondered if he had actually entered the memory-cemetery. Or had he simply stood transfixed in front of it when he saw his visions. His sweat was evaporating in the mild breeze. He was cold-drained somber from visiting the lifeless.

Quantum spookiness is real. EkSeet said.

Reela pinched his ear to focus his mind, pulling KrutChan along. Her retinue followed her, leaving the Memorial area.

The new Zabin-Krils and new Ovals arrived on a raised plateau of the Ovut Area above the memorial with a gaggle of dignitaries. Reela took a moment to assemble the new Ovals and their new Krils around her. All the new Ovals were hugging each other.

The embraces of the Ovals reminded KrutChan of military saluting; more respectful than emotional. He saw the OvalChanHalach embrace Reela first and then somberly approach the crowd of new Ovals and their new Krils.

Moving to a rising platform, Xmucane elevated both her hands skyward in prayer posture.

"You are our new Zabin-Krils. After a Tun year you will embark on your first Uayeb. You will intone this recitation after me." She waited for a beat of time making sure she had their attention after translation.

"You swear upon the Cosmic Egg and her universe she benevolently watches over, that you will defend and forcibly strike down the Dinarchy and its allies wherever and whenever encountered and will cause them to TOTL..."

KrutChan had stopped listening.

The assembled new Krils were reciting the pledge in a monotone in unison with the OvalChanHalach as she droned on.

Reela harshly whispered to KrutChan in irritation. "Swear this pledge to honor your Soothing Oval and her Confederation Empire."

"I've already sworn an oath to my Constitution, country, and my flag." KrutChan whispered back to her.

"That was in another earth dimension. You are in your WHEN at present and you must Service yourself to the Confederation and your Oval."

He frowned. "In my mind I've recited my oath in the proper way."

His EkSeet admonished him. I have advised Reela and my brother EkSeets. They should not believe you are with them in spirit. Why must you be contrary?

I'm in step. The rest of this goddam universe is out of synch.

Reela will use her Oval Device on you!

Won't be the first or last time.  An oath is an oath.  I hardly think she'll draw attention to me; shaming herself.

The rest of the new Krils recited the oath correctly not caring about the finer points.

CheChun and other prior Kril-Uayeb-veterans were exempted by KrutEk.  They had previously sworn the oath.  KrutChan was a Four-Striker.  Keeping KrutChan from punishment was KrutEk's way of singling out the barbarian.  KrutEk wanted KrutChan resented by the Kril Legions as an undisciplined misfit.

"The new Four-Striker Kril sets himself above the Kril brotherhood."  He said to Xmucane, knowing his Stel was transmitting his words to all in attendance.  "As KrilChan I consider that barbarian too dishonorable to be trusted by Ovals or any Krils."

Hearing his commander, KrutChan was weary.  This day is never ending.  He thought.  I been doped since I got here.  "Is that AnticArna drug like LSD-wise?"  He said aloud.

EkSeet sighed.  AnticArna is infinitely more potent, elevating physcosis and dopamine.

Xmucane raised her palms to the assembled masses to silence them.  "The Krils will return to their quarters for further indoctrination and to prepare for our next ceremony at the Ovut area.  We will return later to dispense rewards and Tribunal justice."

Guessing from the look on Reela's face and KrutEk's words, KrutChan figured he was not long for this universe. He was queasy with fear.

Chapter 9


Within the hour the Ovut ceremony began. Hundreds of thousands of troops, along with other millions receiving StelaBalaam transmissions as witnesses in the Kril Legions on planet Arna, were in formations. Columns upon columns of ancient Krils were aligned, dressed in their traditional uniforms.

Wearing armor of shiny bronze or grey iron; many others festooned in feathers, leather, and painted faces of all the colors of the visible light spectrum. They carried spears, rifles, swords, blasters, bows, pistols, whips, and a myriad of other weapons unique to their past timelines.

Warrior ArnaMals bred over a Gryle in time representing the Seeched human species, descendants of the original Zabin-Krils from the earth universe's time eras. Dressed in their ancient combat gear, from Rome, Greece, Judea, Persia, Philistine, Gaul, Egyptian, China, Japan, and the American, European, Asian, Australian, and African Continents.

Other modern era warriors Seeched from later eras of the Americas and Europe; Vikings, Toltec, Mayan, Germans, Russians, and other empires were also present. Many tinier contingents from the Pacific Rim filled in the ranks; many unrecognized and unknown to many of the modern Krils, representing warriors from after the first millennium. The hosts, symbolizing their original Zabin-Krils, were immense and awesome in appearance. The Seech field was indiscriminate; not able to quantum 'fish' for every race from the human universe. Indiscriminate quantum choices.

EkSeet was narrating to KrutChan. <u>These male Kril ArnaMals were ninety-nine per cent bred,</u>

hatched directly from Oval-Kril issue, from Kril-Earth women, and Kril-ArnaVals, mirroring their former-parent Krils lives. The Krils in these formations were not Seeched.

KrutChan was speechless; it was unbelievable this mass of troops had no Seeching pain-memories as he did.

The cost, though the Ovals have much treasure, is money and energy prohibitive initiating the Seeching process. Ninety percent of the Seeched mass received cannot be reconstituted. EkSeet said.

No wonder Ovals enforced Soothing to obtain essence from their new Krils. KrutChan thought.

You are learning, old boy.

KrutChan asked EkSeet. "Okay, how do I fit in?"

After you have completed your Tun year of training you will be on a Uayeb Invasion with countless numbers of these Krils. Along with Seeched Zabin-Krils, like you.

After looking at the multi-thousands of Kril faces in this massed formation of Krils, KrutChan did not believe the Legions were happy. They looked as bored as he felt at the moment.

The Maluayeb Arena was of Kril origin; this Ovut ceremony was a less painful Oval tradition. All the new Zabin-Kril initiates from the Maluayeb Arena were led to the highest platform in the Ovut.

Reela's Nest, first in line was halted; standing in silence around a blue Aura-shrouded corpse. Reela and her entourage of Hortim, Akna, and KrutChan took position near the bier.

A Grey plasma ball of energy surrounded the body. Reela led KrutChan forward. After a moment the blue shroud over a Caucasian face and its body disappeared. Its uniform more modern than

186

KrutChan had ever seen in his earth timeline. Identified by the flag on its shoulder, the deceased was definitely American.

KrutChan was stunned after realizing the identity of the body. More damn Oval magic? KrutChan was concentrating on the black dried blood from the jugular wound in the corpse's neck. He was uneasy and wary. It was the guard he had killed who had been protecting Reela. Was this corpse evidence against me for my Oval and Kril mistake in Arena protocol?

His body grew goose bumps. The chill traveling to the nape of his neck when he looked closer at the deceased. He realized the body was that hooded figure he had seen in the golden graveyard.

KrutChan froze with fear as he stared at the corpse. I killed a woman! Shit…shit…shit…a female guard! No wonder they were pissed at me!

EkSeet softly spoke into his mind. The cadaver was a Zabin-Kril-Oval. Seeched from your future timeline.

KrutChan muttered aloud. "What the hell was she doing in the Arena?"

The Ovals do not allow every Seeched military female to go with regular male Krils on Uayeb Invasions. Unless they are trained infantry. In her case, she became a Palace guard because of her military police expertise.

Shit! KrutChan thought.

Exactly. His EkSeet momentarily paused. You have slayed, not only a Kril but an Oval in Reela's Nest. You will suffer for your actions.

Sensing KrutChan's reaction, Reela tightly gripped his elbow to steady him.

The woman was young, with dull chestnut hair, looking a lot like he imagined an embalmed

187

Sleeping Beauty would have looked. He sucked his feelings inward and put his hand on the butt of his .45 pistol.

KrutChan whispered to Reela, trying to explain. "I didn't know the guard was a woman."

Reela hissed back at him. "Never speak unless I give you permission!"

"Permission my aching ass. I didn't know...goddamit!" He wiped an inadvertent dribble of spit off with his sleeve.

Tense moments before the ceremony, Reela left KrutChan's side and walked forward to the floating corpse.

KrutChan half-listened with anxiety to his EkSeet's translation of Reela's words. EkSeet's interpretation paraphrasing her thoughts as she spoke before the dignitaries. KrutChan heard Reela saying something about Ovals, the Cosmic Egg, the Great StelaBalaam, and the faithful honor of her fallen Kril-Oval, who would be remembered forever in the Memorial.

The crowd began a sing-song melody; different than what KrutChan had heard in the Arena.

KrutEk was smugly staring at KrutChan. His look of disdain telegraphed that he was about to get his revenge. KrutEk's hard edged voice broke the silence. "The Tribunal has decreed this barbarian is to TOTL!"

Until this moment, only the Tribunal members were aware of the death sentence. The massive crowd of rank and file Ovals and Krils were muttering, whispering, and pointing at KrutChan in excitement. Sharks smelling KrutChan's blood in the water.

His grip on his automatic pistol tightened in expectation. Blue-robed guards and the closest Krils gripped their weapons tightly, ready to respond if he did something rash.

KrutChan seemed calm, though a sweat bead rolled down the side of his face. He was not planning on going out in gunfire glory; that option was for idiots. He did not in any way feel invincible. He was feeling nothing. The last words he wanted in his mind. <u>Get it over with...</u>

The OvalChanHalach Xmucane stepped forward and raised her wrist-wand transmitting an ionized bolt towards the body. The grey plasma Orb instantly pulsed and then changed to luminescence gold.

Standing beside the body, Reela beckoned a reluctant KrutChan. They were engulfed in the same glowing gold aura.

Xmucane was intoning a ritual phrase. "Reela is rewarded with her honored Kril-Oval guard's remaining revenue. Reela's Four-Strike new Kril, KrutChan, receives the progeny of this great Kril, as is our custom."

The crowd of Ovals and Krils were shrieking. Some catcalling with disappointment; not all of them happy at the condemned KrutChan acquiring possessions. The majority of others cheering.

KrutChan twitched involuntarily in fear, expecting to be attacked.

The golden aura slowly faded from around both Reela and KrutChan with a humming sound.

The corpse floated into the air still enveloped by the golden plasma and then slowly descended the Ovut staircase towards a sputtering Seech Field in the distance. The body tunneled into the Seech Field; a vortex of spinning particles resembling a whirlpool galaxy deep in space.  The cadaver abruptly disappeared forever.

As part of his role in the Ovut ceremony KrutEk ritually intoned.  "Reela's Kril-Oval has become a Seech-meal.  Her memory will be maintained in the Memorial to the Oval Protectors as a golden symbol.  Her barbarian murderer will be forgotten."  His manner was not of reverence but of tightly controlled anger.

Reela motioned for KrutChan to approach her. He tensed, waiting for an arrow, a spear, a gunshot, or a red-blast from a guard to crumple him.

Reela quietly said.  "The Tribunal was convened.  You were convicted and sentenced to TOTL.  To be executed for violation of the Oval way of life and for perverting Kril Code."

KrutChan whispered back to her.  "Don't rub it in."

"You cannot change your THEN past.  Be silent…and listen!"

Here it comes, KrutChan thought.

His EkSeet retorted.  Cannot you learn to be still, old boy?  Find an ounce of courage in you.

If you're so brave; ya wanna take my place, asshole!

Xmucane is doing her duty as a Monarch. Listen to your Queen!

Xmucane's voice overrode his thoughts; though she was speaking in an alien tongue, she was instantly being translated by his Stel.  KrutChan getting use to the time-delay.

"Your Queen honors the Tribunal."  Speaking in the third person, the Queen of the Ovals said. "There can be no Confederation Empire without Krils."  Her words simultaneously Osil-broadcast to her Empire of Galaxies.  "Kril Legions serve at the

190

discretion of the Queen. The Queen is the supreme ruler. The Tribunal her advisor. Xmucane, the OvalChanHalach Queen, now decrees." The entire empire held its collective breath.

She did not waste words or timeline. "I command a stay of execution for Reela's Kril, KrutChan. Your Queen has fulfilled the Tribunal's decision."

The crowd in attendance shocked outcry was loud.

KrutEk was vehemently shouting in anger. "By the god Mars, do not acquit this creature!" KrutEk felt betrayed.

"Your Queen demands silence!" Xmucane paused for compliance before continuing. "Your Queen has not pardoned this creature."

The crowd calmed down.

Xmucane said forcefully. "Xmucane has her right of modification of Tribunal decisions! She feels the Tribunal decree to TOTL this Kril is too close to the beginning of the Unlucky Five Uayebs. Can anyone predict this creature will survive his first Uayeb through to his fifth? Execution of the Tribunal's sentence is redundant this Ob; serving the agendas of a few."

Those in attendance were intently listening, a few disagreeing, most bowing to their queen's decision.

KrutEk was in fuming-focused hate.

"My KrilChan will have his way on the coming Uayebs, but not the way he plots revenge. KrutChan will 'never' begin his Tun year training of his timeline."

The shouts grew in intensity.

"He will immediately join the Kril Legions. Leaving with our departing Kril Legions, after this Kin day is over, for the first Invasion Uayeb."

There were tremendous shouts of agreement and disagreement from the crowds. The din was deafening, lasting for minutes before the Queen silenced them with a gesture.

The OvalChanHalach continued. "As Queen, Xmucane governs multitudes in our Confederation Empire. This KrutChan will face TOTL innumerable times in the Five Uayebs to come. Your Queen has determined the branching for his NOW future timeline." With a look she admonished KrutEk to remain quiet.

"If he survives the Five Uayebs and performs correctly, the OvalChanHalach will decide in the Fifth Uayeb whether to TOTL him. By enforcing the decision of the Tribunal."

Listening to the outcry, the screaming shouts from some of the Ovals and Krils disturbing the scene; KrutChan heard a majority were chanting praises to the Queen. KrutChan wasn't relieved.

His choking throat was compressed. He saw KrutEk gritting his teeth in anger at his Oval; the veins in his neck and temples throbbing.

In KrutChan's mind, her ruling was a death sentence. To occur in a different place, in another time, with a definite end to his life. The Queen's decision was a minor reprieve; more time for him to squirm. Xmucane had taken away whatever hope he had of possibly surviving.

Reela was smiling to get KrutChan's attention away from his gloomy thoughts by turning him around to face her.

"These progenies of my former Kril-Oval are yours to protect and nurture." Reela ritually said, pointing to a young woman and a young man. The adolescences approached; the girl hesitant, the boy looked determined. The young people standing on each side of Reela. "You will protect and nurture them as part of my Nest."

The young woman's hate bored into KrutChan. The young man was subtler; his feelings hidden. KrutChan's reflexes for self-survival took over his mind.

EkSeet interrupted. <u>Do not refuse to accept them, wanker. That mistake 'will' suredly get you killed.</u>

More words and incantations were spoken by Reela to the crowd.

KrutChan shut off his mind to the praises to the Cosmic Egg, obscuring most of his EkSeet's translation. He personally didn't intend to listen this metaphysical bullshit about his new responsibilities. His survival was uppermost in his mind.

No year-long training to prepare for his first invasion? He would have to rely on his knowledge and expertise from a former universe. Though trying to remain among the living was easy; actually accomplishing survival was moot. <u>I just been hung out to dry before being thrown into the crapper.</u>

EkSeet said. <u>You have finally achieved clarity.</u>

Reela stood forgotten; her part in the ceremonies was done.

Akna was pleased when Reela grabbed her hand when they met.

Reela passed the two orphans into Akna's care. Reela spoke to KrutChan. "Please...I believe that is your earth word...please take Akna and these Germs back to our Soothing room immediately."

The ceremony continued for the other new Krils and their Ovals.

Akna was surprised for she had never heard an Oval say please to anyone, particularly a Kril. Perhaps being polite was a way of controlling KrutChan?

"As a new Consort to the OvalChanHalach, I command you to obey me to the letter." Reela reminded him.

Hortim and Akna were shocked at Reela's promotion.

Reela again spoke to KrutChan. "Let no harm, by you or by others, come to Akna or to the germs in her care. The palace guards will protect me until I require you again. Prepare for your first Uayeb."

To KrutChan, somehow 'please' seemed out of place today. He was rewriting her inferred words in his mind. Please don't shame me by doing something stupid. Just go and die peacefully. Swallowing any smartass remark or 'aye, aye' response; he followed orders and left with Akna and the children.

The celebrations of remembrance and awards would last for the rest of the mid-morning into afternoon, as each new Kril and their new Ovals stepped forward, honored as pairs, with pink orbs radiating energy over them. It was the closest ritual to an earth marriage. The ceremony would be long and different; but just as solemn. The celebrations of remembrance and awards would last for the rest of the mid-morning into afternoon, as each new Kril and their new Ovals stepped forward, honored as pairs, with pink orbs radiating energy over them. It was the closest ritual to an earth marriage. The ceremony would be long and different; but just as solemn.

<u>Wow!</u> KrutChan was thinking on the tense meld back to Reela's pyramid. <u>This Oval-Family-Nest crap is getting out of hand, human-wise. How the hell will I raise children?</u> Children were time consuming, he had been led to understand. <u>I've no father skills or experience at all in raising teenagers!</u>

In sympathy, EkSeet started to say something in KrutChan's ear.

<u>Shut up!</u> KrutChan abruptly thought, in no mood for a paternal lesson.

Arriving back in Reela's Soothing Room delivering Akna and the kids; without a word he stalked to his barrack. Not wanting to talk to or be with anyone, certainly not his awarded children.

Finally, he was alone. He began drinking from a bottle of Jameson Irish whiskey atop the corner metal wall locker. <u>Good old Ovals looked out for their Krils.</u> He wondered if the Mongol PacalMo had fermented goat milk, or whatever the hell Mongols drank, in his area?

He began a mission of his own. He realized since early morning when he was unconscious or alone, an Oval induced time-pause came over him. Particularly a drugged self-induced loss of time.

Drunk, Smashed, Bombed, Blotto, and Plastered covered his later inebriated condition. In retrospect his condition felt great, losing control of walking, talking, and thinking became his reality. Similar to biting a mouthful divot of grass to keep from falling off the planet. He mercifully passed out, entering another dark time-pause, not melding somewhere else, unaccompanied, and alone. Sleep-wise it was great.

Much later, he felt sure he was going to die. Hangovers are great. Hangovers tear at the consciousness, creating a clarity of the mind few non-drinkers would ever experience.

In his hung-over condition, he focused on the here and now. The rough edges were internally felt, broken glass grinding and tearing up his bleeding organs. A Gordian knot pounding viselike just below his head at the top of his neck. He wished he could mount-out immediately for the invasion. He decided dying was a hell of a lot better than this crushing hangover misery. His mind filling with random thoughts.

Who the hell is my ghost wife? He thought with irritation further causing throbbing in his pained head.

He didn't know the youngsters. Hate comes later with familiarity. Getting drunk was his way of coping with this impossible incomprehensible universe.

For the first time his mind recognized he was in another universe. Being AnticArna drugged accentuated his hangover. From now on this place was his reality; not a dream, not a nightmare. In his hangover clarity he sorted out his past earth life and thoughts. EkSeet's subconscious teachings bearing fruit. Preparing to live for the present, what the Ovals called his WHEN, to see what was awaiting him in the NOW future. His earth universe's NOW fading into his past; with its ghost 'Greer' marital-possibility memories becoming his THEN past.

Who the hell is 'Greer'? He thought.

Without his consent, he was melded into the Soothing Room. Hortim, Akna, and the children

196

were there. He thought. <u>Now, what the hell's going on?</u> All he wanted was to be left alone while his hangover beat his brains out. Without a word he stiffly stood, drunkenly weaving in the middle of the room. <u>Let'em think what the hell they want.</u>

Hortim spoke. "Children, this is KrutChan, your new soused OtseVal...your father in his language."

EkSeet chimed into his mind. <u>Oval families are called Nests here.</u>

KrutChan audibly groaned, rolling his eyes at that bit of useless information. "They aren't children." He growled, correcting Hortim. "Young people hate that description."

The young girl's face looked perplexed at his remark, somewhere between happiness and hate.

Hortim spoke softly to KrutChan, to alleviate any gruffness she knew he could display in an instant. "This is your Oval-Germ, BakMeer. She is an ArnaVal bred from a Zabin-Kril and the Zabin-Kril-Oval you kil...." She bit her lip. "She is soon to be a full Oval." Her eyes silently communicated to KrutChan he should behave himself.

His voice drunk-husky and non-committal; he tried exuding warmth. Something he wasn't use to generating. "Glad to meet you, young lady. You're very pretty."

"You are quite ugly." BakMeer said.

"You ain't scared?" His small laugh irritating the girl.

"You gotta be the know-it-all in the fami...Nest. Most families have one in-house." KrutChan smiled crookedly.

"You killed my birth OtseOval, my NOW future of my timeline! I hate you." BakMeer lowered her eyes.

Akna tapped BakMeer's lips to silence her; fearing KrutChan's temper.

KrutChan just stared at the young girl.

Continuing, BakMeer said, prompted by her EkSeet. "I accept you as my WHEN OtseVal. I will listen to your guidance."

"It's better if you listen to your Oval Reela, Hortim, and especially Arna...er...Akna..., not me." KrutChan said. <u>Hell yeah, you don't want a Dutch uncle or a stepfather.</u> "If you need me...."

"You'll hardly have the time, daddy." Hortim said.

He looked hard at Hortim, while speaking to the young girl. "Strong chances are, you'll get your wish, and I never see you again." He said to BakMeer. "Your hate is what you have to deal with." The sum total of his father-knowledge.

The young man startled KrutChan when he ran and grabbed KrutChan, fiercely hugging him. "Not do TOTL in your Uayebs! Come back to us!"

Akna pointed. "This young ArnaMal is called, CauacSky. Not counting their recent OtseOval, CauacSky and BakMeer have lost three Kril OtseVal fathers."

KrutChan thought. <u>Making me Kril number four daddy. Is that irony or more Oval manipulation? What a crock of...</u> "Mom's spread comfort, son. Dad's spread the hooch around."

CauacSky looked bewildered.

<u>You are incorrigible and a bit of a Bugger.</u> EkSeet said. <u>I cannot possibly translate your last idiom for him.</u>

<u>Tough shit, limey!</u>

"CauacSky is effeminate by nature." Hortim said. "He is an ArnaMal artist of some expertise. Destined to be assimilated into the Kril Legions.

198

Service in the Krils is an indignity, considering his vocation."

KrutChan pushed the boy off him. "I plan on coming back, CauacSky." Quit manipulating my words, EkSeet.

My translations are attempting to imply your sensitivity. His EkSeet replied.

"Just remember what I told your sister." KrutChan said to the young man. "Reela, Hortim, and Akna will look after you while I'm gone."

"You aren't rid of me yet, buddy boy." Hortim quipped. "I'll be with you on the invasions. Somebodies got to keep you in line."

CauacSky ignored her and nodded. "I wish I could.... How do earth humans say it EkSeet? ...get to know you better when I join the Legion."

He grabbed the young man's shoulders, holding him at arm's length. "I have to get ready for the Uayeb." He was remembering what Hortim said about CauacSky's assimilation into the Krils. "If you can...stay away from the Legions. War will eat your soul. You don't need that shit!"

"BakMeer and I both honor our OtseVal." CauacSky said.

The kid may possibly be a little queer, but like a politician, he has a great line of bullshit. "Both of you, BakMeer and CauacSky...take off and do whatever the hell you do for amusement." He was feeling a hundred years old. Coping with time, distance, and a raging hangover.

The young people melded, jabbering, joyfully yelling, and pushing at each other. They seemed glad being released from their protocol and their meeting with 'him'.

Hortim snidely said. "My tears are overwhelming me...not! Your restrained emotions
199

must have bottomed out, Sergeant. I almost heard violins."

"I'm a Four Striker. Watch your manners. I ain't a Sergeant no more; name's KrutChan. Don't soak your panties. That was my EkSeet speaking."

"You'll always be an insubordinate enlisted grunt to me."

KrutChan was weaving. "What'd you think I was going to do...punch their running lights out?" He snapped back at her.

"Drunks are sloppy." Hortim smiled. "Knowing you; slugging them would have been more in character."

KrutChan yelled, hurting his hangover. "You weren't in the Arena!"

"I didn't have to be there! Reela told me the facts!"

Nervously Akna came to KrutChan, wrapping her arms around him and after adjusting her aim, kissed him fully on the lips. "Thank you for what you said to them. They are young, needing reassurance they not are alone."

"Sounds like they've had a tough time." He admitted. "Losing three daddies with the fourth condemned to die would make anyone lonely."

Akna nodded and walked away, melding into the area where the adolescents had gone.

Hortim hid her sympathy for the young people. "See you later, tough guy. Go back to your drunk tank." She said. "Reela would have been proud of you." Hortim shook her head. "Me...not so much...I know guys like you better."

KrutChan blearily looked at Hortim, deciding not to answer. He went back to his barrack. Getting ready to go on the Uayeb, meet his TOTL, and his personal combat-hell. He was hung-over and irritated as he was informed by his EkSeet; he would soon be summoned for attendance at tactical meetings.

  <u>What a goddam way to live!</u>

  <u>Be thankful you are still among the living...for now.</u> EkSeet said.

Chapter 10

Planet Arna's galaxy and other galaxies were in celebration. KrutChan forgotten, another Kril on his way to TOTL with the rest of their Kril Legions. Tikumyax was rejoicing with noisy chants, uplifting music, and an immense cheering population.

All the Step-pyramid's and the city's domed habitats occupants were randomly blinking glorious colors; holiday adorned for the festivities. This diminishing Kin day was for loving, intimate embraces and caresses with Nest members. Creating pleasant memories to retain in their NOW, fornicating to spread essence. Singing reverent songs to the Cosmic Egg. Wanting menial and boring tasks to take the resident's minds off the approaching unholy Kins. They would waste no Obets on depressing thoughts or enemies. That would come later.

Akna and KrutChan tensely met in Akna's cubicle. She had invited him. Akna wanted to say goodbye to KrutChan; it was only Ki hours before Uayeb One began for him. She was troubled. "I must speak with you before you go. Can you give me the Obets I need?"

"I'm running out of time." KrutChan shook his head. His AnticArna drugging had been intensified and his hangover was gone. "I don't understand 'Obets'." He closed his eyes, rubbing his forehead as EkSeet's voice whispered to him.

Obets are minutes in your time reference.

KrutChan grimaced at her. "My gremlin just explained 'Obet' to me. Time is short. I have to pack my gear and get ready to mount out…"

202

Cocking her head, with Akna interpreting her own EkSeet's translation, she then tentatively nodded.

KrutChan explained. "...mounting out means getting myself ready to go. When I'm done packing, I'll be back in about twenty minutes...er...twenty Obets."

KrutChan's mind heard his EkSeet say. <u>Bots.</u>

In a silent curse KrutChan said. <u>Fuck you.</u>

Akna also said, "Bots, you mean bots. Twenty Obets are one Bot. I will wait." She said. She could see his building anger. Anger not towards her; there was something else.

He left shaking his head.

Moving back into Akna's area, his shoulder collided with her wall entrance. Not drunk anymore, the Oval drug they had him on skewed his movement. He cursed his stupid blundering.

Akna immediately reached out to steady him. "Please be careful, you not do want to hurt yourself. You not are that huge."

"Wasn't my fault." He was saying. "Your door moved."

"Opening not do move. Oh, I see you were jesting." Her beautiful ebony face looked happy. Her depth perception focused over his shoulder. She grew somber and said matter-of-factly. "You are angry about something?"

"You said not to hurt myself. Where I'm going a lot of angry enemy dudes are planning to put a hurt on me." He griped. "Ain't mad at you. Let it go! What do you want from me?" He said bluntly. He wasn't in the mood for small talk. "I gotta go." He left the cubicle.

Akna followed him into his barrack. Thinking he wanted her to leave, she pleaded with him. "Please promise to come back to me."

KrutChan looked agitated. "Akna, you know I can't make any kind of promises."

She was led by him back into her cubicle.

"My Krils always TOTL. You must return to me."

He said in resignation, though he did not believe his own lie. "Yeah…I'll try." He had to look away from her. "You don't really have a clue what's about to happen…do you?"

"Of course not! I have never experienced an invasion."

KrutChan cleared his mind. "I've been in far too many battles to fool myself." He snorted. "My overwhelming fear is falling on my ass on the Invasion."

"I have never contemplated you in fear of anything."

"You females outta get your acts together. Reela met with me before I came here. She told me not to screw up on the Invasion and shame her. She doesn't give a shit how I feel."

Leaving for a moment, going to his barrack, he came back carrying his gear, throwing it to the floor with a crash; his anger evident.

He got dressed into his field uniform. He pulled on his clothes, his pack, his boots, his ammunition, and his weapons. And at the last, he assumed the veneer that made him a Kril, a frightened and angry old man.

As he was about to exit her cubicle, he stopped before leaving, and slowly smiled at her; his attitude softening briefly. "Get dressed. EkSeet taught me a new Oval word. I have the MishMell wet dream of you I want to remember."

She parted her thighs widely and lewdly, smiling with devilish disregard at his expression and audible man-groan. "You leave the other ArnaVals alone while you are gone, KrutChan."

Knowing full well that was impossible in this universe; but she said it anyway. "Wherever you go…whatever galaxy…whatever system…whatever planet; you must come back to me."

As he left, she heard him mutter in frustrated exasperation and rage, "Goddamit!"

<u>What a way to end a perfect moment</u>. She thought.

Chapter 11

The Krils assembled in formation near the AkSilk elevators. Reela's entire Nest accompanied KrutChan and her other Krils to the Assembly-Burseeosil-Area.

KrutChan unhappy. I never needed anyone seeing me off when I boarded a troop ship.

They kissed and hugged him in their way. Even Hortim was forced by Reela to hug him. KrutChan laughed at Hortim. Afterward she looked like she wanted to spit.

To close out the farewell, KrutChan groaned. "Okay...okay...okay...enough!"

When the good-byes were finished, Hortim's hand cracked KrutChan's butt and she said. "Let's go gyrene." Hortim reminding him who was in control.

Angry at Hortim's familiarity; KrutChan left Reela's Nest behind him. Muttering curses while boarding the AkSilk elevator attached to a TunToob in space.

KrutChan had been dreading this departure, but now it was here; he was ready. This is more like it. He was leaving the Alien and human females behind. His Kril Bot-unit and other non-Seeched human ArnaMal units were boarding.

As the AkSilk elevators descended to load them, KrutChan responded with a huge sigh. He was thinking. Just get me outta here!

Another time-pause began.

KrutChan saw small groups of Krils present from hundreds of ancient and modern eras.

Surrounding the entire assembly, above their heads, was an ebony dome lit by trillions of stars from this universe. KrutChan saw swirling whirls of galaxies: pinwheel types, globular clusters. Towering nebulas of star forming gas clouds, exploding novae, shrinking collapsing suns crunching upon themselves. Spinning disk-searchlights of various speeds from pulsars, quasars, neutron stars, blazers, supernova cosmic ray producers, and pools of deep voids of black space. To make it more confusing the celestial objects were portrayed in rainbow colors; not the black and white he was used to seeing. Never moving, at the center of it all was a radiant white-pulsing light with multi-colored concentric circles within definitive circles, unlike any of the other constellations.

His EkSeet advised him none of the religious groups could see each other at prayer. His EkSeet explained. "That centralized white object is a representation of the Oval's Cosmic Egg. This place is similar to a church, a mosque, or a temple, religious habitats for the earth species on Arna. These creatures hear and witness what their religion requires they see."

"Feels like we're in some kind of Holy Roller convention..." KrutChan mumbled. "Krils gotta get religion before invasion, huh?"

"You are somewhat correct."

CheChun, XibEk, and KrutCheebel wandered over leading an officer from World War II wearing a cross on his lapel. "Chaplain Whalen, this guy is KrutChan, the Four-Striker boot you wanted to meet. CheChun said.

"Stop calling me Padre; I'm protestant, not Catholic." Chaplain Whalen smiled at KrutChan, put out his hand to shake. "I've been informed by the Ovals you were a Catholic altar boy in your past. Do you remember? I could use you at times."

Attempting to be an intermediator, KrutCheebel said. "I think he does reverend."

CheChun's angry look forced KrutCheebel to shut up.

KrutChan kept his distance, without a handshake and said. "I don't remember none a that stuff."

Whalen appraised KrutChan. "Can I help you? Do you have any questions? Your faith is a sturdy foundation you will need to survive here."

KrutChan was staring off into space.

"That expression on your face tells me you're not interested." Whalen joked. "I can see I'm not going to inspire a tough guy like you."

KrutChan was glaring, withholding his inner thoughts from the chaplain. On his second tour in Vietnam he went to confession after a particularly grisly patrol killed three of his friends. After absolution, the goddam priest advised him. "Don't worry about your feelings of hate. Your thoughts of revenge will pass. You have to learn to play the game, son." The priest had said much more. But after that patronizing crack, the young man heard no more. Snake decided the priest was trying to be morally cynical of the human condition. Snake knew the Chaplin priest never went into the bush on any patrols. At the time he remembered thinking. If this 'sky pilot' thinks this war is a game; he's a fucking idiot.

Snake never again went to confession, took communion, or practiced his religion after that afternoon. His religion became as dead as the sea-scrolls they had found in the desert in the late forties. After that incident, his personal religion was in his heart was made up of all the deities he felt could save him.

"Are you listening?" The Chaplin said. "I can help."

Why in hell did the Ovals drop me into this religious place? Makes no sense. KrutChan replied to the chaplain. "You really believe all that shit you're spouting?"

"I'll guide you back to your God. To ask for his forgiveness."

"That's religious claptrap from your seminary days. You hate non-believers as much as you do the Oval's 'Cosmic Egg' here."

"Forget this boot, padre. His specialty is killing Oval-Kril guards." CheChun said.

Reverend Whalen was beside himself, knowing after he departed from this fallen Christian, the correct responses and words would come back to him. Why did that always happen to me?

"C'mon padre, this dude isn't worth it." XibEk said.

KrutChan spit out his long withheld venom. "Tell us! Did you and your other celibate holy men have their essence collected during their Oval's Soothing ceremony?"

The minister grew angry and said nothing.

"I know. The whores of Sodom tied you guys down?" EkSeet stay outta this! "That must have put cracks in any priest's sworn oath of chastity. Not your problem. Ministers breed like the rest of us goats. Stay outta my way! Tell me the truth though, did you enjoy it?"

CheChun's mean laugh broke the tension. "Forget this fellow, padre. Your flocks of true believers need you more than this asswipe."

KrutChan agreed with the Ovals. Timelines eventually branch and change. He would avoid this Chaplin in the future. Wondering why he was constantly in a weird drugged rage, looking for enemies to confront. He was pleased when the Chaplain and his miscellaneous flock wandered off.

The religious ceremonies continued with KrutChan standing aside with other creatures and Krils of his god-ignoring ilk.

Though the celestial fireworks projected in the dome intrigued him; he grew tired after a half hour. His body melded elsewhere.

KrutChan was alone on the Burseeosil EkTsab, other Krils drifting in and out with gear, when an Oval-apparition appeared. His mind flashed on a memory of a beautiful woman, maybe from his future? In his memory she resembled Greer Garson, statuesque with a stunning smile and strawberry blonde hair. So that's why I'm calling my future wife-vision Greer. He couldn't visualize her face-on; only a fleeting memory of someone he should know, someone who would have been in his future life on earth. Trouble was he couldn't remember a woman like that in his past life. Snake, not KrutChan, felt the uneasy connection. Time-paused on it in the moment, attempting clarity.

MishMell was Oval syntax his EkSeet advised. His mind rebooted when he realized his vision was a real Oval. This Oval, appearing in his Burseeosil barrack looked like her; at least how he imagined his future-wife would have appeared.

The melding phantom spoke. "My name is Jaguar, KrutChan. I am sister to Reela." In an instant the Oval changed form with cherry-blond hair.

"Hello, whoever you are." He said.

Her eyes were deep gray and perceptive. "You appear as if I am a poltergeist. I not am supernatural."

210

He noticed a Vietnam-Era Kril seated on a bunk nearby wearing a name tag of Albert. Was he part of this vision? He said to Jaguar. "What are you; more Oval hocus-pocus? Or just slumming; looking for specimens to Sooth."

Jaguar had a pleasant throaty laugh. "You are a blunt Kril. I have my own Krils in my Nest. Your honesty is pleasing to me."

The Albert guy was grinning, studying the ceiling.

KrutChan's brow furrowed. "Have we met before? Someone else, who looked like you when you appeared, is floating around in my head. I don't know who that spirit-woman is either."

"We have never met. Timelines can be confused after experiencing Seech Field dynamics. She existed in your THEN past or possibly your NOW future. There are branches in timelines. Since you not can remember her clearly, possibly you have recalled a NOW future that would have happened to you."

KrutChan raised his hand up. "Don't get my EkSeet dulling my brain with Oval Science."

She was holding his unopened canteen cup questioning its use. "What a strange apparatus." She placed the cup down. "Timeline branching is confusing to many. You will sort it out. You have much to assimilate; remain in your WHEN present for your reality."

The Albert guy wandered away.

He shrugged. "Okay, I got it. What can I do for you; or is the question, what can you do for me?" President Kennedy would have cringed at KrutChan's paraphrasing of his inaugural speech.

Jaguar motioned for him to approach her. "I not will be long with you. I command Oval and Kril Intelligence in our Empire."

She let that sink in.

His EkSeet said. Be careful with this Oval. She is not semantic-driven. This Jaguar, hiding behind her affability, is deadly serious.

"So, what's the scoop? You my executioner? Do I die now or later?"

Listening to her EkSeet, Jaguar nodded in agreement. "Nothing is forgotten in the StelaBalaam. You live because you accepted four strikes and not did excuse your actions in the Arena. That fact, eventually not will protect you from Tribunal justice."

"I fear death." She was laying out ground rules for his continued existence. "There ain't no justice, lady." He growled.

"You are a strange Kril." She smiled. "I appear before you through our Stels. I wish you to recall me later. If you not do meet TOTL, we will meet again, KrutChan. You are an interesting Kril I intend to change." She pulled him in front of her.

Her appearance modified. Her hair was now scarlet; her demeanor all business. When she softly touched his chest over his Stel he felt her attraction. He was irritated by how strong it was to him. The ghost image of his future-wife grabbed at his heart.

"I can't change." He said.

Jaguar put her palms together, peaking her fingers at her mouth. "Males forget they not are always right. Change not will harm you."

KrutChan grinned, thinking. Touché, lady.... "If you don't mind I'll change myself without Oval help." He stopped smiling. "If I don't forget. You know, being a typical dumbass male."

"Ovals never forget." Jaguar said.

212

KrutChan was wondering why Oval Intelligence was interested in him? Maybe it was standard procedure. To let new Krils know who had ahold of their balls in this universe.

Before melding, listening her EkSeet, Jaguar asked. "How do you humans say it? 'Good luck' to you, KrutChan."

He retained that Greer Garson-image of her smile for a moment in time after she left. For now, he obliterated women and female-memories out of his head. His memory of the Albert guy had him wondering if they would meet on the Uayeb planet. Maybe not. Guys drifted in and out of units constantly in the marines.

His Burseeosil quarters was quiet and normal again. He started checking his Invasion gear for the hundredth time. KrutChan was aching to be absorbed into his Kril Bot-unit. Eager to get a healthy dose of gruff cursing male logic without female sensitivity.

Jaguar was finishing her report to Xmucane by secret three-dimensional projection; both of them visible to the other. Xmucane was staring into her Oval-Osil wineglass. "What of your plot?"

Jaguar stared at the wall. "I have met the Kril with the secret Red-AnticArna controlling his mind. He is activated."

Chapter 12


In the Invasion Burseeosil EkTsab, the Kril
Infantry were jammed into their spaces, as the time of
leaving the Arna system approached. The Kril Bot-
units left their personal barracks, assembling in
formation.

Assigned into Malkril creature-compartments
attached magnetically to the Burseeosil, each one
holding two hundred eighty-eight thousand Kril in
forty areas. There were ten Malkrils. The EkTsab
Burseeosil held enough troops to fill a large city. In
all, eighty-five percent of the Krils were support
personnel; the rest actual combat troops.

KrutChan's Bot-unit assembled with their
half-Malkril in a cramped expanse within the
Burseeosil, in time-pause.

The mumbling and talking ceased
immediately when their Leader ZacNaab entered.
The new Krils, including KrutChan, were in
formation nearest their commander.

ZacNaab motioned to his EkSeet, who
reiterated a brief summation of their roles in the
Invasion. The briefing had a familiar ring to it: their
mission, the objectives to be taken by force, the
terrain, the weather, support weapons,
communications, their areas of responsibilities, and
duties. Including the expectations of ZacNaab, their
commander.

Rah-rah speeches given to instill courage in
the Legion troops was ignored by KrutChan. The file
and rank Krils were daydreaming. The EkSeets
wandering through the Krils were spouting the same
party line. Almost before it began it was over. They
were called to attention and dismissed.

His EkSeet grabbed KrutChan's arm and guided him over to ZacNaab. In his past Snake was never comfortable in Officer Country. KrutChan instantly went to attention and gave the required two-fisted Kril salute.

I'm being singled out again. Generals don't talk to anonymous men, just for their health. He thought.

Waiting for the room clearing, ZacNaab stared around and waited a long minute before returning KrutChan's salute. "Do you know who I am?"

KrutChan saw the future Kril NoKoch behind ZacNaab pressing his finger to his lips. KrutChan understood. Either ZacNaab ignored NoKoch or didn't see him. When he saw NoKoch cup both his ears, KrutChan realized ZacNaab was talking to him.

"Did not you hear what I asked? Are you deaf and dumb?"

"Yes sir, rather no sir. My EkSeet educated me as to your rank." He lied.

KrutChan inwardly groaned at his EkSeet's indignation at being blamed. No big deal, limey. A white lie. He thought. This guy's a General-type in the Kril Legions. Better watch my step. Never piss off a General.

Trying to make an impression, KrutChan blandly said. "You are my commander; my General...Sir."

"I am not your General! You are stupid! I hold four ranks above a Tribune! You are nothing but an undisciplined barbarian!"

Remembering NoKoch's signal, KrutChan kept his silence, except to respond correctly; demonstrating his submission to authority. "Yes Sir!"

215

"I witnessed your murder of our female Kril-Oval guard. You received no training for the normal three hundred sixty Tuns because you are untrainable!"

"Yes sir."

"You are mine by order of the KrilChan. I am revolted to have you in my half-Malkril. Assigned to me because you are Reela's Four Striker, whom the Queen has erroneously postponed discipline. I will see you TOTL as soon as possible on this Invasion! You will never survive your first Uayeb!"

Forgiven shit; your Queen condemned me, asshole.

KrutChan intensified his flat stare at ZacNaab. He was used to being tightened up, as the Corps used to say. And this guy's threat was real. He wondered why a goddam General was wasting his valuable time on one obscure minor Kril among millions of Krils in the Legions? Maybe the guy had no second lieutenants to browbeat.

"I will focus on you throughout the Uayeb." ZacNaab whispered.

This kinda guy would have his subordinates drawing his bath, cooking his food, and maybe giving him blow jobs.

His EkSeet said. Be careful, KrutChan. ZacNaab's EkSeet also translates your thoughts.

ZacNaab's bad breath was overwhelming, only inches from KrutChan's nose. "I have noted your opinion of me. You will obey me from this Ob on! Consider me as a vulture perched on your shoulder, evaluating every decision and every act you perform."

"Yes Sir."

"Make no errors! Do no more than I command, if you wish your timeline to continue!

216

You will go where I say! Explicitly following what I order you to do!"

"Yes Sir."

"Cease interrupting me! Be always aware, if you wish to survive, making no deviations from my commands to you. Obeying no other Krils. Obedience to me shall be your watchword. Do you understand me without any confusion in that ignorant barbarian mind of yours?"

Being challenged and dressed down by a General seems to be Legion overkill. I'm aware of his power. KrutChan decided to keep his answer short and militarily correct. "Yes, I do Sir."

Throwing his blue cape over his shoulder, ZacNaab yelled in disgust. "You lie even as you acquiesce! Leave my presence immediately!"

NoKoch was silently applauding, grinning foolishly. Evidently he held no fear of ZacNaab. KrutChan was thinking. NoKoch is in my Bot-unit?

"I said leave!" ZacNaab bellowed. "Do not feign deafness with me!"

That's fucking it? A lousy buck sergeant was better at chewing ass.

KrutChan was relieved as he made an on the spot decision. This General's incompetent. I had a few leaders like him. Worried more about his own butt and comfort than the troops subordinate to him. I survived them in my past. They made lousy field commanders.

Hoping the guy's EkSeet did not translate his last thoughts, KrutChan dutifully saluted ZacNaab. He noticed NoKoch had disappeared.

ZacNaab ignored him and stomped away without returning the salute.

217

Their time, in the EkTsab Burseeosil before harmonies, was in a perpetual time-pause. In a total communication shutdown. Only Stels were operable to let them skim Kril orders, manuals, and military doctrine. No time for frivolities; just work. The races gravitated to their own kind and timeline, feeling more comfortable.

KrutCheebel heard the command over his Stel. "SILENCE! Absorb what is related by your EkSeets!" The EkSeets were cranky; their stress visible, even though they would not physically invade.

I really do not know if the EkSeets are aboard or not. KrutCheebel thought.

"You belong to Makril6." His EkSeet said. "EkTsab is now getting in position to begin its Osil-phasing condition. Phasing will affect each Kril differently. Some will feel slight movement or oscillations."

"Thanks for the warning…." Another Kril said, receiving the same summation.

"You may see other Krils blinking in and out of existence. Do not despair; it is part of the process." He lowered his voice. "We will lose some Krils. But we should get many back once the phasing is complete."

Lose them how? KrutCheebel was thinking.

KrutChan pop-disappeared with a few others amid a general shout of fear.

The Burseeosil crew members among them were busy calming the Kril forces down. "This is expected! You were advised! It is Osil Drive effect! If you are still here, then you have survived."

KrutCheebel did not feel calm; he had lost his leader. What now?

218

"Where did they go...where are they?" The shouts translating on their overworked Stels.

The members of the Burseeosil crew were trying to quiet the new Krils. "No one knows. Many will be back. The unknown is that and nothing more...do not fear."

KrutCheebel had no control over what was happening; too stunned to speak. When KrutCheebel refocused his attention aboard the Burseeosil the Kril's outcries were reduced to mumbling.

Other Krils, who had been on prior Uayebs, experienced with the Osil Drive effect, were helping the Burseeosil crew calm the new Krils.

CheChun was yelling at KrutChan's Bot-unit. "This shit's temporary. You got a hell of a lot more to fear after we Bump! Save your energy."

Reality's a bitch! KrutCheebel thought. Dad was right.

The EkSeets were adding their assessment. "This effect will not last long. We are close to Invasion point."

The Krils in the compartment abruptly felt their skins crawling when the harmonics came over them again; expecting more of them to disappear.

KrutCheebel jumped in fright when KrutChan blinked back into existence near where he had disappeared; but not in the same place.

CheChun roughly grabbed KrutChan's arm to steady him. "Easy boy, you went overboard. You're back. How was the trip?"

KrutChan was groggy, weaving, and blinking his eyes. He tried focusing; cross-eyed and slowly coming around. "Wow, that was great! Why's everyone staring at me?"

"You okay Lifer? You jes got a free Osil-Drive ride." XibEk said.

AhauHuc added. "You disappeared, dude, like a fricking ghost."

"Bad news for yawl. Ya made it back." CheChun said.

Glancing around, KrutChan saw other Krils popping back into the Malkril area. KrutChan began counting heads in his own Bot-unit. "Disappeared? Don't remember nuthin..." KrutChan's head was clearing slowly. "Any our other guys go missing, besides me?" KrutChan asked CheChun.

CheChun snarled. "I counted noses during your R&R. You the only guy went AWOL in this Bot-unit. You won't miss da big show when we Bump."

The EkSeets were advising they were now in another Galaxy, near their Invasion destination planet. The Burseeosil EkTsab was creaking with phasing-down harmonics. The information from the EkSeets was direct and casual; the implications and sciences involved were confusing and treated as trivial facts. The Krils had other things to worry about; their fate was sealed. Not one of them cared how they arrived here.

During the next half hour, the Krils were milling about, trading stories of what they experienced. Their voices resembling a liquor-upped party of fans after a football game.

Then their Stels were transmitting orders. "Retrieve your weapons and equipment and assemble near your assigned Toobs for the Bump. Now! We are at your destination."

Wandering out of the Malkril area, the Krils were not caring the Osil-trip lasted only a few minutes costing some lives.

KrutCheebel went to KrutChan and pointed out the strange guy he had seen on the periphery of their Bot-unit. "Sergeant, you know that huge Kril NoKoch you were talking to?"

"Forget his former rank, asshole." CheChun said.

XibEk pulled KrutCheebel's ear. "Only the strike marks on his forehead count. Forget your old rank too."

Looking where KrutCheebel was pointing, KrutChan said. "All you people are new to me. Never seen him before the induction ceremony. Another time I saw him with General ZacNaab. Looks like another one of the unlucky few going to hell." KrutChan slapped KrutCheebel's helmet. "Forget his ass; worry about yourself."

"Grab yer balls and kiss yer ass goodbye." A non-com was shouting.

"Legionnaires form and we'll meet again in Elysium!" A Roman centurion was shouting.

"Valhalla beckons you; welcome it!" The Vikings were ready for whatever awaited them.

Even PacalMo was shouting to other Mongols. "The Blue Heaven awaits our arrival in this Dragon universe…rejoice!"

KrutChan was never a college 'rah-rah' guy in his past. "Saddle up." He said flatly and anticlimactically. He saw CheChun shrug in agreement.

The Commanding Cunack on Burseeosil EkTsab, through the StelaBalaam was advising Oval Command on Arna the unholy Kin day of Uayeb One was beginning. When the Invasion Force committed, a single word was uttered by BalamEk into the StelaBalaam: "Desert."

Chapter 13

"...These men were born to drill and die.
Point for them the virtue of slaughter,
Make plain to them the excellence of killing
And a field where a thousand corpses lie.... Do not weep.
War is kind."
Stephen Crane 1899--War is Kind

The first Uayeb of the Unlucky days.

A Time hiccup; an alternative-time-lapse beginning for the Invasion forces.

Red Dwarf bleak-dawn above KanBalaam. The red sun had .8 the mass of Arna's sun with a surface temperature of 4,100 K. Red Dwarfs were abundant in this Oval universe having a main sequence stability lasting 15 to 30 billion years compared to Arna's 10 billion. Helping KanBalaam develop evolving life on the planet.

The Oval Science Project Team picked KanBalaam for its high content of heavy metals desired by the Dinarchy and the Ovals. In the murky red light, with silent imposing dignity the Burseeosil EkTsab went into orbit around KanBalaam, gravity-locked over the invasion twilight zone of the planet.

The Malkrils sections attached to the Burseeosil were magnetically uncoupling without sound in space, ignored or unseen by the Dinarchy defenders; surprise possible for the moment. The Biological Burseeosil EkTsab magnetically came apart. Resembling a creature molting. First deploying the larger BakToobs as a perimeter defense for the Burseeosil shell. Smaller TunToobs egressed and took up picket duty outside of them to protect the

223

inner defense of BakToobs. Some assisting the BakToobs in the bombardment of the invasion zone.

Reconnaissance-in-force Toob Bumps were continuously in motion; the first Krils invading. At the same time the Malkrils were dispensing biological invasion Toobs in space crammed with Kril and ArnaMal troops; the spear point of the attack. It was a silent ballet-deployment in space.

Inside the Burseeosil EkTsab noisy activity trembled with continuous thrums of shouting voices, screeching equipment gears of Kril origin, hydraulic thumps, and vibrate rumbling, like gigantic prey monsters. Thudding into place, biomass tendons dragging across flat muscle plates. Muffled Stels emitted a constant monotone in the background, a steady flow of missives and messages.

The Bass-beat of the invasion was increasing. The Toobs, as they were loaded undocked, slowly forming together outside in space into pods of invasion packs. In formations by order of landing, in specific assigned areas away from the Burseeosil. Instantaneously, the StelaBalaam was recording all the bizarre details of the orderly disorder. The troops were heavily loaded down, barely able to move.

The dance of the Toobs in space was an eerie silent impressive sight. BalamEk watched two Toobs collide attempting to get in position, explode silently in balls of ammunition energy, dumping their contents into an orbiting debris field. The survivors of the collision would last brief minutes before going TOTL in the freezing space above KanBalaam. The nearly immortal Dagots were the exceptions who would be rescued. For the Krils, going TOTL in space was an occupational hazard. BalamEk's pity was saved for her Oval Toob pilots; they could not be replaced.

The biological beetle-like invasion Toobs turned into comets upon reaching the upper mesosphere. A shotgun dispersal of tiny asteroids colliding with the atmosphere, inevitable and unstoppable. The KanBalaam invasion once began; could not be stopped.

The Kril's breathing supply, in the Burseeosil and the Toob, was constantly tested by the Oval technicians for inter-species pathogens before they Bumped. One commonality, shared by the Humans, Dagots, ArnaMals, Cunacks, and the Zars, was they could metabolize poison oxygen. They were inoculated through their lungs, or whatever the other species used for breathing, before they ever set foot on the invasion planet. The procedure wasn't foolproof, but it sufficed, if only to quiet their fears.

Strategic planning and Tactical plotting was over. Now it was the Kril Legions mission to capture the invaded planet. In any universe, any planet, any continent, any country, any town, any hill, or in any dirt: objectives were taken by low caste beings. Frightened of death, snarling and vicious, beyond caring or hope. These Krils from different era's, different races and species carrying their weapons would be the chosen ones. Many dying for the elite few.

Within the majority of ancient Kril Legions were the minority modern Zabin-Kril. Zabins did not fight in closed-knit formations emulating the armies of their past, but spread out. Zabin-Kril training consisted of honing skills already acquired. Boring and rugged, physically agonizing work; those Zabin-Kril troops spent KinUts sweating, itching, scratching, and cursing the high Command's formula of war making.

Pressing against KrutCheebel the other Zabin-Krils, not all human, were milling about forming a clot around him near the Toob. He was scared-nervous; in truth terrified. He felt the clumps of bodies moving CheChun, the Kril racist Southern White man, close to KrutCheebel. Unit cohesion took precedence over their hate for each other.

KrutCheebel was moving into their cold-skinned creature Toob towards his assigned web made of AkSilk. Biological AkSilk was manufactured to be as strong as a diamond and elastic as rubber. The humans called it a web, translucent and textured like a spider's trap. Holding the Krils in a standing position for the Bump.

Feeling the bulkhead of the biological Toob when he boarded he imagined the Toob breathing. He had flashes of being in a pink womb.

Once inside the stringy cocoon of AkSilk, he hooked himself with a static line, in case he fell out by accident. He winced when CheChun grabbed his helmet chin-strap and yanked it open, the ends remained dangling.

"Listen up, ya dumbass rookie, I seen guys in the Pacific thrown off a ship have their neck broke when they hit feet first in the water because of those locked helmet straps. Same with concussions from explosions landing too close!"

KrutCheebel had always wondered why movies showed World War II troops with unlatched helmet straps. He thought it was so the guys looked cool on the screen. Now, he had second thoughts.

It was also possible CheChun was screwing with the mind of a new Mulatto Kril. Who knows? CheChun's hand grabbed KrutCheebel's desert tactical belt tugging it downwards, testing its tightness.

226

KrutCheebel choked, unable to speak through his dry caked throat. I wish KrutChan would stop watching me. He struggled to stop his involuntary shaking. He rolled his lips together, tasting the salt from his upper lip. He was pulling his pack tighter; hoping the pinching pain of the straps would give him another thing to think about.

If KrutCheebel's Toob was destroyed on the way in during the thousands of Obsil kilometers they would drop, the Krils inside were secured by the AkSilk thread attached to the top of their Toob. That was not comforting to him. If the biological creature Toob came apart from explosion or structural failure in space, he would silently incinerate or mummify with little time to worry about death.

The AkSilk strands, attached under his armpits, were stronger than steel, supple as spider's silk. An umbilical cord ready to snap the Krils, once they were in breathable atmosphere, to a controlled survivable crash. KrutCheebel had more faith in a parachute.

KrutCheebel was in the middle of the Toob, not the outer fringes of the AkSilk webs. Feeling like a fly caught in a web, he waited in pulse-pounding grimness for the invasion to happen.

The hype of their pre-invasion combat-order ran through his jumbled thoughts. He was vaguely remembering the optimistic instructions after they left Arna. "Confederation's finest troops...Zabin Kril, the leaders of the liberation of Planet KanBalaam."

He had thought at the time. He and the other Krils, were an awesome force. Now he wasn't so sure.

KrutCheebel's body pulsed as if he were tightening, loosening, and then flexing his muscles in isometric exercise. An effect from the Toobs mass-

227

field-oscillators firing at planet targets from their strategic platform quadrant.

The Kril cargo in the Toob were far above Planet KanBalaam. Bland expressions were on many of the faces surrounding KrutCheebel; on the human faces, anyway. It was hard for him to tell on the Krils from the alien races.

The Dagot: eel-skinned-slug-like amphibians with their electric body lights and ultra-violet emissions used for interaction with others. The see-through Zar: grunting gorilla-like beings, their speech more grumbles and whistles and hisses. The Cunack communicators, in case Stels were broken, in the Kril Legions: fish-scaled with beetle hardened backs. Those alien races were impossible for him to read their wavelength conversations.

If he survived many more Bumps, he hoped he would gain subtle interpretations of his alien brother Krils. Easy-going expressions on the packed human beings contrasted the hard angles on their faces. Their tense palpable desperation putting age where youth tried to linger.

This Bump was his first battle.

Moving around him, the Kril bodies were all hollows and humps, with their equipment squeaking. Their grunts, curses, hisses, and sparks emanating from the murmuring Krils. All of them withdrawn into themselves.

Odors were pervasive in the confined area; irritating ozone, with sweat-stink, and sickening sweet perfume. <u>One of the Zars probably farted.</u> Then a strange yet familiar smell: hair oil or after shave?

He directed his trembling voice, mocking KrutChan, trying to joke. "I think I lost my ticket for this Bump." He added with a sneer, to hide his overwhelming fear. "What are you sweating for? Nervous?"

KrutChan's flat hard stare fixed him. "Just about my virginity. I got too many tough guys around me."

"Yeah, me too." KrutCheebel lamely said trying to soak in bravado.

Battle. He had to tighten his quivering stomach and grit his teeth. Every word he spoke came from the hollow of a drum; squeaking like an adolescent. He wanted to spit, but could not.

Nervous laughter in the Toob was contagious, with muffled and crude jokes directed at the new Krils. The sneering jokes were unforgiving and sad.

Reaching up to adjust his AkSilk line, his hand slipped and he punched a face outside his web. A translucent-skinned claw shoved him in exasperation. A moment passed before his Stel translated the curse hissed at him by a huge Zar. "You are a fuckin' idiot!"

KrutCheebel kept his mouth shut. After all, it was my fault, he thought. He yawned, telegraphing his disdain.

Slowly his gaze sorted out the pink dark details in the Toob, reality starting to sink into his mind. All the different dialects and the colors of the different uniforms mottled and drained together into the shining AkSilk webs.

No matter what each Kril wore, if it was different from KrutCheebel's, the common thread was unmistakable. Their purpose was lethality. There was no parade ground strutting. Pretty young women watching them, or glory in this Toob.

The truth jabbed a pain over his heart, twisting off the bile rising from his stomach. Tragically, his mind shouted. God, please don't let me be sick.

Then his stomach jerked...spewing projectile vomit. KrutCheebel apologized between powerful thrusts of his stomach, trying to maintain his dignity. Fighting to control his bowels.

The Zars and Cunacks ignored his distress not interpreting what was happening. Only the human Krils and ArnaMals aboard close to him reacted in disgust.

Nearby a Canuck was swelling up and rippling with a gagging reflex. A sweet odor filled the compartment; signifying a Dagot's anal reaction to the Bump. Helping KrutCheebel somewhat to know inside they were scared too.

The Dagot next to him was chittering his teeth in his round mouth as if conversing with KrutCheebel. He noticed the Toob was screeching with painful muscle-spasm movement and straining fields of magnetic resonance pushing against other Toobs in the Malkrils landing system.

The Bump moved persistently towards initiation. With a vicious crunch, they abruptly exited Burseeosil EkTsab.

The Krils were floating in space. The Dagot next to KrutCheebel sonar-screeched in his ear and KrutCheebel puked again.

Without warning, KanBalaam's gravity well ingested them with a sickening free-fall.

The uncontrolled drop went on for long minutes.

Their Toob didn't land. Mercifully it was a controlled crash.

Alive and outside the Toob, KrutCheebel lay flat on his face. His heart was thumping, his legs spasmodically churning in flight digging a hole in the blue sand of KanBalaam. Scrambling to release his

AkSilk tether, while he was blowing sand from his drooling mouth. The blue stuff was flour-like, smelling of rotten eggs, and covering everything. Keeping his head down, he rolled over and carefully squinted, focusing on his Bump team through acrid red smoke.

KrutCheebel's broken Toob was gnarled debris, biologically destroying itself in a frenzied shower of red sparks; its organic hide blistering in the oxygen rich atmosphere. Becoming another cadaver of the invasion. The sky was light green, heavily oxygenized. Some of the occupants, especially injured Cunacks were glowing red-green in agony, the air attacking their wounds. A simple case of jumping electron shells, KrutCheebel analyzed.

Blood from the Toob was pooling. Cell compartments were leaking. The blood did not react with oxygen. On earth, a gasoline-oil machine would have been a fireball. On KanBalaam, metals and species were corrupted by the oxygen-toxic atmosphere. Humans were immune, breathing poison oxygen since birth.

Cunack were not that fortunate. Not from an oxygen rich atmosphere like the humans. Their planet atmosphere was thinner than the top of Mount Everest on Earth. Their hard-backed shells helped somewhat, if they were not breeched. How do they breathe without getting an oxygen-high? KrutCheebel thought. They were convulsing and having difficulty breathing. Could they acclimatize? Can they adapt? KrutCheebel's thoughts tried making sense of the chaos in his immediate landing area.

He saw other Kril occupants scrambling quickly out of their Toob-AkSilk-wombs; those not thrown out. While Krils from another soft landing Toob stumbled around in a daze. Upon exiting, they

231

collapsed to the ground trying to avoid the incoming blasts of counter fire from the Dinarchy positions.

The Dinarchy were pulverizing the Bump force. Observing the battlefield carnage, KrutCheebel watched a few incoming Toobs awkwardly careening in a light green-copper sky.

Most of the other Toobs hovered, with their Krils rappelling on AkSilk. Then flying off after discharging their loads; avoiding landing. Their Krils moving quickly, attacking in the direction of the fire from the Dinarchy. A minority of crashed Toobs were gutted pieces. Some of those Toobs had dead occupants dangling in their AkSilk cocoons. Other Kril occupants in body-pieces torn apart by the AkSilk meant to protect them. Trails of reddish smoke wisped skyward, in the hazy light of the red-dwarf star.

KrutCheebel saw his busted Toob had landed near the foothills in an outcropping of rock.

Their Toob bulkhead had given way before the crash. KrutChan and CheChun disappeared, sucked out of the hole. Falling through the clouds other Krils he saw were thrown out of gaping rips in their Toobs. The Krils dangling; resembling parachute spiders in the wind.

After surviving the fall, KrutCheebel's Toob-mates were on a billiard table-like mesa full of chunks of rock, mound-craters of sand, and boulders. Modest protection at best; not in abundance.

KrutCheebel tried stifling his vomiting-reflex, smelling the acrid smoke. Then he failed. Bile, in slimy chunk amounts, spurted out black. He had no more contents in his stomach; dry retching.

There was screaming Krils, hissing voices, and ripping sounds all around him; utter confusion. Smells of burnt oxidizing flesh and a rubber-burning-ozone irritant assaulted his nostrils. He recovered from his vomiting, and then wished he hadn't.

The landing area was utter bedlam.

Far across the dark blue mesa, KrutCheebel could see the later waves of Toobs continuing to fall; some exploding, most hovering, dropping Krils tethered on AkSilk, others spilling Krils from ripped hides.

Streaming AkSilk, Krils drifted in the mild wind. Toob pilots were automatically cutting AkSilk guide-lines just before their Krils touched desert. The Toobs were spilling AkSilk cargoes of equipment and men, amid Dinarchy fields of fire. The Toobs landing intact by necessity, against Oval orders, were discharging Krils struggling from their AkSilk cocoons. As stationary targets the Toobs did not last long.

KrutCheebel watched one Kril running out of his Toob, abruptly stop, thrown backwards reaching the end of his AkSilk and crunched to the ground in a blue dust cloud. The guy was loudly streaming curses realizing he had forgotten to release his static line. KrutCheebel laughed at the absurdity, not the Kril.

Blasts of huge zipper weapons from the Dinarchy defenders competed with explosive human ordnance. Roman and Greek fire and arrows, and catapulted siege rocks and deadly huge eight-foot spears were answering. Familiar earth artillery fire from the incoming explosions brighter and more deadly in the oxygen rich environment. KrutCheebel's vision was limited to his immediate area; not seeing the entire invasion area.

KanBalaam, in it's green-copper daylight glory, began to accentuate the craters appearing

233

randomly over the entire plain. The red-dwarf was lower in the sky than an earth's terran star and cast more shadowing.

Blue-tinged faces stared back at him from intermingled humans among the Krils. KrutCheebel with his officer training, kept yelling to the Zar next to him to get down. An Eel-looking Dagot, crawling next to the Zar, appeared calm; but slowly shitting; revealing his fear, blinking his eyes repeatedly.

"Why're you sitting on your ass!" KrutChan yelling, collapsed next to him. "This ain't a war game, Lieutenant! Have those goddam Krils drop their dicks and take off running for the foothills!"

"I do not know these Krils. The landings a failure! I am not their leader!"

"Goddamit!" KrutChan yanked KrutCheebel to his feet. "You are now! Yell, kiss their ass if you have to, swear, bust balls; but get them moving!"

Hate flickered between the former earth non-com and officer. KrutChan dragging the officer by his shirt towards the nearest prone Krils.

KrutChan was yelling over the din as he ran off. "Where the hell is our pilot Hortim from my Nest?" He was not talking to KrutCheebel. His Stel was shrieking from Burseeosil Command. "Your monitoring her Stel! How the hell should I know?" He yelled. "Roger...I'll hunt her down! Jesus Christ!"

KrutCheebel witnessed tiny life at ground level. A KanBalaam menagerie of bugs, ants, worms, and millipedes, or whatever they really were, made KrutCheebel wonder what they thought was going on around them... Do they know war has come to their planet?

"Snap out of it!" KrutChan was snarling from feet away, shoving Krils of different species towards KrutCheebel, as he ran off.

Surprisingly, within minutes the Krils KrutChan and KrutCheebel had rounded up were under a rock cliff overhang.

"Looks like we got the base of the hill area covered." KrutChan said. "Stay here until I get back. Burseeosil Command is hounding my ass! Great way to run a fuking war! I gotta find Hortim."

Feeling like a pet mongrel dog, KrutCheebel was pissed off when he heard KrutChan reiterate a single command. "Stay!"

KrutCheebel thought angrily. <u>Fuck you, white bread!</u> But he did as he was ordered. He felt safe here.

Day passed into twilight quickly after KrutChan went off on his pilot search.

In this area on planet KanBalaam; half-night came with a thud. When the sun flared, the shadows would appear.

In Burseeosil EkTsab, the KrilChan KrutEk, had calmed down now that reports were coming in with more accuracy.

BalamEk, the attacking force commander, told him. "We observe our movement sensors are initiating. Placed by our Toobs in front of the advance positions of Krils and ArnaMals Legions. My Ovals receiving data from the creatures are indicating Dinarchy mass activity in front of our forces."

"Brilliant analysis." KrutEk nodded. "Do you have your eagle-chariots, BakToob and TunToobs, in position to support my Kril? Do they require your eyes to expend their weapons upon Dinarchy forces?"

BalamEk pondered the question for a moment. "Yes and no. A flight of BakToobs have been dispatched to the Dinarchy coordinates. To ensure we not do TOTL our Kril forces, the gunners will have to see Dinarchy before firing. The ships fly at high speed and altitude; their Stels will fire weapons automatically when initiated. The Oval pilots not are able to manually do the firing with complete accuracy. No species is capable at that velocity."

KrutEk said with annoyance. "I leave that in your capable hands, BalamEk." KrutEk had little experience in aerial tactics; his knowledge of air power was nil, except from on his prior Uayebs.

"We have dropped camouflaged explosives around the creature-sensors to protect them. The advancing Krils, when they do move into those specific areas, must be careful the monitoring biological creatures have been deactivated by trained ArnaMal-Kril. This precaution keeps the Dinarchy from disarming them."

KrutEk bristled. "Give intelligence information to my second in command, KrutSeet. He has the responsibility for the main attack. He can advise me of the progress. What is the overall situation?"

BalamEk was cranky in return. "I am aware of your command structure; you must remember mine. An Oval commands this invasion from here." She leaned over a table and a three-dimensional representation appeared.

"You have four Dinarchy…" Her EkSeet was coaxing her for the right words. "…Legions closing upon your exposed Krils on the plain. Your flank

236

Legions on the hills will soon be experiencing minor skirmishes."

"Skirmishes are minor only to observers." KrutEk drily said.

"Do not imply we are indifferent! Ovals care deeply for their Krils facing TOTL!"

KrutEk sneered. "You concern is duly noted. We Krils honor you..."

For emphasis BalamEk whispered. "I will 'immediately' advise 'you' of the most exposed position when the main battle begins!"

"Do not make me wait too long." His hurt pride was evident. "I will add your delays to my Queen's report."

"Kril initiative..." BalamEk's remarks were directed to her Ovals in the command center. "...and KrutEk's vast reservoir of patience is well known to me...and Xmucane."

The invasion of KanBalaam was beginning. In confusion and amid clashing Commander's personalities.

Chapter 14

Hortim had ejected from her Toob just before impact, saved by her AkSilk. KrutChan found her next to a huge boulder decorated with scrub brush. Rolling up the AkSilk as he approached. The incoming fire had moved off behind him.

Grinning, she said. "That was exhilarating…" Brushing blue dust-powder off her uniform; barely succeeding. Hortim had minor bruises and contusions on her face and arms. "…you look worse for wear."

"I'm a grunt; don't fight with in makeup. Thanks for the compliment." He managed a malicious grin. "CheChun and I were hanging on our AkSilk Irish pennants off your mangled Toob." He glanced over his shoulder, remembering the broken Toob near KrutCheebel.

"What's that supposed to mean?"

"We missed out on the crashing-thump you made for a landing. You'll be glad to know most of my Bot-unit survived your exit."

Hortim snarled. "I released their AkSilk cocoons same time as I ejected, wise-ass!"

KrutChan laughed. "We're here to stay. How long? That depends on how the rest of the Toobs. Mosta my Bot-unit barely got off. From the looks of it, we're in a fuckin crap-storm." He said, swinging his arm in a three sixty arc defining the sea of carnage.

His calm appraisal was irritating to her. Hortim replied. "I suggest your attitude should amp up; your coolness is distracting…"

He lowered his head and blew out his breath with a shade too much purpose. "Well, as KertValSaasIch of my unit you have my report, such

238

as it is; you can review it from my Stel later, in the Burseeosil."

"You can bet I'll make time." She said.

"Right on." His red-tinged eyes grew grave. "BalamEk's people are harassing the shit outta me! They want you outta here." He pointed a grimy blue finger at her. "Far as I'm concerned I don't need unnecessary Airdale baggage."

Hortim was still prickling from his crack about her crash landing. "I should be in command. I hold the rank and you should watch your attitude, Sergeant. In my era women commanded grunt units under fire."

"Only thing rank on this planet is the stink. Your commission means nothing to the Krils." Not wanting to get into their usual verbal sparring mode KrutChan did some resenting himself. "We ain't back in your goddam future era. You're out of your element; same as I would be in a fighter jet. Use your head; you've never been in sustained ground combat."

Prideful she said. "As an officer I'm qualified to lead troops."

"My Bot-unit is held together by fear and anger. Not officer and gentleman appointees from Congress." He wasn't in the mood to argue with her.

"You can't order me around."

"Oval Command says they want you up there…" His thumb pointed skyward. KrutChan then pointed at the ground. "…not down here."

"On whose order?"

"Arguing with you is wasting my timeline. My Stel's been constantly screaming at me from Oval Command to find you. I did and told them so. Following their command, you're ordered to haul ass

239

on the rescue Toob they're sending. Get the hell off KanBalaam."

"The Ovals ordered me on this mission, buddy. I obey them."

Fluffing his shirt to flake off blue dust and ventilate his sweat buildup, KrutChan said. "Not my problem. Whine at your female commanders on your own time."

As if someone somewhere was listening, a Toob landed, blowing up a billowing blue cloud of dust, rocks, and broken plants, settling the argument for them. The Oval pilot of the Toob furiously signaling with a blinking yellow laser-like beam at Hortim to get immediately aboard.

Keeping watch around their exposed position, KrutChan gripped his M-14 tightly and said to himself, Dinarchy artillery's gonna spot the rescue Toob soon. I better get outta here.

The landing Toob began drawing Dinarchy attention. Concentrated artillery fire would shortly follow.

He glanced at Hortim with no humor. "It's time for you to rejoin the Airdales and time for me to Terminate my Time Line."

Hortim wanted to regain her command presence before she left. "How do dirty old men say it? Watch your ass, honey." She said.

"You remember what Reela told you. You keep your hands to yourself." His eyes red-sparkled as he ran off, kicking up clods of blue sand, heading back to his Kril Bot-unit.

Before he went too far she shouted over the noise of the Toob. "I'll watch your buns as you leave!"

KrutChan flushed, remembering their last time together in his barrack. "Better watch your own butt, lady." He accelerated his pace, wanting to get as far away as he could from her exposed rescue Toob.

Hortim was muttering to herself, quickly boarding the Toob.

Remember what Reela told 'you' about forgiveness? That's what she should have said to KrutChan but thought of it too late. At least this retrieving Toob is still in working order. She thought, as she caught her ride off KanBalaam.

KrutChan was weaving on the way back, dropping to the ground when he drew fire. Then when it was safe rose again, running towards his Kril Bot-unit. He came upon CheChun; falling with a thump into CheChun's hasty fighting hole at the base of the objective.

"Now that yer here; figure out your plan yet?" CheChun asked.

Irritated, he snapped back. "I ain't in command." KrutChan pushed hard on his temples. "ZacNaab's been hounding me with explicit orders."

"Don't go officer around me, boot."

"How's my Bot-unit digging in coming along?" He asked.

The initial fire from both sides was tapering off. Their view limited to a very small area of men crouching on both sides of them.

CheChun answered. "Like bedbugs under a blanket. Most-a Second Bot-unit's positions are done. We're close to finished."

"ZacNaab's headquarters people are telling me the Cunacks are freaking out; like berserk lighthouses."

"Normal for dem on a Uayeb. Dey's nervous as a virgin in a cathouse." Frowning, CheChun looked angry.

241

They both were veterans but careful about showing the other respect. Their Stels were spewing up-to-date information on their half-Malkril unit's progress; while they ignored most of the information about other Bot-units as useless.

The sounds of the battle were distant, continuous, and faded.

"Shot of whiskey would taste good right now." KrutChan said.

"Ah don't want no moonshine. I gotta keep muh brains outta my rectum. I reckon. Most my guys wanna pull out."

Explosion sounds drifted up from the valley. This area was quiet for the moment. KrutChan's Kril Bot-unit was hunkered down, the Dinarchy gunners had no targets to shoot at as long as the Krils were not moving. Eventually the Dinarchy would begin hosing their Kril's positions again when they got bored or wanted to emphasize they were still on the high ground.

KrutChan was watching CheChun's face. "I don't remember way back then any of you guys in the Pacific asking to be pulled off those beaches. You dudes didn't run when it got hairy."

"Oh yeah? Navy dropped us off. Never came back for us. Yah'll mean da heroes and da scared shitless?"

KrutChan's face was somber tinged with lines of blue-powder on the creases around his eyes. "By the way, you know your EkSeet is translating you with a lousy imitation of southern slang?"

"Don't get high and mighty. He's making you sound like a Jazz-lovin Jigaboo with your dude-shit."

EkSeet piped in. <u>We take extreme lengths to translate the garbage you Krils call language; like your use of 'bummer' KrutChan. CheChun is not</u>

better. Bums mean something totally different to an Englishman.

CheChun was grim. "Fuk him and his Rolls-Royce. My EkSeet spits in my ear when he wants to cram another horseshit rule up my asshole."

KrutChan was relieved he and CheChun could work together. They held the same opinion of the EkSeets. KrutChan carefully asked. "None of my business; but how many landings have you done?"

"Does it matter? Nobody pays you bonuses for invasions." CheChun shrugged. "Ah done my share. Ba-fore I got Seeched."

KrutChan regretted asking the question.

"The Ovals sent me on the last five Uayebs. Invasions on other planets are a son-of-a-bitch. Weren't any easier than the Pacific." He was in thought for a moment. "You'll soon find out more than just the Dinarchy are trying to kill ya on an invasion planet."

"What did-cha say?"

"You heard me." CheChun did not elaborate.

Waiting for orders, they were chatting in a lull in the battle around them, squinting in the drifting smoke; eyes burning. "Maybe our Civil War Calvary Jockey will relieve me for being a chickenshit?" KrutChan asked.

CheChun joked. "Don't try it. KrutEk will crucify you or chop yer head off. Dis is my second set of Uayebs. Yawl the damnyankee who took four strikes in the arena. Youse special."

The Irony in CheChun's voice was not lost on KrutChan. We'll definitely get along.

Eyeing KrutChan, CheChun added. "And keep that Calvary Jockey horseshit to yourself. I served under him in the last Uayebs. Fellow has guts. Looks out for his men. His Civil War tactics are

243

outdated; but whose methods ain't in this screwed up Oval universe? Buluc's the reason ZacNaab's not bitchin at ya. He's okay."

"We'll see." Thinking of ZacNaab, KrutChan snorted a laugh. "Ain't that the true skinny?" KrutChan lowered his head, squeezed with his fingers, and blew blue snot from his nose. "Get the hell out of here and go do your job. Go pester one of your old corps guys."

"Don't get your panties wet, boot. Cudda been worse...dat Calvary mug cudda been a Sambo."

"Fuck off. We all bleed red blood."

"So you say and Eleanor Roosevelt. She jes loved jigaboos. Like we used to say in my Corps, 'You'll be Sorrreeeee...'"

Before arriving back on the Burseeosil, Hortim reported to the Oval StelaVal first; relaying her battle-assessment to the Command Module.

Later she approached BalamEk. Hortim noticed BalamEk was all business, barking orders. She's my kind of officer, totally in her element.

Under BalamEk's beauty beat the heart of a lethal predator. A scientific mind controlling a driven Oval; loyal to her OvalChanHalach to a fault. She was of the consummate Oval breed, a combination reptile and a deadly cat. She was demanding to a fault.

BalamEk drew Hortim to her, gripping the younger woman's upper arm tightly. "I reviewed Stels from the Krils landing in critical areas and the StelaVal of the initial invasion. Things are proceeding as planned, but overall the Kril forces are behind the timeline that was agreed upon. I wish to speak to you."

244

Motioning Hortim to one side. BalamEk was observing the three-dimensional display. Hortim was watching Jaguar absorbed in her tactical screens.

Hortim came out of her thoughts when she realized BalamEk was speaking to her directly.

"Not are you listening? Ovals not do lose their concentration. Hortim, you will assume command of our Toobs. You have prior experience in aerial combat. My last Oval Commander was obliterated before dispensing her Krils. Fly over the areas I specify and egress the other pilots who foolishly lost their attack Toobs as you did."

That was a definite crack against me, Hortim decided. An admonition that angered her. As if she could have avoided being shot down. "I will go immediately."

BalamEk pinched Hortim's mouth in her fingers. "Be advised, the Oval sisters not are to become involved. My Oval commander should have jettisoned her landing forces before she met TOTL. Never become attached to these Krils. You should have released your Kril cargo from a higher altitude. That is the purpose of the AkSilk. Destroying a creature Toob not is what Oval pilots are trained to do."

BalamEk blaming her commander's last act is so wrong; clearly a reference to me. Hortim thought.

"I require those other miscalculating Ovals back on EkTsab; out of danger."

"I'll evacuate them from the invasion mess. Whatever impossible situation they are about to find themselves in...even if they met TOTL." Her teeth were grating with frustration.

"My Oval sister Hortim, be assured I know you will. Precisely why I have given you this new Command. I know the Earth woman in your heart
245

stirs mightily. Ovals not do favor violence. Remember, you are an Oval with an Oval's intelligence and allegiance. Keep aware of that fact. Now leave, follow my missive correctly."

"Yes, ma'am."

BalamEk held Hortim fiercely by the shoulders. "Not do TOTL on me."

Getting killed is the last thing I want to do, honey. Hortim thought. Staying alive is in my plans too.


KrutChan was fighting for prospective; having never made a landing of this size under fire. KrutCheebel was hunched down next to him for comfort. He rudely pointed and shoved KrutCheebel away to a different rock. KrutChan went in the opposite direction and found a huge boulder, split in a 'V' shape where he could observe his Kril Bot-unit.

He was commanding nothing; reacting was more accurate. Even with the Stels, his men weren't puppets controlled by him. His men were making spur-of-the-moment life decisions for survival, ignoring KrutChan.

A huge Zar named Kanix with a red band on his forehead, was growling on KrutChan's left. A Dagot was on his right sparking furiously; both aliens venting their frustration.

He watched in fascination when PacalMo, the ancient Mongol outrider, arrived in a blue sand cloud riding a Mongol pony; not a horse. A fuckin' pony?

KrutChan's mind tried to adjust. A pony in battle not feeling normal to him. Crazy Mongol's name meant Fierce Warrior; seeming apropos for him.

PacalMo was shouting at him. "Your commander Buluc's Stel stopped. He states you must attack to the front and obtain that flattop mound to protect the flank of our other attacking Krils!"

KrutChan peeked out from his vantage point to check the ground in front of him. He felt his gut tighten in fear while his hand shook. Luckily he was gripping his M-14 tight; his quake went unnoticed.

They had to go about a hundred yards in full view of the defenders who were pouring out arrows, spears, boulders, fireballs, explosions, and zippered pulses. The Dinarchy defenses looked resembled a live creature searching for a target to annihilate. A hundred yards sounded small; to an infantryman it looked more like a thousand yards.

He gave a Kril hand signal in acknowledgement to the Mongol. Who galloped off to notify other Krils across the front. He didn't envy the Mongol's job. PacalMo riding that pony in this desert terrain full of potholes and incoming fire.

Deciphering the risks of a frontal assault, he decided to use his own tactics. KrutChan abruptly spoke into his Stel knowing his other Krils in his Bot-unit could hear him. "You Krils with me get ready to follow me up a draw, Kanix the Zar, will dig."

Using Kril sign language, KrutChan sent Kanix off in the right direction. Speaking to his Bot-unit using his Stel communications. "You'll see a ditch appear in front of us. Other Krils in the Bot-unit, under CheChun, trace a route to the left flank. Waste any bastards trying to flank our attack. Whichever of our formations the enemy concentrates on; the others attack and take that flattop hill. Get it done! TOTL here, or TOTL there. Charge when I leave this rock."

Kanix had returned a Kril hand signal he understood. Bared his immense digging claws, went forward and began to excavate a gully for

247

concealment. Followed closely by the sparkling Dagot.

As an afterthought KrutChan shouted to them both, "Watch your balls!" He groaned inwardly at his remark. There's an immortal battle cry for the ages.

"Hold on that order, boys!" A voice of authority bellowed out through their Stels. KrutChan's Cycle leader Buluc wore a new Stel. A Confederate Calvary Lieutenant in the US civil war and a veteran of many Uayeb battles dropped next to him, speaking softly into KrutChan's ear. "Despair not, my tactics will win the good high ground." Buluc's soft speaking was incongruous. Both of them served one hundred years apart.

"Being scared shitless is common in the infantry." KrutChan said.

"Still your worry." Buluc said. "ZacNaab has ordered you back to his headquarters if you survive this attack. He is displeased with you ignoring his tactics."

The Krils in the half-Malkril were hearing Buluc's words through their Stels. "Forming a staggered picket line and attacking to the front will get our Krils there faster. Chances are better if you keep moving; don't stop, move fast. Attack straight into them, boys; aim for the sweet hill above us. Do not skedaddle or skulk; it will only get you kilt. Show yer Grit Krils; make them damnyankees light out! Let's open this ballroom!"

Other shouts and battle cries joined in, filtering through all the noise and bedlam. The Krils were building up their courage trying to jam down their fear of charging over open ground.

EkSeet or KrutChan's Stel was translating on the fly the other Kril's war-cries. "Alala!" from Hellenes. "Hey-Ah!" From Israelites. "Ambrones!"

248

From the Ambrone tribe. "Ker-RUTTTT!" from the huge Zars. "Godamite!" the Saxons. "Deus vult!" the crusaders. "Desperta ferro!" the Spanish. "A Fin!" from the Scottish clan.

Though inspiring to an observer, it was strictly for the benefit of an individual's morale in the moment.

Fuck it. KrutChan silently added his own battle cry for his morale.

Buluc was Stel-talking to the other Krils, changing KrutChan's previous battle order. Dude is a typical Calvary officer. KrutChan thought. This tactic may get us all killed. Or...possibly he might be right. Shortest distance between two points; all that crap. He's in command, not me.

As if he read KrutChan's thought, Buluc pointed his saber at him yelling, "Do as I say!" The horseless cavalryman remarkably charged forward at a walk.

They followed orders, some of them hesitantly; each of them driven by their own suppressed demons.

After they were on line, Buluc yelled while waving his sword. "Forward boys!"

The hill overlooking the mesa dominated the killing ground. Incoming fire from modern and ancient weapons intensified as they moved wheeling to the right directly at the enemy. Because of the terrain they were unable to maintain a straight advancing line. They were breaking up into knots of survivors.

Buluc was yelling to stay shoulder to shoulder. KrutChan momentarily focused on his right and left; battle tunnel-vision enveloping him. The attacking Krils were exposed. Buluc screaming to stay in a picket line shoulder to shoulder. Running and crouching through the murderous fire amid showers of arrows, javelins, explosions, and zipper-weapon pulses; the Zabin-Krils in survival mode, were modifying Buluc's orders.

KrutChan was mumbling to himself, nonsensical rambling with no logic. He was trying to pull his pores inward anticipating a crippling wound.

Exhausted already, his eighty-something joint-aching old body resisting his determination to keep going. He felt as though his legs were made of lead stumps.

An arrow zipped by him, followed by a head-size ball of flaming rock. But those were insignificant to the more familiar explosions and the tracer-like zipper pulse-weapons. Blue chunks of sand and gravel shrapnel stung his face. He noticed sweat or blood flowing in tears down his cheeks. Steel and rock fragments rained down in a continuous buzz. Splattering and ricocheting off hard rock; the firecracker-sparkling flashes of ordinance were everywhere around him.

KrutChan was about fifty yards into the charge when he stumbled over a rock next to a torn bush, sending him face first into the dirt.

Just as a huge shell exploded to his right, a fragment ripping the air and banging loudly off his camouflage-covered helmet.

Buluc groaned, as a fragment from the same explosion hit him. As he fell, he grabbed his left thigh. Buluc crawled over so KrutChan could assess his condition. His trouser leg was ripped in a long slash; his leg bleeding, dark-blue in the light.

KrutChan checked the rest of Buluc for other wounds, kneeling at his side.

The rebel lieutenant mouthed words, but they were broken-speech in the clamor of the battle.

"Keep…boys going…keep moving." He raised his sword and pushed KrutChan on the shoulder with the handguard.

The Zabin-Krils kept pressing the attack. They were sweating with exertion amid scrubby black brush in the blue sand. Their age catching up with them. The hard-scrabble terrain was breaking apart under their feet causing them to stagger-run.

Though they did not move as fast as the younger Krils; they were more determined. None of them had ever seen anyone in their past life outrun a missile of any kind.

Looking behind him KrutChan realized they had only traversed less than fifty yards. Krils were cringing in fear at the tremendous amount of fire thrown at them.

The heat from their exertions consumed them.

KrutChan's was sweating rivulets down his chest to his bellybutton where it pooled in a muddy blue bog. It was strange the water stayed in that pool in his navel without spilling out; weirdly defying gravity. Sweat soaked his socks, squishing in his boots.

This was the worst attack he had ever been in to date. His former attacks, back on earth, were small unit actions, retaliatory responses to ambush, or when he initiated an ambush. This battle felt more like he was back in the States at Camp Pendleton on field problems with flankers out, charging up an objective; trying unsuccessfully to maintain interval and alignment. The big difference now Krils were dying.

On this alien planet of KanBalaam the Twentieth and Twenty-first-century-humans were the point of the spear. After them would come the Zars, ArnaMals, Dagots, and Cunacks contingents. This was Kril Legion tactics. If ancient tactics were used against self-supporting defensive fortifications, the ancients would be massacred. The ancients in practice, formed phalanxes, turtles, and squares. Trained to fight face-to-face, man-to-man, a melee of vicious hand-to-hand combat formations. Huge armies assembling for attack or banding together would take hours to form up; marching in unison shoulder-to-shoulder, onto a battlefield.

After the carnage of the first world war, armies of the Twentieth century had been trained to spread out due to the weapons used against them. Many Katun cycles back in their past the Legions adopted the method, using these Zabin-Krils as shock troops.

However, once these modern combatants fought close in, breath-to-breath; the ancient's formations became a better tactic. They were the optimum choice for gaining territory. Close in, the explosives and gunfire were useless; killing more friends than enemies.

The Zabin-Krils would be stopped to let the ancient warriors filter through them finishing the attack. The Zabin-Krils then protected the ancient's backs with covering fire; using their long-range weapons.

Spurred on by Buluc the Zabin-Krils began lurching forward, cursing in each sentient being's language, tripping, running, trying to maintain their balance.

Attacking up hill was not for the unconditioned. KrutChan was sweating profusely in the heat of his exertions, grunting with each step, cursing his cigarette habit. The oily sweat on his face stinging his eyes; squinting to clear his vision. The ground was crunching with crushing soda cracker sounds under his feet.

The Zar Kanix watched, assimilating what was transpiring. He-She did not understand the concepts of these humans or the Dagots, with their slug-like bodies, or the Cunacks, with their color-twitching-lights. Breeding for self, love, hate, or anger were foreign ideas to Zars. They were hermaphrodites, having male and female organs. Loyalty they understood, but only as it applied to the Zar Cocoon instincts.

Zars had no leader, only evolved instinct made any sense to their many minds. Territorial beings described them the best. Conquering real estate drove them as much as sex drove a human. They communicated by snarls, grunts, and growls spoken sequentially serving as their language. Their intelligence implied by their stature and body language. When they stared, an observer could see the concentration and calculating of their minds. Kanix was paying attention to his surroundings now, understanding his-her task in it.

Zars used Kril sign language in its simplest forms. Zars had slits for eyes, covered by a heavy-furrowed brow, possessing millions of vision rods to absorb low light frequencies; tunneling was not a problem to them.

They were huge, but did not have the mass of a human; they were from a light-gravity world. Their skeleton porous and flexible. Extremely wary of humans, who could harm them with one blow. Not incapacitating them, but give them pain they would avoid at all cost. The upside of their body was it could not be crushed; they could wiggle out of any cave-in slithering like quicksilver, recombining when free of a trap. Making them outstanding miners.

253

When they came cocoon together was the only time their space was violated. Basically they were lone individuals; in a cocoon they were awesome in appearance. The other species, the Ovals, Cunacks, Dagots, and humans never got too close to them because of their unpredictable behavior, and that included the Dinarchy troops. To the Krils, Zars were similar to having a tethered transparent-skinned Alpha-gorilla. More ballerina than brute.

Kanix watched KrutChan's Bot-unit with indifference to their battle maneuvering. He-she heard KrutChan, after Buluc disappeared behind them, change the attack order giving Kanix the Kril sign to again dig a gully on their flank. Kanix instantly went to work excavating.

KrutChan knew if the attack continued to stall they would be obliterated. <u>Get them moving again...now!</u> He thought. Shouting into his Stel for his Bot-unit to go back to his original tactics. KrutChan was glad the Civil War Lieutenant was out of it.

"What are you doing!" ZacNaab was screaming from KrutChan's Stel. "Follow your orders precisely!"

"My orders were to charge!" KrutChan eased his response to his leader. "We are attacking..."

CheChun's flanking unit headed away from KrutChan's base unit. The attacking Zabin-Krils began a running crouch, stopping, and then moving again; most of them silent in gut-wrenching fear. Krils were fearfully shouting in exasperation, in anger, and in rage to someone, or no one in particular. Noise made them feel invincible. A missile smashing into them ceased all feelings.

KrutCheebel showed up with a small plate of plastic-like metal map of the hill in front of him. Small drops of his sweat lubricated his fingers as they moved over the map showing the positions of the Krils. His eyes were wide, the adrenaline pumping in his veins. "Lieutenant Buluc gave me his map to bring to you. He is being carried back by the medical ArnaMals. He's being evacuated to the Burseeosil."

Buluc was useless to KrutChan now and he had no time to spend on wounded men. Glad it was him…not me.

Wheezing, KrutCheebel was asking. "What is going on? I cannot figure it out. The inconsistent incoming is all over the place. Does not seem to be on us right now. Do we just keep going?"

"You horny or something?" KrutCheebel's open-mouthed breathing was irritating KrutChan. "Relax. Hold your breath, then let it out slowly…and calm the fuck down."

His officer pride grated under his noncom's remarks. Clenching his teeth, KrutCheebel shut his mouth and breathed through his nose.

"Stay on the right flank. My dudes here in the base of fire'll head for the draw Kanix dug meandering up the hill. Hang loose."

The red sun rose again, lighting the battle. The high oxygen atmosphere on planet KanBalaam was causing cold drafts. The chilly cool air lowering the temperature of their sweating bodies, making them shiver.

The attack continued intermittently with Krils getting killed and many others being wounded. They slogged forward, protected somewhat from direct fire by the Zar's newly excavated gully.

255

KrutChan's leadership and orders were not genius or made military sense, but at least he was doing...something. At this time, he didn't care whether his Zabin-Krils lived, died, or became wounded. He didn't know any of them...that made it easier, somehow. He was only concerned with his survival.

Even though he didn't want to become a target KrutChan stood. He wanted to make sure all his Zabin-Krils saw him.

He gave a Kril hand signal to attack, the same moment he yelled into his Stel. "Move out! Move! Charge...or whatever you say...move your asses...!"

KrutChan's left ear was ringing loudly with ZacNaab's screams to obey him. As he advanced with the rest of them into the second gully wrinkle up the hill.

For an instant he felt a pang of regret when he saw Kanix, who had dug both gullies; mangled. The back of his white opaque-colored head soup-bowl empty. KrutChan cursed, leaving the gully and continued on up the hill; his lungs screaming for relief.

A Dinarchy trooper literally exploded, inside a shower of yellow sparks, after KrutChan hit the guy center-mass with two rounds from his M-14 and someone else's Dagot blaster.

He saw NoKoch moving away.

The blur of the action was shutting his mind down; his fear gone. He involuntarily stepped outside himself into that peaceful zone where there are no thoughts, no philosophy; only the survival moment.

Moving further, another dying Dinarchy soldier was flat on his back, still alive. Pooling around him was dark copper blood; the ground blotting up the stain. His Dinarchy eyes hooded by a flat-top crew-cut, with crowning bristles. He had no lenses, like a human. There was no look of anger, only acceptance of his coming TOTL. He was small, gray colored, six appendages, with the bottom two used for a bipedal stance. The dying guy was definitely of the Cunack race. He convulsed, squirting from the shredded holes in his body.

KrutCheebel yanked KrutChan's hand away from the dying trooper, at the same time EkSeet yelled into his brain. Be careful, his body could still hold a lethal charge!

KrutChan nodded a silent thanks to KrutCheebel and to his own EkSeet's warning.

To KrutCheebel, the Dinarchy guy was just like them, another poor creature doing what he was ordered to do. He seemed harmless now that he was dying. When his TOTL time came, the Cunack convulsed with his last breath turning ash-white.

"What the hell are you doing?" KrutCheebel yelled.

KrutChan placed his M-14 muzzle where the creature's shoulder-brain was and fired, sparks discharging into the air.

KrutCheebel whined. "He was already dead! What did you do that for?"

"I don't need some Lazarus rising up behind us." KrutChan pointed his rifle up the hill. "Move out."

Blue smoke was scudding thick in the air as they climbed. There would be no surrender by these Dinarchy. The Kril's minds were not on giving up either. It was time for killing and survival.

As they continued to climb they encountered more bodies and wounded from their own attacking force. Some dead bodies were fried with high-

257

voltage efficiency. If human, the wounded writhing in pain.

Other Zar Krils sat motionless with huge horrible burns to their transparent bodies.

The Kril Dagots seemed immune and were far ahead of the attacking force just cresting the hill.

The top of the hill filled with incandescent flickering electricity from Dagot blasters as Kril Dagots were wasting and being wasted by Dinarchy in the defensive line.

Kril hand signals to keep moving were passed along the advancing line as Toob gunships screamed overhead throwing earth ordnance, then laser-pulses, arrows, and ancient fireballs into the top of the hill. The defenders were in a real nut buster.

KrutCheebel wasn't sure who was getting it worse, the Dinarchy or the Zabin-Krils. He saw the attacking Krils getting thinner and thinner.

They continued on the attack, at a snail pace. Most of the attacking Zabin human Krils were in their eighties. The fierce determination of the Krils was contagious. When they ran out of ammo, they pulled out swords, knives, bayonets, pistols, claws, fingernails, and teeth to use in the battle.

A group of Dinarchy defenders abruptly charged over the top of the hill and slammed into the attacking Zabin-Krils, hell-bent with their own determination. In KrutCheebel's time it would've been called a melee but that was too soft a word to him. Hell was a better description straight out of Dante's Inferno.

He saw KrutChan viciously butt stroke an enemy soldier, completely taking off the Dinarchy Zar's head.

In the confusion, brain-numbing violence obliterated everything from KrutCheebel's mind.

258

Except his heavy breathing, his non-existent fear, and his emotional exhaustion.

In front of KrutCheebel, Dinarchy troopers were coming at him from all sides. They were bayonetted by him in slow motion, smashed and shot by him in an adrenaline heightened rage.

The Toobs had left and the other supporting fire had stopped; only the shrieks, the yelling, and the grunting were background sounds all around him.

At the top of the hill, CheChun collided with KrutChan, firing his M-1 rifle at retreating Dinarchy. "Ya stop fer some moonshine? Took yer own sweet time getting up here, boot."

KrutChan's anger arose. Until he saw CheChun smiling at him. "We had to get here before you collapsed and lit the smoking lamp."

CheChun sobered pointing down the hill. "Watch out. Make a hole, boot. The real warriors are coming."

The ArnaMal ancient warriors advancing behind the Zabin-Kril were pushing aside the Krils. Their adrenaline-soaked bodies were relentlessly moving to the front.

This now became their war. They would be no longer followers, but warriors on the line. The screaming hordes of ancient warriors were killing the counter-attacking Dinarchy. Smashing the counter-attack, they formed ahead of the Zabin-Krils, into their unique battle formations.

They had passed through the Krils; a phalanx of death and destruction. Romans, Greeks, Egyptians, and the other ancient tribes, were in their element. Fighting up close and personal, hand to hand, and fierce.

The battlefield in front of the Zabin-Krils became total pandemonium. A death-dance from out of ancient earth history. The modern soldiers were in awe of their ancient Kril warrior's fierceness and guts.

KrutCheebel felt KrutChan dragging him by the shirt over the crest of the hill; both of them ended up in a large, still smoking crater. KrutChan yelled at him to stay put and ran off to the left.

This time KrutCheebel was happy to obey his leader. KrutCheebel shut his eyes and breathed deeply, glad to be alive. I made it! His hands were shaking as the adrenaline was dissipating from his body.

Minutes later KrutChan ran by him, showering KrutCheebel with blue sand and debris as he tripped, then righted himself, heading for another area. As he went by he threw KrutCheebel bandoliers of ammunition for his rifle that he must've collected from the dead and wounded future Krils.

KrutCheebel reloaded his M-16 A2 and scrunched down in the hole with only his eyes above the parapet, waiting. The battle sounds had moved away from him, down the other side of the flattop hill where the Zabin-Krils were now situated.

KrutCheebel muttered into his Stel. "Glad that is over!"

CheChun's voice erupted from the Stel. "That was the first act, sonny-boy."

"We ain't won nuthin' yet F-N-G." XibEk was saying from the Stel. "This here's a fifteen-round fight...keep low and shut up."

KrutCheebel was breathing easier now, his adrenaline-high leaving him shaking, as his body readjusted. I'm alive and well. The next round may be less intense. I hope.

KrutChan dropped into the crater with KrutCheebel. ZacNaab was screaming for KrutChan to report the situation immediately. KrutChan looked nervous. "If they counter-attack now; we're fucked." He figured that summed up the situation for ZacNaab. But, he added. "Buluc told me to hold this real estate until we're relieved."

He turned to KrutCheebel. "I gotta check the lines."

"Just remember to come back." KrutCheebel reminded him.

KrutChan's tired look was obvious. "Yeah...sure. You stay put...and stay alive..." KrutChan ran off, his voice trailing behind him. "...or I'll kill ...".

That was the best order he had received all day: stay alive.

KrutCheebel, took off his helmet, wiping his forehead on his sleeve, the blue dirt and sweat streaking his face. He plopped his helmet back on and waited; night coming upon them. No human Kril without night vision goggles could see much of anything. Hopefully the first Kin day of battle was over.

High in orbit, Hortim was back from her last mission, reporting to BalamEk and Jaguar, standing next to Hortim's tactical screen in the Kril Command Center.

Jaguar asked Hortim. "What of our Kril forces?"

"It's total chaos. From what I could make out on the Toob control screen." Hortim answered.

"Remain vigilant." BalamEk said. "During this Ob, we are blind to individual actions. We not do wish to become involved personally with our Krils."

Hortim said with a touch of peevishness. "I agree. We certainly wouldn't want 'personal involvement' to happen." Her irony was lost on the Oval women.

Jaguar smiled knowingly at BalamEk and said in an offhanded way. "Reela says you have a tendency of removing your Stel when your earth woman urges overtake you." Jaguar was discreetly smiling. "I see you are wearing your Stel now."

Hortim bit her lip, fuming. <u>Their Oval open-mike society was a group of internet social gossipers.</u>

KrutEk's angry voice forced them to look over to the side of the Operations room. The KrilChan commander of the Kril Legions was wildly gesturing; his blistering oaths punctuating his opinion. The Oval-hating male grouch was back aboard EkTsab.

"By the god of Hades, where is the weakest point in our attack? Do you Ovals know what you are doing? I had signals and fires back in my Legion earth era relating better information!"

One of the liaison female Oval Crew said. "We are repairing the slow flow of information as it arrives; attempting to receive our reports from the StelaBalaam as quickly as possible."

"You use Osils when you navigate your Burseeosil bird instantly arriving at an invasion, but you cannot communicate with more haste? Are you a dolt?"

"My lord, we are experiencing the effect of hundreds of thousands of Kril voices talking at once, under great excitement and stress." The Oval explained. "We are monitoring the babbling Dinarchy forces also; there is delay until the information is collated and processed."

"In other words I am as blind as when I was in my Legion back on Earth. The Oval magic is as powerful as the Oracles. The Sirens of Odysseus were more focused."

KrutEk was in stress overload and Hortim sympathized with him. Being in command was not for the faint of heart.

"You experienced delays in your prior invasions!" BalamEk shouted back at him. Her stress-pressure of command mirrored his.

A StelaBalaam transmission calmly stated into the Stels to all onboard the Burseeosil. "AinAcbal has surrounded the Dinarchy fortress on the left flank, regrouping to attack on the next Kin when the Dwarf Sun flares to increase visibility."

"Finally some good news!" KrutEk shouted.

"The Zabin-Krils, led by Buluc, has taken the heights on the right flank. Buluc was evacuated wounded. KrutChan has taken control of Buluc's unit and is consolidating his line."

"Do not report his name! Use barbarian to identify him!" KrutEk yelled at the StelaBalaam.

A one-word answer came from the StelaBalaam. "No."

More information was forthcoming. "ZacNaab's ArnaMal forces on the plain are barely holding their line; the issue is in doubt. KrutSeet's other Malkril Legions are tenuously holding the center of the plain stalemated with the Dinarchy forces."

"Is the StelaBalaam report to your satisfaction?" BalamEk innocently asked.

KrutEk exploded with frustration at BalamEk. "Your truths, your science, and your excuses are a myth! Xmucane will hear of this!"

"If that is your wish...." BalamEk mildly said.

"Send me to ZacNaab's headquarters! Find my Bump Area immediately!" He stomped off, trailing a line of subordinate officers and flunkies scrambling to grab helmets, robes, and other paraphnalia.

On the KanBalaam killing ground the battles were piece-meal, here and there ferocious, other places, strangely quiet. The Dinarchy counter-attack never came that night; the Krils were licking their wounds and preparing for the next Kin.

Nobody cared; the battles petered in and out around the Kril Legions. The Krils hoping a next boring Kin would follow. The First Uayeb invasion had succeeded in creating a stable foothold.

KrutChan stuffed himself between two desert boulders, finally alone for a minute. He felt a vision starting in his mind of 'Greer'. He resisted recalling her. He wasn't in the mood for future history or wet dreams; he was too ornery and tired.

KrutChan jumped in surprise, crouching with his rifle raised, his finger snapping off his safety.

A Kril was kneeling down, scaring the hell out of KrutChan. "Easy man, cool your jets."

"Shit! Where the fuck did you come from?"

The guy dismissed the question. "Don't matter."

"What do ya want, future guy?" The guy was similar, but did not look like the NoKoch; no six fingers on each hand.

He waved. "Forget the future crap. Only you can see me as I really am. I'm not NoKoch."

He let KrutChan absorb the information. "You saw me before when you and Jaguar were groping each other in the Malkril area on the Burseeosil."

KrutChan remembered. "Who are you?" KrutChan had a few hallucinations in his past, but none of them talked to him.

"I'm wearing equipment displaying invisibility to the other Krils. Don't disclose my presence to anyone, Oval, EkSeets, Stels, or Krils."

"That explains nothing." KrutChan flatly said. "Maybe you're a time-pause Oval magic vision?"

"You're crazy." The person scratched his stubble. "Don't play the dumbass with me. I know who you are; we're the same guy's inside. If you don't trust me; you'll TOTL. Who I am is known to the one person who activated me."

"You a spy from ZacNaab…or Jaguar?"

"Don't bother trying to guess. You'll be known only to me, for your protection and mine, a cutout. Security, you know? My name's Albert. Never mind my Oval name."

"What the hell's going on?"

"I'm not NoKoch, but we're the same."

Same as NoKoch or same as me. KrutChan thought.

"Camouflage keeps me secret from all Ovals and Krils. I don't exist to them. When we meet, if I forget, remind me to turn off your EkSeet and Stel. Think of me as an undercover you."

"I don't think I like you Al. I do remember you."

"My name's Albert; use it. You saw me on the Burseeosil when Jaguar talked to you. You don't like me? Good. You ain't my cup of tea either. You

265

already owe me bigtime. Who do you think saved your butt from the Tribunal? My mission is to unmask traitors like KrutEk and hidden Uayeb plots."

"Why should I help you?"

"You'll help me; with or without your agreement. I'll be saving the Oval Empire; maybe you at the same time."

KrutChan was still trying to get his throbbing heart under control. "So, why am I blessed with a messiah? How you gonna keep me safe? What's your primary mission?"

Albert didn't blink with his stare. "I'll kill KrutEk or anybody who gets in your way...or mine. That satisfy you? You're my cover. You got tons of Krils with crosshairs on your head. If I fail and you die..." He shrugged. "I'll get somebody else."

"What if my loyalty to KrutEk overrides your secret mission?" KrutChan said. "Being a traitor ain't my style. You don't really expect me to be a sniper-spotter for you. I hate snipers. I won't dance that waltz."

The glint in Albert's eyes said plenty. "I'll talk to you again during the Uayebs; debrief and help you survive whenever I can. KrutEk doesn't give a damn about you. You owe him nothing."

His face stone cold, Albert added. "I don't need you to help me complete my mission. Just don't interfere. Consider me your Agent control and nothing else."

"Are you one of Jaguar's spies?"

"She doesn't know I exist." Albert laughed.

"You didn't answer my question." KrutChan was not satisfied. "What if I warn KrutEk?" He glanced down between his feet at a flower fighting off a critter with fangs half the length of its body, trying to eat it.

Albert smirked. "My camouflage protects me. He won't believe anything you say without proof." Rubbing his eyes, Albert said. "You may get me closer to KrutEk. Don't give me any distracting ideas. I'll explain more when we meet again." Albert's finger pointed at KrutChan. "Any stupid move by you will be your last."

Before KrutChan could answer, he was alone, leaning back against the blue boulder. If Albert wasn't a vision…that spook scares the crap out of me.

XibEk was standing by him. "Who you talking to man?"

Startled, he said. "Myself." KrutChan glanced around for Albert. "Who else?"

Looking down between his feet, the big jawed bug was carrying the decapitated flower victim away. My future in miniature…

Chapter 15

A week later, Hortim was feeling the strain of flying her Toob in continuous sorties. On KanBalaam she flew more than during her past wars on earth; accelerating her stress. Between flights, Hortim position at her station's console in the Burseeosil Command center, had her enduring a constant buzz of anxiety, edginess, and boredom.

She wasn't happy when BalamEk observed her fatigue and ordered her to sleep before her next rounds of Toob flying. Without warning, knocking Hortim unconscious with a sweep of her Oval Wand; Germs carrying her to her quarters.

The hyperactive invasion Kin days passed into brief lethargy, when the battlefield below was calm.

Burseeosil activity intensified during non-battle on KanBalaam. Material resupply by Germs and ArnaMals occurred, at the same time Toob and BakToob pilots prepared their coming missions.

BalamEk's tactical planning section and Jaguar's intelligence team, were in constant motion adapting to new complications. Too busy to be distracted.

Two Kins later, KrutCheebel heard in his Stel one of the WW II Army Krils sum up the fighting on KanBalaam. "Climb a hill, fight for it, move off the hill, climb another hill, fight for it, move off the hill, and then go on to the next hill. Reminds me of Italy, that country never ran outta hills."

Their Stels kept a running commentary to Krils with similar complaints. What the Stel ignored; the EkSeet voices droned on about in their minds.

CheChun's Bot-unit, which he acquired after KrutChan took over command from Buluc, was on a knoll overlooking a KanBalaam village. The village resembled a mile-long concrete bunker, flat topped with thick walls; apertures, windows, and burrows on the sides for exits or entrances.

Other ancient ArnaMal Kril troops from supporting Bot-units had secured the village a Kin ago, thinking it housed Dinarchy troops in its fortifications.

Their tunneling Zars found no Dinarchy; only squealing, chittering creatures. The Krils herded them into an enclosure until the area was declared free of Dinarchy. Zars and Hoplites had cleared out the indigenous aliens; who would return home after these Krils left. With no visible limbs for walking, the town residents were banded with yellow-orange stripes. Leaving black trails of the slime they secreted when moving.

Some of CheChun's unfortunate Krils had inadvertently stepped into the slime, got stuck, and were devoured by a roiling mass of these creatures.

The Krils were ordered by KrutEk and the Burseeosil not to retaliate. So much for interstellar relationships and ignorance of indigenous hunting and feeding habits.

The villagers were sentient, otherwise docile. Lesson learned: Don't step in black slime trails. The new Zabin-Krils were finding out invading planets were more dangerous than their past earth battles; where they had the enemy troops to worry about.

More Kins passed. The fighting in their sector dribbled away. KrutCheebel was weary. He was intently watching the hand-to-hand struggles between Dinarchy and Krils on a rubble mound a thousand

269

yards away. Fascinated by the ungainly choreography of the death struggles.

Away from his Bot-unit, KrutChan was alone. He was just dozing off when Albert appeared kneeling next to him. KrutChan sputtered. "Will you quit popping in unannounced on me!"

"When you're dead you'll get plenty of sleep."

Albert was wearing his usual dour face; looking rugged and unyielding. Appearing to be in his muscular twenties. He had said, he and KrutChan were the same; which was a lie. KrutChan had never looked like him. The guy was creepy. KrutChan snapped. "I didn't call you. I haven't heard of any plots."

"I told you I'd protect you. But you have to help. Heard anything from ZacNaab lately? No ass-chewing? No threats? No orders?"

"Not in a couple days." KrutChan said. "Heard he was pissed at me; scrounging around for my replacement. KrutEk's been harassing the hell outa him, according to my EkSeet, for some infraction or another." KrutChan squinted. "Why, is his silence a conspiracy?"

"Yeah, you could call it that..." Albert snickered. "Get your radar turned on. ZacNaab put a bulls-eye on you. He's leaving you alone to watch the outcome. I wouldn't sleep if I were you. Look over there." Albert was pointing.

KrutChan walked to get a better view near a couple of boulders. His Kril Bot-unit slightly below were ignoring him. Individual Krils were drifting in and out of the perimeter. He saw nothing irregular going on. He ambled closer to the rocks, shrugging when he saw no more Albert.

When he heard rocks scrunch, feeling a tingling sensation on the back of his neck, KrutChan spun around.

A lone Kril from another Bot-unit appeared, raised his scimitar screaming and charged towards KrutChan. In desperation KrutChan leaned away, but tripped and fell onto his back, looking up at his attacker.

The attack occurred in slow motion.

KrutChan thought to himself. Wrong move! I'm dead. He raised his arms to protect his head.

The rest of the Krils below, seeing what was happening, scattered like a flock of exploding chickadees. Most of them fumbling for their weapons.

A messy hole erupted geyser-wise from the attacker's chest. The wound spraying KrutChan with bloody gristle.

Groan-grunting, the mortally wounded attacker fell on top of him.

Angrily spitting bit-chunks of gore from his mouth, KrutChan furiously pushed the cadaver off him; his fear driving his survival. KrutChan was brushing blood encrusted blue dirt off his face and shirt. His heart pounding. Since the attacker was another Kril, it was obvious to KrutChan he was targeted for a Legion assassination attempt.

PacalMo was standing a few feet away from KrutChan holding a Dinarchy blaster. Absentmindedly inspecting the blaster, turning it over and over in his hands. The weapon slowly dimming down as it deactivated. PacalMo wore a quizzical expression on his face.

"I didn't see that one coming. He was obeying orders?" KrutChan asked PacalMo.

"That would be my assumption." PacalMo said. He viciously kicked the head of the attacker to make sure he was dead. "In battle, Krils see confusing things; enemies where there are none. Or he may have been Dinarchy. My EkSeet says you would understand 'things happen in the moment'."

"Mug was an ancient ArnaMal." CheChun said, after arriving, kneeling next to the body giving his assessment. "If he was Dinarchy, which I don't think so; he's cold turkey now." He grinned. "Poor guy probably had the bejeezus scared outta him seeing KrutChan's ugly moniker."

KrutChan disagreed. "Dude was on a crusade from his boss. I'm the one who was scared shitless."

"Ya piss on angry gods, you be fuked..." XibEk said, lighting a cigarette. "KrutChan looked at that dude like he was gonna put the guy on a shit-burning detail. That'd get anybody mad."

"Nice shot...thanks...but you almost wasted me too." KrutChan was grinning at PacalMo. "You're my friend forever."

PacalMo had a puzzled look on his face. "I do not want to be an associate in your tribe. I do not like you. I am not your friend."

"Why did you save my butt?" KrutChan was wondering.

"I was merely out of arrows. I'm not versed well in this strange weapon." He was staring at the blaster. "A dead Dagot bequeathed it to me. I am as surprised as you it spits out its deadly venom from my fumbling attempts to use it."

"You learn fast."

"I was aiming at the boulder behind KrutChan to distract the killer." He smashed the blaster against a boulder. PacalMo was picking up arrows for his quiver and finding a short sword and a spear. "It was my error. How do you say it? I missed?"

"Well, thanks again...anyway, Mongol." KrutChan said.

"For what reason?"

"For missing me."

KrutChan struggled, not believing the powers-to-be were trying to zap him. He would have to watch his back by himself. I got no friends here to protect me. He thought.

KrutChan heard Al whisper in his mind. Thanks for nothing. I'll remember that the next time... Maybe that was EkSeet? He didn't want any next time.

Spurred by a yelling ZacNaab, throughout the next Kins, KrutChan rallied his Bot-units and the remnants of another unit and led them to a critical position forward of the front lines to the top of another mound. Away from a Saddle on the Hill where they could be ambushed.

ZacNaab was shouting for KrutChan to obey his orders precisely; not adapt them.

Soon thereafter, the Bot-unit came under an intense mortar and artillery barrage, quickly followed by a ferocious ground attack by some forty Dinarchy soldiers.

KrutCheebel saw KrutChan remained at the forefront, shouting encouragement, hurling his grenades, and directing deadly fire. Later KrutCheebel watched as KrutChan reorganized his defensive position with additional Krils in time to halt the third Dinarchy attack.

On two separate occasions KrutCheebel saw Dinarchy soldiers close to within a few feet of

273

KrutChan, but they were killed by CheChun before KrutCheebel could raise his rifle.

Afterward KrutCheebel heard CheChun ask KrutChan. "Trying to win the war by yourself?"

"Naw…I couldn't wait for your slow southern ass to field-strip your fag."

KrutCheebel thought KrutChan at least owed CheChun a thank you.

His Krils repelled the determined enemy.

The battles for KanBalaam were beginning in earnest.

During the second week, ZacNaab was sending blistering commands through his Stel. Screaming for KrutChan to follow his orders to the letter, how to close with and destroy the Dinarchy. KrutChan adapted to each circumstance. In the end able to fulfill the mission of ZacNaab's orders. His commander was not satisfied with KrutChan's performance.

KrutChan had performed so remarkably well that ZacNaab replaced him. With a more suitable Commander who would listen to and obey ZacNaab. He did not want KrutChan gaining more fame. ZacNaab accused KrutChan of murdering one of ZacNaab's own Krils.

Over the weeks the biggest problem facing the Krils on KanBalaam was disease and critters. From microbes and planet-specific-known predators on the planet, even the plant life, became threatening predators against the Krils. The Krils were dropping from disease at a worrying rate.

The Zars and the Cunacks were hit the hardest, getting sick, changing color. They showed indications of tiredness, depression, or dying…just like humans. Puking, crapping themselves, numb with fever, and cross-eyed from exhaustion.

CheChun snarled at KrutChan. "Told ya you got more than the Dinarchy ta worry about."

Like KrutCheebel, KrutChan, PacalMo, CheChun and others, the Legion's aliens kept fighting in the battles by coping. If they left, someone else had to take up the slack of their loss. Keeping them from requesting relief.

On another patrol KrutCheebel saw what the Ovals called 'Loob', used by some of the future Krils to add camouflage to their uniforms for invisibility. Whatever the cost, KrutCheebel wanted to be invisible.

KrutChan acted as if he had never seen or heard of invisible camouflage before. Once you move; you ain't invisible. KrutChan thought, forgetting about Albert.

Chapter 16

Many Krils wondered if they were in a Time-pause. An idea called impossible their EkSeets. Their timelines were scrambled and did not appear forward-linear; but they were experiencing the effect.

Two KinUt months had elapsed on KanBalaam, KrutCheebel sat hunched in his hole, talking into his Stel. He dictated his missive to his ArnaVal, ZocKuk back on Arna. "So far I am alive and well ZocKuk." He decided not to dwell on his chances of survival.

He paused briefly, then continued his missive to her. Minutes later, KrutCheebel decided to tone his missive down and skip the specifics. "Remember how much I love you."

"That is the perilous state of affairs the Kril Legions are facing after too many Kins on KanBalaam." KrutEk was pacing in BalamEk's Burseeosil Command center's conference room, briefing BalamEk's personnel.

"You are correct in your assessment. The invasion is behind timeline and is stretching out." BalamEk said. "The Dinarchy are fanatically holding the fortress surrounded by AinAcbal's forces, which is on your left flank. There is a new Commander on the right flank, assigned there under the orders of ZacNaab. ZacNaab is less than aggressive and furiously berates his former ground commander for precipating actions he does not approve of."

KrutEk thought. ZacNaab again is inactive; he is the bane on my existence. KrutEk retained him not as a tactical leader. Kept him because of his reputation among the Roman Krils as a former assassin. His forte focusing on individuals. Because of his ineptness, ZacNaab solidified KrutEk's power-hold.

"Not did you hear me?" BalamEk asked. "Are you aware of ground commander ZacNaab's intransience?"

"I am monitoring his actions and will deal with him. I do not wish your advice." KrutEk grunted, sighed, turned his back, and ignored BalamEk.

The Burseeosil crew were judiciously absorbed in their screens and duties. No one wanted to be seen or heard, singled out for criticism.

"The Dinarchy are amassing for an attack across the entire front, including your diversionary flanks. We are suggesting ZacNaab be more engaged. Ovals monitor your other ground commanders KrutSeet and AinAcbal are angry at ZacNaab. He is constantly interfering with his Zabin-Kril subordinates."

"You are honored by me." After which, changing the subject, KrutEk angrily whispered. "Once we are committed; do not allow your eagles to tarry. The KanBalaam campaign must be won! The cost in Krils and materiel is becoming too high."

BalamEk's attention was on her 3-D controls and listening to her subordinates and her EkSeet at the same time. "The coming offensive you are planning will not be too expensive to your Legions, we hope. We not do require more Krils in our Memorial to the Oval Protectors."

"My Kril casualties are for me to contemplate." His voice was rumbling with implied threat. "I cannot be concerned about individual Krils in my Legions." KrutEk said, leaving for his Bump.

"Ovals demand life, not suicidal liquidation of our Krils!" BalamEk shouted after him.

KrutEk was shouting back at her over his shoulder. "Precisely the reason why my Legions must attack first!"

BalamEk hissed. "As the command leader on invasion planets, you never understand my meaning."

Melding from the conference, KrutEk muttered to himself. "Your Oval misunderstanding is exactly what limits this plan…and my meaning."

KrutChan was satisfied his troops were ready to begin their part in the new offensive as he walked the perimeter. He saw Albert walking towards him from the Dinarchy side! None of the Krils in KrutChan's Bot-units paid attention.

Probably got his invisible camouflage initiated.

When he got close, KrutChan asked. "How'cum every time you show up I get bad news?"

"Ain't my fault, I'm the messenger." Albert was looking around nonchalantly. "Rain and flash-flooding has stopped. Weather's looking good. I'm on my way to ZacNaab's headquarters to snoop. Can't stay long."

"Thanks for the weather forecast."

In exasperation, Albert looked at the sky. "I'm here to tell you the Kril higher ups are watching you. Your 'loose-cannon' KrutCheebel is sending military information in his missives to his ArnaVal ZocKuk."

"Tell me something I don't know. My EkSeet's been gossiping."

"They're spreading the word your KrutCheebel is daydreaming most of the time. His mind ain't on his job."

"He's my responsibility. I know what's happening. Bug off!"

"Your assassins will try again. Don't go to sleep on me." Albert got serious. "I'm just telling you. Keep letting KrutCheebel pecker-drift and you'll have more like him. With the offensive kicking off, you don't have a lot of time for fixing."

When ZacNaab's voice blasted into KrutChan's Stel with another detailed waltz-card order, KrutChan raised his middle finger at Albert, or ZacNaab or both, as he walked away. When he looked back, in between 'yes sirs and no sirs', Al had disappeared.

KrutEk's new offensive was about to be initiated. KrutChan's Bot-units were sweeping another area above another rock strewn village. Going was slow, except for their ragged breathing and their crunching boots. Nobody had anything to say. They were worn out.

KrutCheebel was so tired he was sleep-walking. He shot and killed an indigenous furry creature in front flushed from his hide. That will teach him to stay out of my way.

KrutChan's grimy hand abruptly vice-gripped KrutCheebel's arm. His furious whisper below his helmet rim. "Get your educated nose outta yer ass, killer. Quit walking in a daze." KrutChan's rifle then swung forward to accent his silent command for KrutCheebel to move out.

KrutCheebel did as he was ordered, glancing back to see KrutChan solemnly watching him. KrutChan's rifle gave him that silent command again to speed up.

Walking carefully, KrutCheebel moved like a crab, his head swiveling on a turret. He was fighting

279

boredom, trying to keep his mind alert. It was a lost cause. He was so tired.

Off in the distance an artillery barrage had begun, following Toob airstrikes.

They came to a huge rock pile, actually a village a Ki hour later. He was on point ahead of his Bot-unit. His dread was increasing, not helping his boredom. Something was going to happen and his nerves were on edge.

A bullet or missile came out of nowhere and sprayed KrutCheebel with dust and rock-shrapnel. I knew someone was out there!

A burst of gunfire from CheChun's newly acquired BAR chugged and killed the sniper.

The other Krils were razzing KrutCheebel's blindness.

"Our professional-rabbit-killer ain't payin' attention on point." Someone growled into his Stel.

The Krils traveled down one rock strewn street of the village after another. Waving for cover fire, as they checked out each hut of stone-rubble. No residents or Dinarchy were encountered. After a while they determined the place deserted, ominously quiet.

KrutCheebel noticed KrutChan was angrily yelling into his Stel; responding to ZacNaab or his headquarters' people. He was arguing again with ZacNaab's headquarters as he set his Bot-units into positions. KrutChan hand-signaled KrutCheebel to climb over the top of a rock-rubble-mound in front. To call it a hill was ludicrous; it was a rubble-pile above the stone village.

As KrutCheebel crouched he saw KrutChan's entire Bot-unit had the same bone-weariness.

Someone was coughing and spitting. KrutCheebel felt alone and isolated in a crowd.

"Move over to the other side of my hole, shithead." Someone was complaining to another Kril.

"This is my hole ya sonofabitch. I dug it. Be nice or I'll kick your ass out."

"Knock it off people!" KrutChan loudly commanded.

KrutCheebel sat up, struggling to get his hose from his Camelbak for water. AhauHuc from later Iraq-Afghanistan wars gave it to him to get rid of his two noisy canteens.

The voice of authority from KrutChan snapped out of their Stels. "You guys aren't here to square dance. Don't get comfortable. We ain't staying long."

The Zar, CoCum, added. "Kril humans talk much. This cocoon is a scrambled..." The rest got lost as the Stel tried to translate what the Zar had said.

"Stick it up your anal hole, ya mole!" An anonymous voice shouted back to the Zar.

XibEk, who had taken KrutCheebel under his wing, plopped next to him. "I'll tell you what the problem is in this here situation. Dudes don't know how to have fun. Any mutherfucker can bitch. Takes a real man to have fun. Fuck and lay the pipe, and then have more fun."

"In this case we are the ones being screwed, you Schvartz." The translated SS Kril said.

"Swallow my black snake, ya Nazi fuck. You ever got tired of sucking on Hitler's dick?" XibEk tossed a cigarette into his mouth. "Dat German's got serious shit down to a science."

"Screw his ass." A voice said, laughing.

"Right on." XibEk said. "Spread your cheeks and love will come."

"Quit skylarking!" CheChun chimed in.

XibEk was thinking. White man says get back to yer killing ways.

They ignored their EkSeets; their attention was refocusing already. They had more important things before them, like staying alive.

Everyone heard the crunching of the rocks as KrutChan skip-stumbled to the middle of the perimeter. Guy was always spaced out on AnticArna was the Bot-unit's assessment.

KrutChan was spitting water in an angry stream; his eyes reddish-blue in the light; after receiving his latest command from ZacNaab. "We're moving. Going back to that fucking hill we took on the first day above the Bump Zone."

He was angry as he screwed the top back onto his canteen.

"Oh...my aching back." A Kril said.

"I ain't walking no more." Another said. "My feets are bloody with blisters!"

"The man says go...you go. Where does it say you got a vote?" XibEk asked.

His hand leaning against a boulder, his rage building, KrutChan said. "I'm gonna give you guys the word...one more fuckin' time! Get your brain-housing-groups cleared out or I'll flush 'em for you."

He was staring hard at KrutCheebel. "Shoot when you have a 'real' target. This ain't no free-fire zone."

"Gotcha, lifer...life's too short to waste ammo on rabbits or chickens." XibEk said.

The others were giggling, nodding in agreement.

KrutCheebel felt they were admonishing him. He got attitude.

KrutChan sneered. "Any daydreamer thinks they're strolling in a whorehouse with their dick trolling for cunt, and that anybody and me are gonna meet behind one of these boulders! Only I'll be coming back!"

The Krils were angrily staring at him. Their hate for him was palpable. KrutCheebel shared their feelings. "Our lord and master speaks! Hail to the chief!" KrutCheebel snarled.

XibEk softly said to KrutCheebel. "Cool your jets Cheeb. Only one lifer's allowed here."

KrutChan yelled. "I don't need your help, XibEk! Keep your mouth shut!" KrutChan spun on his heels. "And that goes for all of you."

KrutCheebel noticed CheChun, PacalMo, and XibEk were staring at the ground or at the sky.

KrutChan turned to KrutCheebel. "You get your head out of your ass and start paying attention to what's around you! You got the word?"

KrutCheebel stared at KrutChan. He was angry at being singled out. He wore his 'I am a combat veteran-face'. "I got it all right...general."

The sounds of fighting drifted into the background. The war had suddenly become personal.

KrutChan was so calm and said it so softly, KrutCheebel almost didn't hear him. "What's the word, Lieutenant?"

"I got whatever the general..." KrutCheebel said with irritation; not knowing what KrutChan wanted from him. "I said I understand...!" He was saying what all of them were thinking. He had had enough of KrutChan's browbeating.

The Krils in the Bot-unit was sneering, some smiling, most of them tensed up in expectation, witnessing the confrontation.

Their Stel's picked up ZacNaab's voice. "Your Krils are insubordinate, barbarian."

KrutCheebel never saw KrutChan's swinging M-14 as its steel butt-plate thudded fiercely against KrutCheebel's Kevlar helmet, sending it flying. KrutCheebel's vision exploded with stars. A moment later he was prone, groggily staring at the dirt-rocks, and drooling. His helmet had fallen back into his fighting hole, knocked off by the blow.

He stuttered. "W-w-why the hell did you...?" His eyes gradually focused, KrutChan's face mere inches from his.

"What's the word?" KrutChan quietly asked.

KrutCheebel stared with hate at KrutChan. "You are absolutely right."

"Sergeant...say it!" KrutChan's rifle barrel was pointed at KrutCheebel's face where he could smell the gunpowder. "You better adjust your fuking Lieutenant attitude right now. You're a day-dreaming cocksucker!"

KrutCheebel mumbled. "I understand. I got the word...Sergeant."

"Anytime you fuck off again..." KrutChan waved his rifle furiously at the Krils around him, "...or any of you other guys! I'll kick your balls up to your ears. Then I'll slice you from your nuts to your throat before you hit the ground."

There was a deadly silence for a tense minute throughout the Bot-unit, thinking in unison. When a killer says he will TOTL you; it is personal.

In the background the thumping muffled explosions announced the war was still going on. This wasn't even a decent skirmish.

"You got a problem too?" KrutChan was glancing at another sullen Kril. "You the Lieutenant's fart-sack buddy? Have something smart-ass to say dickface!"

The Kril kept direct eye contact with KrutChan and he slowly shook his head. "Not me Sergeant."

"Understand Lieutenant?" KrutChan's eyes were boring into KrutCheebel's. "One round up your ass will cure your malingering. Those brown bars of yours don't mean squat here."

Remembering KrutChan's prowess with his K-bar in the Arena, KrutCheebel answered loudly. "Yes Sergeant...you are right. I got the message."

KrutChan spun around to face the rest of his Bot-unit. "Fine...if the 'rabbit-killing-elite' gets it..." In a mock-whining voice he added, "...then everybody understands." KrutChan did not seem satisfied.

KrutCheebel thought. I will kill you. His features glowed with hate towards KrutChan.

Hunching his pack to a better position on his back, KrutChan was addressing his Bot-unit. "Every swinging dick is going to fight and see his TOTL coming right between his eyes!" He hissed. "Understand?"

KrutCheebel's band-of-brothers were looking between their feet or away, nodding.

"From now on there'll be no more surprises, no more wise-ass laxness, no more sleep walkers, or lame-brained screw ups..." He stared hard at KrutCheebel. "...And no more non-Dinarchy goddamn unauthorized kills!"

As he stomped away, KrutChan's Stel was crackling. ZacNaab's high pitched commands were in the background. They all heard KrutChan

285

swearing in between his 'yes sir's'... I understand, sir. We'll be moving out shortly...upon your command, of course, sir."

KrutCheebel sat on the ground, resting his head against a rock-face with his eyes closed. He wasn't hurt badly. A tiny trickle of blood dripped off his nose. He had his bell rung and his boredom was long gone. He stared into the valley-zone, at the dots of Toobs circling above it, and the swirling smoke and explosions.

"You okay, splib?" XibEk asked. "Don't sweat it. Lifers are like that." He was crouching and whispering, cupping his cigarette in his hand.

KrutCheebel rubbed his throbbing head. "He does not have the right to do that."

"Maybe not in ROTC or training; but here he's da man and makes da rules." XibEk said. "KrutChan wanted to get your attention. And ours. You paid the price. Ya piss on angry gods, you be fuked'."

"That bastard will get his..."

"Forget about dat, dude. Watch out for number one. There's lot'sa Krils trying to kill him before ya." His mouth twisted in a grin. "He likes your liberal politics."

"Oh sure he...." He left the thought unsaid.

"As the old salts say, we just got tightened up. Lifers like doin' the Lord's work according ta der bible."

KrutCheebel studied a far-off flying Toob moving towards them, through the twilight and smoke.

PacalMo knelt above KrutCheebel, wearily asking. "Are you normal, little bird?"

"Ya think you're more qualified doing KrutChan's job, butterball?" CheChun asked, referring to KrutCheebel's former rank. "ZacNaab would grind you into punk and piss. You ain't got the experience or the balls."

KrutCheebel was angry. "I could lead Krils better than he can ever...."

Laughter from the Bot-unit drowned out the rest of his sentence.

"Who the fuck would follow you, genius?" CheChun snarled. "You couldn't lead a pissing contest into the wind."

Shaking his head KrutCheebel groaned. "I would have cold-cocked him if I had seen it coming."

Another moment of laughter passed before PacalMo had a CheChun translation he could interpret. "He would have given you TOTL. You deserved what you received. Warriors do not like Krils who enjoy killing. You must learn. The fate of the gods was with you. In my old timeline; you would have been beheaded or pulled apart by horses."

"I need to be Seeched out of here." KrutCheebel moaned.

"You gotta get your head together." XibEk advised him. "Then ya won't need tah be Seeched, shitbird."

"Yew was lucky, boot." CheChun said. "I woodn't been as gentle as KrutChan. In my old raider outfit, I wudda frosted yer N...black ass. Eat shit EkSeet!"

XibEk stood, glaring with hate at CheChun. "You ain't in your past unit, old corps. Why don'tcha try some-a dat raider-shit on me, overseer?" He gripped his rifle and waited for a reaction from CheChun.

287

"Wasn't talkin' to you, XibEk. Remember? We's buddies from the old Uayebs?" CheChun grimly smiled at the challenge, hearing the hesitant laughter that was ringing out.

No one messed with XibEk. The man was a lethal force by himself. Betting started on who would be left standing after XibEk and CheChun locked horns.

In the future, KrutCheebel decided he would be more selective with his killing. How the hell can I stop enjoying it?

Both XibEk and CheChun walked away from each other.

The Toobs flying over the plain were bombing and strafing, thunder-punctuating in a parabolic arc, and then leveling out to resume their ordinance runs.

A Zabin-Kril TzenalAh, the Gurkha from his Sirmoor regiment approached. Sporting his deadly Kukris knife, he strolled by looking for their Bot-unit's leader. Everyone pointed KrutChan out for him.

Later CheChun crawled up, sweat pouring from his long face, his damp beaded forehead furled in thought. "Our fearless leader just got the word from Burseeosil EkTsab. KrutChan got us a ride on Toobs. Least we don't gotta walk back to that goddamn hill. Put on your bloody panties, girls."

Tossing a pack of unfiltered Camels to the asking white German soldier in another hole; CheChun was grinning. "Yawl take care now, jigabooz." He pinched his nostrils. "It do smell some in 'dese here parts. I gotta git me some air."

He wandered off as the smiling German beckoned him with a Kril hand signal.

288

XibEk wheezed. "He's right. This fightin' hole does smell better; now that southern gentlewoman is gone."

"Don't make me laugh; my head hurts, man." KrutCheebel said.

XibEk smiled, which infuriated KrutCheebel, feeling like his brother was assessing him and his fear. "What are you smiling about? KrutChan's an idiot.""

XibEk inhaled deeply, the high oxygen atmosphere of KanBalaam flaring up his cigarette. "Forget KrutChan, he's 'jes shakier than a white boy after dark in Harlem. That dude KrutChan; I seen talkin to hisself with nobody around. Crazy lifers do dat."

He dropped back his head preparing to hack up a green lump. He blew out a line of smoke with the slimy throat butter. He's right 'bout one thang, though. You enjoying snuffin' critters too much."

"So what?"

With a snort, XibEk mumbled. "Where's some pussy when ya need it?"

KrutCheebel smiled at him and jerked his head. "White bread 'makes love'; black man 'fucks' and pounds vagina."

"Forget the college words, nigger. Vagina? Yer education counts fer nuffin' here."

KrutCheebel was sorry he was not like XibEk; nothing fazed him. "Well then, do not go Uncle Tom on me. You are my man."

XibEk furrowed his brows. "Don't shine my ass, boy."

KrutCheebel wanted to puke and swallowed over and over, trying to keep it down. His bowels were rumbling. Nothing made sense to him.

XibEk chuckled and tossed a cigarette at KrutCheebel, saying. "Yer looking a little blue-green, professor."

"All right, all right. There's too many supermen; too many all-American types here for my liking."

"Don't like the neighborhood, boy?" XibEk's tone was tense.

"What is your problem? I never believed in Henry Fifth. Being among the ignorant and the unwashed suckers of this mob don't make them glorious."

"You implying I'm one of those unwashed?" XibEk changed his manner, his teeth set, and his nostrils flaring. "You're saying I never understood or read Shakespeare, motherfucker?"

KrutCheebel's leg was dancing with nervousness. "No man, I didn't mean 'you'."

"Sure you didn't." XibEk wasn't convinced.

KrutCheebel was startled by the swift movement of XibEk, kneeling in the gravel, one finger poking into KrutCheebel's chest. "Don't you ever talk to me like an uppity nigger, boy. I been in more shitstorms than all the times your mother wiped your ass. I'll drag your head inside out through your asshole and make you lick your shit. Fancy talk or cussing KrutChan will never wire up your guts."

KrutCheebel stared at the bottom of his fighting hole.

XibEk grubby hand tapped KrutCheebel's cheek. "Shakespeare and your personal philosophy isn't going to save you. Only lifer KrutChan can maybe do that, along with the great unwashed. And I can call him a lifer because I earned the right. That lifer and me don't kill unarmed personnel. Waste of time and ammo."

290

KrutCheebel thought back, seeing KrutChan blow the head off the dead Dinarchy on the first day. KrutChan doesn't murder the unprotected...?

XibEk gestured. "That's all the reality you got now. Deal with it, college boy, or TOTL. Just don't get the rest of us killed while you deal with your philosophical quandaries."

XibEk is using English as if he did not need his EkSeet. Nothing like democracy in action; the common man comprehends everything. The great melting pot of America. "I'm sorry." His eyebrows lifted in an apology. "I only meant KrutChan is an extroverted-egomaniac leading us to destruction. His ignorance shoves us into danger with no reason; no thinking, beyond his orders."

"That's his lifer job, nigger."

"We will die because of his emotional immaturity and condemn us into his place in hell."

"Dat's quite a mouthful of crap, Cheeb. His orders, my black ass." XibEk jerked a thumb over his shoulder. "Guess what! You're already in Dante's Inferno, my college friend; and with the lower classes."

"I'm NOT wrong!"

XibEk simply shrugged.

Their evacuation Toob was screeching through the twilight smoke towards them.

KrutChan's voice over their Stels got everyone's attention. "This grabassing has gone on way too long." KrutChan walked to the middle of them. His ankle twisted when he stepped on a rock, and he kicked it away, cursing. "Off your asses and saddle up!" He sounded like he was announcing they were moving to another picnic area.

When they headed to the landed Toob to get into their AkSilk cocoons, KrutCheebel felt his bowels groan again. In a panic, he was looking around for any Blue or Green Kril bags anywhere. He hoped his sphincter muscle would hold tight. He dropped his trousers, opened the bag he found, just as his bowels eliminated.

A human from the Army yelled out. "Goddam man...cut us some slack. Jeezuz-Key-rist, don't get any on me!"

His knees were shaking and KrutCheebel was humiliated; his sanity and humanity absent. As they lifted off, his mumbling turned into a silent prayer for his redemption.

Chapter 17


For the Kril Legion forces on the offensive the Kins of the battles wore on. The Oval-piloted Creature Toobs never setting down on KanBalaam. AkSilk dropped Kril Bot-units in the Legion from place to place; landing zone to landing zone. After assembling, the Krils dug and set in. They walked patrols. They did ambushes. They or the Zars dug positions. The Cunacks kept up their incessant body-blinking conversations. Carrying their blasters, the Dagots lumbered along slug-ambulatory.

Another Kin of walking went by in a staggered route march. When they came across a small grove of nine-foot-high pitcher-plant-like structures. Another village? Or a plantation, cultivated by the indigenous farmers of KanBalaam.

"Goddamit...slime-vine!" KrutCheebel swore, steadying himself by touching the closest plant. He tensed when he felt movement. His M-16 sling slipped on his left shoulder while he adjusted his weight.

"What the hell?" The skin of the plant felt tin-can metal to him and breathing. KrutCheebel's hair rose on the back of his neck.

<u>EkSeet, what is going on here?</u>

His EkSeet warned him not to move. "The plant is alive and senses your presence by your motion and touch."

KrutCheebel looked anxiously around for anyone who could see what was happening to him.

No one was paying attention.

He backed off, contrary to his EkSeet's advice, to get away from the living Balaam creature. He wanted to shoot it, but instinctively he knew his Kril Bot-unit would rag on him about prematurely harvesting crops.

KrutCheebel's fear increased exponentially when the creature-plant began to convulse. Instinctively, KrutCheebel fired a three shot-burst from his M-16A4 rifle into the plant, succeeding in wounding his own leg from a ricochet.

The plant blossomed open on top. It rustled with a moan and bent over. Instantaneously devouring KrutCheebel in one swallow.

KrutCheebel screamed!

When he fired, at first the other Krils ignored him, assuming he was taking out another innocent Balaam creature eating grass. His yell got the attention of the Bot-unit.

They quickly formed around the quivering plant in confusion after KrutCheebel was consumed. The milling Krils could hear KrutCheebel's muffled shouts. They attempted to cut through the skin of the plant while KrutCheebel softly begged for help. His mumbling pleas were constant; while his Bot-unit

293

Kril's attempts remained futile. The Kril's swords, spears, and axes were useless, dulled to blunt edges against the armor of the huge plant.

KrutChan was swearing; yelling at the Krils nearest the plant. "Get him out!" He was pushing more Krils towards the alien plant. "Free him, goddamit! Cut through that big bastard!"

Running up close, fiercely slashing his K-bar at the plant's thick metal-like skin. When that failed he thrust the bayonet on his rifle at the plant.

He lowered his voice, enough guys were hyperactive. "Gut the plant; free KrutCheebel."

"What the hell you think we're trying to do!" CheChun yelled back at KrutChan.

XibEk was yelling at the plant. "Let him go, you mutherfucker!"

The other Krils were panting from exertion. They kept probing for a small chink in the plant's armor skin, attempting to create an open gash in the plant's thick hide. The creature-plant resisted their penetration efforts. A few tried shooting at the plant, as KrutCheebel had done, but stopped immediately when ricochets hit some of their own men.

Fifteen minutes passed before PacalMo said the obvious. "He is losing timeline. You must use your noise rockets, KrutChan!"

One Kril climbed onto the plant and was preparing to lay a heat grenade on its mouth. The plant's orifice was clamped shut.

XibEk said the obvious. "Forget about using thermite; you'll kill KrutCheebel!"

Seeing the knives, sabers, spears, grenades, and bayonets were not putting a dent in the creature; KrutChan suggested another attack. "Back the fuk off and shoot at the mouth of that goddamn thing."

Again he was speaking low, as if reading from a manual. "Blow off its top and get an opening in it."

The plant's color changed to a darker hue, hardening its skin in resistance when a furious burst of fire from semi-automatic pistols and automatic weapons tried to dent the mouth of the plant. The Krils avoided shooting in the area of the bulge in the creature they assumed was KrutCheebel.

Again some ricocheting bullets aimed at the roots of the plant from other Krils resulted in slightly wounding a few more Krils. KrutChan ordered the gunners to cease shooting. They were hurting themselves; not the target.

The Kril's Stels were transmitting ZacNaab's voice yelling for KrutChan to explain what was happening, with KrutChan bellowing back, "Not now! We're not fighting Dinarchy!" Seeing his men watching him, KrutChan cooled down. "We got another problem."

The Krils could all hear ZacNaab in their Stels not satisfied, screaming and demanding KrutChan tell him what was happening. KrutChan yelled back. "We're in a firefight with some Balaam plant creature! I'll call you right after we kill this bastard."

Pushing the Earth Krils aside, plasma from a Dagot using his blaster did little more than dimple the plant's armored defenses. Time was clicking forward.

"Get away from this fucker!" A Kril EOD guy from WWII was furiously digging, with assistance, at the roots of the plant. Planting TNT bricks in the hole and damping down the dirt.

An expert, it didn't take him long before he was yelling, "Fire in the Hole....fire in the hole!" He stood and calmly lit the fuses on his charges. He walked away to get distance between him and the creature-plant.

KrutChan fell next to the Explosives guy behind a mound of dirt and rocks. "Won't that much TNT kill KrutCheebel too?"

"I can't stop it." The EOD guy pronounced. "Your man's as good as dead by now anyway..."

The explosion set their teeth on edge. Concussion billowing over them; the shrapnel of rocks hidden in the dust cloud and bits and pieces of dirt rained down on them for a minute.

When the smoke cleared the creature-plant was leaning but unscathed.

There was a lull; KrutCheebel's struggling sounds had stopped. PacalMo, CheChun, XibEk, and KrutChan, with the rest of his Bot-unit stood staring at the plant; mutely recognizing the loss of one of their own.

PacalMo said. "Our heroic efforts failed."

"Concussion probably kilt him. He bought the farm. That boy never could pay attention." CheChun said.

"Shut up." KrutChan said.

"Shut da fuck up, Chun!" XibEk yelled.

Nudging KrutChan in the ribs, CheChun reminded him. "Tol ya fightin planet creatures in the Uayebs can kill Krils quicker than the Dinarchy."

"Save your eulogies!" KrutChan snarled. To change their mood, he added. "He's wasted like any Kril. Move out and keep your distance from these fuckin' plants. Don't even breathe on them."

He had told them what they did not have to be told; on an alien planet anything can bring TOTL.

"And keep your goddamn interval between Krils. I don't need you guys feeding the animal or

296

vegetable life here. We're here to kill Dinarchy; not become coleslaw."

KrutChan's Bot-unit walked away after their EkSeets advised it was impossible to save KrutCheebel.

KrutChan reported to ZacNaab he had lost a Kril to a vegetable predator. ZacNaab's voice in return sneered. "Your EkSeet advised us. As usual, you are late and remiss in your reporting. Continue on with your mission without your usual excuses and weeping."

It would be a long Kin during a boring walk as they slowly forgot KrutCheebel. Their memories would last forever of how he died. There was no time for mourning.

An hour later, one of the Dagot's was translated by their EkSeets. "KrutCheebel body recovered...by disposal Germs. Toobs with other Krils...Burseeosil."

The common thought among the Krils was KrutCheebel would not be buried on KanBalaam. No Kril wanted to be left to rot trillions of light years from Arna. His beloved ArnaVal, ZocKuk, would get another Kril.

Kins passed patrolling with the remainder of KrutChan's Bot-units. There were small sniper fights, no Krils wounded, and then they were digging in again. Their frustration during the absence of action was a blessing. Irritating them the same as the small cuts, bruises, and stinging scrapes on their bodies. Even the bland Zars were getting surly.

The Krils kept hidden their own thoughts and practiced a quiet protocol. They blamed KrutChan for what happened. If only…? Knowing some small slight or word, or minor altercation by any of them could set off an explosion of rage.

The Kril's smoldering hate for KrutChan and KanBalaam festered. Anyone trying to start a scuffle was literally knocked down for insignificant words or deeds. The peacemakers had their hands full. If there was sanity anywhere on KanBalaam, it did not exist within this Bot-unit, among these Krils.

Another kin passed without incident. The Bot-unit taking a break was crapped out, sprawled in a clearing between boulders.

CheChun was incessantly scratching at his itching crotch. If he had been on earth, they would have said he had lice.

XibEk mused. "You just stirring the crabs up, cracker. Quit picking at'em. Dey won't eat much."

Nobody laughed. They all were thinking of KrutCheebel.

"You need'ta pour some gasoline on dem." XibEk offered. "Light'em up. They be gone inna flash."

CheChun had heard enough. "Ah'll unscrew yer head and pour gas down yer throat and make a roman candle out of yawl. Ya fuckin' black genius."

"I tried some mosquito juice on them once long ago." Another Kril offered his solution. "Afterward, I thought my balls would fall off."

CheChun's anger was mounting. "This here's Balaam. These critters ain't crabs, asshole. I had the crabs. These ain't dem."

The Krils had renamed planet KanBalaam 'Balaam'.

A Medical ArnaMal came after XibEk yelled for him. He assessed the situation and pulled

298

CheChun's belt off, then peered down his trousers. "Those creatures are somewhat larger than my Stel advised." He mused. He evoked a manner of medical distance most doctors exuded with patients.

XibEk scornfully said. "Tiny critters feeds on tiny food."

CheChun was embarrassed and not happy. Under his anxious gaze the medical mug took out a glowing purple device. When his device fired, a billowing cloud of black dust erupted upwards out of his trousers, staining CheChun's shirt and his face with the ebony powder of critter debris.

CheChun was stunned. Looking at the ArnaMal, looking at his open trousers, and then betrayed-looking at the ArnaMal with fright. He was sure the corpsman castrated him.

His Kril Bot-unit witnesses fell to the ground and began howling with laughter. CheChun's shirt front and face were covered with black soot.

The Medical ArnaMal stripped the trousers and skivvies down to CheChun's ankles. CheChun was white-naked from his belly-button to his ankles, in stark contrast to his face and shirt. "You are cured! See, no harm...no fowl...your essence shooter is alive and well."

The Krils, who were standing and sitting around, busted out with laughter again, grabbing their sides, some unable to get off the ground.

"So, that's what the Imperial Wizard looks like under those sheets?" An unhelpful black Kril observed.

In the background ZacNaab's voice was demanding. "There is too much strident frolicking going on! What is occurring in that barbarian's Bot-unit? Cease this frivolity!"

"What did he tell us? Quit shriveling?" Someone asked. That caused more outrageous uproar.

Tears were rolling down XibEk's face. He was pointing his blue-dust-covered finger at CheChun. "Sheeee-IT! Shriveling is right! Cover up boy, you embarrassing us real men!"

Another Caucasian Kril added. "Hell yeah, ain't no Oval or ArnaVal bitches gonna AkSilk down ta rape yer black-face ass."

Another Kril soberly said. "I'll bet Ol KrutCheebel is rolling over in his grave, peeing his pants."

The medical guy was fanning with his hand to dissipate the black dust still clinging in the air. "They're gone. You'll be okay." He was packing up his kit. "Think of it like this; if I had used a slightly higher setting, your sperm would have fried. You're Oval would have been displeased for making you a eunuch for a couple of KinUts."

Roaring laughter started again, and pointing fingers greeted that remark. All they had to do was look at each other to bring on more hysterical laughing. The on-looking Krils were rolling in the dirt, fists in their mouths, trying to muffle their giggling.

ZacNaab was screeching for silence.

CheChun ripped his shorts and trousers up to his hips to get decent.

Packing his medical gear, the medic cheerfully advised CheChun. "Remember. You have to keep your powder dry." The Medical ArnaMal then defensively ran away, reading CheChun's expression clearly; getting out of arm's reach.

KrutChan noisily stomped into the clearing. "What the hell's going on? ZacNaab's raising holy hell! Somebody got a feather up their ass?"

"Fuck no!" CheChun yelled at him, as he buttoned his fly, glowering at the Krils still laughing at his plight. "Funny...real fuckin' funny...assholes...eat shit an' die!"

KrutChan yelled. "Quit your grab-assin'! We got new orders to move out. All you giggling-fairy-assholes saddle up and follow me!"

Nobody knew if XibEk was responding to KrutChan, or making a crack about CheChun; but he had the last word. "Twern't nuthin'...ain't no 'big' thang."

They strung out, keeping their professional interval, following their leader KrutChan back up another hill.

As they were route-marching, someone, in a whispering voice under his breath said. "Whooooosh...pecker frying here!" Their chuckling and shaking shoulders could not be stopped.

Some of their stresses and frustrations were stilled at the expense of CheChun, for a little while.

KrutChan's Bot-units spent the next two Kins exterminating a nest of six-inch python-looking things, with large fangs and lousy dispositions. The snake-things attacking the Krils first, killing some of them.

The weeks of stress, frustration, and irritation, at the indigenous life-forms on KanBalaam turned into a mob-execution. Unconsciously KrutCheebel was silently in their minds. It satisfied the Krils even more when the creatures fought back.

Their EkSeets warned them the python-things may be babies and by Oval rules they were not to be killed.

"I saw Dinarchy head-markings!" A helpful Kril said.

One Kril summed up their feelings, yelling. "Fuck these baby-things and fuck the Ovals!"

At the time, KrutChan was immersed in conversation on his Stel, walking away just after the killing started. They heard him reporting to ZacNaab. "We made contact with a nest of Dinarchy spies and infiltrators. We're eliminating em. They won't allow us to take em prisoner."

The EkSeets verified his report.

When no more snake-creatures could be found the Krils lay spent from their killing; they were satiated, a few smiling in satisfaction.

Sometimes, going ape-shit crazy is better than high-priced medication. KrutChan was thinking.

The next Kin they came upon forty acres of wood-like dens filled with more hiding python-creatures in front of KrutChan's Bot-units. His Krils were motionless, indifferently staring at them. Toobs were called in by KrutChan and made short work of whatever structure-dens were in the area, frying creatures to a crisp red-blue blanket.

KrutChan's Bot-unit moved away, never looking back. Their surly moods were increasing.

After KrutEk's new offensive kicked off, each Kin melded into another Kin. ZacNaab was located on the Burseeosil EkTsab with the reserve Malkril. Giving his subordinates, including KrutChan, relief from his constant sitrep whining. He was too busy partying with ArnaVals, sucking on his Roman wine bladders to worry about tactics.

The offensive on KanBalaam started out as a slugfest of ancient and modern artillery duels, in the valley between the hills.

KrutChan's Kril Bot-unit, held in reserve, was milling and weaving like broken dolls; oblivious to everything but their fear growing acid holes into their stomach lining.

KrutChan's Bot-unit had disintegrated into racial groups of snarling cursing animals sick and dead-tired of the weather, sick of smelling each other, and sick of a developing smoldering hate for their all-too-familiar fellow Krils. They had been together in battle too long. Words non-existent between them, their hair-trigger rage too close to the surface.

KrutChan received a terse transmission from the Burseeosil, ending a tense week of observations of the battle on the KanBalaam plain. While his Bot-unit was taking a much deserved break, KrutChan with a small tight smile called his Bot-unit into a circle.

Now what...more bad news? Their Stels, in unison, were reverberating with that question.

They immediately stopped their bitching and held their breath when KrutChan told them. "Our part in KrutEk's final offensive will kick off within hours."

The grousing started right away, summed up by one Kril through his Stel. "ZacNaab's Roman assholes are enjoying watching the rape of Balaam. Guy is really brave holed up in the Burseeosil."

XibEk snarled. "What else is new? KrutEk's sending us all into a shitstorm on his Roman-ass whim."

"Ours but to do or die." CheChun said. "Let it go. Krils always TOTL, like KrutCheebel…for nuthin but angry gods." CheChun's tight smile at XibEk telegraphed he stole XibEk's favorite expression.

KrutChan was grinning with his secret. The others wondered if he was getting battle-happy. And cracking up. Then, he threw an information hand grenade in the middle of his Bot-unit.

"KrutCheebel's alive." He said.

There was a collective blink from all of them as they tried to digest what he had just said.

He had to repeat it. "You heard me. I said he's alive and breathing."

"We all seen him die." XibEk wondered. "How the hell did that happen?"

"He's been on EkTsab getting reconstructed, not reconstituted; that flaky bastard." KrutChan's good humor was wasted on this Bot-unit.

"He'll be here soon." He was shaking his head. "If you can believe it; that dummy asked to rejoin us. He says KrutEk ordered him down. Generals usually don't worry about Kril snuffies. He must be desperate for troops."

Their groans and shaking heads agreed with him. But, their moodiness, rage, and frustration evaporated.

"Told you mugs. Hell, I knew that jigaboo couldn't die…too goddam lucky." CheChun declared.

XibEk snorted, "Yeah, that's true, a white dude would have X---pired; not tough enough."

Another Kril joined in. "He always was a good guy."

XibEk disagreed with him. "You said a couple weeks ago he was an asshole."

"The gods threw him out of the dragon's cave." PacalMo said.

"Yeah, even the fuckin Balaam plant life couldn't stand his superior half-breed ass." CheChun said.

That got a laugh and a lot of nodding.

On the Burseeosil, BalamEk spent her timeline during the Kin supervising the preparations of her Ovals. Tons of supplies were being loaded into BakToobs and the smaller Toobs; ammunition loaded last to be off-loaded first. The scurrying and intense focus of her Burseeosil crews gladden her heart, as the Obs ticked off possibilities in their timeline. BalamEk was thinking KrutEk not could fault the Ovals should he lose this battle.

Jaguar approached her. "We are ready. I require your command to initiate our final support effort. I not do envy the Kril Legions fighting in the next Kin."

BalamEk walked away to the Oval chapel to pray to the Cosmic Egg, saying to Jaguar. "I not do relish telling Xmucane her KrutEk has met TOTL."

Sixty Obets later, KrutChan's Bot-unit went silent when KrutCheebel wandered into the group and sat down. Breathing deeply, not knowing how he would be greeted. KrutCheebel prepared for the worst. He was looking drawn and sickly staring at the ground.

No one knew how to speak to the departed back from the dead.

Finally, someone asked him. "Where you been fuckin' off? Have a good rest, daydreamer?"

Everyone uncomfortably laughed. One of their own Kril had risen from TOTL; creating superstition.

KrutCheebel decided being called a daydreamer was better than the 'Rabbit and beastie Terminator' they had been calling him. KrutCheebel smiled. "I been screwing all your Ovals and ArnaVals."

He grinned at CheChun. "Particularly the Confederate ones."

Their moods changed immediately.

"Hell yeah, all them southern belles who never had a black snake before...tell them, man." XibEk said.

Even KrutChan was smiling. "Doing god's work, huh Cheeb?"

KrutCheebel grinned and stared at his feet. He was feeling better.

"Lazarus, you got point on the next patrol." KrutChan snarled.

KrutCheebel swiftly glanced up at KrutChan with anger and then shook his head, realizing his leader was jesting.

"Why did the dragon gods throw you back to us, little bird?" PacalMo asked.

KrutCheebel explained. "After you guys abandoned me..." He laughed at their shocked faces for a minute. "...I was found by another unit of ArnaMals relieving us. They found me next to some rocks alongside that dead plant-animal that had eaten me."

Their EkSeets explained. The plant vomited him out. His DNA was wrong. The animal could not digest his poisonous DNA mismatch.

"That animal-plant wasn't a Grand Dragon was it?" XibEk asked with a straight face.

306

KrutCheebel mocked XibEk. "Shee-It...wasn't racism...that Balaam moron just couldn't absorb all my college education. Creature had no elitist class."

KrutChan smiled at that crack. "Yeah, he must have been one of those high-class dregs of his society. Too particular what it eats."

XibEk couldn't resist the opening. "Hates dark meat; creature wants only a little white pork and I mean 'little'."

CheChun knew who that remark was aimed at. "Fuck you, Xib, and the slut you rode in on."

KrutCheebel got serious. "To tell the truth. I wish it had been one of you other guys. That was scary being eaten alive. I thought I was a goner, suffocating is not my ideal way of dying. But that would have been better than being slowly digested for hours. I would not have wished it on any of you."

He stared at CheChun. "Even you my cracker friend."

"Sheee...it!" CheChun grinned back and boasted. "Ah wudda ate him from the inside out. I did that wid a gator once-ted."

That got the Kril Bot-unit laughing at his absurd southern bravado.

One of the replacement Krils tried to join in with a profound thought. "Glad to see ya. We need more bodies for the Dinarchy meat grinder."

The angry stares from the older veteran Krils silenced him for daring to join their inner circle uninvited.

Their EkSeets chimed into their Stels. Stop your joking at once! KrutCheebel has not escaped unscathed. He has contracted a lethal virus from his contact with the juices of the plant. Microbes are hardy creatures and not subject to DNA problems.

<u>They   adapt.     They   will   TOTL   KrutCheebel
eventually.</u>
Their silence was contagious.

While the Bot-unit Krils became somber with
their own thoughts; KrutChan took KrutCheebel aside
behind a large blue boulder.
"You okay?   The Ovals turned you around
pretty quick.  You sure you're healthy?"
"They have a first-class hospital on the
Burseeosil.  After they revived me, they said I met
TOTL  dead,  you  know?    I  cannot  explain  it."
KrutCheebel sniffed the air, wrinkling his nose at the
odor.  "Oval Magic like Krils say.  The hospital
smells better than you people."
KrutChan gave him the bad news.  "We're
committed to the offensive.  This time a fight to the
finish."  He eyed KrutCheebel to see if there was a
reaction.  "You didn't have to come back here, you
dummy.  Doc tells me you have a reserved ticket for
the ride back to Arna.  Out of these shithole Uayebs
that are coming."
"I know.  I got the word."  He was staring at
KrutChan, remembering KrutChan banging his
helmet for daydreaming, and then looked down.  "I
owe these guys something after I was eaten for trying
so hard to save me.  I got the word about the efforts
you guys put out, fearless leader."
KrutCheebel's shoulders raised and lowered.
"KrutEk ordered anybody who wanted to go back to
their Bot-unit to let the doctors know.  The horny
female nurses were wearing on me.  KrutEk gave me
a golden out."
KrutChan   peered   under   KrutCheebel's
lowered eyes.  KrutChan was smiling.  "I got nuff

problems without interracial faggots French kissing. You ain't falling in love with CheChun are you?"

KrutCheebel grinned. "No. I did not miss your armpit-smelling gentlemen that much. I do have a few friends watching out for me. Krils on the Burseeosil say more gentlemen are hunting for you."

"Scuttlebutt." KrutChan said.

"Rumors up there say KrutChan is fighting three wars and losing two." KrutCheebel pointed skyward. "Some say you will TOTL during the last battle."

Thinking of Albert, KrutChan said. "I've heard that from a lotta guys."

KrutCheebel kept looking towards the Burseeosil in orbit. "Some say you're already dead."

"Scuttlebutt is always wrong. Do I look dead to you?"

"Are 'you' okay, Krut?"

"Now that you're back..." KrutChan had his hundred-year-old man-face on. "...my life will get easier. Hell, I'd love to send you to ZacNaab to educate that Roman about leadership. I need every swinging dick who can fight on this coming offensive."

KrutCheebel took his remark as an apology. It wasn't. It was a statement of fact. That was why KrutCheebel felt good. It was great for a dreamer-terminator to be appreciated.

When KrutChan and KrutCheebel returned to the group the Krils looked up, glad to get out of their thoughts about the coming battle.

"You forget something, KrutCheebel?" KrutChan asked.

Stunned for a moment, KrutCheebel wondered how the hell KrutChan knew about his gift for his Kril Bot-unit.

Reading KrutCheebel's mind, KrutChan raised his eyebrows. "Forgot you got a Stel on and an EkSeet? Those Burseeosil crews would have donated if you had politely asked them."

Slapping his head, KrutCheebel brightened up. "I was numbed into forgetfulness by this tremendous outpouring of emotions from you bastards."

He pulled items out of his pack and threw them in the middle of the circle. "Oval bottles of booze, the kind that don't leave a hangover in ugly numbskulls. It works on Zars, Dagots, and Cunacks too."

The aliens in the Bot-unit grunted, sparkled, and farted in appreciation.

"I did a KrutChan-stumble over a liquor-pile that was stashed on the Burseeosil. Didn't want some ArnaMal doctor tripping and getting hurt."

"Get rid of the evidence. All of it...and bury the remains." KrutChan told them. "Cheeb probably got it from ZacNaab's private stock. He'll be pissed off and want the booze back."

"What about our EkSeets blabbing to momma?" Someone asked.

We are at rest, as are you. We see nothing out of the ordinary. Their EkSeets said.

KrutChan joked. "We is in reserve, you fart-sack-loving-suckers. Resting and out of this crap for a couple of hours; so consume it...most skosh."

The EkSeets translated that word for the non-Japanese.

KrutChan was walking away. "ZacNaab will be yelling after my ass, in a minute. I'm taking my

Stel and getting the hell out of here before I see something I ain't supposed to see."

"Aw man, you gotta have some hooch..." The Kril Bot-unit said together. Not meaning a word of it.

As he wearily ambled away, tripping on a few stones, they saw KrutChan's hip pocket was bulging from an Oval-Osil wine vessel and they laughed together. The world was all right for a short timeline in their misery. Their anger and frustration had dissipated. They were a functioning Bot-unit again. Krils in relaxation.

Hours later Kril boredom ended abruptly. Fiercely attacking, the Dinarchy were on the two flanking hills, taking the Kril forces by surprise.

The Krils responded.

The Dinarchy was not in its death-throes or close to defeat. Somehow, from out of somewhere, the Dinarchy appeared in strength. With the middle of the Kril lines in the valley holding in place, the two peaks on each side became the objectives for the Dinarchy. Once they succeeded, they could obliterate the Kril forces caught in the valley below.

On the western hill the Viking AinAcbal' Krils took the brunt of one pincer, his Krils recoiling.

Having command by attrition on the eastern hill, KrutChan's Bot-unit and the other Kril Bot-unit forces under his command reacted; absorbing another pincer-attack.

ZacNaab was not happy. When the Dinarchy counter-offensive attacked, KrutEk ordered ZacNaab's reserves onto the plain. ZacNaab had his own problems in the valley, unable to spare any of his officers to replace KrutChan.

The Dinarchy's massive bombardment blasted the entire front with modern artillery, zipper weapons, particle weapons and explosives raining from the KanBalaam sky. The middle of the Kril lines in the valley were attacked by ancient ballistae fired arrows and spears, flaming boulders of pitch-tar, followed by phalanxes and turtles, supported by oxybeles and gastraphetes, belly-bows similar to cross-bows. The attacking forces created an anvil to hold the middle of the Krils ancient ArnaMal troops, keeping them from advancing.

KrutChan was besieged by EkSeet requests asking for situation reports on a constant drumming basis. He was getting too much information, unable to respond to every inquiry.

Even KrutSeet was requiring KrutChan to report up-to-date situation accounts, spurred on by KrutEk. The battle was too fluid to immediately answer the multitude of questions. The breakdown in Stel communication highlighted the flaw in their design capability.

He ordered his EkSeet to stop advising him of every damn possible eventuality, except for imminent peril. This caused the individual Stels on each of his Krils to respond in the same way and freed up their minds for fighting.

Their commander ZacNaab, far below KrutChan in the valley, was the most insistent for information, screaming for obedience. KrutChan, having heard it countless times before, ignored him. He concentrated his focus on the immediate situation facing his Kril Bot-units.

Initiating a defensive-position by midafternoon, KrutChan had CheChun put in charge of the out-most first line, far down below the opposite military crest of the hill.

CheChun was shouting over the din of the battle. "We need support holding onto this goddamn knob!" He was angry at KrutChan. "Them bastards are hell-bent to get up here!"

Yelling back into his Stel, KrutChan was agreeing with him. "No shit old corps! You done it before. Treat the bastards like one of those 'banzai' charges. I need you to stop them as long as you can. I assigned PacalMo to the second line to assist you with advance by fire."

"Who's commanding the safest spot, the third line...are you? That's where you should be to control this shit!"

"KrutCheebel just got promoted." KrutChan said. "I'm coming to you in the first line. I can see more!"

"Bullshit. Commanders belong in the rear, Krut...according to the book!" His Stel transmitted CheChun's Bot-unit returning fire.

KrutChan unceremoniously collapsed into CheChun's fighting hole in answer. "The Kril battle-manual says different. I ain't no fukin' hero."

He glowered at CheChun. "We got XibEk and his assigned Bot-unit with us. You know XibEk. He'd love to crawl out by himself to screw those Dinarchy women. I need you to keep him focused."

"You're full of shit." CheChun snapped. "That spook is more focused than any of us!"

"Let's move before the Dinarchy are butt-screwin' us!" KrutChan yelled over the background explosions.

"You first Sgt York; I'm a cowardly follower." CheChun answered.

The two of them bent over in a crouch, went beyond the military crest of the hill, to the forward fighting hole in the defensive position.

They were hearing ZacNaab shrieking in their Stels for a situation report. KrutChan rolled his eyes.

Once there, KrutChan continued to command by Stel the other piece-meal disorganized units on an eastern peak. Assigning them to tie in into his own positions.

The consolidation took horribly long minutes to get the defensive line accomplished in the midst of continuous incoming-ordinance from the advancing Dinarchy. His Zabin-Krils were taking killed and wounded as they piled up Dinarchy attackers into masses of bodies.

CheChun and KrutChan sustained non-life threatening wounds, too busy to notice the blood or pain. XibEk was uninjured and invincible.

When he looked, the time on KrutChan's watch showed an hour had passed. They still were holding the forward line.

The frustrated Dinarchy charged anew in a series of attacks lasting another hour, gaining ground, coming closer and closer to KrutChan's first line of defense.

On their Stels, ZacNaab was screaming at KrutChan, and AinAcbal on the other hill to retreat back down to his position in the valley immediately! Becoming part of his reserve force. "We need reinforcements! I order both, AinAcbal and the barbarian, to obey me! Do not provoke me!"

Somebody, sounding like KrutChan, spoke out of the Stels. "We only got Zabin-Krils up here. Anybody see naked women barbarians?"

The Zabin-Krils then heard KrutChan speaking to AinAcbal. "We're holding? Are you still in contact? I'm getting static on my Stel. We're both on high ground. Can you hear me, AinAcbal?"

"I hear your whispers. I also have whining noises on my Stel." AinAcbal said. "I will commence my attack when I am ready. My Kril are obliterating the female Dinarchy, sending them to their Valhalla. Are you retreating?"

"Fuck no man; we're having too much fun." He groaned at that insipid remark becoming a quote stored in the StelaBalaam. More bravado for his troops than reality.

Over their Stels, ZacNaab was apoplectic. "My EkSeet insists you are both getting my missives! Withdraw to my position immediately!"

"I'm still getting too much static, AinAcbal. I hope your EkSeet can decipher and translate what I mean." KrutChan was rolling his eyes at XibEk. "We ain't bugging out. This Balaam real estate cost us too much already. I'll be back to you in a minute...er...Ob...a goddamn Bot...or whatever the fuck it's called. Situation's fluid. I gotta go."

All the Stels were buzzing with ZacNaab voice. "Obey me, you filthy barbarians! I am in command! I will have you crucified even should you TOTL!"

AinAcbal cut in. "KrutChan, do what you are capable of until we can make contact with ZacNaab. His ArnaMal Legion is doing magnificent work in the valley."

Sounds of fighting, screaming curses, pleas for mercy, followed by chunky-thumps, were background noise from AinAcbal's position. The Krils heard the Viking chuckling. "My Zabin-Krils have obliterated the fortress. My Legion of Krils is now attacking down into the valley behind the Dinarchy. I use your hill as another fortress we will fight to strive reaching."

CheChun and XibEk collided above KrutChan's fighting hole, saying together. "Dinarchy are overrunning the line!"

XibEk was saying. "Time to saddle up lifer." He was calm but agitated. "Follow ZacNaab's orders and pull us back."

"We'll be frosted unless we scram." CheChun added.

In answer, KrutChan spoke into his Stel. "PacalMo...we are falling back to your position. Cover us with whatever you got. My Kril Bot-units don't need your guys hitting us...okay?"

PacalMo did not answer.

What can he say...no? KrutChan trusted PacalMo had heard. Nothing fazed that Mongol. He thought. That is, if he's still alive.

The Stels crackled with PacalMo's voice. "This is not the time for drinking your fermented goats milk. Are you coming?"

Everybody's a comedian. KrutChan thought. Whistling into his Stel to get attention, with Kril hand and arm signals ordering them, the first line of defense pulled back, assisting their wounded as they left.

Taking unbelievable long minutes, the survivors collapsed into PacalMo's defensive line.

Behind them, in the abandoned first line of defense, the Dinarchy were bludgeoning, stabbing and killing every Kril body they came across. The fire tapering off as they went about their cruel business.

His forces unseen, AinAcbal was continuing his sweep towards KrutChan's forces. Ripping his way into the backside of the Dinarchy forces in the valley; he was not having an easy time of it. He was

316

too busy hacking with his axe to give situation reports to KrutChan and ZacNaab. Doubt began to creep into the KrutChan's Krils mind-set AinAcbal would survive.

During the next hour ZacNaab was ranting continuously for KrutChan to retreat and for AinAcbal to stop advancing on KrutChan's hill. At the same time the incoming fire from the Dinarchy enemy increased ten-fold. Kril hand signals were the only means of communication underneath the noise of battle. The fighting, in the second line of defense, was grinding down to fierce hand to hand fighting.

XibEk watched PacalMo decapitate a Dinarchy trooper. Using his sword to punch a path through the attacking force before retreating back into the line to resume firing his arrows. XibEk mimicked him using his bayonet-studded M-14, clubbing and stabbing any enemy confronting him. On the way back into the line XibEk had seen CheChun and KrutChan, countering the charges of the Dinarchy before withdrawing to the relative safety of the line. They were not brave; more out of desperate necessity.

The entire second line of Krils was bulging outward in the carnage. The survivors pulling back into the line afterward. That tactic would only work once, before the Dinarchy figured it out. The Zabin-Krils were holding the line so far; but not for long. Dinarchy troops occupying the first defensive line of the Krils were swelling in size, pushing forward.

PacalMo was panting and out of breath, close to exhaustion. "We are going to meet defeat if we stay. They are continuously coming over the bodies of their dead." He said to KrutChan.

CheChun nodded in agreement. "Most my surviving fellows got more holes in 'em than a sieve.

317

What now Sgt York? Wanta live for a couple more minutes? Or maybe follow ZacNaab's order to fight another day?"

KrutChan was mulling over his tiny options, his eyes and head scanning the position.

"Tactically we need another defensive line, white bread." XibEk added. "Let's rejoin ZacNaab, man."

"What about AinAcbal?" For a long brief minute, he was in thought, then KrutChan was shaking his head. "You know. I really love it when you guys volunteer to take my fukin' job."

"Don't be a lifer-asshole. Save the Krils you can!" XibEk yelled.

"Custer thought he had the injuns right where he wanted them." CheChun remarked. "We should get the hell outta here, Ace."

"When you guys're right...you're right." He yelled into his Stel. "KrutCheebel, we're coming. Enforce fire discipline." They heard KrutChan chuckle. "And you leave any Balaam critters who wander in front of us alive and well."

CheChun got in the last word almost. "Yer poor relative's issa comin' ta visit yer sharecropper shack Set up da table and I want a room widda view, boy."

"Be advised KrutChan, any indigenous personnel are long gone." KrutCheebel answered back. "Chun...the only room left is in the slave quarters. A confirmed reservation for your fat redneck butt."

Their banter wasn't funny; deadly serious facts. They disengaged as before, dragging their wounded with them.

Long minutes later, when they arrived, it felt like home.

The noise of the battle abated as the enemy went about their grisly task of mutilating fallen Krils still alive in the second line. Krils feigning death got the same treatment.

The Dinarchy were getting more efficient or had less bodies to contend with; within a half hour they had regrouped and began the final assault.

XibEk was singing an old song. "Momma said der would be days like dis...der would be days like dis...my momma said."

"What? Yer 'black snake' asleep? Not inna mood for any screwing?" CheChun asked.

"His working tool's so scared his pecker's peeking his one-eye out of his asshole." KrutChan added.

In his opinion, XibEk said. "Dere issa time fer humpin' and a time fer a love-in'." He cursed at his EkSeet's translation. "Right now I's waiting...till the lifer cuts me some slack."

KrutChan lifted his middle finger in salute to a grinning XibEk.

When the Dinarchy attack came, the Krils bitterly hunched down, forgetting everything except staying alive. Looking into each other's tired exhausted eyes. No more jokes, no more teasing, no more argot, or smartass insults. The time to meet their god was at hand.

They lost track of time in the noise and confusion. They had no fallback position. Kill or be killed was the essence of the moment. The Dinarchy and the Zabin-Krils had been taught the lesson well; needing no critique.

The concussion waves rolled over the Krils from the expended ordinance. Their heaving straining lungs were bellowing in anger. Bits and pieces of gore and animal parts decorating them were ignored.

The two lines crunched together with lethal force. They beat, clubbed, and annihilated the enemy on both sides. The mayhem continued. In a desperate battle to survive, the lonely got lonelier. Both sides together reached a point of total exhaustion.

Then the Dinarchy were backing off. KrutChan guessed regrouping for another decisive lunge.

KrutCheebel heard KrutChan speaking into his Stel with a lot of 'no sir', 'yes sir', and muttered aside curses. He signed off with. "I can't withdraw with Dinarchy bayonets sticking in our dicks!"

The Krils were hanging their exhausted heads in anticipation of the final lethal attack on their position.

When the Dinarchy re-appeared they were dribbling towards the Krils in bunches; about one thousand strong when assembled. KrutChan's Kril Bot-units had cornered them in a blind cutoff of massed bodies. They just stood there, nine hundred plus Dinarchy human and alien; all kinds of sizes and shapes. The Dinarchy must have known there was no route for retreat with AinAcbal advancing.

They were between KrutChan's forces and AinAcbal's Malkril Legion advancing up the hill. They were motionless, not firing their weapons and some of them screaming obscenities at the Krils during the lull.

KrutChan checked on his Krils he could see. Thinking Custer would have mournfully cried. <u>My command is about to be annihilated.</u>

Upon reaching their decision, the attacking Dinarchy force advanced in a walk towards KrutChan's Bot-units; an ocean tide filled with drowning cornered rats. Walking, not running, silent and inevitable.

KrutChan's Krils were stunned, confused by the actions of the Dinarchy troops. These aliens were preparing to die; killing Krils in the process. They weren't charging, just slowly moving forward ominously. Queerly not giving away their obvious intentions, they fired upon any Kril approaching their force to take them prisoner. They were preparing to die. They looked determined none of KrutChan's forces would walk off this killing ground afterward.

KrutCheebel saw CheChun, XibEk, and PacalMo watching the Dinarchy while gritting their teeth, holding their fire. He Stel-ordered his Krils do the same.

ZacNaab ordered KrutChan to cease fire. While in the background, KrutChan was crouching, sliding back and forth along the line, his periscope head rising periodically.

KrutChan was arguing and cursing into his Stel, with someone else higher up, requesting permission to open fire. He was advised to follow ZacNaab's orders. If the Dinarchy had begun a screaming ferocious charge, he wouldn't have asked permission. The Dinarchy were acting strangely indifferent.

BalamEk's voice was heard in the Stels. Ordering Jaguar to stay out of the argument.

KrutEk was silent.

AinAcbal's voice was pronouncing, in a monotone what his present situation was with his Kril Legion coming up from the bottom of KrutChan's hill.

ZacNaab's command to KrutChan was to take prisoners of the advancing Dinarchy. He inferred KrutChan was to obey his order so ZacNaab could take credit for their capture. He threatened KrutChan if he did not want to face charges brought against him; he should obey. If at all possible.

His amendment of that order was as clear as turgid swamp water. The transmission was obvious, ZacNaab wanted to avoid any responsibility for whatever happened.

Bitterly, KrutChan knew ZacNaab wanted him to TOTL, die with his Bot-unit on this goddam hill. ZacNaab would be the glorious hero; KrutChan's Krils the sacrificed.

The Dinarchy kept walking towards KrutChan's positions. The Kril's Stels were quiet waiting for the outcome of the argument going on with their superiors, in the background.

With his Bot-unit subordinates boring holes in him, KrutChan was thinking. I'm running out of options. Surrendering isn't an option. Disobey orders or die by inaction. A nervous or pissed off Kril might break fire discipline. You can't pick where or how you die.

"That's a Banzai attack without the screaming!" CheChun was shouting into his Stel at KrutChan. "Make a fukin decision! Or we're all dead!"

Suddenly, a barking in-the-clear command came over their individual Stels, assumed to be from KrutChan. No one would ever know for sure. "Fuck the Ovals!  Fuck this war!  No goddamn Dinarchy prisoners!  Kill'em all…!"

Fifteen minutes of unrelenting slaughter followed.  No one would or could stop the Krils once they unleashed the maelstrom.  The Dinarchy forces exploded, torn to pieces, broken as they advanced.  Individual Dinarchy blood-mingled in obscene dancing obliteration.  One after another of the Dinarchy were pulverized into gory meat, shredded uniforms, and ripped-off limbs.

The Dinarchy were not fighting back.  The killing of the Dinarchy forces became mass-calculated mayhem.  There would be none left standing.

After the action, came long minutes of stillness.  Numbness and apathy came over the Krils in the line on the top of the hill.

Over their Stels an unknown Kril was asking.  "What the hell they doing?  A mob would be better organized.  Shit…they never fired except when we tried to capture them.  We lost a lot of good Krils trying that brilliant move."

"KrutChan got his body count."

Someone else gave his opinion.  "Who the fuck knows?  They not like us.  They're stupid Aliens."

Another StelaBalaam recording for posterity.  KrutChan thought.

"Maybe they're dumb, but ya gotta admit that took some balls."  CheChun said.  "I wouldn't have made that slow walk."

Three ArnaMal medical troops came to the center of the Kril's line, gazing out at the destruction.  Watching Krils wandering amongst the Dinarchy finishing off the live ones with gunfire, spears, knives, or whatever was available.

The ArnaMal medical teams wandered over to KrutChan. The leader was speaking into his Stel and to KrutChan. "We are of no use here. The Disposal Germs can clean up this mess after we leave."

One of the ArnaMal medics was staring at four Krils surrounded by Dinarchy bodies below the top of the hill near him. The four Kril veterans were crying. Their tears cascading down their cheeks, while many others, along the line were emotionally choke-sobbing.

The young ArnaMal asked. "Why are they crying for the Dinarchy?"

XibEk threw him a tired glance, puffed on his fiercely glowing cigarette, and then threw it in an arc landing among the closest of the dead Dinarchy bodies. "They not weepin' for them Dinarchy. They's crying for us."

The young ArnaMal looked bewildered. "I cannot believe that is the case. I don't understand?"

CheChun threw an arm tightly around the young ArnaMal's neck, squeezing him in a choke hold. "Don't understand...huh? Good for you, judge. You're a sensitive asshole. At least yawl can get sleep tonight without nightmares."

KrutChan snapped at CheChun. "Let him go Chun!" He yelled at PacalMo, KrutCheebel, and XibEk. "Get your goddamn herds back into their fighting holes in the line!"

Despondent and numb with fatigue, the Krils shuffled back to their holes.

KrutChan's anger snapped them all out of their mood. "Shows over; the war ain't!"

ZacNaab's was screaming in outrage at KrutChan. "I ordered you to take prisoners!" His

324

irate voice blasting out from their Stels became untranslatable. The EkSeets were tired of listening to him too.

All their Stels then crackled with AinAcbal's voice. "My legion is arriving. We are sending these scum to Valhalla without a funeral pyre. Do not allow the Zabin-Krils of the Four-Strike barbarian to TOTL us."

KrutChan groaned. "You heard the man.... Friendlies coming in through the wire." KrutChan whispered into his Stel. He admonished himself. Concertina wire was non-existent in the Kril Legions.

There were no cheers from KrutChan's force. They had massacred any surviving Dinarchy within sight; anymore killing would be redundant. After they collapsed to the hard scrabble-blue ground that was KanBalaam; they were waiting for AinAcbal's troops to arrive. All of the remaining Krils were stunned they had somehow survived. Now beginning to feel the pains from their wounds, they realized in horror what they had done.

They inwardly rejoiced watching AinAcbal, wielding that battle axe of his, leading his legion up to their position ripping a path through dismembered, screaming, and dying Dinarchy his force was killing. When he broke through, AinAcbal paused briefly when he came upon the Dinarchy forces KrutChan's Bot-unit had massacred. And then continued to the top of the hill.

KrutChan rose to meet AinAcbal and they saluted each other correctly as Krils, bumping their foreheads informally. "You are one fucking Viking killer..." KrutChan said. "...for a short puny dude."

Huge AinAcbal punched KrutChan's arm in response, rattling KrutChan's teeth. "It is pleasing to see your men are more violent than your pretty female rump."

His EkSeet had cleaned up what he could of the translation.

AinAcbal was puffing in exhaustion, his face blood-spattered and sweating. "KrutEk has his Legions charging and killing the fleeing Dinarchy on the Balaam plain. Capturing a few while pursuing the survivors, sending them to Valhalla. KrutEk does not condone mercy."

They hugged each other, glad to be vertical. The commingled Kril Legions on the hill were bumping foreheads. The Krils grew pensive as their broken-minds absorbed and crammed-down the sight of the grotesquely mutilated dead bodies of their fellow Krils. Most looked away after a while; their nightmares would conjure up the horrific scene as their timelines traveled to their end time.

Anticlimactically, their Stels crackled with ZacNaab's voice. "We will meet again on the Burseeosil EkTsab. You are a disobedient barbarian! You will face beheading or crucifixion for disobeying my commands!"

"Sounds like ZacNaab isn't full of mercy either." KrutChan said, knowing ZacNaab was addressing him.

"If only his courage was as brave as his mouth." AinAcbal whispered and grinned at KrutChan. "You must have Viking blood."

"My ancestry is Irish." KrutChan shook his head.

AinAcbal laughed. "My raiders plundered much gold from your churches." He was quietly assessing KrutChan. "ZacNaab will bring you TOTL for disobeying him."

"He's in a bind." KrutChan shrugged. "Damned if you do...damned if you don't."

KrutChan started laughing when AinAcbal roared at the EkSeets to make his Stel and the others be silent from ZacNaab. "Being one of the surviving damned is glorious, is it not?" AinAcbal said to KrutChan.

AinAcbal gave XibEk and CheChun bear hugs, roaring loudly in laughter at their embarrassed grins.

XibEk was complaining. "Let go, motherfucker!"

CheChun carried an equal body mass to AinAcbal and squeezed harder matching the Viking's bear hug.

PacalMo and KrutCheebel retreated to a safe distance away from the celebrating huge Viking.

Seeing AinAcbal lumbering back to his victorious Malkril, KrutChan was reviewing his battle decisions from his Stel, on his flexible 3-D scanner. He looked up when he saw a World War II Kril from CheChun's Bot-unit approaching. Thermometer thin, his head resembling the blade of a zipper, stopped before CheChun and XibEk.

KrutChan eavesdropped.

"You ordered me to report to you, CheChun." The skinny guy was loose and calm. "We did good, huh?"

"Why do I get all the numb-nut ten-percenters?" CheChun asked KrutChan.

KrutChan could see something was bothering CheChun. "Anybody that got Seeched is a ten-percenter." He offered.

CheChun's finger bounced on the thin guy's chest. "Ya see those ArnaMals picking up those

327

dead Krils." CheChun pointed and was wearing his 'old corps' irritated face. "Go help em and carry your rifle to protect them."

The skinny Kril's demeanor changed instantly; he was nervous and unsure of himself. "I don't understand." CheChun didn't have him report just to assign him to a dirty detail. Removing dead bodies wasn't his job. He was wary; something else was up.

"Oh you don't understand, huh? You heard the order to open fire and take no prisoners. When we had those Dinarchy bastards cornered; you didn't understand?" CheChun's helmet was cocked to one side; his hard stare boring holes into the skinny guy.

"I fired double the bullets as the rest of the guys, maybe more." The Kril bragged. "I did as I was ordered. What's your problem?"

CheChun reached out and slowly put a finger down the front of the guy's shirt, pulling the Kril with a jerk towards him. "Keep being a wise-ass. And you and me're dancing the 'fist city waltz'. Ya fired twice as many rounds as the other Krils in the line, did cha?"

KrutChan saw the flustered man didn't know how to answer without arousing more of CheChun's anger. KrutChan stayed out of it. "The man doesn't understand what you're asking, Chun." KrutChan was trying to help. "Get to the point."

Looking over at XibEk, CheChun asked him. "Did you tink dis Kril was 'confused' about his orders, Xib?" The menace in his voice was clear. He was building to something.

XibEk shrugged and nodded. "Seems ta me he understood. Must be a lifer, making his own orders." He glared at KrutChan."

The skinny Kril's choking question had a squeak in it. "What'd I do wrong? What're you accusing me of...?"

CheChun shove-slammed the guy to the ground. The skinny Kril jumped back up immediately. CheChun waited until the skinny Kril was standing in front of him again, at a respectable distance. "Killing is easy. We were shooting lined-up ducks on a pond. You was burning holes in the sky, asshole. I know the U. S. Army taught you better than that. You're lower than whale shit at the bottom of the ocean."

The man's eyes widened in recognition. "What do I do after I help the ArnaMals with the dead?" By his grey face it was obvious CheChun had him cold; all he wanted was to get the hell away.

"Those frosted Krils were killed by the Dinarchy. The enemy yawl didn't shoot. I don't give s shit what you do. Go with the Germs, stay on the Burseeosil, but 'never' come back into this Bot-unit. If ya wanna keep living."

The guy tried one more attempt to pacify CheChun. "I don't think KrutChan will allow you to do this to me."

CheChun exploded. "Move your ass picking up your friends you killed! I own you! Get outta my sight!"

The man saw KrutChan turn his back and walk away. The skinny scared man ran away to help the ArnaMals with their grisly duty.

XibEk murmured to CheChun. "A few guys were like him in the 'Nam. Never could trust them again. We shipped them out."

CheChun was looking to the sky watching Toobs come, land, and leave with full cargoes of dead Kril. "We didn't always rotate them out. A lotta dem

329

never survived their next patrol." CheChun did no further explaining.

"Same with us." XibEk agreed. "Ya piss on angry gods, you be fuked. We never relied on the Staff NCO's and Officers ta do da right thing. He's lucky."

Their unwritten rule of infantry: You had to trust the guy alongside you. No excuses, no buts, no ifs, or no after-action excuses or logical explanations. Just the way it was.

On the plains below, the ancient Krils were routing the Dinarchy, punching through the middle of the forward line and destroying everyone in their path. They were relentless and unforgiving, sparing no Dinarchy soldier. By the end of the Kin, victory was achieved, the planet KanBalaam theirs. The Kril survivors had prevailed.

On KanBalaam, in the last days of mopping up, a weird incident occurred, recorded by the Stels into the StelaBalaam. A furry-stiff-snake-like creature from KanBalaam that glowed in the dark, attached itself around KrutChan's left arm for an unknown reason of its own.

KrutChan warned the medical ArnaMals coming to assist him the creature was a mean little bastard and would not get off. "Maybe he's joining my Soothing Nest." KrutChan joked.

It reacted aggressively when the medical ArnaMals attempted to detach it. The medical ArnaMals determined it was not microbe laden and causing no harm to KrutChan; advising him to cope. They assumed the creature would leave when it wanted.

KrutChan shrugged. The furry creature was like having a flat glowing snake-cat-tattoo
330

combination that moved. The little bastard is cool. He thought. What the hell should I call it, EkSeet?

The EkSeets threw in their usual analysis. The Styx; in Ancient Greek, or Greek Mythology was called the river of death, forming the boundary in the Underworld, named after Hades, the reigning god. The rivers Phlegethon, Acheron, Cocytus, along with Styx, converged on a great marsh at the center of the Hades underworld.

"What da fuck do rivers haveta do with that fuckin' ting?" CheChun asked.

"Ain't gonna try pronouncing the big Greek words." KrutChan snorted in acceptance. "I'll name it Styx; River of Death...fits its mood. Styx is one ugly-wise looking dude. I'm starting to like ugly."

"He does look like your son." XibEk said.

"On the bright side, Styx'll keep you guys away from my liquor." He smiled inwardly, thinking to himself. Reela's about to inherit two-for-one Snakes to put up with; she'll hate it. Serves her right for dropping those kids into my lap.

Two Kins later, they secured the next objective without resistance. KrutChan was approached by his Zar, CoCum. The Zar was billowing up his fur and shaking like he was in a cat-fur-ball-coughing fit.

He is laughing for joy, KrutChan. His EkSeet said.

CoCum's large lethal claws held huge chunks of shining nuggets. EkSeet told KrutChan the Zar was holding gold. The Zar was pointing to a huge excavated mine entrance on the hill KrutChan's Bot-units had secured. The ancient Kril ArnaMals were scrambling, pushing, shoving, and climbing over themselves to get into the mine.

331

KrutChan was mystified; oblivious to the excitement. It was a quiet time and he could care less about the geology of KanBalaam. Hopefully, I can relax and be anonymous again back on Arna.

He heard Albert whispering in his brain. Fat chance of that. Hopefully you wise up, you mean.

KrutChan moved on to check the lines, letting his EkSeet report the gold find. KrutChan wanted to find some place to fall asleep. I gotta get away from the Kril herd.

Eventually, on a blue-green Kin, the Stels reverberated with the order to stand down and evacuate KanBalaam.

While in their AkSilk bindings, KrutChan's inherited units boarded their Toobs leaving KanBalaam. KrutCheebel remembered a passage from his past he had read on the internet about ground combat. He could not remember word for word. He had lived it. 'Plagued with creeping crud and diseases; living like an earthworm or wallowing like a dung beetle in the rubbish of a battlefield'.... KrutCheebel recalled reading. '...not even mentioning the hate that develops, not only for the enemy. But with the grunt's all-consuming contempt in their growing familiarity with each other'.

KrutCheebel had not understood what hate meant at the moment he had read the page on the internet. The passage was clear as a window in a house to him now.

Recorded in the StelaBalaam, the battle of KanBalaam officially lasted for fifty-five Kin days. Inside the biological evacuation Toob the Krils were somber and silent trying to sort out their survival. Unit by unit the survivors were pulled off the planet, back into the Toobs, flying off KanBalaam to reform into the Burseeosil.

These surviving Krils would never miss Balaam. Tucking deep into their minds; they would never leave it. These Krils had three more Uayebs to go on, looking forward to surviving. That was the rub; could they? Nothing had changed for them. All of them questioned if they would possibly get lucky again.

As suddenly as it had started, it was over. Nobody was celebrating.

KrutCheebel was a combat veteran now; even KrutChan was treating him better. Could be KrutChan was mellowing; but more likely KrutCheebel was cynically maturing. KrutCheebel carried a Balaam death-sentence-virus in his body. I want to enjoy my time I got left with ZocKuk. I love her.

CheChun was sleeping, full of wet dreams of mating with his female ArnaVal until she squealed in pleasure.

XibEk was moody, an introspective tallying of the cost of losing his friends. Krils and aliens alike. Burying them in his mind until tossing sleep came to him to bring them back. Hatred growing for KrutChan.

KrutChan's nightmares had him dripping sweat. When awake, KrutChan was satisfied; but he knew survival came at a cost. By Stel and KrutSeet, KrutChan was informed he was facing a court-martial; initiated by ZacNaab. KrutSeet would summon him when to appear.

This shrinking timeline of mine is getting old fast. If he survived Xmucane's death sentence, if he could survive ZacNaab's punitive court-martial verdict; he wasn't home free. He was still an assassin-marked-Kril-target.

# Chapter 18

After leaving KanBalaam, when the Burseeosil phased, the Krils in the Malkrils experienced the popping in and out Osil effect. Many of the lost Krils from the Osil effect were never seen again. Other Krils twinkled back into existence from previous Uayebs long gone by, confused and out of time-synch. Statistically probability was factored in their favor.

Osil harmonic phasing back to Arna was an excruciating ordeal. The majority of the Krils collapsing where they were, in their racks or on the deck squirming, moaning, and cursing their treatment.

Long ago, in Arna's past, a KrilChan decreed they had to endure Maluayeb Arena pain in order to cull out the Krils who could not endure the Burseeosil Osil return abuse.

Early morning on Arna.

Egressing from the Burseeosil EkTsab, the Krils assembled on the causeways of Arna to be dismissed.

They recognized the Tribunal pyramid overlooking two gigantic flat-top pyramids, a bright-yellow Ingress Seech pyramid and a vibrant-purple Egress Seech pyramid on the Ovut Plain of the Ovals.

Another smaller-green-rounded pyramid, offset from them was the Ovum, functioning as a supersensitive telescope periodically observing the entire spectrum of energy, including the torn from subatomic inter-dimensional cosmic radiation.

In between the pyramids, massively laid out like a Pre-Columbian ball court was the Ovut. The receiving area, a portion delegated where the Seeched humans were to become Krils in the adjoining Maluayeb Arena.

From the huge pyramids outward, the city sprawled to the horizon with its blister-bubble habitats, similar to pox, between the Oval edifices. Stark yellow-gold against a green canopy of jungle in the distance. The returning Krils remembered the Capital city of Tikumyax on Arna by its sheer massiveness and grandeur. They were home.

Above them, their barracks were redeploying on biological AkSilk into their Oval's pyramids. There was no Kril support from the Ovals, except for mass-melding them out of their formations to their respective pyramids.

Germs left the Krils where they melded into the pyramids, close to their Soothing rooms. Surviving Krils would crawl, walk, or stagger by themselves to their personal warrens.

Their treatment was purposeful, on orders from the Kril High command. Making the Krils, coming out of the last Uayeb, reorientate their minds. The Kril High Command did not want their Legion Krils to get soft receiving pity. The Kril Legion Commanders knew exactly how to discipline-tighten their Legions.

The High Command's edict was drilled into the Krils. "No one cares how or what you survived. Disregard self-pity. Reinvigorate your discipline and begin acting as Krils."

KrutChan was sweat-dragging a small-slime-trail of clothing and crap from EkTsab's Burseeosil purgatory, down the hall to his destination. A Germ behind him was cleaning his debris. He was passed by strolling Ovals, oblivious to him.

The male Germs, non-breeders of the Ovals, were scurrying about cleaning after the Krils. Germs were considered slave-class by the Ovals. Those not eunuchs were allowed to breed with indigenous Arna woman. Both bred from each other's gene pool to keep their species in existence.

The Krils disliked Germs. Under Oval and Kril rules Germs could not fight as Krils. Germs reciprocated by loving each other; they hated everyone else. A class among the classless.

KrutChan was lightheaded. Deep in his mind, conflicting voices were whispering for his attention; Albert, future-vision Greer, the unknown Oval, and his EkSeet. Too exhausted to pay attention, he concentrated on moving forward.

He was in a fog when he arrived at his destination. Germs formed a wall in front of him guiding him to Reela's Soothing Room. KrutChan melded through the wall, falling to his knees; palms on the floor. Wearing shredded clothes irritating his skin, wafting ozone-cleansing vapors.

He was deciding what was the worst. Being Seeched-tunneled from his former earth universe and then being beaten in the Maluayeb Arena, or returning to Arna in a Burseeosil nutcracker with its Osil-drive.

"Can you hear me?" Reela was fifteen-foot-tall, looking down at him when she spoke.

He nodded in reply.

"I need you to be aware of my words. Are you conscious? Your mind is working? How are you feeling?"

KrutChan answered to screw up her female EkSeet's translation. "I'm peachy. Simply hunky-dory." He croaked, and then wise-cracked. "How's

337

your life been on Tikumyax while I been gone? How you doin', honey…?"

Hortim snapped at him. "I see you haven't lost your petty arrogance, KrutChan."

Reela's back stiffened as his words were translated in her mind. "You are babbling nonsense."

Akna cautiously moved towards KrutChan in concern. "Please, may I help him?" Akna asked. "He is hurt. Allow me to assist him."

Hortim held Akna tightly to avoid her touching him. "Get real and keep your distance. Obey the Oval-Kril decrees."

Taking control of her Nest, Reela commanded both of them. "Not do touch him. He is crawling with infestations; biological contamination. The invisible field around him will disappear once he is disinfected. He must attend the Oval Memorial Ceremony to the Krils who met TOTL."

Her father's Kush warrior blood had Akna snapping at Hortim. "My helping him not will abuse him more than he has already endured!"

Arna'd hug a leper if she thought one needed support. He said to himself.

Sitting sprawled against the yellow walls of the Soothing Room while they assessed him, KrutChan experienced a queer sensation looking at Reela, Akna, and Hortim. The three of them had changed in appearance. Their usual Aura-radiance from prior to the Uayeb was softened, as if a painting of them had aged in a good way. Their stunning looks had faded into a more realistic picture of them. More mature, less hyper-sensual. The effect of visual sexual-signals toned down; female directness and intelligence came to the forefront.

His noticed his own hand held up in front of them looked different. His age spots on the back of

338

his hand were not gone, but much tinier. Was this from the Osil drive? Or the faster-than-light effect of ageing backwards that the EkSeets had spoken of during his time-pause? He decided his body had been affected by both.

This Oval universe keeps getting weirder. KrutChan was grimacing and blinking. "Why am I spaced out on dope again?" He asked.

"When you start time-pause, blue AnticArna drug is re-introduced into your system heightening physiological reactions." Hortim explained.

"Enough! He not does require information." Reela said to KrutChan. "You will learn to cope as your timeline progresses. Your response is normal."

KrutChan hung his head and scoffed. "Haven't felt right since I woke after Seeching."

Her eyes widening with shock, Reela exclaimed. "KrutChan your arm is writhing!" She held her cheeks. "By the Cosmic Egg, is that Styx?"

Smiling, KrutChan thought. That'll shake your complacency, sweetie. Now you got two ugly monsters in your Nest.

Styx's huge green eyes were bright.

Reela was neither afraid or disgusted. "I was advised by BalamEk that you acquired him." She cooed like a woman in the presence of a newborn.

"Styx hates everybody." KrutChan warned. "Be careful!"

"He is adorable." Reela said to Hortim and Akna.

If he wasn't so tired KrutChan would have laughed. I'll never understand some females. They could cuddle a muscular wild baby gorilla calling it cute, yet run like hell from a spider.

Akna was observing Styx with wonder. "The KanBalaam creature has so many bright colors; his furry skin is inviting, urging touching."

He was used to that reaction towards Styx from a few females when he came aboard the Burseeosil. "I'm warning you. He's got a menagerie of bugs, same as me."

"I not would be harmed by him." Akna said.

KrutChan sighed. "He's more Komodo Dragon than a huggable kitten. Those colors are his emotions. Stay clear. Styx isn't mine or anybody's pet."

Arna swiftly approached and reached out for Styx. "Is he your friend?"

Styx responded by turning pitch-black with red flashing eyes as he puffed up; his open mouth full of slime-dripping fangs. He roared with disagreement while Akna jumped back.

"Little pissant has a mean streak in him. He took a shine to protecting me. I guess I smell good to him. Maybe I remind him of his mother or father. Who knows?" KrutChan carefully reached behind Styx's eyes and scratched. "He's the only thing I have in this goddam universe that doesn't want something from me. We ain't buddies. We co-exist. I think we're alike in many ways. We're ugly brothers."

"I do see the resemblance." Hortim gravely said. "By chance, did you happen to donate your essence to an ornery Anaconda on KanBalaam?"

"Stop it!" Akna said. "KrutChan is exhausted. He is in pain and must recuperate."

"He not is as hurt as you think he is." Reela said.

"Pain is relative." He waved them off with both hands. "Stay the fuck away from me too. We ain't nobodies' pet."

Hortim laughed. "You see; KrutChan's his own sweet sensitive self. You better watch your tongue, KrutChan or..."

KrutChan challenged Hortim. "Or what? You females will ship me off on another Uayeb Invasion?" He shook his head, clearing the cobwebs from it. "Oooops...I forgot. The next Uayeb starts at twelve-dark-thirty. Too bad...that screws up Hortim's thoughts of discipline."

"Hortim obeys me! I command this Nest." Reela took over. "You will accompany us to the Oval Memorial to honor the Krils who met TOTL. You must do this. Not do you care about the Krils who met TOTL on KanBalaam?"

He made two fists on his face. "Don't lay any guilt-shit on me. You gonna moan for all those dead Krils who can't earn no more shekels for their Ovals?"

"Be still!" Reela commanded.

KrutChan pushed himself off the yellow-lighted wall and stumbled towards his barracks room, saying as he went, "I ain't goin' nowhere til I clean up my scuzzy ass."

"You will obey me!" Reela warned him. Her Oval wand on her wrist initializing.

KrutChan was wearily staring at his Oval. "I'm being court-martialed. Hortim can explain." He snarled. "I don't give a shit what you do."

"Be careful KrutChan, you're bordering on mutiny." Hortim said.

"Being knocked out by Reela'll have benefits; at least I'll sleep for a while." He glared in defiance.

"Cease confronting my words!" Reela hissed.

341

He calmed. "You're wasting our timelines. Who knows? I may be court-martialed and dead before I'm able to attend your horseshit Memorial."

Deep in thought, Reela was modifying the setting of her Oval Device glove on her hand. Inwardly frustrated again by his overdose of AnticArna. She wondered if an Oval or Kril was increasing his dosages.

"For a guy that's twenty years younger; you haven't gotten any smarter." Hortim chimed in. "She's serious buddy, don't push her."

KrutChan icily answered. "Reliving any strafing or bombing runs? How're you sleeping Hortim? Going to share any war stories with your NEST?"

Akna eyes pleaded with him to listen to them both.

Hortim noticed KrutChan's manner was somehow different, cagier or maturing. His sixty-year-old body was in terrible shape, but his willpower was stronger.

Pointing at his Oval, KrutChan said. "Reela, you can't command me in Kril matters. Read your own Oval manual." He understood what was expected of him. "Your Queen and her Confederation Empire defers to her KrilChan in military matters." I'm getting sick of EkSeet and his dull mis-translations of my words.

"My EkSeet is tactfully reminding me this Ob of your correctness." Reela said, not sounding happy.

KrutChan continued. "Before we left Balaam, all of the EkSeets explained our pre-Second Uayeb duties. The Kril High Command decided we'll be appropriately dressed for your Oval Memorial function. Our KrilChan's no stranger to pomp and circumstance. We 'obey' him. I'm under your command 'after' the Uayebs are over, not before."

Reela was seething, wondering if Xmucane suffered such disobedience from KrutEk.

As he approached, his barrack door went translucent. "If you guess wrong...at least I'm out of the Uayebs and you lose fortune and fame. You wouldn't like that. When I'm dressed as a respectable Kril to do honors to our fallen; I'll accompany you all."

Goddamit, EkSeet, I don't talk like that...too southern. "I meant...all of you." He said and walked into his barrack.

Reela shook her head in frustrated bewilderment. "My KrutChan has returned an angry Kril. Let him be alone while he rests, Akna. I will deal with him in his NOW."

The blinking stars behind KrutChan's eyelids and his hazy vision cleared as soon as his barracks door tunneled behind him, though he was stumbling again. I wasn't so clumsy before I got here! These AnticArna drugs are a pain in the ass!

Akna following behind him bent in instinct to assist.

"NOT NOW!" KrutChan bellowed, viciously shoving her away. "GET AWAY FROM ME!" He had fallen onto one knee.

When Akna slowly left the room; he struggled to pull himself up, leaning against the green-glowing bulkhead, taking a moment to catch his breath. The voices quieter but still swirling in his mind. He was blinking, taking baby steps as he shuffled around, attempting to regain his composure.

KrutChan had finite number of hours before he would be on another Uayeb again. He stepped

343

fully clothed into the Oval shower, cleaning his weapons. He stripped and threw his discarded clothes into a slot.

Maybe the Pyramid Oval creature-walls eat discarded clothes for dinner? Who knows?

For long minutes, he was hanging his head. The second Osil-dry-shower he had entered burnt a layer of skin off. Refreshing his body and spirit.

Keening, with a high pitched sound, vibrating Styx was self-satisfied the dry-shower was eliminating stink and pathogens off his body too.

KrutChan slowly refreshed himself, attempting to get his head together for the next Invasion Bump. There would be no days off, no rest and recuperation, for any of the Krils.

He analyzed his reflection in the 3-D projection above the sink. KrutChan was in his sixties, had fist size whitewalls highlighted above his ears. His scar-rippled ugly face was swollen and wrinkled, with a new Uayeb mark denoting KanBalaam's invasion.

I sure ain't looking any prettier. My monster face looks crueler. His tired bloodshot eyes, tinged with red-blue highlights, stared back him.

KrutChan was daydreaming about the 'woman' he would have been married to, had he not died in Vietnam. She floated into his mind now and then. The longer I'm in this shit-hole universe. His feeling of the loss of her becoming an obsession.

He thought. Remembering is 'not' the correct word…that word implies a past. He had 'no' past with her. I 'daydream' of her is more accurate.

He referred to his future wife as 'Greer' because 'she' reminded him, in appearance, actions, and presence of the movie star Garson; representing his ideal woman in his earth youth.

344

KrutChan got himself presentable. KanBalaam receded into his past. He was putting on new utilities, before he heard and noticed her. He motioned Akna to enter his barrack; moving hesitantly in her near-blindness.

He apologized, sort of. "I shouldn't have yelled at you. You didn't deserve that."

Akna stared off-center at KrutChan. "Are you in pain? Are you feeling well?" She hollowly sounded like his 'Greer-Vision'. "Must you drink so much?"

"Yeah, you're right." He glanced at the glass in his hand. "I gotta cut back..." He shrugged. "I'm not in the mood to talk...about anything. Okay?"

Her knitted brow showed her EkSeet translation was incomplete. "You appear much younger in appearance. I hate the Uayebs because what is only a Kin day here is many Tun years for you. You look exhausted." She was fidgeting with her sheer gown and the lacing on it.

"I am." KrutChan noticed her huge belly pregnancy. It's strange my youthful time-dilation isn't affecting Akna. Hortim and Reela weren't with child, as the saying goes. Looks like my standing stud had no effect on them. Akna's child ain't your kid, remember dummy?

"Not can I help?"

"No." He felt better seeing Akna's glow. Her eyes were soft with hormonal tiredness. "You've grown more beautiful in my absence."

You're a sly old devil, EkSeet.

"I not am KeechPan...meaning beautiful in your language. I am a walking Maluayeb Arena-sized ArnaVal."

Come on EkSeet, help me out here.

345

"I'm seeing your KeechPan from your inner self, radiating from your beautiful outer appearance. I am full of admiration."

That's true. I think I've found your poetic calling, EkSeet.

Most EkSeets do not treat women who are Up-the-Duff like a Knob Head.

Not needing a further translation, KrutChan shook his head at EkSeet's wise-crack.

Akna's smile was luminous. "Your thoughtfulness is pleasing to me. You are KeechPan to me." Then she quickly asked. "Do you want to talk about KanBalaam?"

His vision wisp of his Future-Greer's pregnancies flooded into his mind. He was talking to her as well as Akna. "I already said I don't...." KrutChan gruffly said. "There's nothing to talk about...it's over." He crossed the room, consciously wiping out his 'Greer-Vision', and washed his hands in water, like Pilate, wiping them dry under his armpits. He repeated. "It's over."

Akna's expression indicated she was doubtful of his assessment of himself. "Perhaps not?"

Falling for her would be easier if her other dead Kril wasn't between us. Maybe he's her past 'vision'?

"In my earth timeline we would get married. This Oval-mating puts an edge on my upbringing." His drugged AnticArna red-blue-tinged pupils stared into her moist eyes. Akna was crying. He felt like a monster.

She swallowed emotional gulps and then pulled herself together. "I thank you for unshaming me. Akna is not worthy; she has Beekav love for you."

She's mixing me up with her dead Kril. KrutChan wiped her wet cheek. "I like who you are. I'm just ugly old KrutChan. Even Ogre's have feelings deep down." He quickly held up his finger. "But don't tell anybody else that Ogre crack, okay? Don't wanna screw up my legend."

His EkSeet was hysterically laughing. Bloke, you have a bit of the Irish blarney in you, for a colonial rebel.

Stuff it, you bloody Tory!

I have achieved the impossible! I have taught you proper English.

KrutChan sighed at EkSeet's words. "I'm thinking I need a change."

"What do you wish KrutChan? Another Oval?" Her unhappiness at his wanting to leave Reela's Nest was clear.

"Hell no, I don't want another goddamn Alien." Why can't I get through to her?

Her eyes were downcast. "You wish another ArnaVal?" She was heartbroken.

KrutChan softly and tenderly touched her cheek, wondering if his EkSeet was controlling him. "Never do I need another ArnaVal." 'Never do I need'? Quit making me sound like I'm Oval, EkSeet. I'm better at being me. "I've been too busy surviving."

"Your NOW I not am part of will never be your fate." Akna was somewhere else, seeing someone else. "I not could exist without you."

Walking past Akna as she composed herself, he left his barrack. Entering the Soothing Room, he sneered at Hortim's attitude and followed Reela; her back straight in patient resolve.

Reela, Akna, Hortim, and he melded. To another goddam revival. He thought. Where's my

347

Court-martial defense officer? Typical bureaucratic Kril bullshit.

The KrilChan spent six Bots two hours, reviewing the KanBalaam Invasion with his commanders; Jaguar present. He debriefed his commanders on the mistakes and victory. For an hour KrutEk made sure no one was spared criticism. He berated BalamEk's Burseeosil and Toob command misinterpretation of Dinarchy entrenchments and poor intelligence.

Jaguar came out of the shadows. "My Kril agents reported constantly to your Commanders on the ground, including yourself. Our intelligence was ignored."

"My Queen ordered me to have you present as an observer!" KrutEk was livid; nobody questioned his motives in his own meetings. "Do not speak! Observe!"

Jaguar smiled, mockingly putting her finger on her lips and withdrew to the shadows. She had made her case in front of his Commanders and him.

Previously AinAcbal had critiqued his KrilChan of the battle for the hill with the gold mine. "What of your Zabin-Krils, ZacNaab?" KrutEk innocently asked.

ZacNaab jumped to attention. "Acting like freebooters. Roman Mob behaving, mindlessly following a barbarian." ZacNaab said. "I ordered him to my headquarters in reserve to be part of protecting the KrutEk's flank."

KrutEk's question appeared to be a shrewd test and moved on to another subject. "ZacNaab what have you to say about that Four-Strike barbarian? I see he did not meet TOTL; still a thorn in my side. Did he listen to you?"

ZacNaab saw his chance to exonerate himself of any consequences. "He was the most uncontrollable. After his previous Commander went disabled and evacuated to the Burseeosil; he led the rest of the Zabin-Kril units. His Bot-units followed him like sheep. Because of his constant insubordination, I replaced him with another more loyal Kril Commander. However, that Commander met TOTL also. The barbarian assumed command again. I thought he met TOTL numerous times, but he survived...the ignorant worm."

"And you let him go unchecked?" KrutEk was shrewdly hiding his real point.

ZacNaab's hands were shaking. "I ordered him to retreat...I mean...withdraw to the valley as a reserve force. He disobeyed me!"

"I have seen your court-martial charges and I have received reports from the StelaBalaam and Jaguar." KrutEk said.

AinAcbal related. "ZacNaab 'also' ordered me to withdraw. 'Me' to retreat along with the barbarian. He dares not add my name to the court martial for my actions. He knows I would kill him. As an honorable Kril and a Viking, running away is not a tasty meal to digest."

Red-faced with anger ZacNaab shouted. "You admit you both ignored my order!"

KrutEk yelled. "I am well aware of what occurred! I do not recall ordering you to send me any reserves."

AinAcbal had an opening. "I never received a command from KrutEk to withdraw. I retain warriors who do not submit prematurely. I do not TOTL them by court-martial drivel!"

ZacNaab explained his position. "One of my most experienced Krils tried to TOTL the barbarian

349

for disobeying me. KrutChan was modifying my commands. His Mongol Kril broke our code and killed my Kril!"

"NEVER USE THAT BARBARIAN'S NAME IN MY PRESCENCE!" KrutEk screamed. "I DO NOT WISH TO HEAR ABOUT DISHONOR!"

His eyes met Jaguar's standing off in the shadows observing them.

AinAcbal saw his leader's glance at her, saying quietly to defuse the atmosphere. "The KrilChan told you before ZacNaab, the OvalChanHalach did not want the barbarian assassinated. By Kril Code of Honor and on the Command of our KrilChan you were to obey her Decree. Perhaps you need to refresh 'your own' obedience to orders?"

KrutEk took the initiative from his favorite Viking. "Exactly!"

KrutSeet added. "Perhaps Jaguar could arrange a time-pause for ZacNaab to teach him obedience, in the Oval Pavilion?"

KrutEk was scowling at Jaguar. Her smiling face indicating she would relish witnessing the Pavilion Ceremony.

Except for KrutEk and an embarrassed ZacNaab, the rest were laughing.

"I did what I thought was correct." ZacNaab whined.

"We have discussed this before. Thinking seems to be one of your failings." KrutEk said.

KrutSeet looked knowingly at KrutEk. "Your Half-Malkril Commander, ZacNaab, stayed where he was in a safer position."

"I DID WHAT I THOUGHT WAS CORRECT!" ZacNaab lowered his voice when KrutEk narrowed his eyes at him.

KrutEk said. "That barbarian will 'not' be court-martialed. Not face TOTL for rallying his Zabin-Krils." KrutEk was tapping his white baton on his thigh. "I retain much disfavor in my heart for this barbarian. My Kril Legion's bellies will not accept their KrilChan punishing a Kril for successfully helping achieve our victory."

He was looking, one by one, at all his Commanders. "The assassination plot brought dishonor to my Legions."

"I understood you wanted to be rid of him." Was all ZacNaab could manage to say.

"It is only necessary I understand…" Making up his mind on the spur of the moment, KrutEk said. "What should ZacNaab do on the future Uayebs."

ZacNaab was near tears. "What could I do with the disobedient barbarian?"

KrutEk turned his back on ZacNaab, staring cruelly at his Commanders. "ZacNaab should be concerned more with what the barbarian is going to do with him."

Jaguar was sitting quietly away from them, keeping silent, gaining intelligence. She immediately left after KrutEk said what his plans were for the barbarian. As she left, they were forming a plan for the next Uayeb.

Xmucane was meeting with Reela and Hortim in her quarters, while KrutChan stood guard. Outside, Jaguar melded into the corridor, motioning with her finger for KrutChan to approach her. "You look fitter and younger, after your first Uayeb. Your time-pause is going well?"

KrutChan shrugged. "I'm spaced out...feeling well is relative."

She looked concerned. "Do you always have to add another scar to your face? In your earth idiom, it is not dishonorable to duck. Younger is something you wear with distinction."

"Sixtyish wasn't exactly young-wise in my old universe."

Jaguar tingling-bell laughed when his words were translated.

"Pregnant Akna's carrying my child. Too bad I can't see the kid born."

"I not did know Akna is carrying your hatchling." Jaguar frowned. "She is typical of her human species. They earth-love whoever they Sooth; it is one of their failings."

"Too each his own..." KrutChan said.

My EkSeet has translated your words." Jaguar was correcting him. "Akna has a 'germ' in her, not a 'child'. She not is an 'earth woman'; she is an 'ArnaVal'."

"Blame my EkSeet. What the hell. Means the same." He said.

Jaguar then said something strange-sounding to him. "Akna will pay the price. This universe has no need of another KrutChan."

KrutChan remembered CheChun's guy on KanBalaam the Ovals wasted for having a doppelganger in the same Invasion. He dismissed her thought as female Oval pettiness against a subordinate. KrutChan promised himself. "Just so you know, 'this' Kril intends paying the price 'for' her." He answered in the third person parroting an Oval tendency. <u>That's my job. Protect the innocents; Ovals aren't virgins.</u>

352

Moving KrutChan off into a secluded corner, she asked. "I gather intelligence about the Uayeb that not is from your Stel."

He was wary of her more than Reela, Hortim, and Akna. I wonder if she's using her Oval magic on me right now.

Halfway down the hall, he saw Albert leaning against the corridor wall. He was doing his usual thing, with Kril hand signals, silently advising KrutChan to be careful.

KrutChan remained cagy with Jaguar; the Arna barrier of Me-Oval-You-Kril-and-subservient-to-me-shit not at all between them. But for his own protection he decided he should hold something back, not revealing Albert.

"You live more than one day during your Uayebs; you need the rest..."

"What relief?" KrutChan saw Albert turn his back, laughing. "We're back here for only a few hours. Bringing us back between the Uayebs is nuts!"

"I not do understand your connotation. Your food word is not translatable."

She took his Stel from him. putting it with another and hers into a hall hidden drawer. After their words became secret Jaguar recruited him to 'spy' for her. She called it reporting.

KrutChan considered it, 'covering her ass'.

"Do you wish to survive and not TOTL?"

Silly question. He thought. Or was it? He surmised Jaguar had layers of deceit in her.

"I have assigned assistance in your mission for me..."

Albert was shaking his head, hand-signal warning KrutChan.

Jaguar beckoned a future earth Kril-Oval standing near an opposite wall from Albert. When

353

the female Kril approached them, Jaguar said. "This is Kelel. She has Bumped on eight prior Uayebs. She is your Control, one of your Bot-unit commanders in your future headquarters group."

He recognized her from his first time-pause in the training area. They remained adversaries. "We've met."

Kelel was smiling. "You can pick your jaw off the floor, KrutChan. I'm a grunt."

He was thinking about how he could ever fit Kelel into his table of organization. What about a Kril-Oval's male buddies wanting to protect females during combat?  Wanting to impress her, screw her or ...worse.

"Relax, KrutChan. I'm not one of your assassins." Kelel said.

KrutChan wondered if she was telling the truth. Her attitude in his first time-pause came back to him. "In other words everything is hunk-dory; all the Krils from KrutEk on down treat them like one of the men?" KrutChan asked.

"I'm not a 'them'." Kelel peevishly said. "And I don't like you either."

He shrugged. "Hope you ain't sensitive. Infantry vulture-feed on the weak."

"I castrate vultures." Kelel was speaking to KrutChan.

"Like Hortim, huh? Just the male ones?" KrutChan turned to Jaguar. "I don't think she's a good idea."

"Quit discussing me like I'm not here, goddammit! And I'm not Hortim." Kelel said. "Judge me by my performance; not by your gender-biased old Vietnam notions about women in combat. The Ovals don't trust Kril-Oval grunts any more than their male Krils."

354

KrutChan nodded, fully understanding what it meant to be singled out. He spoke to Jaguar. "I was told Ovals can't land on planets? You're saying that's bullshit?" He held up his hand to Kelel. "That's a rhetorical question; not about you."

"You're talking about Ovals; not Kril-Ovals." Knitting her brow, Kelel sneered. "In practice, the Kril-Ovals answer to KrutEk and his discipline. Kril-Ovals are stained, kept apart from real Ovals, except for service to them. Jaguar is my Oval. I'm in her Nest."

KrutChan was shutting his mind off about 'essence' gathering from Kril-Ovals and their breeding habits.

Kelel was brunette, big boned, a beautiful woman officer from his future. Her smile disarmed KrutChan for a moment. "If you could see your mutilated face. Kril-Ovals breed Krils after the five Uayebs are over and have our hatchlings during the Tun year between Uayebs. Our children are raised by the ArnaVals. We're bred by our Ovals best Zabin-Krils. She was laughing deep. "You look like you're going to implode."

KrutChan lost his cool, stammering. "Never met...never trained with.... I never saw an infantry woman officer. How the hell did you?"

"Become a Kril-Oval?" Kelel was smiling. "Same as you, Sergeant."

"Not do underestimate Kelel. She is exceptional. One of many in the Kril Legions. She has much ground-fighting experience, leading earth men, in your THEN. And Krils in Uayeb battle. Her Bot-unit volunteered to stay with her due to her prowess. Ovals not do favor her."

KrutChan was recovering from his shock. "This is a bummer." He stared at Kelel with hope,

355

looking at her former rank of Captain. "You going to be my leader?"

They ignored his question.

"Intelligence you acquire for me, verbally tell her. Her report secretly related to me on the Burseeosil." Jaguar said. "She will refrain from contacting you about intelligence matters for security reasons. Treat her as a fellow Kril commanding one of your Bot-units, with no reservations or favors."

Kelel said. "Skip over the macho-bullshit about why women can't perform in the infantry, okay? I've paid my dues, ten times over in my past. From what I hear you don't need more conflicts of interest in your timeline."

"Agreed. With reservations..." KrutChan said. "...we'll see how it goes. I got one Bot-unit now. Can't use you unless they promote me."

Jaguar and Kelel locked eyes; saying nothing.

Al was shrugging.

Kelel tightly shook his hand. "Fair enough; neither one of us knows which end of the rattle snake we're gripping." She eyed Styx. "Keep a short leash on your snarling pet wrapped around your arm and we'll get along."

It was KrutChan's turn to frown. "We'll both try to get along. Styx has his own mind. Nobody controls him."

Jaguar warned. "Not do let KrutEk hear you speak thusly. I have personally lost some of my Krils to..." She was listening to her EkSeet. "...suicide."

He blurted his wise-ass answer. "Ovals enjoy taming Krils."

Kelel said. "In my timeline here, I've never seen one tame Kril."

Jaguar touched his shoulder. "Many of the Krils I have debriefed are from your ancient past.

356

They say their descendent Krils are gods. Your KrilChan is considered by them to be Zeus or Odin or Jupiter."

"I'll bet he loves that."

"You misjudge KrutEk." She said. "These same ancient Krils relate events about you. They say, on KanBalaam, the Four-Striker KrutChan, dropped out of the sky on a bolt of lightning. Their story; KrutChan single-handedly continued into the battle, obliterating your enemies and you alone saved the ArnaMal Army."

Kelel and KrutChan snorted at that idea. Jaguar had to be teasing him.

"That's baloney and you know it." KrutChan said. "I fuckin' hung outside the Toob with AkSilk saving my ass! I wasn't fighting by myself. The Kril Bot-units I was with done much more than I did."

Knowing he was telling the truth, Kelel discretely stared off into space.

Jaguar's brow wrinkled while her EkSeet translated his answer. "In your changing timeline I would say you have acquired a following of fans…or is it disciples? Are those the right words?"

"A better word is morons." KrutChan shook his head.

She laughed at his translation. "I should not be the one to tell you but you will find out very soon…"

"Jaguar wants you to act surprised when Reela tells you." Kelel said.

Jaguar got serious. "Let it be 'our' secret…for now."

"What secret? What're you guys talking about?" KrutChan did not like where the conversation was leading.

"You have been promoted by KrutEk to be a Sub-Nacom of a Malkril. Another reason I have placed Kelel under your command."

His knees unconsciously started shaking. He fought to control the effect. "How many Krils will I lead now...four hundred? Too goddamn many."

"She benefits, so Reela will be pleased, as will Hortim, Akna, and your Bot-units. It is a great honor."

"Some honor..." KrutChan said. "...me sending more meat to be slaughtered."

"Not do be fearful." Her hand waved in dismissal and she turned her back. "Four hundred is far too few in number. One third of a Malkril is forty-eight thousand."

"Hell no!" KrutChan felt his body heating and freezing, at the same time. He exploded. "Goddam Ovals and KrutEk's ass! I'm a lousy squad leader, not a Multiple-Division Commander!" His fear was overwhelming.

"You will, because you've no choice in the matter." Kelel said.

"I ain't fit to command that many troops." He was pacing up and down, his arms flailing. "Jesus H Christ!"

He looked down the corridor for help from Albert. Al was gone.

"Calm yourself. You cannot overrule KrutEk."

"Goddam everybody, including the Kril Commanders! I ain't gonna do it!" He was scared. "I'll quit. I swear I'll desert!"

Seeing his flaring red eyes from his overdose, Jaguar relaxed, stroking his shoulder. "Not do blame we Ovals. The Ovals not did decree your advancement. KrutEk controls the Krils. It is his

358

authority. You not will dishonor the Kril Legions by running away. Not is your way."

KrutChan pointed at Kelel. "She's more qualified than me, goddammit! Hortim's enough of a pain as an officer. Let Kelel take command!"

"I'm not like Hortim, in any form." Kelel spread her hands. "I'm not the one who took Four Strikes in the Arena. Don't do me any favors."

KrutChan turned back to Jaguar. "I can't stick pins on a map wait for reports and communications to see what happens." He was pleading with Jaguar.

Jaguar placed her hand on his back. "KrutEk intends to put you in more precarious situations. His order states your former Bot-units from KanBalaam will go on the next Uayeb with you."

He groaned in disagreement and frustration. Oh, I understand too well. I'll be more of a high-value-target for an assassin. "KrutEk has a bigger hard-on for me than I realized." He moaned.

"I'm not patronizing." Kelel said. "I've never commanded Divisions, but I know the 'G' command structure. I'll be available, should you need me."

Jaguar said to Kelel. "Your words are foreign to my EkSeet. Clear your thoughts and be more specific."

"Clear? Clear you want? It's very clear to me…" KrutChan's voice was cold and resigned.

"My EkSeet not could translate Kelel's last sentence." Jaguar looked at Kelel for help.

"Don't get me involved." Kelel said. "I'll be with KrutChan."

Jaguar thought for a moment, picking her way through the translation. "My EkSeet states tactics are tactics, perhaps the semantics are different, but my EkSeet says a flanking movement is similar to envelopment, or end-run, or whatever the term is called by you."

KrutChan tried logic on her. "What I'm arguing Jaguar, leading a division is more complex. Now KrutEk wants me to lead four divisions! My Divisions in the Marines had competent officers, G-1's, G-2's, G-3's, and G-4's, and their subunits that handled the beans, bullets, bandages, intelligence, and other logistics."

Kelel nodded. "Names are different, but you'll have a similar headquarters group."

"There's another goddamn thing, Kelel! I never did drugs of any kind back in my past life. I even avoided aspirin. Ovals have doped the piss outta me. I'm stoned right now! Get it? That ain't right."

Jaguar smiled and shook her head. "Again, that specific translation is lost on me, but my comprehension is better than you think, KrutChan."

KrutChan thought. How the hell can I make her understand I'm in way over my head?

"You will have assistance." Jaguar's demeanor was serious. She noticed his unique red AnticArna reaction. He was correct; his overdose was kept at a high level per Xmucane's plot.

Kelel reminded him. "You'll take command on the next Uayeb." She wagged her finger at him. "Don't ask me how. I've never commanded that many troops either. But, I won't leave you stranded."

"Thanks for nuthin, Captain." He was realizing what Jaguar was 'not'_saying. That fucking KrutEk was putting him in position to meet TOTL. Welcome back survivor. He felt like an M-60 tank had rolled over him and stopped to grind him into the dirt with its tracks.

"You'll be golden, buddy." Kelel said. "I'll help if you get confused with the nomenclature."

He spoke to Kelel. "Yeah...I know...give my soul to god, 'cause my ass is in the KrilChan's hands. I know the drill. But this coming Uayeb is not a fuckin training exercise!"

"Not is the way you were trained, to assume command when your leaders suffered TOTL?" Jaguar asked.

<u>She just doesn't get it!</u> KrutChan rolled his eyes to the heavens. "Listen to me! Fucking Buck-Sergeants don't inherit command of an Army, or a couple Divisions by attrition!" He pointed at Kelel. "She knows!" He was pleading with Kelel to help him explain. His anger was surfacing and he fought to control it. "It's not fair to the Krils crazy enough to follow me." He tried one last plea. "I'm not qualified! Kelel, you know I'm right!"

Opening the palms of her hands, mimicking a football wide receiver who just dropped a touchdown pass. "You're preaching to the choir. I'm glad it's you and not me. I'm truly on your side, Sergeant."

"Your Bot-units from KanBalaam are very loyal to you, as are all your subordinates." Jaguar said.

"Horseshit! They hate my guts!"

"Hortim will now command a sub-BakToob. You will have familiar faces with you when you next Bump. That should be some comfort to you." Jaguar touched his cheek with a soft caress.

KrutChan was shutting his eyes for a moment. He sounded in pain. "I know I'm going to regret this."

Kelel said, in a serious voice. "That's the worst burden of command; having to remember the ones you send off into battle."

"Don't give me that shit! We both got too goddamn many dead guys to forget already."

Grinning ghoulishly at Kelel, KrutChan said. "What the hell, I ain't coming out of this alive anyway." His leer was deadly. "Fuckin brain-damaged Krils led by the ultimate dumbass."

"Must you curse?"

He ruefully winked at Jaguar, smiling at Kelel.

Kelel had humor of her own. "Don't make me tell Hortim about your potty-mouth." She said, with a smile.

Reela and Hortim came out of Xmucane's quarters, saw Jaguar with KrutChan and a Kril-Oval, and walked towards them.

Whispering, KrutChan dropped his own bomb on Jaguar. "Any Four Strike Kril ever commit suicide in the history of the Ovals or Krils?" He asked.

His EkSeet cut in quickly. Do not even think about it, KrutChan. I can and will stop you.

"Sorry Sergeant, leaders don't have that option." Kelel said. "Ours not to reason why…you know?"

Jaguar was not the least fazed by his question. "Be advised, your dishonorable act would make KrutEk a very…and I stress…very happy KrilChan. Solving all his problems with your self-inspired TOTL. Do you wish to help 'him'?"

Joining with his Nest, KrutChan had heard Jaguar's last words loud and clear.

Chapter 19


After the Memorial Ceremony Reela invited her former Oval-trainees, Krils, ArnaMals and ArnaVals in her Nest, to her Soothing room. Which had hugely expanded to hold them. As Consort to the Queen, demanded by protocol she be the host. KrutChan had unearthed the gold mine on KanBalaam; her prior Trainee-Ovals becoming wealthier and sharing the cost of the dinner. The banquet was held in a pink-red brightly lighted, flower-bedecked dining area.

Hatchling-progeny of the Ovals and ArnaVals were invited. The younger ones were boisterously scampering about, dancing, fighting, meeting old friends, and making new ones. The teenage Ovals coyly flirting with possible Kril mates. Their educational classes postponed by the Five Uayebs of mourning. This Kin unleashed them from parental restrictions and traditions; but not entirely. When the banquet began Oval discipline ruled; the young would be strictly sent to another room.

For the benefit of the Uayeb Krils, time-pausing prevailed.

Reela followed Oval tradition celebrating the end of the first Uayeb which had given her wealth and honor. Gathering the non-military in Oval Nests before the Krils and ArnaMals left for Uayeb Two.

There were no ostentatious celebrations outside in the city of Tikumyax, or on the planet Arna; celebrating was reserved for Oval Nests.

They partied with a meal suiting each of her Kril's and ArnaMal's ethnic heritages. The Krils resumed drinking upon arrival, previously soaked from potent Osil Wine supplied by their Ovals. The

363

Oval's plan was to keep them reasonably under control; protocol breaking was not pleasing.

The veteran Krils and ArnaMals nodded indifferently at each other, saluted in their Kril way, and exchanged mumbled greetings before sitting down. The KanBalaam ArnaMals and Zabin-Krils formed their own pockets and groups. Their familiarity with Kril Legion members grated on them. They wanted to be alone with their own women, not each other.

The ArnaVals and Ovals were carefree and lighthearted, striving to cheer up their KanBalaam Krils. The noise and cheerfulness in the banquet was brain-numbing. The females succeeded in relaxing a few ArnaMals and Krils; those with supporting roles in the invasion.

The combat Krils getting drunk. Unspeaking to each other; they were throwing sullen glances at enemies, old and new. They were speaking, more demanding than civil, in monotones requesting seasoning or liquid refreshment. Scars obtained from old enemies were scratched open; in battle their animosity was hidden, here not much.

The assembled combat Krils ignored any references to Uayeb Two. Consistently snubbed any misguided accolades for their actions on KanBalaam. No Krils of Reela's were in the same Bot-unit; they were kept apart in battle for Reela's benefit.

KrutChan's Bot-unit commanders were an exception. Over their many objections, they had been summoned by her to attend this banquet. KrutChan, like the others in his Bot-units, was sullen throughout the meal. The surly Zabin-Krils acted as if they wanted to be anywhere but here.

The festivities were as joyful as the Oval and ArnaVal women could make it.

The meal lasted without incident for forty minutes. Forever time to the bored Krils, who were drinking to excess.

After the meal, to offset the drunken solemn Krils, a tipsy Hortim arose and offered a toast. "Here's to the finest bunch of Krils I have ever known. I salute all of you." She waited for the EkSeets to translate and continued. "You have brought much honor to Reela, her Nest, and to your own Soothing Ovals."

The support Krils who were present, were cheering the tribute and saluting Reela, the other Ovals, and their Nests.

The Zabin-Krils, who had survived KanBalaam, raised their mugs and glasses, drinking one small sip to Hortim's toast.

Hortim getting tight, drank some of her liquid refreshment in displeasure, and then held her half empty glass in front of KrutChan. "Here's another toat...er...toast to 'Grumpy' KrutChan and his other friends like him."

Tapping Hortim's other hand, Reela silently admonished her not to incite these Kril guests.

One of the Kril veterans of KanBalaam muttered. "There's nothing for us to celebrate."

Reela loudly sighed and answered him. "Kril attitudes never change from one Uayeb to the next. I follow Oval tradition, even though the celebration is for Reela's loyal and ungrateful Krils." Her demeaning emphasis on 'ungrateful', referring to KrutChan, was not lost on anyone present. "My sister Ovals and ArnaVals want our Krils to understand; we appreciate your service to us."

Downing the rest of their drinks; the combat Krils slammed their empty mugs and glasses onto the floating table, indicating their opinions. The Krils were border-foolish, soon to be mean-drunk. Civility was not a strong Kril mood.

The following silence was thick and full of hidden meaning. The drinking vessels were filled again; including the Krils wanting to abstain and leave.

"We're here to honor Reela and her sister Ovals, who Soothed you all." Hortim thanked Reela.

"I have learned to ignore squawking crows in my past. Women belong in the kitchen." A weaving-drunk Wehrmacht soldier from the nineteen-forties said, precariously standing to make his toast. "Here's to my dish-honored brother Krils insulted by female whining. In my old timeline women knew their place."

The party was slowly degenerating into a gender war.

Attempting to forestall a male-female insurrection at her feast, Reela quelled the Kril uprising with a wave of her glowing red Oval device. Reela arose. "We raise our Oval wine goblets to honor the Four-striker KrutChan. He has brought the most wealth from the gold mine on KanBalaam."

"I submit...urp... the Barbarian was one of...urp...Odin's god-heroes." A Viking Kril hiccupped in irony.

He was ignored.

"From his Uayeb command we received much wealth." Reela said. "I honor him, my ArnaMals, and the rest of my Krils in my Nest." Reela raised her glass.

Fidgeting in his invisible chair at the invisible table, KrutChan was speaking to Reela. "Don't pile it on." KrutChan was sober and internally doped-angry at his EkSeet's lousy silence. "I didn't find the goddamn gold mine. My Zar excavated and later

366

showed the gold to me. You should be honoring CoCum."

"The Zar received his share of the gold for his species." Hortim said. "You were in command. In the Confederation Empire commanding officers on the Invasion get the credit."

"Likka…da…dude KrutChan say, 'lifer-officers' suck." XibEk grumbled. His beautiful black ArnaVal, OwlSmoking, punched his arm to be still. XibEk was on his way to being completely soused.

"ArnaVals present are permitted by their Ovals to speak." Reela said.

"Do not you have more to say?" PacalMo asked KrutChan.

KrutChan gestured to those in attendance with both of his middle fingers.

Immediately KrutCheebel's ArnaMal ZocKuk assumed the role of peacemaker. "Krils not do have much love bestowed on them from their Ovals. I understand. My Kril's heart is full of much Beekav for me. He is tender and considerate of my needs."

KrutCheebel wore a silly grin, his head rolling with drunkenness.

KrutChan felt his hormones stirring inside him. Much more statuesque than the other aura-body-enhanced woman he had seen on Arna. He envied KrutCheebel.

KrutCheebel was weaving and rubbing his drunken head against his beloved ZocKuk's hip after her statement.

"As you see, KrutCheebel is overcome with liquid gratitude." ZocKuk said. KrutCheebel's smirk lasted briefly before his head crunched the banquet table and he began snoring.

Reela did not like the topics of discussion. "I see KrutChan is standing to be heard."

KrutChan said what his EkSeet was feeding to him. "I honor Reela first, for her thoughtfulness and her beautiful banquet. She's a great Oval. Second, I honor all the surviving Balaam Krils present, your Ovals and their ArnaVals." About time you showed up, EkSeet.

The combat Krils reasonably stood up after his words and growled the Kril salute. "Ker-errrr-RUTTTTT!"

KrutChan stared intently at Reela. "We honor you Ovals for your generosity and your concern for our internal brooding. You're appreciated by all of us."

His EkSeet speaks. He is a lying... All of the EkSeets said into their Oval's minds.

KrutChan heard the EkSeets comments, not wanting to shame Reela. "I intend to get some air." KrutChan had not wanted a drink since coming back from Balaam and certainly not at the banquet.

"What-isss your pib-lem?" Hortim slurred. "Erp...mad cause...ain't not drunk...."

"The alcohol fumes from these Krils are stifling." After relating his EkSeet's correct speech, KrutChan continued. "I got other places to go." He left his destination hang in the air.

"I would appreciate your words more if they not were your EkSeet's." Reela nodded. "I wish KrutChan to leave in his contrary way." She wagged her finger at him. "Not do wander far."

KrutChan, cold sober, walked out to Reela's veranda.

Reela took control of the banquet. "I also wish the Ovals and ArnaVals spend this time-pause following ZocKuk's example and give us their thoughts."

Akna was summoned to Reela's side so she was not alone. An intimate joyous celebration continued sharing fellowship amongst the non-incoherent part goers.

Enjoying the silence in the black night on Reela's flower and tree bedecked veranda, KrutChan was somber. Alone with his terrible thoughts of what was coming tomorrow. His new promotion crushing his spirit.

A cough from behind him, near a clump of bushes bedecked with sprays of flowers, caused his head to swivel in a startle-response.

A woman's voice said. "My sister Akna advised me to announce my presence to KrutChan; for my protection."

ZocKuk came out of the shadows.

KrutChan was in adolescent awe. The same reaction he had at the dinner watching ZocKuk. I think the Sicilians called the effect 'Venus-struck' or something.

He replied. "If you fight like Akna, I'm the one needs protection."

KrutCheebel's ArnaVal was wearing a seven-layered hoop gown of deep Azure from Civil war days. KrutChan was swallowing the lump in his throat before he made more of a fool of himself. "Great dress...er...clothes...er...gown." He lamely said. "I meant...southern women floor me. Shit!" Scarlett on Arna; I'm fantasy lost.

Her lace half-glove covered her smile. "You are a gracious gentleman. I am flattered." She peered at his lowered eyes. "I think?"

Her EkSeet was eliminating her southern accent. She added. "My OtseVal father was a Zabin-Kril; Seeched from the Confederate War of Northern

369

Aggression. My dress is a present from him to me. My OtseOval is BalamEk."

Holy shit. BalamEk's daughter? Wonder if they get along.

"BalamEk has observed you since the Arena. She confers with me while you are in the Uayebs. Her information about you is more accurate than your Stel and the StelaBalaam."

Christ! More people keeping track of me to add to my list. Is everybody part of the Assassination plans? No wonder I'm going nuts.

ZocKuk whispered. "KrutChan seems troubled. Did I misspeak?"

"You're too beautiful to correct." KrutChan cleared his throat and changed the subject. "What's your job? Your mother shouldn't waste you on decorating banquets."

ZocKuk laughed. "My mother is training me to Osil-command a Burseeosil. We are much alike, though she is more qualified."

"You got class..." He stupidly said. "I mean you're awesome...with all the wrestling...qualifications."

He shut up.

Her smile was radiant. "I not do Akna fight or beat Krils. Why are you staring?"

His voice was husky. "You don't look like your mother at all. You're gorgeous...er...have your own beauty. Not saying she ain't..."

He shut up again.

She tried to ease his nervousness. "May I stay for a minute? I not am bothering you?"

KrutChan was reining in his schoolboy hormones. "Well...I am pretty busy." KrutChan said. Indicating the empty veranda. "No, don't go; my lousy humor." He held out his hand.

"My curiosity made me come here. KrutCheebel said you were a monster. I wanted to see who you were inside your ug...gruff exterior." Her smile was warm. "I apologize for my forwardness."

"I am ugly. Ask any kid on Arna. My orneriness turns people off."

"You are fierce looking. KrutCheebel related how you treated him. After hearing that; I did not like you."

He was happy her EkSeet stopped translating her without Oval transposing verbs. "I ain't been invited to any plantation balls. No social graces; get the drift? I'm harmless. Just not comfortable around women with class."

She was listening to her EkSeet. "I think my Kril KrutCheebel is mistaken about your hated persona. You are very humble inside your exterior. I perceive your sadness betrays you."

"Your EkSeet is mismanaging your southern-belle speech...er...charm." KrutChan said, wanting her to stop appraising him. "I'm disappointed by your female EkSeet's betrayal. My EkSeet constantly tries improving my speech. In my time I remember southern women called men 'honey' or 'sugar'. Implying stand-off politeness or a come-on to a future lover. Their innuendo blew my mind back then."

"Many at the feast did not understand your distress. Your words in the banquet touched me deeply." Her smile lit up the veranda.

"I don't remember EkSeet's speech." He said. "And I'm gentleman enough not to ask you to explain."

"I do not require flattery. I recognize pain."

"How's KrutCheebel?" He wanted to bite his tongue for bringing him up. In his past, KrutChan

371

was tongue-tied talking to stunning women. They mystified him; usually avoiding his clumsy approaches.

She was sighing. "My EkSeet informs me KrutCheebel is blissfully in a stupor. I will have to meld him back to his barrack cubicle."

"Don't be pissed at him. Cheeb saved my butt. He's been through a lot. I recognize you. His 3-D projections don't do you justice. He shares his missives to you with everybody. Guy loves you. I think your love for him helps him survive." Yeah sure it does. Survival requires hate, not love. My own EkSeet's mis-translating my answers.

"Beekav is not always shared." ZocKuk cocked her head at him. "Akna tells me she has much Beekav love for you."

"She loves someone else...not me..."

Misunderstanding, she waited a moment with a discerning look on her face. "Do you love her?"

Avoiding her gaze, KrutChan wandered to the invisible railing and leaned over. He refused to lie to her. "I don't know. I've only known her for two days. I get the feeling she sees me as someone from her past. Akna's half blind, you know."

"Since hatching. She sees the world in blurred gray-tones; no color."

"Akna loved that Kril. I'm a poor substitute. Visions and memories ain't simple."

ZocKuk leaned over the railing next to him, mellowing him out. "My problem is similar. KrutCheebel was engaged to be married in his past earth-life. He died before...you understand. I too am a replacement. We are both dealing with ghosts."

"Right on." He sullenly made direct eye contact. "You're way too beautiful deep down to be a stand-in."

372

"And you are blunt." She met his intense stare with her own. "My EkSeet warns me to be careful not to incite your anger. I believe your true emotions are tender and deeply hidden."

"My rage is part of my bullshit legend." KrutChan felt awkward with ZocKuk so close to him. Uncomfortable near her. KrutCheebel was hers. "Thanks for the insight. I feel like a creep around Akna. What happens when she figures out I'm not her true love?"

"KrutCheebel needs me." ZocKuk sounded more irritated than prideful. "Not many understand how stifling that kind of need can be. If I walk away from him, he follows like a cuddly puppy. Being someone's whole world is suffocating."

"I know what you mean." I can relate to that feeling.

KrutChan's family, except for his father, always came to him for solutions their problems. That attitude strained his relations with his older and younger siblings. He matured at seven after getting sprung from the County home.

"I have my own need for a strong equal Kril to share my timeline." She said.

"Nobody ever gets what they want." He was thinking of his 'Greer' vision.

She read his expression. "I have said too much; night shadows and breezes loosen my tongue. Please forgive my outspokenness. My mother told me Ovals should not be inquisitive."

He lightly touched her arm. "No, forgive me for embarrassing you with my whining. Spirit-visions popping in and out are my specialty."

"As are mine..."

He kept his distance. "Expectations put us on the spot. I feel the same way as you. If I went

373

ballistic in one of my rages, Akna would hug me, standing between me and any attackers."

"That attitude is stifling." She looked over the railing at Tikumyax. "Akna is protecting her protector."

"I've never been the cuddling type. In my past life I've been known to show affection in private; not in crowds." He shook his head. "I didn't say that right. I'm way off base pouring my past feelings all over you. Guy in the corps told me once, 'ugly hides a lot of beauty and beauty hides a lot of ugly'. The woman he said that to slugged him."

ZocKuk appraised him for a minute. "You are not ugly to me. KrutCheebel is wrong."

KrutChan's stare hardened. "No he's not; he pegged me right. He hates my guts like everybody else. Don't interrupt. I've given him and others plenty of reasons."

"If it is any comfort; they do respect you."

"Respect don't pay the bills." He looked away, going morose. "In the second Uayeb, I'll inherit more Krils to add to my hate-list."

Her small palm turned his face; her smile dazzling him. "I will never be on that list. May we share our love-proxy status?"

"I'd like to; if you don't mind."

ZocKuk is part of another Oval's Nest. Be very careful, old boy. She is KrutCheebel's ArnaVal; in training to be a high-echelon Oval, like her mother BalamEk. EkSeet warned.

She touched his forearm, sending a tingling sensation up his arm. Her EkSeet told her the same thing. "I must go before Reela becomes aware of us together. Protect my secret, as I will protect yours."

KrutChan thought she was feeling their heightened attraction for each other. "I never saw

374

anyone out here." He smiled confidentially. "Only my ghosts are with me." He grinned again at her.

ZocKuk walked back inside. Leaving him empty of feelings. KrutCheebel would kill him if he knew of KrutChan's love-at-first-sight feelings. His attraction to her was strong; more so than a Greer vision.

KrutChan left the Veranda and re-joined Reela's Nest.

Reela's Soothing Room banquet guests drifted away, most struggling with inebriated Krils.

Back in his barrack, his rage obsessed him and he stood in his firing range squeezing off twenty rounds from his M-14 downrange, followed by eight rounds from his forty-five pistol. The concussion sounds were echoing off the walls in the range. Minutes later, he stupidly stared at his warm-to-the-touch weapons.

<u>Goddammit, now I have to clean them again!</u>

He took off the clear bracelet containing hair, given to him by Akna, CauacSky and BakMeer at the banquet, fitting it under his Stel where no one else would see it. Some kind of Oval talisman to keep him safe on the Uayebs. Hiding the bracelet didn't help his mood, but hidden from view it helped him focus on his coming mission in the approaching early hours. Maybe the bracelet could help him with his immense battle-fatigue on the next Uayeb.

KrutChan was planning on leaving alone, forgetting them while in his second Uayeb. His way of conditioning his mind.

Once on the Burseeosil he was to be briefed in his duties as a Sub-Malkril leader; the rank he hated. <u>Here's hoping I don't get too many guys killed.</u> He said to himself.

Chapter 20


Viking AinAcbal, met KrutChan exiting Reela's Soothing Room.

"Our KrilChan commands your presence."

Shit!  I forgot about that fuckin court-martial! The time-pause his mind was replaying with ZocKuk evaporated.

Together they melded to KrutEk's quarters.


Kelel and Jaguar were together in the Intelligence Oval's quarters before they left for the Burseeosil.  Kelel had never seen a more austere Soothing room.  They were staring out at a darkened Tikumyax from Jaguar's veranda.  The city had shut down for the night, except for the bright yellow lights and activity at the Ovut embarkation area where the Krils were loading onto AkSilk elevators.

"Are you listening?" Jaguar asked.  "I wish to be assured you understand my intelligence-mission orders.  Regarding KrutChan and other agents accompanying us on the Invasion."  She was nodding her head.  "Repeat them to me."

"I'm paying attention."  Kelel said coldly. "Treating me like a male Kril imbecile doesn't inspire me."

Jaguar was waving her red-activated Oval-wand between them.  "Then humor me, as your earth species says.  What will you relay to my agents?"

With a sigh, Kelel said by rote, as if she was talking to one of Jaguar's agents.  "This is ultra-top secret.  Divulging this order, to anyone, will bring about your immediate TOTL."  Kelel shook her head in exasperation.  "That TOTL phrase sounds silly."

Jaguar glared at her.

"I'm telling them what they already know."

"They not should inquire about the over-all strategy. KrutChan not is trained to spy. He not will bother interrogating you, as the other's might. Continue, and refrain from commenting."

Kelel frowned. "In my day spies were...forget it. My missions never fail.""

"You retain Kril insolence."

"Isn't that exactly why you trained me?"

"My choice was made because of your intelligence. Not do test my will."

Seeing Jaguar's eyes turn into slits, Kelel continued. "I'll tell them they might cross paths with an operative, your opposite number from the Dinarchy, on the next Uayeb. The enemy operative will divulge a message your agents are to keep secret; without Stel or EkSeet knowledge."

"That is correct."

"I'll pass on the order to eight agents, including KrutChan, before we Bump. Kinda flimsy a mission, if you ask me."

"Brief and concise is more secure. What is the coded counter-sign they must look for? What is the contact code words?"

"The Dinarchy agent will say, 'we are in a stalemate'." Kelel held up her hand, crossing her middle finger, entwining it over her index finger. "This counter-sign is silent and indecipherable to most observers. I've never seen it used flagrantly."

"The Dinarchy interpretation means, 'I hate Aliens'." Jaguar said. "Why do you smile?"

"KrutChan will love that. On earth, we hid that crossed-finger sign behind our backs. To us it meant, 'what I just said is a white lie'. Superstition so God wouldn't strike us dead for lying."

"Ovals not do lie." Jaguar was grim. "I prefer you keep your earth-ghosts out of our discussion."

Having KrutChan drop his weapons and gear on the floor in an anteroom, AinAcbal led him into KrutEk's inner villa. The room was constructed of biological white Italian marble glowing brown in the yellow lighting. The room located near a rectangular water filled pond in an atrium.
   This guy KrutEk sure lives in luxury. KrutChan thought. But, what the hell, no Commanding General sleeps in shelter-halves.
   ArnaVals wearing see-through negligee gowns, with bouncing breasts and bare bottoms, were scurrying about, setting places. A dinner table filled with food befitting a king. KrutEk was not a king in name, but he was the closest thing to one on Arna.
   I've never gotten used to Oval invisible furniture and the floating stuff on em. At least here in KrutEk's villa everything is visible. Akna would look enticing in one of those ArnaVal uniforms. He lewdly thought.
   KrutEk wasted no words on preliminary protocol. "You were ordered here by me. AinAcbal wishes me to speak to you." His manner of a guy that just ate a handful of slimy-white wriggling slugs.
   It figures. This Roman dude would never have met with me unless the Viking pushed him.
   KrutEk bluntly got right to the point. "I understand you disapprove of my promoting you?" KrutEk waved his ivory baton. "Do not look surprised. Stels and EkSeets do not hide your thoughts."
   KrutChan thought about what he could not say to his leader. I'm the kind of guy who mistrusts

378

higher promotions. The EkSeets and Stels are nothing but useless piss-ant spies.

His EkSeet was warning him to watch his demeanor, thoughts, and language with KrutEk.

AinAcbal said. "Do not forget yourself, barbarian; your thoughts are shared."

KrutEk laughed. "For once the barbarian and I agree."

"I apologize." KrutChan said. "In my old universe, my thoughts were my own."

Thanks EkSeet. Now get in here and say the right things.

KrutChan continued carefully. "Contrary to rumors, I have no ambitions to attain your high position, my KrilChan. Most of the Krils I have met think I am plotting against 'you'. That's bullshit."

"Possibly not or you hide your nefarious plans."

Clearing his throat, KrutChan said. "Others say you're scheming my TOTL. I don't have the inclination, the skills, or temperament to be your kind of leader. You are more militarily qualified then me." KrutEk can read that remark either way.

AinAcbal said. "You do not act as a deserter nor command cowardly. Your KrilChan wants you to command a half-Malkril upon my recommendation. You are announcing then that I am wrong and your Kril leader is also in error?"

"Let him speak! Answer my question!" KrutEk bellowed. "You appear to need leader discipline when obeying orders."

The Marine Corps would get a laugh out of that statement. Lack of discipline was not something the Corps condoned. This guy has a hard-on for me, looking for any excuse to get me killed. Report that, you EkSeet spy! "I ain't afraid of dying. Strike

379

that...I'm not looking forward to death again. Having thousands of your Kril Legionnaires TOTL under my inexperienced command is to me, not winning..."

"Being in command and obeying orders is reality." AinAcbal said.

KrutChan could hear the water dripping in the Atrium pond. Reminding him of Chinese water torture he had heard about.

"Should you contravene my orders; I will have you join those thousands." KrutEk said.

"I was trained that burying our dead, or the men we command, is not what any leader strives to accomplish." KrutChan said. "Leaders cope when it happens, sure. They don't hold prayer meetings and burial rites during battle." KrutChan added. "Winning is as subjective as losing; the end results are just as painful."

KrutEk snapped. "Do not lecture me, barbarian. Apologies are not accepted for poor performance. None in this villa would disagree with your analysis of winning versus losing. It is the way you relate your philosophy that is the stinging nettle."

"Yes sir." The extent of KrutChan's response. Arguing with Generals was not healthy.

"You have not explained your philosophy sufficiently to me. If you are not a coward, and you dislike seeing so many dead Krils, then why did you survive Four Strikes?"

"I had no other options...."

"Do not interrupt me! Why did you disobey your commander ZacNaab? You were perfectly aware you were risking your Kril Bot-unit's lives. Why did you hold your position on KanBalaam when ZacNaab clearly ordered you to withdraw to save yourself and your Krils?"

He felt safe scoffing at that question. "ZacNaab wasn't trying to save me or my Zabin-Krils. We wudda been creamed if we bugged out. And if we survived, he would have accused me of cowardice."

You're on a roll, EkSeet.

"Are you saying ZacNaab did not inspire you?" AinAcbal teased.

"I adapted to battlefield conditions. Again, my options were limited." KrutChan was feeling slightly better. He was remembering many combat choices he had struggled with in his past. "There were a lot of confusing things going on at the time. I was trying to survive. I wasn't getting clear information from my Stel. My EkSeet was rambling."

"Your EkSeet advises us he was receiving the order in a flawless fashion." KrutEk said. "He is not authorized to interfere with command communications, as you are well aware." KrutEk raised an eyebrow. "You commanded him to be silent."

"There are times my EkSeet overloads my mind with constant trivia. Yes, I ignore him or tell him to shut…be silent…at times."

Thanks a lot, EkSeet…that should piss-off this Roman. KrutChan ironically thought.

KrutEk was thinking for a moment. "In my earth Legions many people defensively maintained, that they did not see signal fires, did not see communication flags, or did not have the order correctly transmitted by messenger-riders. Ignorance was not sustainable or condoned then and is not tolerated now."

I don't think he gives a shit what I think or what I did or not do.

"You are not promoted to command a half-Malkril of half a million Krils; the term is half-Nacom-Malkril, about fifty thousand. You were misinformed by unmilitary amateurs."

Wow...those table of organization figures make me feel a whole lot better. KrutChan was squirming inside, forcing himself to remain respectful. "In my military units on earth, discipline was ingrained in us daily. We obeyed orders, especially in battle...."

"Then why did you disobey ZacNaab?" KrutSeet asked before KrutEk did.

KrutChan tilted his head, anticipating a blow. "...Unless field circumstances changed. Both of my former Bot-unit commanders on Balaam ordered me and my Krils to take the positions; holding them at all costs. As a Four Striker I inherited command. Wasn't my choice."

AinAcbal grinned. "I am the last one who ordered you to maintain your position. Why do you not include me in your defense?"

The old squad leader training came out. "I never give orders to my troops saying it's not my idea. Telling them a higher commander orders them to fight is weak and my Krils hate a weakling's way of avoiding responsibility."

"Authority begets responsibility." KrutEk said.

Knowing he was putting his boot in his mouth, KrutChan said it anyway. "In my past life...and here...I've been under a few officers who have the authority but never take the responsibility."

KrutEk laughed. "I doubt any commander could achieve reaching your high standards of expectations."

AinAcbal picked up his glowing reader-sheet-device. "Your rank as a Four-Striker is precisely the reason you have been promoted."

"Your promotion will alleviate you blaming others for your failures." KrutEk added. "AinAcbal requested I give you to his Malkril. I refused. I do not wish two barbarians operating together."

Thank god. That Viking has too much blood-lust in him!

"Your new commander KrutSeet will order you as he sees fit." AinAcbal said. "He has battle-time against the Dinarchy and is more experienced. KrutEk is magnanimous in promoting you."

KrutEk asked. "In your past life, did you promote one of your men, over his objections? In your past timeline, did your former Legion commanders allow you to choose where and how you would fight?"

Thinking back to his time in the Corps; he had to admit KrutEk had a point. KrutChan ran his hand through his hair and rubbed the deep scars on his face. "Sir, I understand and I respect my KrilChan, as I did all of my good leaders..."

Keep on rolling EkSeet.

KrutEk scoffed. "The barbarian honors me. How amusing."

This General's Mast is not going well. KrutChan added. Might as well go down in flames, kicking and fighting. "...I've realized, since I arrived here, our KrilChan leader has the worst job of all of us."

Oh yeah, I'm acutely aware of how much KrutEk is looking out for my welfare.

KrutEk snidely remark. "You see AinAcbal, this barbarian acts like he is in the Senate of Rome. He disagrees and in the same breath, he agrees. We
383

punish disobedience after the fact; should a crime against our Kril code be committed."

Breaking out in a sweat, KrutChan nodded. Here it comes. The death before dishonor shit! "If I survive I'll keep that in mind." KrutChan said. "The Kril trying to assassinate me needed no punishment from anyone. He died a coward."

AinAcbal snarled. "KrutEk did not condone that act or order it done! Are you accusing him?"

KrutChan was hoping his EkSeet would pass on his thoughts. I may be stupid, but I am well aware of 'plausible deniability' in commanders. Generals motivate others to carry out those kinds of mission; they don't do it themselves. Careful dummy, keep your street-wise remarks out of the discussion. If you want to survive.

KrutEk grinned. Listening to the translations.

"I'm not accusing him of anything!" He yelled, going on the offensive pointing at KrutEk.

"Do not raise your voice!" AinAcbal shouted.

In a more moderate tone, KrutChan continued. "Sir, I said I honor the KrilChan and respect his rank. Things go wrong in battle. I've seen it happen before. I can't blame anyone but the assassin."

"Be clearer in your thoughts." KrutSeet said.

"On review of the circumstances, that assassin 'could' have been after me. And he 'could' have been after PacalMo…or some other Kril. He 'could' have been battle-crazy. I don't know for sure."

"You do not sound sure of anything." AinAcbal said.

He paused for a moment, deciding his head was already in the hangman's noose. "If an assassination attempt happens again; then maybe, just maybe, I'll have less 'could's' to deal with. Happening again, I'll find out from the assassin who

384

gave the order." KrutChan said. "Then I'll kill the sonofabitch. I didn't at the time and I've no time now to dwell in my THEN past."

KrutEk's crocodile grin looked deadly. "For the first time I see a glimmer of reason in your inner thoughts, barbarian. Although I will never call you intelligent..." He was listening to his EkSeet. "...in your language street-wise is unwise."

Evidently, my EkSeet passed on my thoughts.

Changing the subject, waving his Roman baton, KrutEk said. "ZacNaab's court-martial of you is voided."

KrutChan's shoulders dropped in relief.

"He has been demoted. You will assume command of his half-Nacom-Malkril as soon as we are aboard the Burseeosil. My staff meeting is within two Bots of our boarding. That is forty minutes in your language, my EkSeet informs me."

KrutChan wheezed, like a balloon deflating. He kept his silence. Does he think I'm going to argue with him? From their non-reactions, he judged his EkSeet didn't pass that thought on to them.

AinAcbal continued. "Your EkSeet will appropriately meld you to the Staff meeting on time. It will be a tactical conference."

KrutChan said. "I understand completely, sir." He decided he sounded entirely too happy.

Getting KrutChan's attention by throwing his wine goblet; clanging against the wall. Pausing for effect, KrutEk said. "Barbarian, you must try to avoid any more assassination attempts." KrutEk insinuated, without sympathy. "I would mourn mightily should you prematurely TOTL. How do you say it...be dead-wise? What further demands do you wish to make of me, barbarian? Does the Four-Striker require lustier ArnaVals?"

385

<u>Akna would kick their ass and mine. Combat is safer.</u>

KrutChan heels clicked to attention, locking himself in place, muscles taunt. "I never 'demand' anything of my leaders I respect, sir."

Hoping that would satisfy his General, he could not resist adding. "I'm looking forward to my KrilChan calling me KrutChan someday."

EkSeet sighed in exasperation. <u>My boy, just once I wish you could resist your insane desire to have the last word.</u>

AinAcbal inwardly winced at the barbarian's request while he watch KrutEk's anger rise on his face.

But KrutEk held his composure, softly saying. "AinAcbal said you have huge stones. I will call you by name when you earn 'my' respect and not before, barbarian."

Knowing instinctively not to retort, KrutChan backed away, stopped at the exit, clicked his boot heels together, and stood at attention.

KrutEk waved his white baton. "You are dismissed. I do not want to see you again unless you reside as a golden statue in the Memorial of the Oval Protectors."

Meeting KrutEk's hard glare KrutChan used the Kril salute with his fists under his chin, before he departed.

Outside and alone, he breathed a sigh of relief at his survival and luck. KrutChan hoped his luck would hold through Uayeb Two.

Chapter 21

The second Uayeb on Nihberu.

A Time hiccup: an alternative-life beginning.

Nihberu was located in a nine planet system surrounding a Class K Orange-white sun, eighty per cent in mass compared to Arna's sun, with ninety percent of its radius. With a sixty percent luminosity of the Arna sun and a surface temperature of 5,000 Kelvin.

The fifth planet out from its sun in the 'goldilocks' zone of habitation. Compared to the heightened solar flares and coolness of KanBalaam's sun, Nihberu was more earthlike, with an elevated oxygen level atmosphere.

Internally Nihberu was rock-core-similar, but different from KanBalaam. The exo-planet had a minor tilt to its axis, was globally temperate heated, with volatile weather.

Though fierce when they occurred; the lightening and wind storms were Kril survivable, laden with a peroxide mix. The battles would occur on a massive island continent, surrounded by a single and huge Peroxide Ocean. Unless metallic elements or surfaces were hot, which can precipitate rust, the peroxide had no effect on the Kril's biological weapons. Human Krils ingesting large amounts of the ocean water were subject to vomiting, internal tissue burns, and depending on the amount swallowed, death could occur. Dagots enjoyed peroxide.

That Ocean, along with the human necessity of needing water within three Kin, factored a science hurdle to overcome. Meli's science team would have to purify the ocean water for drinking while on the planet eliminating the extra oxygen atom. Along with killing any indigenous oxygen-reducing microbe soups the Krils ingested. Their medical remedies are not absolute. Meli's forensic exobiologists prepared for every Uayeb based on methods and data accumulated from previous experiences and studies.

Non-pathogen diseases were difficult to control. Many Earth Krils smoked plant fibers, causing heart problems, had cancers, vital organ failures, and pulmonary problems. Ancient Krils like the Romans, used lead metal in their drinking cups, water fountains and other building materials. Those habits factored in their TOTL. Zars had a type of non-pathogen in their DNA turning their system off when it reached its zenith. Cunacks met TOTL on occasion, by not being able to communicate with its species. Dagots, in some instances, succumbed to massive dehydration.

Hundreds of thousands virulent pathogens existed on home planets or any Invasion planets. Indigenous races endured the infestations by evolving immunizing resistance.

Pathogens existed and reproduced on compounds found present in their home-world habitats. Impossible to stifle all of those interactions. The Oval scientists knew they would have to TOTL all life on an invasion planet to be safe. Annihilation is scientifically impossible. Even if possible, the Oval Confederation would never allow total extermination, no matter the size of the creature. How could Ovals ever colonize? Pathogens mutate constantly at a prodigious rate. Oval Scientists were in constant battle on every Invasion.

The invisible nasty pathogens present on Nihberu were dangerous to foreign species. However, Meli's team felt they could reasonably manage. Because it was an alien planet, the Krils would have to contend with the indigenous flora and fauna and possible pathogens in them. Meli's scientists worked diligently in the pre-Kins to manipulate, not wholly eliminate or control, the environments the Krils would face.

In the final analysis, Nihberu, like KanBalaam, was as potentially lethal to the off-world Krils as any stable oxygen-poisonous atmospheric environment would be to any non-indigenous life forms. Dangerous but survivable. Off-world creatures landing onto a non-home-world planet would be comparable to the hostile environment a species from Nihberu would encounter after landing on Earth; with similar challenges to survival.

Their biological Toob KrutCheebel's Bot-unit was in was scarred and crumbled in places, but held together. Being a biological creature, the Toob had the ability to repair itself. With no Dinarchy resistance, AkSilk repelling was nominal, not like the last bump on KanBalaam.

Repairing what major constructs the Oval pilot and Germ crew could, after AkSilk dropping their cargo, abruptly screeched into the air. Minor outer pieces of skin dropping off, drifting debris heading back to space. The Toob and Oval pilot had other things to accomplish.

KrutCheebel swore and dragged some of the other indolent shocked new Krils off their AkSilk lines. Getting them away from the ocean; heading them for the lush-green sparse jungle, resembling broken teeth that ominously awaited them.

389

Deadly silent in the warm tropical mild breeze, the brush-tangle of green was quiet. Running towards unknown numbers of enemy forces in this new Uayeb.

PacalMo and KrutCheebel both watched KrutChan, moving away from them trailed by his headquarters'. His huge force unseen. Yelling into his Stel to his Kril Bot-units to get inland off the high water line of the deadly peroxide ocean. Herding them into the sparse shrunken growth of the island-continent.

They glanced at each other, believing KrutChan was in his element again; sober and concentrating on his mission. KrutChan looked to be adjusting to his new rank, in an anxious hesitating way in this first Kin.

He was talking not to the Command headquarters, but by Stel to his subordinates Bump-landing elsewhere. Half of his assigned Toobs went amphibious eliminating themselves as airborne targets, approaching the island from oceanside.

KrutChan, kept his fears of making a mistake in check. He was an immovable rock relentlessly pushing himself harder than his Kril forces.

KrutCheebel was now more seasoned. He still thirsted for piling up kills. His way of coping with his inevitable virus lurking within his body. KrutCheebel was stronger in character after his bout with the infestation he caught from the devouring plant on KanBalaam. He was not forgetting his responsibility to his own Kril Bot-unit. He led by example, as he had been taught by KrutChan, CheChun, XibEk, and PacalMo.

PacalMo, though a gifted Mongol from the steppes not used to island warfare, came into his own. On this island of low jungle-brush, his prowess with his bow, his spear, and his other weapons would serve him well. He did not miss his pony he lost in the beginning of their last invasion of KanBalaam.

One of the landing Zabin-Krils was standing outside their Bot-unit, assembling a weapon. PacalMo observed his strange inverse haircut; a brown forest with a firebreak in the middle. He was standing near KrutCheebel's Bot-unit; alone but a part of them also. <u>Was he a dragon creature summoned by the Ovals?</u> <u>The Oval women had strange magic.</u> PacalMo was thinking.

This huge Kril had numerous scars from many fights and was dressed like no warrior PacalMo had ever seen. His skin was yellow-tinged, but not like the Mongol. His hands had a dragon's six fingers. All the Krils avoided him like water in a stream roiling past a large rock in a current.

KrutCheebel got PacalMo's attention. "I see NoKoch survived Balaam. You saw him before, my Mongol friend."

PacalMo bristled. "Do not speak KrutChan's words. I am not your friend."

KrutCheebel shrugged. "That six-fingered-mutant guy is always around; he's harmless to us." He pointed to the jungle. "KrutChan is demanding we keep our Krils moving inland."

For a Kin day, KrutChan's half-Nacom-Malkril Kril force landed without making contact with Dinarchy defenses. Having no enemy in sight or incoming fire, unnerved the Krils.

The Zabin-Krils encountered a variety of purple masses of Nihberu creatures attaching themselves. The invading Krils brushing and pulling them off their bodies as they ran. Irritating, but not lethal. Not yet; the Krils remembering KanBalaam.

Discarded by every Kril species, the biting, sucking creatures departed, save for a hidden few, back to the beach.

Skidding, slipping, and sliding in the sugar-cane-consistent sand had the Krils heaving and breathing in gulps, while sweat drenched their uniforms.

CheChun's voice crackled through his Kril's Stels. "Pay attention! Flush out yer brain-housings. Our first wave on Iwo landed without any firing from Jap positions. Oncet the first wave went inland, the gates of hell opened."

A Japanese Imperial Marine spoke right after CheChun. "On Iwo Jima, we were ordered not to fire until commanded. We let the marines stroll into our guns. Our fear was as great as theirs. We were destined to die gloriously for the Emperor. The invaders spit hell-fire back at us with their flamethrowers."

"Jap's right. That stinking sulfur-smelling island made the Nips and us forget about living. Like dying in a cesspool; both sides ignored and pushed aside the dead, and kept moving. Both a us died there."

The Krils spent two Kins organizing and sorting themselves out on the island, that technically was a continent.

KrutChan was constantly on his Stel straightening out his lines, assembling them for future movements. Assigning his sub-Nacoms handling the logistics. As he grew more certain of himself, he issued his orders, and unlike ZacNaab, he left the details in the hands of his subordinates. Though, he did follow up on them and correct any tactic or tactical position not to his liking. Getting credit for his successes, but accepting the responsibility for errors or failures.

He had decided a long time ago on earth, whether he was going to get the blame or the honors; he would do it his way and no one else's. After the fact, to his Kril Bot-units he became a disembodied voice from their Kril Stels, of their always-watching-listening leader.

It was sweaty boring work during those two Kins, with no contact.

Meli's science teams published an edict for the Krils to immediately don protection for their respiratory organs. Humans covered their noses and mouths with Oval-supplied clear masks. The Zars wore chest sweaters. Cunacks looked ludicrous; their body-bag covered by their masks. The Dagots, with their Osmosis membranes, ignored the order. Straining out atom sized pathogens or larger.

Each Kril's EkSeet explained the Oval doctors had dead and dying Krils carrying hosts in their bodies. Meli's teams determined the fungal hosts were from the purple creatures they had encountered. The fungi sprouting stalks of beaded branches within hours in the moisture. Heavy rains would accelerate their growth and puff-ball explode releasing more spores. Krils were to wear the masks until Meli's team could find a cure for the malady.

The Krils were thinking. <u>What a way to fight a war.</u> Moving around in gas masks was not comfortable. Breathing or sight restricted not advantageous if they got into a firefight.

On the third Kin, darkness had fallen when they came into contact with the Dinarchy. The firefights small unit actions, with snipers, and isolated incidents leading up to small battles of attrition.

393

CheChun's Bot-unit made first contact. "Yawl see 'em?" CheChun, hearing the exchange of fire, said into his Stel.

"Not individuals...but I see movement." Kelel answered.

Not wanting the Executive Officer with him, CheChun was grumbling to himself regarding the uselessness of female ground pounders. Maybe KrutChan's extra-eyes? She was not giving any orders to him. He was happy when she wandered further up the line.

CheChun located himself twenty yards forward of his men. He discerned movement to their immediate front. Under the eerie moving lights of the weird flying Oval Toob-flares KrutChan had called for, a couple bunches of Dinarchy were seen working their way toward CheChun's position. Nudging his dozing companion, CheChun whispered, "They're coming out of that road-cut into the brush. Open fire!"

His Kril XocYax, was a US Army PFC from World War I; Seeched from Le Champy Haut. XocYax was expertly bolting round after round into his weapon's breech until he emptied his M-1917 rifle.

CheChun quickly joined him using his Thompson submachine gun and fired at the three remaining advancing figures. XocYax quickly reloaded and began firing again. But after three rounds, his weapon jammed from purple spores.

XocYax and CheChun had cut down a dozen of the infiltrators. During the lull, discarding his jammed weapon, XocYax left CheChun to get a replacement weapon.

As XocYax disappeared into the darkness, a grenade rolled near their fighting hole and exploded.

394

The detonation showered CheChun with a cloud of sand and hot grenade fragments. The concussion stunned CheChun momentarily; feeling warm blood seeping from small prickly punctures.

"You alive?" XocYax had rejoined him. But before XocYax could see if CheChun was ok; four snarling Dinarchy charged them. The first guy swung his large bayonet-attached rifle at CheChun. XocYax quickly parried with his rifle. Lost his grip, grabbing the enemy rifle with its bayonet still attached.

Ripping the man's weapon out of his hands to assist XocYax, CheChun forced the assailant to impale himself on his own bayonet, collapsing with a hair-raising scream. The dying Dinarchy had inflicted a slashing wound across XocYax's hands and shoulder in his attack.

Mowing down more Dinarchy, CheChun realized his submachine gun was out of ammo. CheChun yanked his pistol out, firing point-blank at more intruders. Another Dinarchy attacker abruptly stalled when a .45 round penetrated his forehead, hurling him backwards.

Around them, the entire front of CheChun's unit exploded with activity. Firefights were picking up all around them; grenades 'whumphing'. Blasters, arrows, and lances joined the automatic weapon fire and hand-held mortar explosions sprinkled with the fire-crackling sound of rifles. The fighting dwindled off and away from them.

CheChun ordered XocYax back for more ammunition and to get his wounds attended. CheChun was driving his bayonet studded rifle again and again into the sand to clean the blood off it.

Wouldn't be kosher ta kill some asshole with a germ infested dirty blade. CheChun thought, breathing heavily through his mask. His fighting-hole-buddy gave CheChun more rifle and pistol ammunition from his own gear. Vowing to return.

Upon XocYax's departure, CheChun was exposed and alone, sucking in air to recuperate. He had his Bot-unit's Stels giving him head counts and ammunition reports. His blood slowly trickling from his numerous pin-prick grenade wounds.

Instantly a Dinarchy was charging his fighting hole, throwing a grenade. It landed at CheChun's feet. He kicked the grenade into the sump hole and bailed out.

Instinctively, CheChun fired his pistol at the attacking enemy. The Dinarchy trooper collapsed, mimicking a broken rag-doll. CheChun grabbed his submarine gun and rammed a clip in and cocked the weapon.

Two more magazines and then I'm up shit's creek without a paddle. XocYax better hurry. He then reloaded his pistol, half-cocked the hammer and replaced it in his holster.

Slumping back into his fighting hole, while he had time he cleaned his K-bar in the dirt. Finishing that task, CheChun waited.

His Stel engaged, KrutChan was whispering to him. "Your Bot-unit holding, Chun? You okay?"

"How yawl think I'm feeling? My unit's passable, but a mite ornery." CheChun said. "One KIA, three, no, four WIA…walking wounded. Ammo's tolerable; I've sent for resupply."

"It's coming. Can you hold?"

Cranky, CheChun snarled. "Yah'll think yer talking to da enemy? Lot'sa dead Dinarchy around us."

"The ArnaMal doctors have evacuated some of my Krils because of those beach-bugs we ran through." KrutChan said. "Probably want more specimens to poke. You know how the Ovals are."

"Yeah, they got a few of my Krils too. Buncha horseshit! I'm fine."

"You sound muzzled, but normal...don't take it out on me. Ready for some bad news?"

"Ain't had no good news lately."

"You're fightin' hole buddy XocYax, got scarfed up by the Burseeosil crowd. They killed him."

CheChun was pissed. "He was sent for ammo. He wasn't deserting!" Having to keep his voice down did not help his anger. CheChun's rage was coming through the Stel. "What the fuck for? Why they killing our own?"

Whispering back, KrutChan's voice hissed from CheChun's Stel. "The StelaBalaam thing screwed up...or hiccupped...or had a quantum meltdown. The Burseeosil dudes were correcting the StelaBalaam's error. Ask your EkSeet. I don't understand either."

In his mind CheChun yelled to his EkSeet. You heard him, you freak...explain!

The StelaBalaam designates each Kril to go on each invasion. His EkSeet said. No reconstituted Kril of any individual is to be on the same invasion. The StelaBalaam will never allow two individual Krils to occupy the same invasion timeline. You are the only CheChun allowed on this planet in your timeline. Only one CheChun or one XocYax is allowed here.

"Well...ah'll be a 4-F asshole. Why did they TOTL the guy?"

"Sounds like another Oval rule...old Corps." KrutChan hissed from his Stel.

I just imparted what occurred. No Oval, Kril, BalamEk, or even the OvalChanHalach can change the StelaBalaam function.

Over his Stel KrutChan said. "So that StelaBalaam prick made an error...then corrected it with another error? Why not send XocYax to another planet? Who is StelaBalaam?"

The StelaBalaam is not a who species. Transferring XocYax would be cost-prohibited. The StelaBalaam is a quantum entity, too complex to explain to you. It was not an error. The StelaBalaam was realigning the Invasion timeline.

"In other words I lost my Kril XocYax." CheChun was muttering. "Great system yawl got here. I lose a murdered Kril and yawl say toughski-shitski?"

No, that is not correct. He has been replaced with the original XocYax from the other side of the continent, who was assigned to this invasion. He has the same memory as the XocYax who met TOTL by the Ovals. He will rejoin you.

KrutChan snarled. "In other words, the Table of Organization, or Order of Battle did not add up. StelaBalaam guy erased the mistake?"

I did not cause XocYax's expiration!

CheChun spoke at the same time, into his Stel. "What a fuckin' crock of...!" CheChun had the last word. "Thanks for nuthin', EkSeet...you're a fucking fairy faggot."

"Cease this gibberish on your Stels!" KrutEk's voice was unmistakable.

Near dawn CheChun was surprised when XocYax returned with a resupply of ammo and ArnaMal stretcher bearers, hoping to find CheChun still alive. CheChun thought XocYax looked different; but the new-old XocYax didn't act different.  CheChun decided to keep his mouth shut.  CheChun was always prepared for battle losses of his men; but being frosted by their own Oval commander's Quantum Entity seemed a futile and ignorant policy.  With two XocYax's, CheChun felt his Bot-unit would have been twice as effective.  What a snafu war.

"You okay?"  CheChun later asked XocYax when they sat upon the edge of their fighting hole.

XocYax nodded.  "I'm fine.  The ArnaMal medics repaired my wounds.  I'll be better when we get out of here."

"What I meant was…"  CheChun grabbed a fistful of XocYax's uniform.  "…do you feel different?  Are you the same as you were…before?"

XocYax's hard stare fixed CheChun.  "Aren't we all the same?  We're all different every day."

CheChun wondered what the hell that meant.  It sounded like XocYax was telling him either to shut up, or to forget about the StelaBalaam transformation.  It irked him, in this universe there were possibly other CheChun's out there on other planets that he would never meet.

His EkSeet advised him.  Out of the mass ejected from the Seech Field, the reconstituting Ovals have trillions upon trillions of quanta raw material to work with.  Most killed by the Seech Field.  Surviving 'you's' went to the Arena.  Afterward, the others were disbursed to other planets.  Rare occurrences, at best.

In disgust, CheChun said.  "This me of my me's thank them fer nuthin."  What if his doppelganger and he did meet?  Would I die or the other guy meet TOTL?  CheChun was thinking, and then shrugged. <u>The hell with that noise, forget about it.</u>

After attending his minor wounds, the medical ArnaMals declared to KrutChan CheChun was fit for duty.

Through the sheets of rain, the Bot-unit commanded by CheChun could see the Dinarchy hauling away their dead and wounded.

Kelel's voice whispered to KrutChan. "The Dinarchy are full of masses of spores on the dead and wounded I see through my binoculars."

"Guess they don't get Oval masks." KrutChan said.

Kelel added. "Fungal stalks are growing out of the eyes and entire bodies. Must have drove them nuts."

After a while they could hear the Dinarchy opposing units digging in about one hundred yards away, in tandem and in concert, with the Kril Bot-unit's sloshy burrowing.

Over the Stel, KrutChan advised CheChun. "Dinarchy are consolidating in front of you, Chun. Not much of a force, but don't get cozy. The entire TAOR, tactical area of responsibility is seeing and hearing the same thing as you are."

"You mean the Main Line of Resistance?" CheChun asked. "Don't worry, we ain't gonna party wid dem. We cancelled our hootci-cootchie girls for this Kin." CheChun noticed his EkSeet was cleaning up his English. "It's stable here for the moment."

KrutCheebel chimed in. "Roger that, only thing in front of us is wet Nihberu jungle life."

"Ain't none-a-dem Dinarchy dude's female?" XibEk's EkSeet evidently was having a harder time fixing language when mis-translating XibEk's idioms.

400

"Quiet here too…wasta time waiting. How 'bout wasting dem dudes widda airstrike?" He suggested into his Stel to KrutChan.

Kelel's voice announced. "I've assumed command of one of your Bot-units, KrutChan. Their leader met TOTL. CheChun doesn't need me."

The silence indicating nobody had a problem with her volunteering.

KrutCheebel chirped in with his opinion. "Yeah, these planet-critters are always up to no good. I am still itching from those purple critters on the beach. I vote we eliminate them."

"You would, Cheeb." KrutChan relayed. "Xib ain't talking about your kinda critters. He's getting a hard on for more advanced evolved animals, like willing Dinarchy Ovals. Kelel will waste you guys if you try any shit with her."

"Roger. I'm golden with that. Count on it." Kelel said.

KrutChan ignored the updates. "I'm busy monitoring my half-Nacom-Malkril. There are more intense battles a couple of klicks…er…miles from your units. I'll get back to you. Alert me if you're attacked in force."

Immediately, KrutSeet's voice transmitted. "You belong in front of those Bot-units of yours in the worst fighting, barbarian. I transmitted their location to your map."

"On my way…." KrutChan's voice sounded grave and pissed off. "I'm half way there…."

Five Kins later, KrutChan was doing the math, when KrutCheebel's Stel chirped. KrutCheebel wondered what had stirred the Dinarchy up and reported to KrutChan immediately. "We got

401

movement and contact coming right at us. Damn the rain!"

"Kick their ass, Cheeb." KrutChan answered. "Kelel, support Cheeb's flank. They're testing us all over the line. Let me know if you meet TOTL."

XibEk firmly muttered into his Stel. "That's not funny, man."

KrutChan was new at this half-Nacom-Malkril leader stuff, but he had his subordinate logistics people working at laser-speed. He couldn't stop the Krils under his command from meeting TOTL but he could make damn sure they had the tools to fight with when they needed them.

He remembered back about the bad feelings generated when his patrols he led were out of touch and ignored in the Vietnam jungle. That would not happen under his command here. He endured a lot of bitching from ZacNaab, AinAcbal, and KrutSeet that he was taking their supplies. He figured they could court-martial him if his Krils got overrun; not if he wasted a lot of Dinarchy. Besides, those other Commanders did the same to him every Kin they could get away with it.

KrutEk's rasping voice crackled into KrutChan's Stel. "Do not disrupt my supply line, barbarian. Heed my orders about maintaining discipline with your troops."

"Yes sir. You honor my Krils, sir." That seemed like the proper response.

"You are an adequate battlefield Centurion. Do not become enamored of your prowess."

KrutChan was properly chastened and silent after that. Don't fuck up is what he's telling me.

The Dinarchy disappeared from his sector. KrutChan's forces were at rest for the next week.

402

The ancient ArnaMals were absorbing the brunt of the attacks elsewhere.

The lightning crashing and soaking rain on the island atmosphere stunk from the Dinarchy bodies piling up over the last five Kin.

The non-fighting medical ArnaMals had carried Krils to evacuation Toobs and loaded them without differentiating between dead or dying. The fallen ancient ArnaMal's and Kril's appendages were species-color-crusted with dried blood. Let the Burseeosil doctors sort them out in triage, was their creed. The Krils fighting had no time for processing.

Krils left alive, smelled of sweat, urine, feces, and sweet-sickening death; the survivors could get no relief from the stink.

No wonder the infantry back on Earth called themselves 'grunts'. Wallowing pigs would be more apt a description. KrutCheebel thought.

For a half-Kin, the ArnaMal medics arrived and sprayed disinfectant on the Krils, advising them to not throw away their masks. Meli's team had come up with a fungal cure. The Cunacks were Christmas tree flashing their happiness.

Mercifully, the downpour ended with clear skies.

Finally, the time came for them to be reinserted back into the advancing line by their half-Nacom-Malkril-Commander. The second Bot-unit of KrutCheebel's Kril Bot-unit, attached to KrutChan, were assigned by him to point.

Kelel was roving again.

KrutChan was angrier than usual. KrutCheebel was faring about the same; a constant refrain running over and over in his mind. Why are we on point again?  Why us?  Probably, we are not as willing to die as KrutChan.  KrutCheebel thought.  "I have 'my' Bot-unit under control.  Do not need Kelel."  He angrily advised KrutChan.

"You said you served with women grunts.  What's your problem?  She goes where I tell her.  Just like any Krils under my command."

His Stel, relaying XibEk's voice chimed in answer.  "Stay focused, Cheeb, one of your new Kril recruits KinikXoc, is watching you."

Seeched from Salerno Italy, KinikXoc, as an untested private, had died during the first four days on the beachhead of attrition against the German Wehrmacht.  He played the harmonica incessantly to the pleasure of most of the Krils in the Bot-unit.

"I am aware."  KrutCheebel wearily said out loud to his own EkSeet.  After waiting a minute, he gave a Kril hand signal to XibEk who ambled over.

XibEk stared into KrutCheebel's eyes.  "You okay, mutha-fuc…?"

His EkSeet angrily said.  Stop your infernal cursing…you bloody twit.

Ignoring the remark, XibEk continued talking to KrutCheebel.  "You lookin' little gray around da ears.  You got the Balaam bug agin?"

"Not complaining.  My Balaam viruses slayed my fungus critters.  My blood microbes from Balaam are always with me in my timeline; however long that will be."

They were both saying the unsayable.  They were in this crap for the duration of their meager lifetimes.  Happy the rain had dribbled away.

"We gonna call ya 'chuck' iffn ya get any whiter, splib."  XibEk said.

KrutCheebel decided XibEk was good-naturedly jiving him.  "I am fine."  KrutCheebel pointed with his thumb. "Insert your guys on point. Put KinikXoc up front."

"Thanks fer nuthin', dog." XibEk yelled to the new guy, KinikXoc and pointed him to the front of the Bot-unit.

The man crumpled over his own feet, righted himself, grabbed his pack, and almost dropped his rifle.

XibEk was shaking his head in disgust. "Dat guy's gonna get lot'sa Krils wasted."

KrutCheebel was not in the mood and poked back. "Take the point yourself then."

"Like hell I will, dat newbie gotta learn sometime. Ya piss on angry gods, you be fuked..."

His mind swimming, KinikXoc waited to be told to move.

The Bot-unit was unscrambling, forming into a route march formation, single-filed and spread out.

Back in my time, knee-grows were invisible; weren't near us or in command. He never served under a knee-grow before or even saw one in the Army. KinikXoc was thinking. I don't like this Stel. I write in my journal. KinikXoc was uneasy.

My grandma used say if you don't write it down; it didn't happen. Her great-grandpa had been in the Civil War and he had kept a journal. When the Confederacy gave up and he got out, his great-grandfather had packed his journal away, where his grandma found it after his great-grandma died. She adored great-grandpa even more after reading 'his' journal.

My mom and pop raised me to be God-fearing and respectful. Grandma gave me character and the backbone to strive to be a better person. I would still be alive if she had not stressed having gumption so much.

KrutCheebel was yelling at some Krils to hurry up and keep their interval.

This KrutCheebel guy's a veteran and I'm going to stick to him like glue. KrutCheebel used to be an officer in the US Army. Probably hates enlisted GI's; maybe that's why he rides me so much.

KinikXoc didn't know his leader well; that seemed to be the story of his life, a perpetually ignored replacement.

I'm not even sure I won't crap my pants when I do get into battle. Our Malkril leader KrutChan is a scary fellow. I want to stay far out of his sight.

KrutCheebel walked up to KinikXoc, grabbing his shoulder, pinching it for effect. "You keep your eyes open and pay attention. You are not going to a square dance looking for girls."

Quietly admonished himself in his mind, to stop parroting KrutChan. "If you see something suspicious, halt my Bot-unit and alert us." He tapped the new guy's shoulder to reassure him. "Every Kril here were new to this. We survived and so will you."

Then he pinched KinikXoc's face to get his attention. "Listen to me! Don't stop too many times because you are nervous." KrutCheebel pointed to XibEk. "Or he will kick your butt if you hold up the patrol for nothing. XibEk has been known to TOTL new guys that screw up."

That little exaggerated lie would keep KinikXoc on his toes. His EkSeet says this new guy is overloading on daydreams about black men and their straight razors. KrutCheebel thought in anger. So much for leadership according to KrutChan's book of discipline.

406

XibEk played the role. "Listen tada man. I'll wat'cher back; you pay attention to the front and sides." His scowl added meaning to his words. Glancing at KrutCheebel and rolling his eyes, XibEk said. "I guess he ready to go."

KrutCheebel bellowed. "Saddle up...charge your weapons...and move out."

His Bot-unit straggled to their feet, waiting briefly for KinikXoc to start off on point.

CheChun wandered by with his Kril Bot-unit lining up behind KrutCheebel's and muttered. "Toughski-Shitski said the chaplain to the knocked up virgin." He frowned at scared KinikXoc. "Dis lashup sure be going to hell in a hand basket."

Mimicking KrutChan's voice KrutCheebel said. "Keep your own legs crossed...old Corps."

"Up your black-ass, genius." There was no humor in the remark. CheChun slid his middle finger up the bridge of his nose in a salute to KrutCheebel.

They moved for an hour without incident. KinikXoc, seeing ghosts and misidentifying movements, stopped them four times. They began crossing the flat boulder-bridge, with KinikXoc in the lead, keeping their interval on both sides.

Starting across the bridge, in an instant, KinikXoc was obliterated by a mushrooming explosion. He disappeared, the target of a hidden gun. He met TOTL.

"Keep your interval, goddammit!" XibEk was yelling into his Stel and at the same time his eyes were trying to find out where the firing was coming from; so they could maneuver. "RUN! Get across the bridge. Then get your fucking-asses down and dig cover for yourselves!" His EkSeet was cleaning up his translation as XibEk was yelling into his Stel.

XibEk's mind was racing with the alternatives and the risks. Without knowing where the enemy was, he didn't want to take his point further away from the ambush.

At the same time running directly into the ambush could get them out of the killing zone. He was thinking of the old timeworn axiom. <u>Charge into the fire!</u> He ran from Kril to Kril in a crouch, expecting a buzzing bullet slamming into him at any time. He moved when the enemy rounds went searching for another target.

"What's the hold up?" KrutChan yelled.

He yelled into his Stel to KrutChan. "You comin' man? We gonna get wasted here if we can't see the dude pretty soon."

"We're on the knoll above you to your right and starting to draw fire. There's more than one bunker, Ace."

"Ah don't give a shit! I got my own problems."

The incoming fire from the Dinarchy was increasing as more targets popped up the Dinarchy gunners could see. XibEk's Krils were pinned down.

"You got bunkers about ten o'clock in those bushes, atop that small incline." KrutChan was spotting for XibEk. "Looks like three rock bunkers with interlocking fields of fire. Maneuver and bring fire on them. Cheeb's coming!"

KrutCheebel was calmly speaking into XibEk's ear. "We are on line. We'll lay down a base of fire so you can maneuver."

XibEk hand-signaled his Krils to return fire and to speed up the tempo, laying on suppressing fire. Returning fire would make him and them feel better even if they couldn't see the Dinarchy.

Through the din, a future Zabin-Kril added, "We are so fucked."

XibEk agreed wholeheartedly. He inwardly wished he was back in 'Nam, lost in the confusion, where nobody in a deafening firefight knew where anybody else was, or what they were doing.

He could barely make out KrutCheebel and the rest of his Bot-unit moving on the right flank attacking a bunker with smoke and a flamethrower. Blast-flashes and concussion waves breezed past him from the satchel charges thrown to finish the job on a bunker.

Kelel was maneuvering one of KrutCheebel's squads on the left. XibEk watched her Krils from Cheeb's Bot-unit cut down trying to take out a bunker. One guy charged the bunker with grenades, but he was shredded and never made it. Another guy was firing from the hip with a futuristic M-60-like machine gun.

That's gotta be wonna them Saws, I think they call them. XibEk thought.

His EkSeet concurred with his observation.

He wished he had an M-60 right now. That machine gun was some piece of work and he was more familiar with it.

The Kril carrying the Saw lost half his face and then got chewed up; dead before he hit the deck, never getting to the bunker.

His heart sank when XibEk realized there was two more bunkers to his front. XibEk's men were getting ground down.

The battles to eliminate the hardened bunkers seeming to last an hour. He kept yelling at his men to keep throwing ordinance and be careful they did not hit any of the men flanking. These Stels would have been great in 'Nam. Once the shit hit the fan, vocal communication was impossible in the Vietnamese jungle. He thought.

KrutCheebel's voice squawked then from XibEk's Stel. "Move your guys forward and advance on those bastards...draw some of their fire! We're almost on you!" XibEk was cursing KrutCheebel, at the fates, and his men to get moving. But it had to be done; the job of the mobile smaller units.

When they got to their objective, it was a smoldering burning bunker someone else had fried and blasted. They dropped against its side for cover, sweating with fear; gasping in adrenaline-sucking gulps of air.

Unconsciously XibEk was doing a mental head count. Sure as hell, as if reading his thoughts, KrutChan was asking him. "You got enough guys left? How's your ammo count? Can you hold?"

XibEk was angry because they had walked right into the ambush. Shit never changes. He thought. "Ammo's okay. Looks like...we can hold unless a Dinarchy platoon wants this real estate back. I lost three for sure, including the point man...couple guys missing, though. They'll catch up or will wander into you."

"Roger...we got one of yours already. I'll keep him with us for now to help out on our bunker." KrutChan sounded as busy as XibEk was; he was handling more than XibEk's unit. The rest of his half-Nacom-Malkril were holding the line waiting for attacks in their areas. "KrutCheebel get to you yet?"

Dirt, debris, and broken plants announced KrutCheebel's falling arrival. "Tell him I'm here...I'm spreading my guys out to fill your line." Immediately he went behind XibEk's tiny fighting hole.

XibEk grunted, relaying the word to KrutChan. "Blood has arrived like Audie Murphy's tan brother. The situation's well in hand." Whatever the fuck that means? "Not for long."

"My Krils have overrun two bunkers and setting in." Kelel said.

"You done your job." KrutChan hissed. "Let somebody else lead. Get back to my headquarters."

KrutCheebel was hiding behind an orange thin bamboo-looking tree to XibEk's rear. There was more KrutCheebel than tree and XibEk laughed. "Git your ass behind the bunker, Cheeb! You don't wanna get that poor Nihberu plant kilt."

They were trying to make themselves as small as possible waiting for a counter-attack.

"Were the hell did CheChun go?" KrutCheebel was yelling over the noise around them.

XibEk was shrugging and then pointed.

CheChun's men were getting closer to the last bunker. CheChun was yelling furiously to his flamethrower guy to get the fuck down. The guy was hosing the slit in front of him when his throat exploded with a red spray of mangled meat and blood. XibEk saw CheChun curse and looked directly at XibEk. CheChun was tight lipped and shaking his head. XibEk frowned and shrugged; they could do nothing else. No time for wise-ass remarks.

Running and landing on top of the last smoldering bunker, XibEk and KrutCheebel saw CheChun slide down to a backdoor which wouldn't open. Taking it personally, he was cursing and furiously firing his M-1 in frustration, and then he stopped.

They saw Kelel take over CheChun's spot

CheChun ran around the bunker, rolled under the firing aperture, was pulling pins, one after

411

another, and threw a satchel charge into the aperture of the bunker.

KrutCheebel and XibEk vacated their spots atop their bunker, taking cover. Kelel ducked.

Rolling away to his left, away from the gun port, CheChun pulled his helmet down tight and scrunched up into a ball. The blast muffle-chugged inside the bunker and the stone-door in the back of it blew off.

CheChun charged back around the smoldering bunker, screaming a rebel yell, and disappeared inside the bunker. Kelel covered him. The cracking sounds of his M-1 killing whoever was still alive.

XibEk offhandedly said to KrutCheebel. "That redneck sure do like his work…that crazy muthafuka is…."

His words were lost as explosions after explosions around them grew to a crescendo. The teeth-crunching concussions filled their existence; they thought of nothing but surviving.

Within minutes the noise of battle was slowly dropping off to individual Blaster-pulses or rifle shots.

XibEk elbowed KrutCheebel in the ribs. "Ain't much like a movie…is it?" He knew of KrutCheebel's dad's Hollywood background as a director.

"Not when you are in the middle of this shit. Full of concussion blasts and smoke. With screaming and dying people wanting to live." KrutCheebel said. "It is more exciting within the safe perspective of Theatre objectivity."

XibEk was frowning. "That's what I said, nigger. Don't get your superior genes inna uproar."

KrutChan was speaking into his Stel to Kelel, ordering her back to his position. And gave XibEk a Kril hand signal to get his men down. As XibEk dropped to huddle closer to the bunker, out of the corner of his eye he saw a phalanx of ancient helmeted Spartans, with their eyes hidden, marching steadily over and past his Bot-unit; continuing the attack.

They were followed by Romans in armor with short swords, alongside feathered Zulu warriors looking fierce and deadly. The ancient warriors kept their alignment and formations. They were impressive.

The explosives and small-arms fire dwindled away. The sounds of ancient warrior shouts and curses were followed by the screams of Kril and Dinarchy dying in agony. The ancient troops were locked in combat, shields clanging and thumping, and going about their hand to hand mayhem; killing according to their trained customs.

The three attacking Kril-Bot-units from KrutChan's half-Katun-Malkril could relax. It was over for them.

KrutCheebel was crouching beside XibEk, watching the scene with him. "It is kind of nice those ancient troops are part of our 'civilized' team of killers."

"You got that right. They's civilizing the shit outta dem Dinarchy. Ain't no barbarous shit going on here." They grinned at each other.

Both of their EkSeets said. <u>You two Wankers Are Twits.</u>

Out of the billowing smoke of battle, KinikXoc slowly wandered up to them, disheveled, dirty with dust, his right pants leg missing, his shirt

413

shredded and he was staggering. He was in a daze muttering something to himself, when KrutCheebel yanked him to the ground with a thump.

KrutCheebel and XibEk, both began laughing. "Where you been, Chuck? We thought you was wasted. You been hidin'?" XibEk asked. They were clearly relieved that they didn't have to replace him with someone dumber.

"You guys see me? I got a Dinarchy machine gun nest by myself; killed three of them!"

"Easy dude, yer Stel recorded what you did." XibEk turned to KrutCheebel. "Ah guess war's over."

KinikXoc was smiling back at them with relief. "Second time is the charm. That's the second time goddamn artillery searched me out."

"Maybe you said something to make him mad. You need to correct your syntax and not piss those cannon-cocker-gentlemen off." KrutCheebel offered.

XibEk added. "That t'was clearly a butt-fuck round. You almost lost yer panties. That love-round was aimed for yer ass, boy."

KinikXoc decided he disliked knee-grow marines right now. "That's not funny. The artillery blew me off the bridge is all." He dramatically formed his bicep and stared at it. "Shazam! Captain Marvel has rejoined his motley crew of survivors; feels great!"

KrutCheebel snarled to get him re-focused. "Get your daydreaming butt back to help those other guys move the wounded Krils. Who were doing their jobs; instead of malingering like you."

Chastened, KinikXoc headed for one of the bodies, deciding he disliked knee-grow Army officers too.

"Mite hard on the white boy, aincha'? You got another body to use because he survived."

"I do not want his mind up in the clouds thinking he's a hero."

"Yeah, yawl know all about daydreaming, Cheeb." XibEk said.

"Shut up. He could have got us all killed! He should have done his job."

"That's bullshit!" XibEk was irritated. "He fired at the ambush to warn us before Arty nailed him! Ah seen him myself." XibEk sneered. "Jes saying…you're turning into a fucking KrutChan lifer, nigger."

"Cheeb, get over here!" KrutCheebel walked away heading for the assembly area and KrutChan's wrath. He felt he never could please that guy.

KrutChan's angry voice shouted over their Stels. "Quit playing with yourselves and assemble by the bridge. Goddamit, move out…this ain't R&R and I ain't your fucking mother!"

Later on when they set in again, in the distance XibEk and KrutCheebel could see KrutChan was intensely arguing into his Stel. His face was red; the veins were sticking out in his neck. Through the smoke, the tail end of his conversation drifted to both of them. "Know that! I'm collecting my walking wounded as we assemble. ZacNaab, don't…me…my fucking …! You ain't my…anymore!"

When KrutSeet jumped into the conversation, KrutChan was dramatically bowing like a knight at court. "Yes sir…right sir…of course sir…I understand, sir." As he wandered away into the battle-smog, KrutChan was giving the 'bird' to the sky.

415

They were losing Krils to sickness, to indigenous flora and fauna. Liaison screw-ups killing anonymous Krils by intermittent fire, explosion accidents, and other screw ups continued. Allowing Krils to kill, giving them lethal weapons and explosives and sleep deprivation, was an old story in the armies they came from.

Conflicting emotions and morality added a bad combination for rational control. They had had no contact for a week, with plenty of ammo. Water supply was okay, but not in abundance, and the food was non-existent since one Kin ago. KrutChan's Kril Bot-units were in dire straits with no resupply of food or water in their near future.

KrutCheebel slid into the fighting hole next to KrutChan, Kelel and XibEk.

"Are you glad you bumped back in here from the Burseeosil a week ago?" KrutCheebel asked. KrutChan had been evacuated for too many wounds he had taken over the weeks.

"I got dry. Tired of chasing ArnaVals." KrutChan said. "Even Hortim was getting jealous. I came back down to rest my balls with the stinking herd."

"Uh-huh…bet KrutEk assigned you to find some Nihberu pussy on this island continent." XibEk said.

Kelel looked down at them. "I'm going to check the lines. Male bonding gives me hives."

Before she walked away, KrutCheebel yelled. "Don't grope any Zars, Captain!"

"What's your problem?" KrutChan said to XibEk. "You pissed KrutEk'll find your stash of nubile maidens?"

Kelel said as she left. "Only sex you're gonna

get is humping one of those dead female Dinarchy." Kelel pinched her nostrils shut. "No live female with a Vee-Gee will get within ten feet of you boys."

KrutCheebel decided even their gallows-humor was flat and not funny. They were in a food and water bind...all of them. "KrutChan, if you stop talking like an illiterate asshole, I'll give you my report."

KrutChan listened and watched Toobs dropping supplies attached to AkSilk. Then told his Krils they were going hill-humping again, uphill of course.

Three Kin elapsed again with no contact with the Dinarchy. They struggled up a hill, down a hill, up another hill, and sliding down the debris of another hill.

On the fourth Kin they were in desperate straits.

Solar flares and auroras killed communications. Again no food, and no relief. They were on their latest patrol for a couple of hours, sweating with exertion, and out of breath; hoping they were not spotted.

The emergency packs of recycled water were consumed without thinking about where it came from. At their second stop the Dagot bags of excrement honey were passed to the human Krils; the human Krils bags to the Dagots.

Cunacks and Zars swapped theirs.

The Dagots had no problems; seeming to exist on whatever critters were in the air, and what passed for Dagot grass.

CheChun refused his ration; he wasn't hungry he lied. He wouldn't eat anything he or his mother didn't cook. He was not alone. On his first Uayeb, many years ago, an ancient ArnaMal coached him to cannibalize dead Krils. "War has no morals; but feed covertly."

They were out of touch with the Toobs, with the BakToobs, with their Command, and with the Burseeosil.

Normal for a Bot-unit stomach-growling hungry on their way to another objective. Their timelines came down to staring at their feet, taking one step at a time, gravity pulling at their tired limbs.

The Cunack named CaanHop was furiously changing colors on his hide as he turned and met CheChun, arriving from the top of another hill. "Any word from above? I'm a designated runner. KrutChan wants to be resupplied with food and water."

Both their EkSeets were chipping teeth trying to translate them.

"What is a fleeing human asking us?"

"I ain't runnin' away, asshole. What do ya mean, 'us'…ya gotta turd in your pocket?" CheChun was absentmindedly scratching his scrotum, looking around, picking small bits of plant life off his gunsight. Dagots, he hated. Cunacks were second on his list. He hated any species or race not Caucasian. I know that fuckin KrutChan gave me this detail on purpose. I hate that sumbitch too.

"Cunack are many, and speak concurrently. Your tribe is scattered. Speak numerous dialects…as blinking stars in the universe, illogically all at once. No comprehension…you only babble." CaanHop settled on a grey appearance, the Cunack color when around humans, with emerald rhinestones flashing.

What else could he expect from a walking beetle lit up like a flickering Burma-shave sign.

418

CheChun swore under his breath and assumed the Cunack was waiting for information. "Listen, you walking twinkling garbage can. I need to contact our support, the EkTsab Burseeosil, the BakToob, the Toobs, or any other goddamn Oval commander for resupply."

"Cunack speak to Cunack. We have no …connections…EkSeets or the StelaBalaam…. This will change."

"How'cum we understand each other if yer EkSeets haven't a connection?" Balls…I'm talking to a blinking fuckin' lamp. This lashup is getting' stranger than a bull with sucking tits. "Yawl mind not speakin' with forked-tongue?"

"Human Stels translates we Cunack. You Osom photon…not audio wavelength concussion and repeats to your language. We…waste discuss on…such minor speeds."

"Yer as useless as a tick onna dog. I'll tell KrutChan yer a deaf-fucking-mute."

Blinking lights and rolling sheets of greens, blues, and yellows flowed over CaanHop's skin.

CheChun figured the Cunack guy was back in his original thoughts, forgetting the humans. He left to meet KrutChan in the alien Bonsai-looking trees on top of the hill. Our Krils are up a creek made of shit. Great pun, gyrene. I might be a poet.

He struggled to get through the tangled foliage. I'll be lucky if I don't run across no Dinarchy couple fuckin' in the brush. Hmmmm…come ta think of it, I'll shoot the Dinarchy dead and finish up his job on his Oval. He was grinning in fantasy.

When CheChun arrived at the top of the hill the Krils were dividing up the resupply material CaanHop had requested.

Chapter 22


In the Burseeosil Command Center, Jaguar and BalamEk had concluded their morning Kin intelligence update. Meli reported on the pathogen outbreaks and how her medical-scientific teams were handling the spots of infestation. Her teams were analyzing the data and deciding on how much information they gleaned.

Updating KrutEk on matters he should know. BalamEk and Jaguar were contemplating strategic counter-strikes on Nihberu. Their 3-D projections, intermittent and garbled, showed a peaceful planet below; hiding the war that was continuously breaking out hidden from this view.

The Nihberu operation was in flux. It was not on schedule and running into problems. The Dinarchy were subterranean and fortified. Communications were spotty at best, from jamming interference, corona mass ejections, huge electrical storms, or unanticipated Auroras. Dinarchy troops were unexpectantly popping up in areas thought secure.

Four weeks into the invasion everything was going wrong for the Kril Legions. They were making grudgedly slow progress.


"What's that?" Hortim's flickering tactical 3-D panel from her station in the Burseeosil caused her to ask. There was a green-dotted mass of movement roiling like an ocean surf with six foot swells. She did not recognize the breaking up signals. The Dinarchy and the Kril did not display in green. She immediately summoned Jaguar.

421

A couple of Obs passed as they watched the intermittent green movement. "I believe you should be aware of this phenomenon." Jaguar was speaking to BalamEk.

BalamEk approached them. "My other Oval commanders have acquired the curious projections. Meli's science teams have also attained the information."

The wall to wall projections in the command area was abuzz with verbal readouts, interspaced with StelaBalaam projections, analysis, and crackling voices crosschecking incoming information.

Obs slid by, the command module was focused and speaking to each other about data and analyzing readouts. The Burseeosil Germs were scurrying about transferring 3-D projections and cleaning biological debris littering the module. They were unaffected by the anxiety.

Hortim fixed on her projection. "This huge mass is located on a plain between KrutChan's half-Nacom-Malkril and ZacNaab's reserves."

Jaguar agreed with Hortim's assessment. "The phenomenon appears to be a lava flow. Any phenomenon was likely on an exoplanet. It is strange...very strange."

She was looking at Hortim. "What is your opinion?" Jaguar asked.

"I question whether the incident threatens our forces on the ground...or not."

"Could it be a natural process?" BalamEk asked her scientific crew. "As Hortim speaks, we must be concerned for our Legions."

Jaguar's suspicious nature was on full throttle. "I wonder if the Dinarchy is behind this event."

"I agree." Hortim said. "Though, I've never gotten those kinds of readings when I've piloted my TunToob over the battlefield."

An extremely excited ZacNaab was shouting into his Stel. "They are attacking us! Thousands upon thousands of Dinarchy creatures are massing towards our positions! We need assistance!"

BalamEk's calm voice cut in, speaking to those in the command module and to KrutEk, in response to ZacNaab's warning. "I have sent urgent missives to our affected Kril Legions on the ground to report what they are seeing." She was oblivious to ZacNaab's hysteria.

"Do you want my Toobs to investigate? Hortim said.

"No. When you alerted me, I already dispatched reconnaissance Toobs to the area. They should be over the affected zone within Obets."

KrutChan's Stel reported, using EkSeet's British translation. "My Krils are reporting moving masses of single furry indigenous animals in some kind of frenzy. Their intense growling can be felt in our feet, earthquake rumbling and low-toned. We're attempting to get further information. So far, they're moving on a lateral course to us; making no attempt to close with my Bot-unit's lines."

AinAcbal added. "My Krils are not juveniles. They are advising our positions are quiet. They are not seeing any massed Dinarchy. In the far distance the phenomenon behaves as wind blowing through barley fields. There is no perceived threat of any kind."

"I want immediate access to what you gain in intelligence, BalamEk." Sounding irritated, KrutEk broke in. "ZacNaab, contain your emotions! We need information. We do not need fables; not everything on this planet is Dinarchy perpetrated."

Meli's second in command added. "KrutEk may be correct." She said to BalamEk. "However our past invasion data experiences indicate some of the indigenous creatures on a planet are as dangerous, in some circumstances, as the Dinarchy."

BalamEk's voice was aloof. "KrutEk is incorrect in his assessment?"

Meli answered the query. "My aide is simply pointing out KrutEk may be overly optimistic in his assessment of ZacNaab. If creatures are approaching his lines, ZacNaab has a unique perspective and is prudent to be concerned."

Hortim spoke to Jaguar. "That possible threat is more like waves crashing on a shore. KrutChan says they are individuals. I wonder..."

"KrutChan's on-site calmer perspective is better than ours." BalamEk said. "In your earth measurements, we see the entire battle line hundreds of kilometers in length. His vision sees the terrain in meters. I trust uninflammatory reports."

"What about ZacNaab's concerns?" Meli asked. "Are we to ignore him?"

BalamEk was thinking for a moment. "Meli, assemble an ArnaMal science team with medical personnel and yourself, down to ZacNaab's command post and report to me directly. I need facts on what we are dealing with...."

Jaguar added. "Protect your Ovals down there. Leave immediately if they or you sense Dinarchy danger." Jaguar then instructed Meli. "Take one of my Intelligence Toobs. They are faster and more maneuverable than the standard Toob."

Meli nodded and left the command center.

Jaguar grabbed Hortim's arm to stop her from leaving. "Stay aboard the Burseeosil EkTsab. My Oval pilots are more qualified with those particular Toobs."

BalamEk agreed. "Hortim is needed here at my side to coordinate the BakToob units."

Jaguar was smiling. "Our Oval Hortim has a Kril tendency of letting her blood-lust for battle forget her responsibilities."

Hortim remained silent, hiding her displeasure. Hortim always plotted ways to get back into her cockpit.

"KrutEk, are you receiving the reconnaissance Toobs perspectives?" BalamEk said into her Stel after a few minutes.

"They help me not. Not when your demon gods grow silent at inopportune times. My experience is of attacks over terrain I can see. Your Toobs-visions are from a god's perspective. I cannot deduce any tactics or what the purpose or intent the green masses have towards my Kril Legions. I will move my present headquarters to ZacNaab's until the threat is determined to be a legitimate attack."

"It is too early to determine if there is hostile intent involved here." BalamEk said.

"Enemy intent is hidden, as they are when your signals are interrupted. If you cannot repair your signals, resort to our old ways of communications." KrutEk laughed without humor. "You realize it will be my Krils meeting TOTL who will determine their intent. If we wait too long?"

"You are correct." BalamEk concurred. "Our scientific teams are presently analyzing the situation and our communication problems. Your Cunacks not do need our Stels."

"Your Cunack aliens do not respond well to my human queries. They are not forthcoming."

"Not do mislay your patience. I will contact you as soon as we reach any conclusions. Order your Kril Legions to 'observe' only and not do take any hostile action unless a perceived threat clearly becomes hostile."

"I have already done so." Speaking drily, KrutEk responded. "Inaction is 'not' one of my Kril Legions preferred battle precepts. At the present I am monitoring ZacNaab's medical and scientific teams who are supporting him. I ordered them to directly report to me; not to ZacNaab." Diplomatically he added. "Of course, after they report to you first."

"You are honored by me..." BalamEk used diplomacy in her answer, though her irritation came through to him.

Trying to help, Jaguar spoke to BalamEk. "I suggest we bring Meli's scientists still aboard the Burseeosil to observe this phenomenon on special monitors to assess the exobiology ramifications. I should be down there obtaining intelligence."

"Information will be obtained by Meli as I commanded her. You Jaguar, are not an Earth-Oval like Hortim. You will stay with me where you can be more of use." BalamEk said."

Hortim was smiling at Jaguar's put down by BalamEk. How does it feel to be left out...honey?

Speaking through her Stel to Meli and her scientific teams, BalamEk continued. "I not do wish any Ovals on Nihberu should this event turn hostile. Meli, you and your Ovals leave as hostile intent is confirmed. Not do tarry for an Ob."

Not requiring an acknowledgement to her order, BalamEk was concentrating on the overall situation. She expected compliance instantly to her commands, not inappropriate deference as KrutEk demanded. Human male commanders were full of vainful pride. As an Oval female she did not suffer from either malady.

CheChun and the other commanders in KrutChan's half-Katun-Malkril, along with XibEk and KrutCheebel, were huddled in KrutChan's claustrophobic headquarters. A musty cave, dug by KrutChan's Zar, CoCum.

KrutChan's Zabin-Krils were observing a flexible 3-D projection sheet he had thrown against the pebble-strewn dirt wall. Sticking in place between orange roots cut by the Zar when excavating. Styx's nose was on the map, leaving a slime trail.

When KrutChan pulled Styx's face out of the way, his intense blue eyes went from the map to KrutChan, and then back to the map. His thoughts were his own, though he seemed to sense the problem.

KrutChan glanced at XibEk concerning Styx's behavior. XibEk shrugged his shoulders. "That little dude's from KanBalaam. He doesn't know nuthin. Maybe he screws maps where he's from…?"

"Anybody else got any…bright ideas…on what the fuck these furry critters are outside? Anybody ever see them before, or knows what their intentions are?" KrutChan asked. He noticed Styx was affectionately nuzzling his hand.

"I'm going to the bulge nearest our lines." Kelel waved before KrutChan could stop her.

After she left, there was silence.

"Come on you Zars, Cunacks, and Dagots! This is your universe, for Christ's sake. Are they dangerous or not?"

AhBacao, a Dagot was translated first. "These creatures not sentient." The EkSeets interpreted. "Other planets...not ours...many species terminal...similar but not known...loving fond creatures."

Blowing up his cheeks, KrutChan exhaled loudly in exasperation. "That's clear...if we were Dagots."

"Loving is my idea of how ta live." XibEk grinned.

"Shut up if you can't help, Xib!" KrutChan answered. "Stick to the issue at hand, these rampaging creatures!"

CoCum, the Zar said. "Hive none experience with them in tunnels."

The Cunack, ButsChan, was furiously flashing multicolored, his thoughts on the creatures, talking to his other Cunacks on the Burseeosil at the same time. He did not share his ideas with the Zabin-Krils.

The consensus of their EkSeets? ButsChan or the other Cunacks know anything helpful.

The Dagot, AhBacao, as the oldest species in the Oval universe added. "We know...purpose is wondering...fierce competition...overly aggressive, angry...killing many."

XibEk asked the slug-like Dagot. "Dey pissed...angry at the Krils?"

"Ah seen rampaging Wart-hogs. Dose hogs were crazier than dem little fuzz-bastards; I tink lemmings." CheChun said.

"I think they group like African Army ants." XibEk said and held up his middle finger in CheChun's face to quell any racist comment. "Ya don't wanta be in an Army Ant column's line of march."

428

KrutChan sneezed twice, wiping his nose on his sleeve. "You guys aren't much help. These dudes are larger than feral pigs or ants. Concentrate and analyze. KrutEk and BalamEk are both hounding me for answers."

The Stels of the Krils were cracking a constant rain of overlapping confusion and suggestions from the Krils nearest the phenomenon.

KrutChan walked to the entrance of the cave and waved for the others to follow him. "Let's get the hell outta here. Sound off if you get any clues as to who these goddam creatures are..."

Outside, two Germs, assisting the Medical ArnaMals, were carrying an invisible hover-bed between them. On it was a blood-encrusted Kril corpse. The medical personnel were on their way to an injured-TOTL station. Struggling with the body, trying to keep their feet on the uneven ground.

The Germs had not completely covered the corpse. They were too busy compensating for the terrain as they carried their shifting load. Picking up pieces of body parts, another Germ was following them, struggling to routinely keep inventory of the human remains as the foursome traveled in their macabre duty.

The Ovals demanded the mass from TOTLed Krils be gathered to feed their Seech Field. Picking up debris, there was a finger here, a bright white bowl from the corpse's head with chunks of brain-matter in it. There was a ninety-degree fulcrum, either of a foot or elbow joint, and a quarter of the face of the corpse. Projecting from what was left of the head, was one eyeball hanging by a thin tendril of gristle out of the grotesquely agape socket.

Yelling into his Stel for his Chief Medical ArnaMal; KrutChan was angry. "Get to my headquarters immediately!"

KrutCheebel yelled, though he was right next to XibEk. "I have to go!"

"What's up, blood?" XibEk asked.

"My EkSeet told me this casualty is one of mine! I am going back to my Bot-unit."

"KrutChan's gonna be pissed you left."

"That is his problem. The responsibility for my Bot-unit is mine, not his." KrutCheebel was scrambling away.

"Dat lifer's gonna have yer ass, dude!" XibEk yelled after him.

"KrutChan owns all dese Krils, includin' yer's, Cheeb!" CheChun shouted.

The Chief medical ArnaMal arrived. He was standing in front of KrutChan, who was snarling. "I told you before and I'll tell you one more fucking time Doctor. Set up your collection point at least two hundred yards 'behind' my goddamn headquarters!"

Calmly, the ArnaMal answered. "As I have told 'you' before, I need to be where my ArnaMals can get to my position quickly without having to travel a kilometer. As Senior Medical officer in this Malkril, I command the Germs and where my ArnaMal medical teams will situate. I decide where my collection point is to be..."

KrutChan was coming to a slow boil. "I'm tired of your people taking wasted Krils past us in a Memorial of the Oval Protectors procession. I don't want a parade of dead or dying Kril marched past my surviving Krils!"

"Your Krils have seen it all before. My ArnaMals and Germs have the grisly work!"

430

"My Krils memories are bad enough without you making them keep tally!" KrutChan's voice was rising.

The headquarters group was silently observing the confrontation. They began laying bets on who would win.

The ArnaMal doctor was saying. "My ArnaMals and Germs are striving, as much as your Krils, to keep their minds right. My ArnaMals and Germs more so."

Lowering his voice ominously, KrutChan said. "Just do what I said. Move your goddamn collection area! Now!"

Everyone in the headquarters was listening and watching them both. Krils, Zars, Cunacks, and Dagots were taking sides as to who was right. In this area of the line, this was the only show. The green furry horde was forgotten. Tense minutes of staring obstinacy between the doctor and KrutChan accumulated.

The Chief ArnaMal doctor stood his ground. "Maybe your headquarters should move forward two hundred yards."

XibEk saw KrutChan's face harden; he had to step in. He got between them and put his hands on the ArnaMal doctor's chest before KrutChan exploded. "Why should you be baiting each other? Never tell da man you don't like his orders, doc. There's a time for principal and a time fer retreatin'. Lifers don't like to be challenged."

More minutes of tension grew while the ArnaMal doctor and KrutChan mulled what they were going to do. The onlookers were either grinning or looking away in embarrassment.

CheChun saw XibEk's head-nod cue him and CheChun approached KrutChan. CheChun was grabbing KrutChan's elbow, turning him away. "I wanna get my brain-housing back on that furry mob. I gotta ton-a-questions about dese critters."

Blowing out a sigh, the doctor ArnaMal hung his head after he heard CheChun. He addressed XibEk. "You are correct and honored by me. I am here to save Krils. I cannot help the Krils who are on their way to the Cosmic Egg." He was staring at KrutChan. "Or leaders who…" His words trailed off indicating what he thought of KrutChan.

"Way ta think, doc…way ta go…life over death, huh man?" XibEk said. "We all gotta do our job…right?"

XibEk was pulling the doctor with him. "Hey, doc…me and my Bot-unit guys will help you move back. Ya piss on angry gods, you be fuked'."

As XibEk and his picked 'volunteers' walked away with the fuming Chief medical ArnaMal, the headquarters tension was evaporating.

CheChun watched KrutChan's eyes throwing hate into the expanding space towards the medical people.

"Hell of a thing, ain't it? Ever-body wants ta be da general." CheChun said.

Nodding, but keeping his stare on the doctor's back, KrutChan ignored CheChun. He was someplace else in his mind. "Doctors see only the meat, not the individuals…" He mumbled.

"Bullshit." CheChun was saying. "Ah nev'vah met a corpsman ah didn't like. Dey put up with a lot of crap. They gotta lousy job. Ya gotta 'mit it…dose ArnaMals and Germs gotta lotta balls."

Remembering he also had never met a cowardly corpsman, KrutChan came back from where his mind had gone to, and cocked an eyebrow at

432

CheChun. "When're you going to stop talking like a redneck-fuck and talk English so I can understand you."

"Now my feelings are hurt." CheChun was grinning. In his best EkSeet manner he continued. "I am crestfallen. I am lower than Leeseanna mud in the Mississippi Delta."

KrutChan grabbed CheChun's shoulder affectionately. "You ain't got no feelings. Get back to your Bot-unit like Cheeb. And..."

CheChun bent his head towards KrutChan in a conspirator's manner. "And what?"

"Thanks. Even commanders get on the rag sometimes."

"Bot-unit commanders too, me darlin'. May the devil find you after ya enter heaven...or some kinda shit...?"

KrutChan groaned. "Jesus, don't go Irish-Southern-boy on me and make me puke. I wasn't mad at the doc..." KrutChan was staring off into the distance. "I just..."

"I know...I know, Hoss."

Abruptly pushing, KrutChan shoved CheChun away. "Take off your apron, mother. Grubby as you are...you're giving the Ovals a bad name."

"As Cheeb says...we're good to go...ya mud-marine?"

Laughing, KrutChan was his old self again. "While you're EkSeet is working on your syntax...have him teach you some respect for your leader."

CheChun said. "Yes...oh mighty exalted half-Nacom-Malkril commander. Ah'm tighter den da bum hole onna whore on Saturday night."

In answer KrutChan spit as he walked away shaking his head. Minor irritations always spurred on by the major glitches in wartime.

KrutCheebel was agitated when he wandered back to his Kril-Bot-unit's positions. <u>KrutChan wants answers. I'll give him answers. Not one more of my Krils are going to die because of these mysterious furry creatures.</u> His exo-critter-hatred building in him.

Reminding himself of being swallowed on KanBalaam by that plant, KrutCheebel's mood was darkening. He had developed an overwhelming fear and hate for all indigenous life on other planets. He felt he was not the only one.

Coming back to his Bot-unit area he was looking out over the low terrain in front of him. The rapidly approaching green squiggling mass of creatures caused him to ask the closest Gurkha Kril leader to him. "Your people have orders from above to only observe those creatures. What the hell happened to my dead Kril?"

TzenalAh, the Gurkha, spoke softly. "The creatures are in frenzy."

"My EkSeet says you state the obvious." Drily, KrutCheebel said. "I can see that. What riled them up against our man?"

Close up, the sight of thousands upon thousands of furry, snarling creatures big as black bear yearlings was frightening. They seemed to fill the ground to the horizon, emitting deep low-pitched growling. They were snarling, spitting, and tearing each other into pieces. They were out of control. Jumping on each other pell-mell; they were vicious and frenzied. Hinting no indications of their lethal intentions towards the Kril forces.

Off in the distance, on another high ground, the Dinarchy had paused with inaction; as puzzled as the Krils were at the sight of the writhing animals.

The growling creatures had huge forward sabre fangs in a mouthful of large teeth and were shredding each other.

"Our dead Kril saw an albino furry creature mewing for assistance." TzenalAh was saying. "He picked it up removing it from the group." The slow speaking Gurkha seemed more curious than anxious.

"And did he succeed?" KrutCheebel sneered. "Come on...what happened?"

"You notice there are many white creatures within the mob." Rubbing his forehead in memory, TzenalAh continued. "His Kril friends shouted for our man to return to us. He did not listen. When he picked up the frightened white creature, he was instantly enveloped by many of the green-colored creatures, slashing furiously at him. One of our Mayan Krils said it resembled an attack he had witnessed from piranha fish in one of his rivers on earth."

His stomach was roiling, KrutCheebel's personal memory of being eaten alive was still with him. "Stick to the facts."

Even though TzenalAh spoke English, it took a moment for the Gurkha's EkSeet to translate. "We killed the ones attacking towards us and clubbed the others off of our fallen Kril. We were tearing the creatures off in handfuls. They are of lightweight mass. We threw the snarling animals back into their mob. They ignored us after they rejoined their pack. Our defensive actions never bothered them. They are unstoppable."

The frenzied green furry creatures were getting uncomfortably close to the perimeter of

435

KrutCheebel's Kril Bot-unit. A few were spilling into his lines biting some of his Krils, and were broken in two.

"Unstoppable? We will see about that." KrutCheebel yelled at three Krils with flamethrowers to come to him. They were standing around observing the creatures.

"What are your intentions?" TzenalAh asked.

Staring long and hard at TzenalAh, KrutCheebel softly said. "No more of my men are going to be chewed and eaten by those bastards."

"The Burseeosil and our KrilChan are ordering us to remain immobile and observe what is occurring."

Seeing KrutCheebel's expression, the Gurkha shrugged. "I am reminding you of our orders. We have only ten Krils with bites on them who went to the aide of our man. You see the current Krils who are being bitten, throwing the creatures back into the mob. Our man made a fatal mistake choosing one when going to its assistance."

"The last blunder he will ever make." KrutCheebel turned to the flamethrower Krils. "Spread out parallel to our line. After I fire; fry each and every one of those bastards into bacon who even hints at approaching our lines." He used his Kril hand signals to bring up the ancient Greek Fire troops to get on line with his flamethrowers. His area of responsibility would be covered.

"I believe your error compounds," TzenalAh said, trying to calm his leader. "Our orders are..."

KrutCheebel ordered his mortars, machine guns, and rockets to prepare to support the flamethrowers. "I didn't ask for your opinion!" KrutCheebel's lip curled in anger. "This is my call."

He yelled into his Stel. "When I open fire, aim at the closest mass of these creatures to you. Lay down grazing fire out for one hundred yards. Do not allow one of them to penetrate our lines. Kill every one you have in your sights. And be careful you don't hit any of our own Krils."

The transmissions in his Stel were phasing in and out, overlapping with BalamEk, KrutEk, Jaguar, and KrutChan as the voices cut each other out. The substance of the garbled questions was: "What the hell are you doing!"

In the distance KrutCheebel could hear and see the Dinarchy's intermittent Greek-fire and blasters were taking the same action as his against the creatures. None of the Dinarchy weapons were directed at the Kril lines.

Minutes later, when the green bearlike creatures were tumbling towards the Krils, KrutCheebel started firing; using his Krils weapons on the creatures. Followed immediately by the rest of his Kril Bot-unit and 'whoosh' from more from his Greek-fire and flamethrowers. Down the line other Greek fire weapons and flamethrowers were belching following KrutCheebel's orders.

The green creatures were exploding, twisting, growling in agony, and turning into screeching furry-fire-balls. The creature-survivors were flipping away from the Krils, landing amongst themselves, continuing their ferocious melee. The Kril weapons cleared a dark-green charcoal-bulge around the Kril positions. The high oxygen content helped the conflagration. KrutCheebel grabbed a flamethrower that was not being used and ignited another area of the creatures.

437

Abruptly coming from nowhere, KrutChan ran up to KrutCheebel grabbing the flamethrower nozzle. Yelling into his Stel. "Cease fire...cease fire!" He turned off the valve and lowered it, shouting at KrutCheebel. "What the fuck are you doing? I commanded Cease Fire!"

KrutCheebel was as loud. "I am doing my job! Protecting my Bot-unit, goddammit!"

"You don't need your EkSeet to know what cease fire means!"

"I didn't hear you!" KrutCheebel was screaming back at him. He calmed down when he realized it was KrutChan and quietly repeated. "I did not hear you..."

"Don't give me that crap! You hear what you want to hear!"

"It must be your wonderful on-the-job training, Krut!"

Calming down, KrutChan grabbed KrutCheebel's shoulders to get him focused on what KrutChan was saying. "BalamEk's science people say the creatures are mating!"

KrutChan ripped the flame-thrower off KrutCheebel and threw it into a fighting hole.

"What?" KrutCheebel was stunned. "What are you saying?"

"They're in rut; in their annual season. The males are finding females. Fighting among the males to be the one who mates with her."

KrutCheebel's eyes were flat and uncaring. "Tough...they killed one of my men. They wounded my other Krils. They sure as hell were not making love."

In the background, KrutCheebel could hear KrutEk yelling over the Stels for all the Krils involved to stop the killing! Ordering them to only

438

use Greek fire to keep the creatures out of their perimeters.

BalamEk was screaming just as loud the creatures were harmless.

About as harmless as a stampede. KrutCheebel told himself.

In the distance the Dinarchy explosions and ordinance was raising in tempo and ferocity.

A Hindu Kril's mind was lost in the sight, his Stel paraphrasing the Bhagavad Gita, mumbling incoherently, "We have become the destroyers of an alien race. We are doomed."

Eventually, all the firing petered out, then stopped and a deadly silence overwhelmed the Krils. The creatures were still in frenzy, avoiding the Kril lines, ignoring what had happened to their dead or dying brethren. The stench of mangled burnt flesh hung curtain-like, green-smoking over the scene.

A few Krils were puking from the smell.

The Zars began digging holes, the Cunacks flashing in all their colors were clearly disturbed, and the Dagots retreated into their hardening shells with indifference.

KrutChan pulled KrutCheebel away from the line, to be alone.

KrutCheebel was in a daze; in another place. The unbearable pain in his eyes as he stared at KrutChan said it all. "None of my people are going to be suffocated. My man was eaten alive...ripped apart like a piece of liver. I paid those bastards back."

"Combat vengeance isn't compatible." KrutChan said. Shut the fuck up, EkSeet.

KrutCheebel snarled. "I have seen you do worse. In my time we called it payback."

439

He grabbed the back of KrutCheebel's neck. "I've never killed for personal reasons."

"Good for you. I am not as well-trained as you."

KrutChan gave up, remembering his argument with the doctor. "Disappear...vanish...and pull your head together." KrutChan took KrutCheebel's Stel off him. "I'll send XibEk for you when I want you back. I don't care where you go...just get out of my sight."

Turning to leave, KrutCheebel picked up his gear.

KrutChan's shoulders slumped. "I really expected more of you."

"Why, because I am so elitist...better than a common Kril?"

"No...you used to have a soul." KrutChan shook his head. "I liked you better the way you were..."

"I wish I could have killed more of the bastards." KrutCheebel was fighting his conscience in his mind. "They asked for it."

"Just go hide. I got a lot of explaining to do before this shitstorm's over."

Their Kril Stels crackled. "The old 'black critter killer' is piling up indigenous corpses again." Some anonymous Kril was saying.

KrutChan was shaking his head as his Stel squawked. "KrutChan, report to me immediately!" KrutSeet ordered, followed by BalamEk and KrutEk demanding explanations.

In the Burseeosil afterward BalamEk, Jaguar, and KrutEk were meeting. Hortim was back at her 3-D viewer half-listening.

440

"Your Krils and the Dinarchy troops were obliterating an entire species!" BalamEk shouted at KrutEk. "Killing them because they were fulfilling their biological imperative to reproduce! That is abhorrent to the Ovals!"

"Do not Oracle-preach to me!" KrutEk yelled back. "The Dinarchy were killing more than we killed! It was a natural mistake for survival. That irresponsible Kril is under the command of that barbarian. He is not used to discipline." KrutEk finally said. "It is done and is unchangeable."

"You condone it?"

"I command. I do not condone."

"You are extremely benevolent when one of your Krils is unjust." BalamEk said. "The Cosmic Egg teaches us better morality."

"The gods have nothing to do with this!" KrutEk shouted. "War tragedies have occurred long before I joined the Roman Legion in my old universe. My EkSeet advises me my descendants followed my example. Battle sometimes cannot separate the kernel wheat from the husk."

BalamEk sneered. "How convenient for you. What happened is morally repugnant to Ovals."

"How conveniently you, the OvalChanHalach, and the Confederation Empire attempt to control timelines." KrutEk was smiling.

Hortim whispered Shakespeare's line. "Cry 'Havoc,' and let slip the dogs of war..." It didn't justify the act. She thought. Oval morality or any moral code does not function during war-killing. There are too many probabilities and philosophies in play.

"This discussion is over." KrutEk said. "That barbarian will be taught to control his Kril Bot-unit commander by making an example of the culprit. Not

441

by having him meet TOTL but by berating him in front of his half-Nacom-Malkril as a stupid Kril. You Ovals should take away wealth from the irresponsible Kril's Oval. There is a war at hand and is not over."

Later, speaking to his Krils, KrutChan had taken full responsibility for his subordinate's actions. KrutCheebel knew KrutChan did what he was forced to do; maintain discipline. Inside KrutCheebel's conscience was hurting him.

Afterward, his other Krils in his Bot-unit, had come to KrutCheebel in commiseration to tell him they supported his actions. Hated KrutChan was covering his own butt. Their words did little to comfort him. KrutCheebel swore to himself he had been right; no matter what anyone else thought. He would never have any regrets.

Before rejoining his Bot-unit, his EkSeet gave him a visual missive from ZocKuk. She was fined, along with her Oval for KrutCheebel's errors. ZocKuk was disappointed he had changed into a Kril she would find hard to forgive.

Kins later the Krils were watching the ravine-pass between the hills exploding with Dinarchy troops on the attack. The lines were drawn. The feinting, the patrols, the searching for each other's forces were in their THEN. The battle for Nihberu was now joined. Victory or defeat hung in the balance. Both forces collided-crunched in an offensive death-grip.

Between a hill and a river bank containing liquid purple pond scum, KrutCheebel's Zar, CaanHop, was wounded. His Kril Bot-unit had cleaned out a pocket of resistance. In the slow-motion aftermath KrutCheebel was counting noses and could not find CaanHop. He had seen his Zar absorb two or three projectiles that entered CaanHop's translucent gelatin body.

Eventually finding him, KrutCheebel recognized the yellow headband in his hair, turned his Zar over onto his back expecting the worse. The wounds did not look too bad, but the Zar was out of it. "You and a few others are taking a ride to the sky. Hang in there, CaanHop. I called for an E-vac Toob."

CaanHop was wheezing and choking, unlike the stoic Zars KrutCheebel was used to watching. "Must find...need my cocoon...need to dig...survive."

In their shared past his Zar friend had strived mightily to explain to KrutCheebel where his asteroid-pummeled world was, how he had grown into his cocoon, and what his dreams of life after the Uayebs meant to him.

Even with the help of his Stel and his EkSeet, KrutCheebel couldn't understand everything of the alien's philosophy or lifestyle. He did relate to CaanHop's individual loneliness when not among his own race.

"The ArnaMal medic says you will be fine. Just relax and follow his orders. You were not hit by an asteroid, my friend."

"You are...Zars...few friends...Me alone...stay until me go...no cocoon with..."

KrutCheebel knew that was as close a request for help from a friend that he would ever get from a Zar. He was honored. Sad he could not help his Zar more. "Your cocoon will be with you soon."

CaanHop was evacuated. As KrutCheebel watched the Toob shriek off, the sound sent shivers up his spine. KrutCheebel forgot about CaanHop, gathered his troops and continued his mission.

The next Kin, in their shared fighting hole, XibEk was with him when a huge sparkle of strobe-light lit up the dark sky. It was noticeable to all in KrutCheebel's Bot-unit. Most of them did not see it or ignored it.

XibEk surmised. "Burseeosil EkTsab just farted."

KrutCheebel had a feeling of dread. "Too easy an answer. That looked too large to be a Toob hitting the atmosphere."

"The Burseeosil creatures gotta let off gas too; for relief." XibEk laughed at his own wit.

"I guess so..." KrutCheebel mumbled. "...you may be right."

The next afternoon CheChun approached KrutCheebel, who was enjoying the quiet on the front; away from his Bot-unit.

The sky was drizzling again.

"KrutChan said I should talk to you." CheChun said, without a preamble.

"He send you here to train me in indigenous customs or interracial relationships?" Seeing CheChun did not react, KrutCheebel had an ominous feeling. "Why are you here...what for...is KrutChan's Stel broke?"

"There ain't no good way to say this." CheChun's face was granite. "CaanHop is dead."

"What? What are you saying?" KrutCheebel was confused, explaining to CheChun. "He was fine when doc airlifted him. What the hell happened up there? You know I liked that Zar. The StelaBalaam's numbers not add again?"

444

"He was one of the better Zars." CheChun agreed.

"You mean he was not a colored one?" KrutCheebel snapped.

"Don't get your hackles up with me! Ah didn't kill em!" CheChun glanced skyward towards the Burseeosil. "I'm saying he bought the farm."

KrutCheebel's bitterness showed. "You must be enjoying this, Chun."

CheChun was deadly somber. "I got assigned to this shitty detail cause KrutChan's too busy and he hates my guts. He figured you would accept it better if you take your anger out on me."

Gritting his teeth, KrutCheebel nodded.

His untranslated words sounded like his EkSeet was silent. "My EkSeet said the Ovals got CaanHop back to the Burseeosil and into their medical unit...alive and well." KrutCheebel said.

"You don't wanna know."

"What the fuck happened, goddammit!"

"I don't know for sure!" Looking at the sky, CheChun said. "My EkSeet told me Zars can't stand being alone." CheChun coughed. "The Ovals announced they think he detonated some ordinance he had on him...a kind of Zar suicidal hari-kari, I figure."

KrutCheebel's veins were bulging in his neck. "Bullshit! CaanHop was better than that!" His mind was reeling. Maybe I should have gone with him. He asked me not to leave him alone. He knew KrutChan would never have allowed him to leave; needing KrutCheebel to lead his Bot-unit.

"Don't start blaming yourself." CheChun was trying to help. "I don't know shit about Zars, but wounded guys get plenty of crazy thoughts after they're hit. Who knows?"

445

"I know! Ovals are trying to explain away their own incompetence is what happened. I been a patient in the Burseeosil hospital section."

"You ain't alone, buddy-boy."

"I 'know' the Germs always disarm the wounded before they bring them to the medical unit. The Ovals Seeched CaanHop, because of a StelaBalaam Quantum fart. How do you know what happened, Chun?"

"I already said I don't." CheChun was staring at the mud, thinking about his own WWI guy the StelaBalaam screwed. "You may be right."

"Being correct does not bring CaanHop back."

CheChun pushed his rifle butt softly against KrutCheebel's shoulder. "There's another Kril Legion theory; scuttlebutt says someone else committed sabotage."

KrutCheebel showed a glimmer of hope. "Who?"

"The Germs have a hate-hard-on for the Ovals. The Ovals may want to keep a Germ revolt quiet. CaanHop wasn't the only guy to die. Other Krils, ArnaMals, Germs, and Ovals bought the farm too."

KrutCheebel said bitterly. "Doesn't change the fact it happened while under the Oval's protection."

"Before you go off half-cocked; it 'is' possible it was an accident...and not planned. KrutChan says BalamEk has counterintelligence and maintenance crews investigating the incident. But you know they are going to be brief; they got a war on their hands."

"You could be right." KrutCheebel snarled. "I am not the brightest Kril on this Uayeb."

446

"If any of us were bright; we wouldn't be here." CheChun said.

"In my day it was said, 'shit happens'. I never understood how accurate that phrase was before now."

CheChun locked eyes with KrutCheebel. "I don't like this post-mortem-crap no more than you like me telling you...okay?" He thought back to Tarawa and Iwo Jima and the survivors having to be the ones to carry the bad news. "I never did. We have to just saddle up and move on."

KrutCheebel grabbed CheChun's dungaree shirt fiercely and glared at him. "If I thought both of you were fucking me over to get a rise out of me...I'd kill you and KrutChan."

CheChun relaxed his body, remaining motionless and stared hard at KrutCheebel. "If we were jacking you around; I might let you try to frost me."

Releasing his shirt, KrutCheebel stared into the rain.

Calmly, CheChun asked. "Do you know how many dead guys I got in my brain-housing group? How many gyrenes I would have gladly changed places with in my past?" CheChun looked very old.

KrutCheebel was silently mulling CheChun's words over in his mind. "You know as well as I do this is fucking bureaucratic bullshit over a dead alien Kril. A Zar who I considered my friend." He wore a grim smile. "I think I like you better when you don't speaka the fuckin King's English."

"Yawl use-in barracks cuss-words now?" CheChun asked. "Dickens said. 'It was the best of times; it was the worst of times.'"

"I am astounded you know Dickens?" KrutCheebel said. "As if you knew the difference between an ice cream flavor and Dickens."

"Even a poor southern cracker can read...you fukin genius."

"I meant you reading classic English literature."

Laughing, CheChun said. "I be go to hell. The professor is jes like the resta us poor mugs, dragging his dick in the dirt, over a casualty."

"I may be cleaner in mind than some of you, but I'm as grungy as the rest of the Krils." He picked up his M-16 and sling-shouldered it. "I have to go hunt up the rest of my Kril Bot-unit and break the news to them."

"If XibEk was here he would tell ya 'Ya piss on angry gods, you be fuked, dude'. Some things ya can't change. Hang in there..."

"I think XibEk would cut your balls off if he heard you plagiarizing him."

"Hit your Krils right between the eyes with CaanHop's death. That's da best way." CheChun advised him.

"Your sensitivity has underwhelmed me."

They both laughed at the old Bot-unit joke.

They walked off together, for a moment in their timelines, as war brothers.

Chapter 23

The Nihberu invasion had gone wrong from the start. KrutEk created problems, not consulting BalamEk, adjusting his tactics, without consulting her. BalamEk ignored KrutEk making her own Command corrections. They both stayed out of each other's way; operations did not go smoothly.

As their timelines evolved together; their dual invasion strategies conflicted. Their individual animosities infected the invasion with problems that could have been avoided with a more unified command structure.

Now they contended with crippled Burseeosil EkTsab.

Hortim's BakToob and her sister BakToobs were surrounding the disabled biological Burseeosil EkTsab, at a safe distance should more detonations occur.

Other smaller Toobs, Malkril Kril compartments, and maintenance bio-creatures were ordered by BalamEk to remain apart from the Burseeosil. The standing-off beasts had been notified there was an internal problem developing in the basic configuration of the Burseeosil.

BalamEk was speaking through her Stel to KrutEk on Nihberu. "I received a missive from Xmucane. Why are you resisting the command from your Queen? There was an armed insurrection on Arna by unknown culprits. She orders your return to Arna immediately. She needs your protection."

"Where was your security forces on EkTsab...asleep?" He growled, changing the subject to the Burseeosil.

"The protective forces in your Legions on Arna not were prepared." She countered. "Security is reactive, never proactive. Xmucane requires your return to Arna as soon as possible."

Visually stifling his anger, KrutEk stated. "I cannot leave at present. Your creature is broken. I have contacted the Palace Guards on Tikumyax." KrutEk said. "Ordering them to seal off Xmucane's pyramids, public and private. My concern is the Tribunal buildings."

"I understand." BalamEk said.

He was swinging his white baton. "I want the Tribunal isolated; the Germ's insurrection may be orchestrated by them." KrutEk raged. "Speak my words correctly, EkSeet!"

"I not do need EkSeets for my translation!" BalamEk shouted. "He has correctly spoken. Xmucane commands you to leave Nihberu to aid in her defense."

"Changing command will weaken our forces during an unnecessary transition to another commander."

BalamEk sweetened the tone of her voice with an ironic twist. "I fail to understand your hesitation."

KrutEk was gloating. "Your Burseeosil is in a state of disorder! How am I to leave if that flying beast of yours cannot be used by anyone?"

BalamEk held her temper. "Calm down KrilChan, not do infect your Legion forces with your petulance."

"What is your present situation, Oval? Are your forces also in disarray; broken like your sky-chariot?"

"Upon a two Kin investigation we have determined the medical unit was sabotaged. I not do want to further elaborate in this Stel conversation."

450

"Your loss of one unit should not shut down your entire operations."

"Of course not! When Xmucane's missive arrived, we immediately began sustaining other incidents I not will discuss with you without secure communications. Prepare to use your own Cunacks."

There was a pause before KrutEk answered, as if he was struggling to remain civil to her. "I have no trust in these Cunacks. However, you are acting correctly." And then he added. "Preparations for the change in Kril Legion command has begun."

"I entreat the Cosmic Egg you will be forthcoming with your progress."

Pausing again KrutEk continued. "I wish to leave the Dinarchy confused as to when or if I depart. Depending upon...your Flying-Colossus of Rhodes."

"You are truly a great...and intelligent...Kril leader." BalamEk said drily. "I honor your magnificent patience under these circumstances."

"I am not happy going back to Arna in the apex of these battles. How do you know the Dinarchy have not infiltrated your Flying-Chariot?"

"We will discuss such matters after you arrive on board. I not will break security! Prepare your Legions."

He said. "I will. You are honored for your unpredictability."

BalamEk did not need him reiterating to her the various possibilities her crew had already considered. "My intelligence apparatus, security forces, and my crew are coping with all possibilities. You will stand by for my egress orders."

She cut off his Stel so she did not have to listen to his cursing.

451

Within an Ob of her discussion, the Burseeosil EkTsab's damaged biological bulkheads were finally sealed. Smoke was hanging in the air after they had biological hull integrity. The creature could rejuvenate itself faster than a mechanical device. The displays were blinking off, then on, as Ovals corrected them, and then they snapped off again.

Jaguar approached BalamEk. "My security forces have taken control. Your Germs, ArnaMals, and Ovals onboard are valiantly fighting to contain the damage."

"Never refer to my Ovals fighting anything in my presence." BalamEk said. "Report your progress."

"We have found cadaver pieces of the Germs who committed self-serving TOTL during their sabotage. The main Osil drives are damaged and are non-functional at this Ob. Communications are less than optimal at present. We are stationary and maintaining orbit, but there is a degradation in our orbital position."

"I am well aware...."

"As one of our human Krils has summed up our situation, 'we are dead in the water'. One of our ammunition units has exploded. The area of destruction was biologically sealed and is contained for this Ob. All non-essential personnel have been evacuated onto BakToobs and smaller Toobs. However, the Burseeosil EkTsab soon will begin falling into Nihberu's gravity well."

On the same Kin, the Kril Legions were advised by KrutEk, through on-the-ground messengers, of the change in their command structure. Because of the worsening battlefield conditions, the haters of KrutEk thought it was about

452

time he was relieved. His allies were wondering what was happening. KrutEk's emissaries did not elaborate.

Hours later, peering at an undamaged 3-D monitor, seeing the darkness of space twinkling with a million-multitude of stars, BalamEk was concentrating. "The Germs certainly knew where and how to cripple us. KrutEk may be right; though I doubt the Dinarchy is behind this. Someone else is the instigator."

Jaguar sighed with concern. "We have captured one of the treasonous Germs before he could implement his explosive device. He has confessed the plot."

"Not is the Germ alive?"

She held up her hand. "Not do be concerned. We used a drug on him; not torture. Your Krils aboard the Burseeosil are loudly begging to TOTL him."

"Protect him from Kril exuberance. Something is bothering you...what are you hiding?" BalamEk asked.

"My intelligence teams have captured an Oval co-conspirator. The captured Germ accused her."

BalamEk did not look surprised. She hid her disappointment well.

"She has remained mute about her part in this mission. The Germ was in her Nest. What are your intentions about her?" Jaguar said.

"The Tribunal will decide her fate." She did not say more.

Jaguar paused for a moment, and then asked. "What are you going to do about this?" She waved her arm at the disabled EkTsab Burseeosil.

BalamEk looked worried. "As our Earth Ovals say, 'Somehow, I intend to fix the problem'."

A half-Kin passed. Squawking in unison, the Stels from the BakToobs Oval pilots were asking her for updates on the developing situation. "You will get information from me at the proper time." BalamEk replied.

"I can silence their Stels…" Jaguar muttered.

"Stay in formation as long as your life-support holds. Transfer your crews and de-activate your Toobs before you evacuate. We will reactivate them later. Not do Bump onto Nihberu. Not do attach yourselves to our Burseeosil EkTsab until you are authorized."

The Oval groans from the protesting pilots were audible in their Stels.

Hortim's calm voice came from Jaguar's Stel. "I've already taken other personnel aboard my BakToob. We await further instructions from our Six…" Hortim used the earth military jargon for the leader of a specific unit. "If there is any way we can help you, ask us, we will do it immediately."

Jaguar sounded irritated. "For an Oval, she still not has discarded her human pilot idioms. She must be controlled."

"She 'is' in control." BalamEk admonished Jaguar. "I am surprised and happy. She is the only one of our thousands of Pilot-Ovals who has asked to help us…instead of asking for our help. Oval coldness not is always the best solution."

Another half-Kin passed with the Burseeosil EkTsab leaking material into space and rolling softly, while his adjusting vapors corrected his attitude.

Oval syntax always referred to their biological creatures with the male pronoun.

Their untenable position becoming a matter of the gravity well from Nihberu dragging them down.

Jaguar stared at her 3-D monitor. "We are stable, but in critical condition. The BakToobs, Toobs, and other crews are watching our timeline expire. If you not can fix…what do you intend to do…land on Nihberu?" Her scorn of that solution indicted her thoughts.

BalamEk waved her hand in dismissal. "Ovals will never put foot on an Invasion planet." That seemed to satisfy her. "We will maintain this crippled Burseeosil; helping to repair him. ZocKuk would have been helpful, if she were present.

Jaguar did not know of whom she was speaking.

"We are nearing the planet's atmosphere." Jaguar said. "Repairing EkTsab would involve four to six KinUts of timeline. We not do have that long. The Legions down on Nihberu not do have the timelines. Evacuating the Kril Legions only postpones the moment of their TOTL."

"Should we meet TOTL, the Krils will remain on Nihberu until we dispatch another Burseeosil to rescue them. Timelines cannot be controlled but they can be altered by the right decision." BalamEk mumbled.

"How will you control this tragedy-crisis?"

Calmly BalamEk maintained her inner thoughts and risk-calculations. "I am managing the situation."

Another Ki hour passed with the EkTsab Burseeosil's timeline nearing expiration. The command center was tense. All eyes were on

455

BalamEk, waiting for her order to abandon EkTsab. Uppermost in their minds was the question, where could they evacuate when the time came?

Nihberu was looming large, serene peaceful and gravity-beckoning them to a fiery TOTL. If they rode their Burseeosil on a comet trail of destruction, slamming into the atmosphere, they were doomed. The Toob creatures, away from the Burseeosil, could not absorb the remainder of EkTsab's occupants. If they evacuated into space, they would die when they ran out of breathable air. If they met TOTL by their own hands, they would have to face judgment by the Cosmic Egg. They had no more options, except waiting for the inevitable.

After an hour of boring work in his headquarters, KrutChan slipped unobserved out of his Zar-dug cave overlooking the sparse jungle to observe the battlefield. He saw a four man Kril unit meandering to him coming off patrol.

He raised his binoculars, looking into the distance, trying to make out his defenses. He saw the lightening-flashes of the explosions from the modern artillery, the fireballs from the Greek fire, and the catapult missiles. The concussion waves were slight aftershocks rolling towards him. His force's Stels were routinely giving him reports in the background, from out of the cave and on his personal Stel.

After a few Ob minutes, Styx was constricting on his arm, his eyes glowing red as he looked at KrutChan, then around them.

Alerted by Styx, KrutChan noticed the four-man patrol was on the level with him. Strangely, they were approaching him from four different directions. That wasn't normal. Usually, the leader of the patrol reported and the others scattered. Two of them were ancient warriors, one with a bow behind him and the other one on his left was expertly balancing a spear in his grip.

A half Ki hour later, BalamEk announced to the Command Center. "Now you will see my skills." Determined, BalamEk said. "We shall manage this crisis, as the Krils say, by Oval magic."

Watching her monitor spewing data, the Burseeosil EkTsab was in the process of approaching the upper atmosphere of Nihberu.

Jaguar was gloomy. "We need Oval magic and soon."

BalamEk pointed to her 3-D screen and pulled Jaguar towards it. "Peer out into the darkness." Those were her only words.

When Jaguar looked she saw another Burseeosil had appeared within close distance from them, in basic configuration, with no Malkrils attached.

"How did they know?" Jaguar wondered aloud.

"I ordered them to come, during the cascading crisis." BalamEk smiled.

Doing a mental analysis of the condition of the approaching Burseeosil, Jaguars' mind was confused. "But he is an ancient Burseeosil. He not is up to our Invasion standards."

BalamEk sniffed. "Not be critical of Burseeosil GryleTunToob with the oldest Osil Drive in the Oval Inventory. He will suffice. He has survived countless Invasion Uayebs. He has a history timeline of half a Gryle. Not do dishonor him."

In wondering awe, Jaguar whispered. "I thought he was disassembled, biologically assimilated into parts, existing only in our StelaBalaam."

457

BalamEk pointed at the 3-D screen. "Dinarchy intelligence thought likewise." Her pride was significant.

Jaguar was in admiration of BalamEk who had existed, it was said, in the same timeline as GryleTunToob. Jaguar knew, from her intelligence information being processed now, the THEN past history of GryleTunToob. BalamEk was its original Engineer-Commander.

"That Burseeosil is much valued and honored by me." BalamEk whispered.

Jaguar snickered. "The human Krils would call that a scrap heap."

BalamEk ignored Jaguar. "If you are so fond of human Kril listening, then you must learn to believe like the human Krils, in Oval Magic."

"The Cosmic Egg smiles upon us." Jaguar said.

"I have ordered EkTsab's BakToobs, Toobs, Malkril Kril compartments, and Malkril maintenance crews to immediately magnetically attach onto the GryleTunToob Burseeosil." BalamEk said.

"Our Oval pilots will be pleased."

"Before they become attached, GryleTunToob will tow our EkTsab Burseeosil to a safer orbit. I will transfer my command to GryleTunToob as soon as repairs are finished on this EkTsab Burseeosil."

"May the Cosmic Egg watch over you."

"We are in control." BalamEk was speaking to her Command Center entourage. "Our Kril forces on Nihberu are safe."

Jaguar watched the Command crew swiftly run to their assigned stations.

KrutChan was alone; the Krils in his headquarters were busy with their duties in the cave.

He was the target again. Two ancient ArnaMals approached. His neck was itching and sweat began rolling down his back.

The other two were modern warriors, one with a 9mm Beretta pistol; the other carrying an M-1 rifle with his bayonet fixed.

KrutChan decided to meet the modern Krils head-on because he was more familiar with their hand-to-hand tactics. He shouted into his Stel for help, knowing rescue would arrive too late.

The guy with the pistol fired three shots, while KrutChan was firing back at the same time with his .45.

They both missed. They were cursing loudly at each other's bad aim. KrutChan grabbed the pistol-carrying man's forearm and spun the Kril with the pistol around to mask KrutChan from the guy with the rifle bayonet.

KrutChan smelled the sweat on the guy he was struggling with. He viciously rammed the other guy's M-1 rifle bayonet into the back of the neck of the 9mm pistol-carrying guy. The blade exited from the guy's mouth, gushing gore. KrutChan then fired with his .45, exploding the pistol-carrying guy's face.

The thudding sound from an arrow, released by the ancient Kril behind him, pierced KrutChan just under the right side of his flak jacket. KrutChan felt the numbness surround his wound and groaned as he fought to stay conscious after he went to his knees.

He had no time for thinking about his dangerously exposed position. The guy with the bow was readying another shot and fired. That arrow clanked off KrutChan's camouflaged steel helmet. KrutChan's .45 pistol jacked open when he fired his last round into the bowman's crotch sending him sprawling in screaming agony.

Time slow-motion ticked in those seconds-minutes of confrontation.

Styx had dropped off KrutChan's arm, exploded in fury; swelling to ten times his normal size. With his enormous steel-toothed mouth viciously snapping, he attacked the Kril with the spear; shredding that Kril's neck and upper body. Styx was loudly grunting in hellacious rage.

With Styx's diversion, KrutChan went after the guy with the bayonet on his M-1 rifle, who was trying madly to free his bayonet from his friend's neck. KrutChan parried the bayonet as the man lunged at him after breaking his rifle free. KrutChan drove forward, under the man's M-1 rifle, with his .45's magazine butt-plate, crushing the attacking guy's larynx. The choking assailant dropped his weapon.

Following up, KrutChan immediately raised his K-Bar, ramming it into the gasping man's neck; spurting blood. The assailant was already dying when KrutChan picked up his M-14 and fired above the man's heart after he was on the ground.

Styx's screaming ancient spear-carrier was cursing and pleading loudly with KrutChan to pull Styx off of him. Styx was too far into his killing-mode to quit, and tore the man to pieces, leaving parts of the guy and his gore littered around him.

KrutChan staggered, trying to catch his breath.


Observing GryleTunToob magnetically attaching the Toobs and Malkril pieces from EkTsab, BalamEk said to Jaguar. "I will need the Cosmic Egg's help dragging KrutEk up here. He not will be thrilled traveling back to Arna on our crippled EkTsab Burseeosil."

"We will all endure his wrath."

460

"I may have to bind his body with my Oval Control Wand. He will be sent back to Xmucane, whether he agrees or not. Sometimes he needs his attention refocused as to who is in command here."

"Perhaps KrutEk requires some of Oval Hortim's attention." Jaguar grinned. "Our Oval Pilot Hortim has her way of disciplining."

As she walked away, BalamEk looked incensed. "I not do resort to obscene Willow ways, as Hortim does, to control Krils." She was checking data streaming. "BalamEk teaches the OvalChanHalachs how to be strong. We will obey our Queen and so will her KrilChan by using...my methods."

Lowering her head in submission to BalamEk. "I honor your wisdom." Jaguar said.

BalamEk added as an afterthought. "I consider Hortim an honored trainee, with an overindulgent Oval commanding her Nest. I will teach both of them how to follow our Oval ways."

When timelines converge; events happen spontaneously. She thought.

For KrutChan, minutes of slow-motion had passed. The lethal melee had lasted under a minute. He was screaming at a dying assassin. "Who sent you? Who? Tell me, ya sonofabitch!"

Seeing the four Kril assassins were dead or dying, KrutChan slid down. Sitting on his butt, cradling his M-14 in his arms, gasping and heaving. His adrenaline pumping. His fear was overwhelming. He struggled to control himself. I was an easy mark for those goddam assassins.

He was consciously breath-sucking in and out to ease his hyperventilating. Al must have been slumming somewhere.

461

He was past blaming anyone for this. He thought it was his own fault; he had let down his guard. <u>Better not let that happen again...ya dumb shit.</u> He was still a marked man to the Kril hierarchy.

Styx slithered up to KrutChan as he sat there. KrutChan's hands and his whole body were trembling with fear. Styx coiled with contentment onto KrutChan's forearm.

<u>I gotta piss.</u>

KrutChan stood, relieving himself on some of Nihberu's plant life and rocks. The instant relief was exquisite; his adrenaline and fear were dissipating in a frothy pool between his feet. He would catch hell for not using one of the elimination bags. Looking down, he foolishly wondered if the plants would collapse-die or thrive on his urine.

His Krils had erupted from his cave headquarters and were running towards KrutChan. They held back when Styx dropped to the ground and inflated himself to full attack-size facing them, protecting an again-seated KrutChan; hissing and snarling a warning. Styx's fur was shiny and blotted in places, with congealing blood.

"Are you normal?" KrutChan's Zar, CoCum asked. His red hairline band was glowing scarlet.

"Fine...I'm good." KrutChan pulled his Styx creature by his tail, down to his level and was softly touching Styx's head, calming down his little symbiotic animal. "I'm good. I'll make it. Thanks to my one-man fire-team here."

"Your friend...angry is he."

KrutChan was talking to himself. "Styx isn't my friend; we just travel this timeline together." <u>You sound like an Oval.</u> "You guys stay away for a minute while Styx reins in his mad."

One of his headquarters' Kril said. "I notified your staff. CheChun is going to kick my ass. He told me to protect you."

"I'll calm Chun down. Don't worry. Old corps has too much mother in him. I'm fine."

"You are not fine." The ArnaMal doctor was kneeling near KrutChan assessing how to get the arrow out above KrutChan's hip. "If that fetid-mouth Styx bites me; I will drug him and put him to sleep forever." He proceeded to open his medical kit and leaned over to attend the wound, ignoring Styx's hissing.

KrutChan was gently holding Styx's head. "Easy buddy...doc is one of our friends." He glanced at the doctor knowing they had no love for each other.

Stroking Styx more briskly. "At ease...you little killer. You don't have to be an asshole all the time."

You could be talking about yourself. EkSeet said.

Except for the few Krils still attending the 3-D monitors in the cave; the rest of the headquarters Krils were surrounding KrutChan. As they stood there kibitzing the doctor, most of them were relaying messages to KrutSeet, KrutChan's staff, and the Burseeosil."

CoCum was heaving; his Zar laughing amusement evident to all.

KrutChan thought. Goddamn Zars always laugh like they're coughing up a hairball.

"My excavation cave...would have...safer for you...not healthy here...above."

KrutSeet's voice broke into KrutChan's Stel. "You need not report. I have the doctor's prognosis. You have survived more dishonorable Krils. It is getting to be a habit with you. We are investigating."

463

Knowing full well the StelaBalaam would be recording, KrutChan said. "As he was dying, one of these bastards whispered who gave the order. I won't forget."

"Concern yourself with your WHEN!" KrutSeet barked.

KrutChan trivialized the incident now that his emotions were under control. "They weren't Krils. They were amateurs. I had a secret weapon."

"You are running out of secret armament. And are delirious. My doctor and I agree with your doctor on site; you must be evacuated to the medical unit on the Burseeosil GryleTunToob."

"What happened to our own Burseeosil EkTsab?"

"That is not your concern."

KrutChan could not resist, insubordinate or not. "Tell our KrilChan KrutEk his jerks have failed again."

"This is being recorded by your EkSeet and your Stel; choose your words more precisely. If you wish to remain a survivor. Our KrilChan is…busy."

"I'm sure he is…."

After a pause, KrutSeet said. "I believe you are delirious or the doctor has overmedicated you." There was again a pause before KrutSeet continued. "I now command all ground forces of this invasion."

Forgetting KrutEk's messengers had notified the Legions; KrutChan did not know whether to cry or laugh, but he was hoping. "KrutEk meet TOTL?"

"Worse for him…he has Bumped to another battle zone."

Before KrutChan could answer, KrutSeet said. "And not for the reasons you are fantasizing. The good doctor will transport you to the Burseeosil medical unit."

With no reasonable answer left to him, KrutChan replied. "As soon as my staff arrives. Screw the doctor's assessment of my condition. I'd rather TOTL here than under a blue sheet on one of those Oval invisible beds."

"I hear your delirium is passing." KrutSeet said. "Be advised, if you come back down to Nihberu; I will not be as humane as KrutEk. I do not suffer fools gently."

One of the Krils from KrutChan's headquarters unit was expounding to the small crowd of other Krils, his view of what had happened. "Those guys lying over there were here to frost KrutChan. He took 'em all on and left them to rot." The Kril either did not see all of the fight, or left out Styx and his protective actions.

Another headquarters Kril said. "KrutChan even pissed on them. I saw him...yeah, I ain't lying...he did. KrutChan is one killing sonofabitch."

A yelling, cursing XibEk coming into their headquarters brought them to attention. "Who the hell screwed this detail up?" When he stared at the man CheChun had left in charge of headquarters security; the man beat a hasty retreat.

He yelled after the man. "Run! You and lifer CheChun will meet again, asshole!" He spun around, glaring at the rest of them. "What're ya standing around for; one goddam mortar or grenade could wipe you dudes all out! Spread out! This ain't a wake!"

In a frantic hurry they disappeared back into the cave.

XibEk knelt down next to KrutChan. "I can hear him now, Chun leaves the Lifer alone for one goddam minute...jes oncet...and you turn into a daydreaming FNG."

465

"Chun taught me how to fuck off, Xib."
KrutChan was fading away; the doctor's medication
kicking in.

CheChun arrived and knelt next to XibEk.
"My ass has been transferred to KrutEk's
headquarters."

Reaching out to Styx, who hissed ominously.
"Shut up ya little fuck. Ya covered Krut's back
good." He flicked his finger with a snap against
Styx's saber fangs. The creature seemed pleased.

CheChun addressed KrutChan. "Keep your
balls covered while I'm gone, okay boot?"

"Go play with yourself someplace else."
KrutChan groaned. Poison from the arrow was
effecting his system. He was going cross-eyed when
he said. "Where's Cheeb when I need him?"

KrutEk was furious as he boarded the crippled
Burseeosil EkTsab with his new coterie. With her
own expertise, BalamEk had initiated EkTsab's Osil
drive to operating mode again, in a record time of
three Kin.

Speaking to CheChun, KrutEk told him. "We
were together four Uayebs ago. I am ordered back to
Arna. I will explain our mission and your place in it
after we have departed. You were a great Kril
bodyguard then; do not fail me in my NOW."

CheChun felt great KrutEk remembered him.
Those Invasions were hairy; KrutEk had a habit of
being fearless. CheChun had had his hands full
protecting the guy then. They had barely survived.
He rendered a proper Kril salute to KrutEk to signify
he understood.

KrutEk stopped CheChun before he went to
his quarters. "You will command a Kril sub-Malkril;
cleanse yourselves, dress in proper attire, and look

466

like respectable Krils before we disembark on Arna."
He gently shoved CheChun on his way.

BalamEk sent the crippled Burseeosil EkTsab
commanded by ZocKuk back to Arna to finish its
inner repairs, with a fuming KrutEk, She was too
experienced to let a few malcontent saboteurs or a
disgruntled KrilChan to stop her mission. At last I
have become the Invasion commander again without
his interference. I not do envy Xmucane. I have no
doubt he will blame me, but she will admonish him.

Near death from his poisonous festering
wound when he was evacuated, KrutChan had been
unconscious in the medical unit for two Kins.
Melding, Hortim visited KrutChan.
"We were told you died. I wanted to tell
you...." She was wearing a half-hearted grin.
"...never mind. I see the Oval doctor's worked a
miracle."
He was leery. Hortim and he had a tenuous
relationship. Was she was hiding something? Or
maybe he was imagining things. Reela probably sent
her to check on me. Wanting to make sure I didn't
die on her and lose her wealth. Hortim could care
less about me.
"How are you feeling?" Hortim said. "Are
they treating you well?"
KrutChan remained non-committal. "I'm
recovering...and feeling just fine."
"I'm reading some irritation in your attitude.
Are you angry they made the attempt on your life or
mad you can't kill the guy who sent them?"
KrutChan decided to play her game and hide
his real thoughts. "You didn't come here for a bible-

thumping-prayer vigil. You angry they missed? What're you really here for?"

Three ArnaVal nurses approached him on their rounds.

Their appearance next to him gave Hortim a chance to wander away from him. Reela had ordered her not to divulge the Nest secret to him.

Checking KrutChan's systems thoroughly, he was joking they leave his essence penetrator alone. The lead nurse was the only one smiling as she melded.

Hortim approached him. "Up to your old tricks, I see. Is sex all you ever think about?"

"I don't use tricks. Like a gopher, I only burrow when I'm excited."

"You died on the Evacuation Toob. The Oval doctor's recovered you."

"Dying is turning out to be a habit of mine. I pay attention to the doctors."

"You never listen to anyone." Hortim admonished him. "I strongly suggest you forget about mating with that ArnaVal head-nurse without Reela's permission."

Feeling evil, KrutChan snapped at Hortim. "How's our baby doing?" His self-satisfied smile was more of a grimace.

Hortim flustered. Then she realized he was not clairvoyant; he was jabbing at her about Hortim's egg in the Oval hatchery. "My egg is fine. Don't worry; you won't be burdened with the baby. You got two other youngsters to protect."

KrutChan said. "Why are you here? My morale is fine, so let's skip the officer-visiting-the-wounded-patient-bullshit."

"I'll keep it short." Hortim continued. Then she got angry knowing he took her words as a pun

468

after the head nurse had fondled him. "Reela was advised by BalamEk when you transferred up here."

KrutChan believed she was holding something back. Maybe attempting to arrange her words he could not misconstrue. "And...?" He asked.

"I...simply...wanted to relieve Reela's anxiety. Wipe that attitude of disbelief off your face. · It's a common Oval trait. She cares for all of her Krils in her Nest."

Sure she does. Hortim did not look sincere. What're you hiding, Lady?

Hortim continued. "Reela wanted me to verify you are healing. She ordered me to assess you. In their Nests Kril leaders are valued by their Ovals."

KrutChan locked eyes with Hortim. "You mean valued, like KrutEk, for instance. I heard they shit-canned him. Or did the StelaBalaam TOTL him?"

Updating him on ground progress, Hortim said. "KrutSeet has won Nihberu with ancient ArnaMals. Without KrutEk...or you." She pushed a tendril of hair away from her face. "KrutEk was commanded by the Queen to go back to Arna."

KrutChan felt immense relief he was out of the insanity. "Oval plans are always self-serving."

Irritated, Hortim yelled. "Clean out your ears! Nihberu is conquered! The surviving Krils are golden."

"Our guys find more gold mines?"

"This conversation is over."

"Liar." KrutChan said. Something else unsaid was bothering her.

"Believe what you want."

As Hortim stiffly walked away, KrutChan said. "Let's be candid and cut through the crappy-official line. I ain't earning Reela anything lying on my back in this hospital! That's her only concern. Providing KrutSeet doesn't warehouse me. Boy, having me on the sidelines would really crank up her shame."

Hortim never turned around; waving her hand in the air, acknowledging his message to Reela.

KrutChan shouted at her retreating back. "And tell Akna I can't wait to see our baby!" That crack ought to endear me to Reela.

Hortim stopped in mid-stride.

KrutChan added, to dig a little deeper. "Tell Akna that I miss her."

Thanks, EkSeet, for cleaning up my Beekav message to Akna.

Hortim abruptly turned to face him. She had watery eyes.

That jab got to her female pride; one point for me.

Hortim spun around, shaking her head, and melding from the medical unit, keeping whatever thought she had to herself.

KrutChan came out of sleep seeing the AnticArna blue rain stop cascading around him. My goddam Greer-vision dreams are getting too real. None of that future time has anything to do with me. It never happened.

He was looking at the upper bulkhead of the medical unit to readjust to the present time on the Burseeosil. His Greer-vision had brought him full circle. His thoughts were of Akna again and how he had treated her when he was with her. A minute later, he conked out.

KrutChan was dozing, breathing deeply, and relaxed. He startle-jumped when an ArnaVal nurse appeared over him.

"You are conscious. That is good." Then she said automatically. "How do you feel?"

"When my heart starts again; I'll be fine." KrutChan said. "You spoiled one hell of a wet dream, doc."

Her EkSeet was translating for her. "I not am a doctor; but a nurse. I believe your systems are functioning. Is that the correct earth-phrase, EkSeet?" She drained the blue medication from his tub. "Your release from this unit is imminent."

KrutChan took her hint and slowly sat up, while coughing and wheezing. "The same old shit starts again."

"Stop cursing...."

He crawled out of the tub and shakily went to a module in the wall to dry off and kill any nasty bugs he had accumulated in the hospital.

The nurse melded while he went to another module to get his uniform.

As he flapped up his combat boots he was thinking again of Akna and anticipating their reunion on Arna in her cubicle. He had to treat her better.

KrutChan was not surprised to see Albert leaning against the other wall. "Your visits are getting old, asshole. Bugging the hell out of other sick Krils?"

No one in the medical unit was watching them.

["You survived another kill-attempt. I'm getting bored watching out for you."] Albert pointed to KrutChan's Stel from his personal effects. ["There's more coming your way, buddy."]

471

"Where were you when those four dummies showed up? You do know how to fight, don't you Al?" KrutChan was doubting himself about who this guy was. "Or is your weapon invisible, shooting vision blanks?"

Albert was smiling. ["Why am I always the one you blame? I'm trying to help save your ugly butt."]

Who is Al, KrutChan? EkSeet asked. Whom are you speaking?

Albert was dressed in his invisible iridescent suit, now wearing a matching fedora hat. He looked highly amused. ["You going to answer your Oval snoop?"]

You blind?

Don't be a Nutter. I perceive no one.

Laughing, Albert adjusted his suit, wiggled into position, posing for a snapshot portrait. ["My camouflage makes me invisible to others. Not even Ovals with their full spectrum electromagnetic vision can see me."]

He wandered to the other side of the room. ["You better give your EkSeet a good explanation, not the correct one, or he'll have you committed to a loony bin."]

Are you all right, KrutChan? EkSeet asked.

Don't be an English jerk. One of my friends back on earth was named Al. You ever talk to yourself? A lie is as good as mangled truth.

I question your veracity.

EkSeets needed constant information to share. KrutChan not wanting to be an informer. Albert was the only one he could speak to frankly without fear of recordings.

I know your reconstitution history. I do not recall an Albert or Al, in your past.

472

["Told you he wouldn't be satisfied. You better come up with a better story."] Albert was spreading his arms, flicking his upright palms, indicating more explanation was necessary.

KrutChan walked to an invisible cart, reading his Bot-unit reports, buying time. He sat on an invisible stool he knew was there.

You're not the only voice in my head, EkSeet. I got plenty of them whispering to me. You're the idiot. Leave me alone.

KrutChan could hear his EkSeet sigh. As you wish. If you absorb additional AnticArna and drink more Osil wine, you will hear a thousand voices. I leave you to your delusions.

A few minutes passed while KrutChan made sure EkSeet was gone.

["See? Was that so hard?"] He came closer to KrutChan. ["My fault, forgetting to shut off your Stel and EkSeet."] Albert was drinking double-shots of whiskey. ["Acting crazy will fix the problem in the future, in case I don't remember again."]

"I gotta get outta here." Taking his EkSeet's advice, KrutChan poured himself Osil wine. His renewed red-AnticArna high increasing with the combination. "I must have more pull as a half-Malkril Commander. You got any strings you can use to get me back to Nihberu?"

["I'd lay off the liquor in the future."] Albert said. ["KrutSeet doesn't need you. He won the Invasion. You're a diversion."]

"Then give me the short version of why you're here? Without the Oval-Magic displays and hocus-pocus invisibility crap."

["Still don't get it? Short version?"]

"You're never at a loss for riddles."

473

Looking gravely at KrutChan, Albert said. ["You survived Nihberu. The assassination attempts have moved to a higher level. Higher than ZacNaab."]

"Tell me something new, asshole."

Albert's face was serious. ["If you make it back to Arna, something's real wrong there. You saw how Hortim acted?"]

"Didn't think nothing of it; she's the same old Hortim. Too full of herself and her Officer Rank." KrutChan rubbed the stubble of beard on his chin. "Course, she 'was' more evasive, I'll admit."

["A lot more KrutChan-killers have been recruited. Don't look so suspicious; I ain't one."]

"Why don't I believe you? They've shown up every time you do."

["You're getting paranoid. Hortim might be a new addition; maybe the Queen, Reela, Jaguar, Akna, and even BalamEk's daughter. In addition to the assassination plot; another plot has been initiated, I think. You've been ignoring my advice. It's time to stop, listen, and trust no one, except me."]

Akna being involved was too ludicrous to think about...yet, it was a possibility? As for the rest of the Ovals, he wouldn't put it past them. Scorned women have a lot of mean in them. His Kril enemies had a lot of hate in them. Including his Zabin-Krils, like CheChun, KrutCheebel, XibEk, and even PacalMo.

The alien Zars, Cunacks, and Dagots; had plenty of other chances to zap him. Albert had given him ideas in the past. The need to listen to him was imperative to KrutChan's survival.

KrutChan would be careful, warier of the people surrounding him. His advantage was in being alone. He was used to feeling he was a Seal on a floating broken iceberg waiting for a hungry Orca to show up. Lonely was miserable at times, but safer. "Go pester some other poor schmuck, Al."

Albert was gone.

Throwing his Osil-drink against the bulkhead, KrutChan screamed into his Stel. "As a half-Malkril leader, I want my doctor! Get her, now!"

An ArnaVal nurse immediately melded into his cubicle.

The Kril's Malkril compartments were silent and moody; there was nothing to share or talk about between Krils. Critiques briefly occurred in some quarters.

The Zabin-Krils in KrutChan's headquarters briefly summed up their analysis. "My dead Kril's number-count you already know about..." KrutCheebel said. "...four times higher wounded in the hospital; including twenty with Post Traumatic Stress."

"Shit. Show me any Krils aboard this Burseeosil who don't got battle fatigue." KrutChan said. "Those wounded are full of shrapnel. Plenty still dying from the spores. A quarter of them have bloated faces from blast-concussion effects. You ought to see the drunk-staggering guys with brain damage. Trying to make daily choices, concentrating on difficult decisions whether to leave or enter a room, sit or stand. Lot'sa them unable to speak a paragraph coherently."

"You a doctor now, lifer?" XibEk said.

PacalMo reported. "My Kril's numbers reflect much of Cheeb's. Half of my wounded Krils have Dragon-fever. Many have broken spirits."

"After you evacuated, my Krils were close to breaking after that shit at the river." Added XibEk.

475

KrutChan said. "When I came to, my brain regrouped. My fault. From the hospital, I ordered ZacNaab to have his Roman engineers building that bridge to take boats to secure the other side of the river. Made sense at the time."

"Water-Dragons killed most of them and destroyed the bridge." PacalMo said. "ZacNaab ran away."

"Those weren't Water-Dragons." CheChun said. "More like Channel cats the size of whales with dinosaur crocodile teeth. Planet creatures, I told you, are worse than the Dinarchy. Mopping up is always deadly." CheChun added. "You should court-martial ZacNaab. Tit for tat, Hoss."

"ZacNaab followed my orders. No fucking way."

"I got twenty-two Krils screaming, crying incessantly, preparing to kill anybody touching them. Raging with post-traumatic stress." KrutCheebel wearily waved his hand. "Who gives a motherfuck?"

Hearing KrutCheebel obscenely curse, brought thin smiles to the headquarters' group.

Silence reigned in the Malkrils after Osil harmonics began. Tunneling upon the dark matter-dark energy Graviton rail was painful for them. The senior Kril Commanders never spoke to each other as they bore the effects from the Osil drive. Even the Zars, Dagots, and Cunacks had grown pensive remembering the Island Planet Nihberu.

The old Burseeosil GryleTunToob was adequately carrying a load of now younger Krils. Acting-Zombies going back to Arna. These shattered Krils would never return to Nihberu. Garrison assigned Krils would, but not them. The twenty-year cycle for these Krils, in the Second Uayeb, was over.

Chapter 24


Arriving on Arna, KrutChan's half-Malkril Bot-units were watching him. Leading them off the Burseeosil GryleTunToob from Uayeb Two on Nihberu. His mind was reeling. The other Krils filled with their daily diet of administered blue-potion-drug aerosol keeping the Krils healthy and in a mental state of floating-happiness.

Except KrutChan, who was more fuzzy-headed physcotic. Al's warnings were in his mind. In his case, he believed the Oval's AnticArna drug had another purpose. The Ovals wanted him hyper-active on-edge.

Tendrils of smoke were rising skywards all over Tikumyax; battle-torn ominous. The Ovut area was crammed with milling Krils looking for exit causeways as he melded.


Entering Reela's pyramid, following the Germs, KrutChan not doing his usual bug-like crawl to Reela's Soothing room. Entering, he stiffly shouldered past Reela. He was in control of himself until he stumbled into his barrack. He was angry at his unsteady gait; three days and forty years of drug-tripping over his own feet.

KrutChan's face collided with his barrack room floor after wobbling past the door jamb. Collapsing in a heap of sweat-dripping body parts. At least I didn't leave a slime-trail like the last time. He thought. Reela's Germs won't have to clean up the crap.

He was grateful Reela's Nest was leaving him alone for a few minutes; which was strange. KrutChan recuperating from the Osil-effect.

As he lay quietly, Kelel's secret missive from Jaguar flitted briefly into his mind. Never was approached by any Dinarchy agent on Nihberu. He forgot completely about that spy-agent crap. Being back in the city of Tikumyax felt like he was home.

After a few minutes, he ripped off his clothes and inserted himself into the Oval Shower cubicle; scouring off a layer of his skin, cauterizing his abrasions and cuts. The healed wound from the assassin's poisoned arrow on Nihberu a minor dimple on his flank. My aching face in the 3-D mirror device looks younger. When he first arrived in his eighties, he constantly had the urge to pee. He was glad that went away.

His EkSeet corrected him. Device is wrong. Every biological creature is controlled by its function.

He kicked away some debris accumulated on the deck. Well, tell their biological critters not to molt their skin onto my Barrack floor. By now they outta be housebroken. In Vietnam he was in his early twenties but had a forty-year-old mind. Time had finally caught up with him, he thought.

He began cleaning his weapons and gear, preparing for his third Uayeb, working alone for a half hour.

Feeling a presence behind him, he snarled with a joke. "Goddammit Akna, I told you before, not to sneak in for a grope from me."

"I not am Akna. I am of your Nest. I am BakMeer."

KrutChan was surprised and briefly looked at her. "Well, go away. I want to talk to Akna." He squinted at BakMeer. She had grown into a beautiful woman with the poise and carriage of a true ArnaVal. "You ain't hardly a kid no more, are you?"

479

"I am less of Akna's age when you first met her." Her eyes were downcast; so unlike her.

"Okay. Tell Akna to come in. I want to talk to her first."

BakMeer walked out, shaking her head; going to Akna's cubicle. Her stiff posture indicating she never could talk to him.

KrutChan saw Reela meld just as BakMeer disappeared. Her appearance and clothes were adjusted, reflecting a more mature, prosperous Consort to the OvalChanHalach. He wondered if her impressive muted-down look reflected more of her reality without any Oval magic. She seems more human? Don't pass that on, EkSeet. I don't need her Oval xenophobe-temper surfacing.

EkSeet answered immediately. I agree. She has more on her mind than controlling you.

He was happy to be back, even if for only hours. "Jesus Christ. We having a Nest Convention or something?" KrutChan stood in case Reela was going to blast him with her biological wand for being a wise-ass. "I honor you, my Oval. I'm a little cranky. I need my second wind to get normal again."

KrutChan was puzzled when she did not remark after receiving her translation, asking where did his 'first-wind' go?

Reela was speaking in a deadly serious tone. "BakMeer was correctly doing her duty. She wanted to help you adjust to coming back after your twenty years...is 'years' the correct word?" Her EkSeet was helping her.

She was in stifled pain. Her eye ducts were dry; no crocodile tears streaming down her scrunched up face as she maintained control. The effect was strong on him; like when his father told him, 'you're in deep trouble, boy'.

480

KrutChan tried to joke. "Damn, I can't look that bad. Hell, I'm twenty years younger." He placed a sleeping Styx in his transparent module. "Maybe you ought to buy yourself a younger stud-Kril." He saw from her attitude something else was wrong and his humor was unwanted.

Reela held up her palm for him to be silent. She looked devastated.

"Shit, you are sad." KrutChan said. He tried again to joke; to lighten up the atmosphere. "I know I didn't earn you as much wealth as you wanted. Hell, my half-Nacom-Malkril unit was in a quiet area most of the time. And my time in the Burseeosil's medical unit didn't help."

"Not do speak!"

"I couldn't do much for your treasure chest." He could not help himself. "I'll do better next Uayeb, I'm sure. I kind of got busted out on this Uayeb. Hell, you can Sooth another richer Kril, honey."

"Will you be silent!" She snapped. Reela was nervous, ignoring him by turning her back to KrutChan, and then revolving around to face him again; looking up to the projection of the universe on the star-filled ceiling.

I think I pissed her off. Hortim is behind this, I'm sure.

Reela was either listening to her EkSeet or she was deciding to come to grips with something else.

"Hell, I survived. Nothing can be a bummer today!" KrutChan got serious and tried to figure out her mood. "What's wrong? I know I irritate you. Should I call Hortim to help?"

"Hortim not will again enter here. She not is of my Nest. She earned her own Nest on the first Uayeb." Reela's voice was cracking. "My Nest ArnaVal Akna not will be greeting you."

481

He felt that deadly pause before bad news is given. "Why, was Hortim awarded her too? Akna will hate that turn of events. Did you Ovals banish Akna over some fucking rule I broke?"

"Must you ramble! Stop babbling!" Reela was definitely in an emotional state; struggling to compose herself.

KrutChan politely wandered away, to save her embarrassment from her outburst, storing some of his gear into a pack. He held onto an ominous feeling.

Speaking to his tee-shirt, Reela softly said. "Akna nearly met TOTL on the Kin when you left for Nihberu. Human's call it childbirth. She survived dying, but the germ...her child...met TOTL."

He fiercely gripped his wall locker, rattling and banging it against the yellow wall. For a couple of minutes, he slammed his fist into his locker.

Akna was dead? He had survived Vietnam battles. He had endured Two Uayebs. How could she be dead? She had been left here, where it was supposedly safe.

"Smashing my creatures will not change Akna's timeline."

"How could that happen?" He softly asked. "Births are normal SOP. One of your doctors screw up?"

"TOTL happens; life is uncertain." She was listening to her EkSeet to make sure she spoke correctly. "You have my condolences for your Hatchling-child. Akna is a great ArnaVal."

"You know goddam well the kid wasn't my child!" KrutChan's rage flared. "Skip the Oval horseshit-grief and tell me what happened. How did Akna die!"

"I speak. But not do you listen!" Reela yelled. "Let your blue-potion calm you."

<u>Dope?  That's a joke!</u>

Speaking slowly, with enunciation, to prevent him from misinterpreting her translated words. "Jaguar said she was in attendance. Akna was calm at her hatchling's demise."

"Yeah sure.  Nobody's calm when they're dying."

Finally defining what his problem was, in afterthought Reela clarified. "I stated Akna 'nearly' met TOTL.  In her present state of condition, she may as well be TOTL."

"Then she's still alive?"  KrutChan's hope was in his voice. "Akna survived?"

Reela remained firm. "You not can observe her.  She not is with any of us in her mind.  Akna recognizes none of us.  She has closed her mind off to her timeline reality.  Meli's doctors advise me she will never recover."

"Is that wishful thinking on the part of the doctors?  Or is it medical-techno-talk, meaning they don't know what to do?"

"You consistently blame the Ovals!"  Her head was shaking.  "When an ArnaVal loses a Hatchling, a part of themselves, they sometimes mind-withdraw forever.  That is a human, not an Oval trait."

"Akna ain't no Alien."

Reela touched his forearm to make a connection.  "Her emotions and thoughts are now purely her own. Her only function in her NOW is her loss; unfeeling other beings around her."

KrutChan was fighting for control, staring at her hand on him. "I want to see her."

Reela's Oval acceptance of timeline branching occurred.  "No...the doctor not will allow your presence. You will be leaving for the next Uayeb."

483

"Maybe she would recover if I see her?" Feeling guilty, KrutChan was pleading with Reela.

Reela shrugged. "The doctor foresees more pain for her if you persist. You must stay away to protect her timeline sanity."

KrutChan's eyes narrowed, swearing an oath. "I will see her...one way or another." His persistent anger was welling up in him. He walked away from Reela to distance himself.

"You will obey the doctor and me!" Reela asserted. Her Oval-device on her hand activated with a humming sound.

KrutChan glanced back at her, suspicion growing at his next thought. "Jaguar was on the EkTsab Burseeosil with us on the Invasion."

Reela spoke carefully, not wanting to incite him more. "Jaguar was here with Akna. After the hatchling met TOTL, she joined you to go to Nihberu." It was as if Reela was reading travel orders; bland and official and boring.

KrutChan was accusatory. "Why the hell didn't Jaguar tell me after she boarded Burseeosil EkTsab?"

"Ovals not would have mentioned timeline-branching before an Invasion. All the Ovals on Arna and in the Burseeosil knew about her mind-condition simultaneously through our Stels."

"Hortim knew about Akna losing the child too?"

"Yes. She wanted us to send you a missive beforehand. Before the invasion Bump began. The Queen commanded me not to let Hortim tell you. Hortim would have grievously suffered for breaking our secret."

KrutChan was grimacing. "When Hortim visited me in the Burseeosil medical unit, she was busting a gut wanting to nail me with what happened to Akna and the baby."

Reela angrily approached him and held his arms. "Not do say that about Hortim! She not was happy to keep it from you."

"Right on. I'll bet." KrutChan felt his old familiar distancing from death. Inwardly, Akna's child joined the ranks he had seen die. According to Reela, Akna was as good as dead. He added Akna into that dark pit in his heart keeping only the memory of her. Deep down her image would fade, like his future wife's image, and become faceless.

["I told you something was going on, didn't I?"] Al was leaning against the Barrack door jamb.

"Relax and try to regain your perspective, my Kril." Reela said.

"Don't preach to me about mortality."

Reela grabbed him from behind. "I want you to be alone for an Obet. Afterward, come into my Soothing room and partake of some of my Osil-wine with me."

"I got a shi…er…a pile of blue Ak-dope on the Burseeosil. Don't need no more."

"We must go to the Memorial of the Oval Protectors ceremony to honor our TOTLed Krils and our Ovals, ArnaVals, and ArnaMals who have joined with our Cosmic egg."

"No way, Jose. I don't need a goddam ceremony to revive my 'Ghosts and Demons'. The hell with that noise!"

"You are out of control!" Waving her glowing Oval-wand in his face, Reela said with determination. "I command you to attend." She initiated her device to a higher setting. "If you continue to resist, you will attend unconscious. I will

have you carried by my other Krils to the ceremony!"
She was a more serious Oval than he remembered.

Albert said. ["She means it, ace. You better
think of something to talk her out of it."]

She's holding all the cards and power.
KrutChan mumbled in resignation. "I'll be there. Let
me rest for a little while."

Satisfied, Reela melded.

KrutChan was alone, staring at the wall.
BakMeer entered with CauacSky. Her to give
condolences and he to proudly advise of his departure
on the next Uayeb as an artist recording for Kril
posterity. KrutChan exploded at both of them, yelling
for them to get out! They left in anger.

Lord almighty, save me from people full of
pity.

Moody at the Ovals for purposely letting
Akna semi-TOTL in childbirth, for some goddamn
reason. He had spent an hour sorting out his unseen
enemies. Conspiracies were growing. Al was right.
KrutChan's whiskey bottle was beckoning him, but
he refused to drink from it. He would never again use
liquor as a crutch. He viciously threw his Bourbon
bottle into his rifle range, smashing it into pieces.

Xmucane was meeting with her Tribunal,
reviewing the status and the innumerable problems
that were demanding she address, in the thousand
galaxies Confederation. Jaguar, Dirva, her anti-
Uayeb members, and BalamEk were present in the
flower and jungle atrium of the OvalChanHalach.

"Every Uayeb we continue this constant
dislike of our war." Xmucane said. "Your Queen
tires of the jungle-bird chatter from our opposing
Ovals; noisy and contributing nothing."

One of the Anti-Uayeb Ovals said. "We not do experience hateful words with our Krils. We have much Beekav for our Krils. Protect our galaxy. End this war."

"Not do insult our Queen!" Jaguar shouted.

Dirva interrupted. "Exorcise your immature attitudes, my sisters, before the Queen banishes you to a life of shame that will exist for a Tun! We must honor her."

In the silence that followed, BalamEk spoke. "Have any of these dissenting Ovals ever controlled their Krils?"

Jaguar answered BalamEk. "Their Krils are useless for Uayebs, according to KrutEk, because of these Oval's political views." Jaguar said. "The StelaBalaam and my intelligence department indicate these Ovals Sooth a Willow Nest for their pleasure."

"That is enough infighting." Xmucane held up her hand. "I honor BalamEk and Jaguar. Although he not did want to come back; KrutEk not did abandon me. His Beekav for me saved all of you! He protected all of the Ovals with his Legion guards and his ArnaMal Legions not on Uayeb. Where were your Soothed Krils?"

Dirva's Ovals were somber and silent as their Queen asked the questions. A few were squirming.

Xmucane cupped her ear dramatically. "Where are your professed words of honor for KrutEk or those Kril Legions of his who disbanded the Germs? Who here can heal the shame he feels in his heart because he left his Uayeb Legions on Nihberu."

"He did as you commanded." Dirva said.

"Do any of you feel his shame? Something he has never before done in his Earth lifetime! You are safer because of him and your Queen's foresight. Did

487

any of you suffer loss of treasure disabling the Germ insurrection?"

There was no answer from the Ovals; their pride hurt.

"Your Queen did, suffering much wealth loss. Our Confederation is safe, for now. Glory in that. My KrilChan protects me. I protect you. Your Nests are safe because of us."

BalamEk spoke. "Our Queen knows your feelings. However, she not does trust the Dinarchy. She must protect our Confederacy by challenging the Dinarchy Willows far away from Arna in our universe."

"My sisters and I are well aware of the threat." Dirva calmly added. "We are pleading we should protect our own species; which was the original intent of Seeching the Krils. Any prudent OvalChanHalach would seek peace with the Dinarchy."

"I am certain that is how you would act; if you were Queen again." Xmucane's voice was low and menacing. "Consider your thoughts, Dirva. You are dangerously close to receiving my wrath."

"You would be wise to listen and obey your OvalChanHalach." Jaguar said.

"Some of us think, some of you, may have plotted to instigate the Germ revolt." BalamEk said.

Another isolationist took the floor. "Since our first OvalChanHalach Seeched the first Kril to invigorate our species bloodline, we have been at war." She said. "Has anything changed in this monstrous war against the males of our species? Xmucane is the fifth OvalChanHalach. KrutChan the Fifth Four Striker. Five is an unlucky number."

"Do not blaspheme the Cosmic Egg with your superstitions!" BalamEk was bristling. You can speak because our Queen allows and protects your rights with our Tribunal system."

Another one said. "And what have we achieved? We should stop this war madness."

Xmucane arose, her hands on her hips. "This Queen not does advocate stopping our counter-aggression, using our Krils against the Dinarchy. I not do have the patience or the timeline to spend on theoretical political nonsense." She said.

The room was silent.

"The next OvalChanHalach will do as she sees her duty to be, and not what you think should be done." Xmucane said. "If any of you were ever to succeed me by Tribunal decree and become the OvalChanHalach; then you can command the Confederation. You are all dismissed!"

KrutChan was in a Uayeb Three strategy meeting where he met CheChun during a break in the discussion.

"Heard you was TOTL before doc did her Lazarus thing. Ah'll break a few gourds for putting you at risk. How yawl doin', Hoss?" CheChun asked.

"Leave those guys alone. Hell, I didn't see it coming, either." KrutChan grinned, glad to see the old marine. "How you doing, Chun?"

Waving his arm in dismissal, CheChun said. "Wasn't much trouble...same as always."

CheChun bumped foreheads with KrutChan after grabbing his elbow. "Sorry to hear about yer kid dying, Ace. That's a tough row to hoe."

Looking down at the deck, KrutChan said. "XibEk is pissed at me. The Ovals are to blame."

CheChun gripped KrutChan's shoulder and shook it. "C'mon, the Ovals had nothing to do with it. Kids die in birth. It happens here and happened in our past."

"Not to me, old Corps."

"We seed em. We don't birth em." Shrugging, CheChun said. "Put it behind you. Nothing can be done about it now."

KrutChan walked away. Something else was bugging him, and then he came back. "Strange, isn't it, with all their 'Oval Magic' and their superior medicines, and their elite doctors; they couldn't save one small kid?" He grimaced. "You know their attitude. Heaven forbid they should change a branching timeline."

"You could do better?" CheChun asked.

Paraphrasing the usual Oval philosophy. "Ever notice the Ovals do nothing for their handicapped? Let them make personal adjustments, is the Oval way. Did you know Akna was born technically blind."

"Ever since we met, you always were a guy that saw subterfuge in everything." CheChun said. "What're you laughing at?"

"Is that a new Southern word...subterfuge?"

CheChun snorted. "That's my EkSeet guardian-angel-asshole. He insists my vocabulary should include ten dollar words."

The Command meeting was getting louder with laughter and genial kidding among those present. Drinking Osil wine and celebrating their survival.

"Listen Krut, I gotta go. KrutEk is throwing me the 'evil eye'. He fired me when the Germ rebellion was over, and transferred me back to your half-Nacom-Malkril. I ain't supposed to be here among the elite. I came in during the break to give you my sympathies and to tell you to hang loose."

KrutChan gripped CheChun's dungarees with his fist to hold him for a moment. "Why in hell did KrutEk take you with him when he left Nihberu?"

CheChun said. "We go way back. I rode shotgun for KrutEk back in the old Uayebs. He called me one of his proletariat guard. I never saved his ass from TOTL back then, but he thinks I did. He treats me like his good luck charm."

KrutChan inquired. "Why here? Why did he bring you back with him to Arna?"

"You ain't heard? Thousands of Germs on Arna went berserk and were lighting up Tikumyax, looting, and fuckin' over the Ovals. Ah never seen Oval real estate on fire before. Did you know these Biological Oval pyramids can burn?"

"I ain't a Fire Marshal."

"I landed with KrutEk to protect his keester while he organized the Kril Legions and guards left here. I'll tell you what, that Roman fellow is no stranger to instigating blood baths! Cheeb wudda been right in the middle of the action."

"Germs ain't critters, man."

CheChun frowned. "I've seen lot'sa killing in the Pacific; but contrary to the usual bullshit, we left the civilians alone on the islands whenever we could. This fellow, KrutEk, had his Legions frosting anyone who even looked like a Germ."

"I read somewhere those ancient Roman Legions re-invented the process of 'sacking' a city'." KrutChan said. "Kinda like a search and destroy mission on dope, huh?"

"Hell no! KrutEk was already pissed about being relieved on Nihberu. His Germ mission gave him an excuse to stomp on someone. He kept muttering throughout the massacre, slaves should know their place. He was too supreme to dirty his hands wita mob of slave-scum. How do you like that shit? KrutEk was as cold-blooded as I've ever seen."

"We're all the same, I guess."

"He was fire-directing occupation Krils on other planets, massacring other revolting Germs out in the Empire, at the same time. His rampage on the Germs wasn't revenge. It was a lesson in power. The Germs won't try that shit again...the ones who survived, I mean."

KrutChan looked at KrutEk who was staring into his eyes. "Glad I missed it. You better head out before our Roman leader decides his 'favorite' barbarian is plotting with you against him. You take care of yourself. See ya on Uayeb Three."

"Fuck him and the chariot he rode in on..." CheChun said and melded, as KrutSeet was approaching KrutChan.

KrutSeet grabbed KrutChan's arm. "Return to the Laktavil Invasion projections to review your half-Malkril objectives! Will you never learn? You are constantly being observed for incorrectness."

KrutChan saw KrutEk point at him and say something to make AinAcbal and the other Malkril commanders look at him. ZacNaab and NoKoch, BalamEk's blond six-fingered Kril, were ominously watching him.

Being observed, my aching ass...more like being targeted.


XibEk, CheChun, and KrutCheebel met together with their Ovals and ArnaVals in a Tikumyax pavilion close to the Arna Ocean. They were attending another ceremony called, 'The Nesting Rituals of Renewal'. Unofficially, it was named by the Krils the 'One-Striker Adjustment

492

Center'. Ovals used the area, around the pavilion, to give their Kril warriors rest and relaxation.

In the case of the pavilion, the edifice was used to modify One-Strike Kril incorrect behavior their Soothing Ovals deemed unacceptable. The structure for the coming ceremonies was near the restricted shielded wall-field of the dinosaur jungle area. The white-brown and black striped pavilion had a huge clear dome top with open sides rolling like surf around driftwood.

Looking as if the structure could rise into the air above the sea gave atmospheric buoyancy to the effect. The smoothness of the construction was pleasing to the sight and feel of it. On the ground floor were hundreds of hard pink marble benches laid out in concentric circles, with pathways between them. What looked like Maluayeb pillars were sectioning off the benches.

The pergola sprouted imaginary wings from its base, into a seawall of the same biological material, extending from both of its sides. Relaxing Oval Nests were strung out around its marble sides, enjoying the fresh breezes from the ocean with their returned Krils. The Ovals loved the pavilion because it had none of the distant crude granite look of Tikumyax.

CheChun, XibEk, and KrutCheebel were attended by their Nests of Ovals, Krils, ArnaVals, and ArnaMals. Avoiding any discussions about Nihberu while at the pavilion with their Krils. KrutCheebel saw Reela's Nest with KrutChan was absent.

He said to XibEk and CheChun. "I was not happy when I found out after I arrived back on Arna, my Nest's been sold like slaves by my Oval into

Dirva's Nest." He swung his arm indicating his Nest mates.

Their EkSeets chimed in at the same moment. <u>The Queen, Xmucane, commanded the realignment of the Nests after the Germ insurrection. The Ovals caused, but did not execute the rebellion. Your quibbling is useless.</u>

KrutCheebel had to admit, the best news he received was his ArnaVal, ZocKuk, was transferred with him.

When Reela, as the Consort to the OvalChanHalach melded into the area, near the pavilion with her Nest, the ceremony officially started. Xmucane directed Reela to oversee the traditional ceremony. Reela was somber, keeping her thoughts to herself.

The pavilion Ceremony began. Hortim, newly promoted to Oval was in charge of the ceremony. "The Nesting Rituals of Renewal now is in session."

Following her announcement, in lines of one hundred, One-strike male Krils and selected ArnaMals, were taken out of their groups by determined individual Ovals advancing from the pavilion.

The Ovals were dressed in functional black short toga-style skirts, bare legs, with leather Willow-like spike-heeled boots. They walked into the male groups, leading them with AkSilk on the wrist. Their chosen male charges were guided to a single pink marble bench in the pavilion; others to a Maluayeb Arena pillar.

The male selectees were all dressed in blue shifts; the hems ending mid-thigh on the males. The selected males were naked underneath and were positioned in front of the Oval or ArnaVal who selected them for the ceremony. The Krils and

ArnaMal males heard some of the female Ovals and women ArnaVals in the attending crowd, hooting and cheering; waiting for the coming ceremonial sights.

Maluayeb-style Pillars in the pavilion, were attended by similarly dressed Ovals carrying Willow-whips. The Ovals were slowly raising to the shoulders the blue shifts of their blushing male charges. Causing a spattering of applause from the onlookers. The circles of benches were occupied by sitting Ovals and ArnaVals carrying their choice of Willow paddles or thin leather quirts, with their selected males tensely standing in front of them.

Waiting for what was about to happen was an interminable time for the chosen males. The Ovals in the pavilion, taking part in the ceremony, were instructing their chosen males. Their words unheard by the boisterous females in the crowd. No Stels were permitted on Ovals, ArnaVals, ArnaMals, or Krils participating. The Ovals and ArnaVals were softly and sternly lecturing their selected males.

When the majority of the Ovals and ArnaVals, nodded they were ready, Hortim walked out of the pavilion to a group of anticipating males, briskly selecting her ArnaMal. Leading him to her own designated pink bench in the center of the pavilion.

Her selected ArnaMal was CauacSky.

KrutChan was loudly hissing at Reela, his Stel fully functional. "I thought you hated this kind of violence?"

"What you will witness not is abuse." Reela was calm. "You are witnessing Oval Tradition. Our Willows were more vicious and bloody with Ovals in their tradition."

Hortim was speaking. "The Nesting Rituals of Renewal Ceremony is proscribed to remind these One-Strike males and our ArnaMals of Oval discipline. Of what the cursed Willows did to the female Ovals for disobedience long ago in our THEN. This Ceremony reminds these males of Oval control when they shame their Ovals."

She was picked for command of this ceremony by the Queen herself. She looked stern and up to her task. "I control these males in this pavilion."

An eager cheer rose from the observing crowd of watching Oval females and human women. The StelaBalaam was transmitting the ceremony throughout the Empire.

"Do something!" KrutChan snarled at Reela. His Stel transmitting to the attendees outside the Pavilion. "Whining about Hortim's twisted ways ain't going to change her! She's your goddam trainee!"

"Hortim not is from my Nest. She has her own. I not do control her."

Dirva was whispering to her newly transferred Kril KrutCheebel and his ArnaVal ZocKuk. "You deserve to be part of this ceremony, by your disobedient actions on KanBalaam and Nihberu. You should be happy not to be one of the chosen males. Dirva and her political Ovals not do participate in these rituals. We have a more ancient Oval nature."

KrutCheebel, as a One-Striker Kril, sighed with relief. ZocKuk was whispering to him not to look so glum; Dirva's nature would have KrutCheebel on the giving not receiving end when his Oval got him back to her Soothing room.

Abruptly, KrutChan pushed off Reela's restraining hand on his arm, bounded up the marble steps into the center of the pavilion. He was the only uniform-dressed Kril in the ceremony. As he stomped towards Hortim, his anger was growing.

The entire assembly of spectators were wondering what the Four-Striker was doing.

KrutChan stormed forward, weaving through the pavilion. KrutChan reached Hortim, who was smiling, diligently preparing to begin on CauacSky, who was over her knees. <u>Her child is dead. Akna is a vegetable…and the Ovals condone this crap!</u>

In controlled fury, KrutChan roughly grabbed Hortim's wrist holding her Willow paddle. He yelled at her. "This is my adopted Son, you fucking pervert! You know goddamn well CauacSky has never taken one strike of service to your crazy Ovals!"

He twisted Hortim's arm until she dropped her paddle.

There was a pause in the proceedings. The participating ArnaVals and Ovals stopped to watch. Both Hortim and KrutChan glared at each other in a test of wills.

"I'm a Goddam Malkril Commander with Four Strikes!" KrutChan was shouting at Hortim. "Try it…go ahead!" His drug-induced physcotic rage was in play. "If you hit him, I'll wipe that shit-eating grin off your face and break your fucking jaw!"

His words were translated by his Stel to every spectator.

The crowd in attendance stared in awe at the furious Four-Striker Kril. There was no doubt in their minds he would do as he promised.

Hortim kept her smile on her face. She spread her arms with her palms out, cocking her head in disdain at KrutChan. Her actions indicating, as usual, KrutChan was making his own rules.

KrutChan yanked CauacSky off Hortim's knees and dragged him, through the pavilion and the startled Ovals, back to Reela's Nest.

Hortim beckoned with her forefinger to a One Strike Kril from the waiting group. The Kril ran to not keep her waiting. As KrutChan left, a few of Dirva's Ovals in the crowd came forward into the pavilion and following his example, collected their males; taking them away from the pavilion.

Following Hortim's example, the remaining Ovals sitting on the benches slowly laid their male charges across their knees. Raised their male's hems, and began briskly applying the implements in their hands with stinging results. Copying Willow justice from their past. Blow after blow, without let up. The punishing Ovals were lecturing their charges and punctuating their remarks with briskness. The Ovals attending to the males on their pillars, in unison were whipping their males; matching their sitting sister's actions.

The surrounding spectators observed the proceedings. A few females yelling out to the Ovals in the pavilion to increase the tempo. The punishers were concentrating hard on their respective males wriggling on their laps intensifying the pace.

KrutChan pushed an astonished CauacSky into Reela's outstretched arms.

Reela was shouting at KrutChan. "What have you done!"

An angry CauacSky was screaming at KrutChan. "I am an ArnaMal! Hatched and raised by the female Ovals and the ArnaVal women on Arna! I have much Beekav for the Oval Confederation and their society! Why did you interfere with their ritual? You are a fool!"

"Stop whining!" KrutChan yelled.

CauacSky was livid. "Why do you always respond without thinking? I would suffer no hurt from Hortim! This Renewal Ceremony is expected and honored by all of our normal One-strike Krils! ArnaMals in our Confederation obey! The Ritual is, as it should be, during this Nesting Rituals of Renewal Ceremony!"

"You like it too much." He had a bitter taste in his mouth.

CauacSky furiously shoved KrutChan away from him in disgust. "You have shamed me forever in our Confederation society!"

KrutChan was taken aback. "You'll get over it."

KrutSeet's voice came into KrutChan's Stel. "Disobeying rules as usual, barbarian?"

"You condone this shit? You believe in it?"

"If I did, I would seek KrutEk's approval to tie you to one of those pillars and have you scourged fifty times by one of our palace guards. As an example of Kril discipline. This is Oval discipline. You have been enough of a disruption. Leave!"

Reela tightly hugged CauacSky to calm him down. "Think of your NOW and forget what has occurred. KrutChan hates our customs. You will soon be made 'unashamed'…by me." Reela was speaking clearly into her Stel. The spectators would not be confused by her meaning.

"Aliens would blowjob a lizard." KrutChan said. His EkSeet and the Krils knew what her really meant.

Speaking in a quiet voice, her arm was around CauacSky's waist, whispering. "When we arrive back in my Nest, I command you to appear before me in my Soothing room dressed as you are now."

His eyes to the ground, CauacSky said to her. "You will give me back my respect?"

She held his hand in hers, in reassurance. "I intend to re-engage the Ritual, with the session witnessed by BakMeer.  I will finish complete what Hortim began before she was interrupted..."

BakMeer was smiling.  "You are honored by me, Reela."

His thin lips, with grey-faced fury, had KrutChan stifling his objections.  After the death of her child, Akna's comatose problem, and all the other Oval mothering-nonsense he had witnessed in this timeline; he despised Oval cultural differences.  He vanished.

The crowd was joyous, praising Reela.

Reela grinned.  "You will experience I am more expert than Hortim in my renewal ritual session.  For many Tuns before you were hatched, my timeline was spent seated on those same pink marble benches in the Pavilion.  I have unshamed more naked Kril's and ArnaMal's than Hortim.  Be assured, not will you forget receiving my Beekav."

The females in the crowd were cheering and applauding loudly.

She was grinning aloofly.  Xmucane will be proud of me.

Before leaving with Reela, BakMeer was giggling and wagging her finger under CauacSky's nose.  "Our Oval relates this Kin I begin my training to be an Oval."  Her eyes were twinkling.  "Reela tells me I not will be only a witness, but actively taking a part in your unshaming session."

CauacSky was blushing recalling his former OtseOval's lessons as he matured.

KrutEk was holding another after-the-last Uayeb organizational conference with his Command staff.

In a dark corner, KrutChan was staying out of his KrilChan's sight; still smoldering about being dismissed at the Pavilion charade and the consequences. After completing her task on CauacSky, Reela took vengeance on KrutChan in her Soothing Room for shaming her in front of the Empire. When he came to, his nose was broken…again.

KrutEk waited for KrutSeet to finish speaking, analyzing the final tactics he employed on Nihberu before speaking. "Does any Kril here have anything to add or have a question?"

There were none.

"Our ancient Krils won the field." KrutEk said. "The barbarian and his Zabin-Krils were kept safely in reserve." KrutEk was speaking quieter, with his hands behind his back as he paced the room. "I believe your moderns appreciated their time of scratching their genitals. They must have been full of admiration for the ancients fighting their battles for them."

Here it comes! KrutChan thought. He didn't like what KrutEk was insinuating.

Tapping his white baton into his other palm, KrutEk was uncharactiscally soft in his question. "And that is why KrutSeet has rewarded the barbarian with another promotion. Because of his obsolescence in this battle."

Damn! More Kril dummies I gotta send to their deaths. My own fault. Shudda quit after one strike, stuttering or not, in the Arena.

KrutSeet voice was bitter. "When I arrived on Arna, you and I discussed promoting the barbarian to command a full Malkril. I did not act unilaterally."

"Again must I apologize? I bear the sole responsibility." With a benevolent sneering voice, KrutEk said. "I am certain ZacNaab will be overjoyed to be further below the barbarian in rank;

501

needing no apology for his cowardly actions at the bridge."

KrutEk gave KrutChan a hard stare. "ZacNaab will now have more reasons to eliminate his Four-Striker competitor."

<u>What a sonofabitch!</u> KrutChan thought. Not decided whether he was referring to KrutEk or ZacNaab.

KrutSeet eyed ZacNaab. "Eliminating other Krils seems to be ZacNaab's total ambition."

"You do nothing to shame ZacNaab. Hades cares not. Even the barbarian forgave his incompetence. TOTL is the destiny of all Krils." KrutEk sighed.

The debriefing about Nihberu ended with the final conclusion that it was a terrible miscalculation but a mission completed satisfactory.

Afterward, when the de-briefing concluded; KrutSeet approached KrutChan as the others were melding.

"Do not fail me in your NOW. Promoting you has put my head under the executioner's axe. As a Malkril commander, before, during, and after the Uayebs, you are not under your Oval's control. You can meld to any place in Tikumyax without your Oval's permission. Be mindful, however, that you are not free to act as you please. You have taken a Kril's vow to protect her; that has not changed."

KrutSeet had given KrutChan a small opening for his plan.

Later, walking from his barrack, KrutChan met Hortim and Reela in an anteroom, and asserted his rights as a Malkril Commander. He asked Hortim, as a fellow earth human, to explain to Reela he wanted to see Akna. "Explain to her it is an earth thing."

"Screw you. I don't owe you anything." Hortim said. She thought for a moment. "But, I'll do it for Reela's sake. And Akna's.

He swore to Hortim he would follow the doctor's orders and not speak to Akna. But he must see her and what condition she's in before he left on the next Uayeb.

After listening to Hortim, Reela did not relent. "I have been re-invigorated after my ritual with CauacSky. My status has increased with your promotion of full Malkril Commander."

She initiated her wand to red. "You not will leave for Akna's medical pyramid. As penance for your unconscionable act at the Pavilion."

KrutChan snarled. "I'm leaving then. Going to the Burseeosil for the Third Uayeb. I won't be back."

"That is your Kril privilege!" Reela shouted.

Before he left, KrutChan brusquely told Reela and Hortim. ""This isn't some fucking biology assignment."

"Whatever are you talking about?" Hortim asked. "What're you going to do?"

KrutChan was deadly calm. "I don't care what you two Ovals do. What the Queen commands. Or what the doctor does. Whatever you have to do…fix her…just bring her mind back." He smashed a holograph mirror with his fist. "Use your Oval Magic…."

Hortim yelled. "And if we can't? You're threatening to kill us?"

His face hardened into stone. "Zabin-Krils don't threaten. Words are a waste of time."

Reela was furious. "Ovals not can change timelines!"

"You still don't get it? I don't care if she never remembers me! Or the kid! Restore who she was; don't resign her to limbo, goddammit!"

"He's in testosterone overload." Hortim said. "Ignore him"

"You think my hand is shit...?" Leaning into both their faces, KrutChan snarled. "Call my bluff!"

Melding, KrutChan had a new Uayeb to go on.


A half day elapsed while the Kril Legions and the Burseeosil crew's timelines were converging. The afternoon wore on. Uayeb Three was approaching.

KrutChan had a brief moment of happiness when he met KrutCheebel.

"ZocKuk sends you her condolences. She wishes you to return from the Uayeb. Akna was her friend."

His heart soared thinking of ZocKuk. Cheeb doesn't know what he had lost. I'd kill to have her in my arms. Cheeb's lucky I ain't an assassin.

"ZocKuk has changed. She is more concerned about you. Since I got back, she has been cold towards me. Angrier than I have ever seen her. I cannot do anything right after Nihberu. She lost an Oval and a pile of money."

"Krils are the ones that lose in the end, Cheeb." KrutChan said.

Tikumyax turned into a ghost town again with the wind picking up as twilight approached. Flotsam of minor debris, not cleaned up by the Germs, was swirling around Jetsam of broken carts or machinery

504

that littered the streets like broken parts of a ship on reef breakers.

Signs from the Germ revolt was still visible. The biological structures still mending. The city was eerily quiet with muted lights in the Arna individual's living areas. The sing-song sounds of chanting, loud arguments, and religious singing were wafting on the breeze in the background of the city.

The broken, gray clouds revealed patches of an ebony sky full of potholes of a bright swirling paintings in the Oval universe. A universe the Krils were about to Osil-drive into to meet their timeline splits. They were to TOTL, be maimed, or survive, by chance and nothing more.

Chapter 25

The Third Uayeb of the unlucky days

Laktavil was located in a Brown Dwarf binary system. The low mass Dwarf star was magenta-purple-red. At one time it had a fusion core. Over billions of years the fusion diminished, generating reduced heat and radiation, eventually arriving at its present state. The Dwarf had eight planets orbiting but only Laktavil was close enough in the life zone to sustain water and vegetation. A companion White Dwarf, much smaller than its giant sister, supplied visible light and heat to the Laktavil planet orbiting around both suns.

The flora and fauna on Laktavil adjusted to the climates. Atmospheric conditions enabling the Dinarchy and the Krils to cope. The habitable twilight zone was less cold, livable for indigenous life, and Dinarchy and Kril survival. Because of the massive gravitational pull from the Brown Dwarf, Laktavil had a hot iron core to create its own magnetic field. Constantly groaning while flexing its crust.

The purple and black vegetation on Laktavil adapted, though sparse on its numerous mountains, craggy passes, and geothermal created riverbeds. Laktavil evolved vegetation and animals under cold inhospitable conditions.

In the glow of its suns, the Krils dreaded the Laktavil invasion on its ice and frost covered terrain; constituting the landing zones for the third Uayeb Kril invasion.

The surrounding mountains on Planet Laktavil were snow-capped in the biting cold. The mountains were dark angled, with hidden crevices, and crammed with purple blankets of snow, looking peaceful.

The Kril Legion invaders thought the Dinarchy troops had to be huddled down and waiting for their approach. The Toobs had placed some of KrutChan's Malkril Bot-units between two peaks, vacantly forbidding; definitely not where they wanted to defensively set in and mingle. The old Pacific island saying of 'get off the beach' changed now to 'get up the hill'; both sayings just reiterating the obvious. They had to move out of the valleys.

Laktavil was numbing cold. The wind chill was creeping from within themselves; only their pace kept them a little warm. Moving out in a staggered formation in route march, without resistance from the Dinarchy. The sounds the Krils heard, besides the howling wind, was the crunch of snow, ice, rocks, and dead brush under their feet, as the Kril force snaked along, following trail-breaks made by the Dagots in the snow; alone with their individual thoughts.

The Kril invasion force had been advised on the Burseeosil, they would encounter weather-worn beaten-down garrison Dinarchy forces. Predicting they would achieve victory in a couple of weeks. They had been told battles would be over in a half a Uinal.

KrutChan's Malkril meandered upwards to their assigned mountains. Cold, bored, and cursing the idiot who picked this planet. The modern Zabin-Krils complained their compasses were crazy. They were informed by the scientists, Laktavil's magnetic poles were reversed; they had to adjust their compasses to establish true magnetic South.

KrutChan's Krils reached a huge body of glacier water, similarly potable to earths, named after the planet Laktavil. They formed a U-shaped defense, facing Laktavil's southward pole, and established their lines around the Lake. Scuttling around by the Kril units during the first Kin, they waited for the Dinarchy to make contact. They sent out patrols to flush out the enemy.

Eventually, the ArnaMals of Ancient troops, with patchwork leathers and basic winter clothing, were on KrutChan's Malkrils western flank by nightfall of the third Kin. The modern Zabin-Krils with KrutChan's Bot-units, wearing more suitable winter apparel, occupied the high ground nearly in the center of the Cordillera. The northern area of their defense held reserves of ArnaMals-Ancients.

At the very tip of the U-shaped defense was the Kril Legion headquarters, with their overall Kril commander, KrutSeet, and his five subordinate Nacoms. KrutSeet's headquarters' group in instant communication with Burseeosil Central Command. Many of the modern Krils thought KrutSeet's position was tactically unsound. How could they protect a commander who was in front of them?

As his ancient engineering troops built their Castrum barricades, KrutSeet ordered a small Oval Toob-area be built so they could fly out casualties and bring in badly needed supplies. KrutSeet was a virtuoso minority of one, who adopted Oval biological equipment to meet his needs during Uayebs. His tactic could be considered a KrutSeet strategic blunder, or he could turn out to be a genius.

Predictably, the KrilChan KrutEk, high above on the Burseeosil, waiting to go to the most critical

508

battle, thought Toob zones was a waste of time and effort.

Remembering back, on board the Burseeosil, prior to the Bump, KrutCheebel had been summoned to KrutChan's area, along with CheChun, XibEk, and his other Bot-unit commanders.

KrutCheebel had been promoted. His Kril-Bot half-Katun Unit had all new members assigned to KrutCheebel. He was not happy about that state of affairs; feeling his sins on Nihberu caused the transfers.

He was aware KrutChan assigned him Krils that were considered intellectually challenged by their former Unit commanders. Although his new Krils were competent; he had heard they were difficult to control.

The upper-class Kril Legion leaders did not appreciate freebooters. KrutCheebel's new Bot-units were fierce, were unwavering; but not very reliable at following off-site orders, making their own on-the-ground battle decisions.

KrutCheebel had asked KrutChan after the meeting. "Why am I getting command over every dimwitted Kril in your Malkril?" His deep down distrust of racism fed his frustrations.

Stifling a joke about numbskulls needing educated officers, KrutChan lit a cigarette. "Have you met with them to see who they are?" KrutChan reasonably asked.

Irritated KrutChan would ask such a question, KrutCheebel said. "Of course I have!" He felt his past anger that KrutChan considered him a rookie. "I know my job."

"They need seasoning." KrutChan said. "You're an officer; a lieutenant. So train them."

KrutChan rubbed his neck. "You, CheChun's, and XibEk's constant bickering were hardly icing on my cake."

"Am I supposed to feel grateful me, Chun, and Xib fueled your racism?" KrutCheebel was feeling picked on and said so. "Thank you for nothing. Half these Kril-people cannot read or write."

"You three guys were a pain in the ass when I took over. I have a reason for putting them under your command."

"Is it an Oval and Kril secret...?" KrutCheebel said peevishly. "Are you going to acquaint me with your grand strategy?"

"Detach your Stel." KrutChan removed his Stel and placed both of them into a hidden drawer.

Seeing KrutCheebel's wary look after taking off his Stel, KrutChan said. "Don't worry about it. Removing Stels for security reasons is one of my perks as a Malkril leader. Too many gossipy ears around us."

KrutChan was quietly reviewing his Malkril order of battle with KrutCheebel, on a 3-D hologram in front of them.

Skeptical, KrutCheebel said. "You are informing me higher headquarters will use your secret plans to plot against you. You are paranoid."

"Be thankful they're not plotting against you!" KrutChan snapped back. "They'll find a way to stop me."

"I am eagerly awaiting your explanation." KrutCheebel said with sarcasm.

KrutChan was frowning. "I'm ordering you. Not offering advice."

KrutCheebel went silent.

"Any decision I make is scrutinized for deviation from Kril Code. I need a secret weapon. You're it."

"Am I supposed to be grateful?"

Studying his 3-D map, KrutChan said. "I tagged you and your Bot-unit's as my special ace in the hole. The Kril Legion's command hierarchy doesn't trust me; in fact, they hate my guts."

"Really? That is hard to believe."

Slapping his hand on KrutCheebel's shoulder. "You've been facing irrational hate since you were born. Use that now."

KrutCheebel silently understood he had inherited an additional burden. Besides racism on his shoulders since KanBalaam, he became a carrier of the lethal virus. He sighed. "What is your plan?"

"The Kril Legions are following the tactical party line of the KrilChan; not wanting to piss off KrutEk. Those Krils I assigned you have repeatedly demonstrated they will follow a leader they respect. Like them, you have a tendency to be a little rash in your judgment. Those Krils like that attitude."

"Really? Is that what you think of me?" KrutCheebel had wearily said.

KrutChan whispered. "As Malkril commander I ordered our EkSeets to wander off and ignore our conversation." He looked up at the sky. "And they better fucking do it!"

KrutCheebel wheezed. "You hardly believe they will be silent, do you?"

"You are fearless when you follow your heart, without distractions." KrutChan was shrugging; not caring anymore if he was disobeyed by the EkSeets. They knew he would deal with them if they blabbed.

"As I said, what are your plans then?" KrutCheebel wanted to get back to his units.

"I want a strike force within my Malkril units completely dedicated to you. I want Raiders. Men capable of total focus on a mission; specially trained by you and ready to meet TOTL achieving it."

KrutCheebel was thinking of the consequences. "Why me? Xib could do it. Chun was a marine raider in WWII."

"They could, but I think you're unique; an unforgiving asshole."

"You are describing yourself in detail, Krut."

"Shut up and listen! You fear death, of course, all Krils do. At the same time, you're begging to TOTL because of your Balaam virus in you. Don't deny it, I've watched you. The Krils I've assigned to you have other reasons to not want to live."

"In other words, you want to use me to manipulate their tendency for suicide, against the Dinarchy?"

"Hell yes!" KrutChan lowered his voice and calmed down. "But I don't want a one-man army."

"I misunderstood you. For all your playing dumbass; you really do have a cold-blooded mean streak in you."

Holding up the 3-D projection in front of KrutCheebel's eyes, KrutChan said. "I'm going to throw your Bot-units into the worst shitpiles I can locate on the remaining planets we invade when I feel the need."

"I think I understand. You want me and my Bot-units to take the pressure off you facing TOTL."

"I don't care what you or anyone else thinks." KrutChan sneered. "You and I can't be saved."

"So why designate me?"

"You'll want payback on every indigenous plant, animal, and Dinarchy we encounter. Those

512

Krils you got have their own reasons, but they possess the same anger."

"You want killers."

"I want your anger and theirs directed at the enemy Dinarchy forces. I need you. What the hell more do you want? A white cottage in the sticks with ZocKuk."

"After Nihberu, ZocKuk is frustrated with me. And shows it when I am around. I get sick of hearing how she admires you."

KrutChan sneered. "I don't give a shit about your Nest problems with ZocKuk." Remembering beautiful ZocKuk on the banquet balcony made him lose his focus for a minute. Am I setting Cheeb up to be killed? So I can get ZocKuk; like in the bible?

"What if I don't want this job? What if those Krils don't want it either?"

KrutChan dropped his helmet onto the table, with its 3-D projection; a hollow thud for emphasis. "In that case I'll make sure you and your Krils die keeping my secret. Krils get mangled every Ob we are in battle. At least with this plan you got options."

"I understand more than you know." He locked eyes with KrutChan. "ZocKuk keeps telling me...never mind."

"Focus Cheeb. Your ZocKuk's Beekav will be there when we get back."

"Want to bet?" KrutCheebel was hiding his anger. "You want me to be you."

KrutChan had made a tight fist. "I'm ordering you to cool your emotions down, grab them by their nuts, and make them follow. I know you can handle the assignment, Lieutenant."

"I was born for command." KrutCheebel chuckled at his absurd comment. "As XibEk is constantly reminding me, you are turning me into a

513

'lousy lifer', Krut'. I won't like it but I'll do my best."

KrutChan was laughing as he retrieved their Stels and they put them back on. "That sounds like the superior ingrate I need to lead those Bot-unit Neanderthals. Your second-in command is a grunt from my future. A woman with combat experience, named Kelel."

"Thanks a lot. I saw her in action on Nihberu. She's golden. I remember her from our first time-pause. She'll TOTL any Kril staring at her tits or ass instead of paying attention to her orders."

"You don't really know Kelel. Use her and don't 'ever' underestimate her abilities. You'll need her to control those misfits. She doesn't take crap off anybody. CheChun and XibEk will be scratching their heads, if that's any comfort to you."

"Shove it, Krut. Now I got to call breaks for sanitary napkin changes; thanks for nothing."

KrutCheebel melded while listening to KrutChan roaring with laughter.

Laying now in a Laktavil freezing snowdrift, KrutCheebel was watching his Raider Bot-unit.

KrutChan is turning into a goddamn general. KrutCheebel understood, by being in charge of his Bot-units, he would face a lot of racist crap when he gave an order. The only good thing was the woman combat soldier had her own crap to put up with when she led her Krils. Two peas in a pod.

Sounds like she knows how to straighten out any Kril disobeying her.

KrutCheebel would handle each situation as it arose; up to and including culling a disobeying racist by fire.

That is one thing KrutChan, XibEk, and CheChun taught me. He pitied his new secret Bot-units, more than himself. Almost.

After those early frosty Kins in position, awaiting further orders from KrutSeet, a cold front blasted in upon Laktavil Lake. Creating bone-numbing temperatures dropping to −25 °F.

KrutCheebel had never experienced that kind of cold in his life. There were a few old veterans in the ancient troops, who had climbed the Alps with Hannibal and his elephants, and other Krils, back on earth, from different frosty climates in their eras. A few others who had fought on the Korean peninsula in 1871 at the Ganghwa Citadel. Their descendants, who fought in the Antarctic Circle, saying back in their day it was worse. Most of the listening Krils ignored their bravado.

KrutCheebel was told by one Kril. "I was at the Chosin Reservoir in Korea. This here place is different. KrutChan wanted me to tell you guys, how to handle cold. Shit, it was miserable." The man, one of his new Raiders, was nervous, chewing his nubs of fingernails. "If I knew the secret of surviving in the cold; I forgot it. I'd say, once you're in it, watch the guys that look the warmest and find out why."

"Just for the record..." KrutCheebel said. "...the Kril Legions I have seen have never complained about traveling to the ends of this universe to fight. Not one word. We must be getting used to invading other planets; nobody cares about the science anymore."

"Queer isn't it?" The new Raider said. "No one talks about home like we used to."

KrutCheebel thought. Being miserable was the main focus of infantry and we poor bastards are all getting it up the keester, as Chun would say.

KrutCheebel's Krils were in different postures of misery. To be as small as possible against the biting wind. Coping with alien purple snowflakes. Huddling for heat amongst their miserable brethren. The humans were oozing clouds of vapor, steam engines panting, wanting to move to keep warm. They scrummed together, griping, cursing, and muttering at their gods. Nothing they said or did alleviated their chilly misery.

The Zars were cold. Their huge fur coats wrapped tightly around their hairless auburn skin. Their grunts and whistles translating wonder. Why they were here. Their tuneling efforts using six fingered-claws and opposable thumb was slow. Producing sparks from snow-covered rocks as they excavated. Frustration among them was non-existent; their easy-going nature prevailed the more they were stressed. Their anger, if they had such an emotion, was constant, controlled, and unemotional. Therefore, unseen by bystanders.

Possessing no ears or nose, the Cunacks were the worst off of the Kril force. Their slimy scaly-fish body with withdrawn stems of binocular vision were most affected. Beetle-like backs held their octopus arms tight to their bodies against the cold. They communicated with different colors changing on their skin, which now was snow-purple-white. Flickering charges of blue-static were the only indications of life in them. Their Chameleon ability was not very helpful, except as camouflage. They were normally bipedal, sentient beings, very intelligent, but highly emotional to a fault. Now they looked rock-like in appearance, laying rubble-strewn in the bluish-purple snow.

The Dagots would never freeze to death. With their Tardigrade-like body, three pairs of legs pulled in, no joints and anti-freeze blood, they were better off than the rest of the invasion force. They could survive from --459 up to +304 degrees. Absorbing 9,000 rads more than any human. Sonar charged signals were constantly sending out pulses of information to each other.

They might have even been happy, KrutCheebel thought. The bastards looked like they were in a sunny, warm garden pool on earth.

KrutChan crunched up to KrutCheebel through the crusty snow, blowing on his hands. "How ya doing, Hoss, you gonna be okay?"

"You look like a frozen popsicle."

Wearing a cold tight-skinned taunt face, KrutChan grunted. "On my way here, I seen a couple of those Zars looking your way. I can't tell when they're in heat, though." KrutChan grinned. "You're shaking so bad; you might be turning them on, Cheeb."

KrutCheebel snapped back. "Well, if they can get me warm, I ain't gonna resist. Xib says Lifers have their power to keep them warm."

"Xib's got a smartass answer for everything. He drips hate like a rubber tree."

KrutChan's attitude grew sober, staring off into the mountains. "Word just came there's movement up ahead in the passes. Kelel will help keep your misfits under control? The Burseeosil transmitted that they're coming, so watch yourselves. And don't give me any Xib street-jive."

"Control? You sound like an Oval. XibEk's the only sane one."

KrutChan looked concerned. "You okay, man, you're gonna shake yourself to death. Still got that fever from Uayeb One on KanBalaam?"

"Fever? I'd welcome that. It's the usual incurable crud. Going to snuff my young ass."

"Your vocabulary is deteriorating. Since when do you use the words 'crud and snuff'?"

KrutCheebel attempted to imitate CheChun's jargon. "Ah been hanging around youse guys too much; snuff, kick the bucket, buy the farm, cash in my chips, or get frosted. Like Xib says, in the long run it means nothing."

"Well, all the same philosopher, watch yourself."

"KrutChan, shove your sympathy. I'll be there to throw your TOTLed butt into the Seech field. Keep vigilant for your herd of assassins."

"Vigilant? Now who's sounding like an EkSeet?"

"I do not want CheChun leading your Malkril. No one is ever going to be okay."

"Thanks. Leadership and command will make you a better man." He grinned at KrutCheebel. "I'll have to pass on to KrutSeet your offer."

"Take your white-bread lifer-bullshit someplace else. I 'vont' ta be alone, as the WWII guys say."

Running off, KrutChan was cursing; slipping and sliding in the blue drifts. He looked spaced out to KrutCheebel. He was having trouble running with snow up to his hips. Checking on the rest of his Bot-units, in his Malkril, by Stel.

Close to darkness, in that two-star system; when the Dinarchy attack came, it came in the thousands, hurdling straight at KrutCheebel's Bot-units.

*I bet KrutChan assigned us here for a reason. I would not put it past him. He needs to test my Raiders.*

Nothing would stop the Dinarchy; they were screaming and focused on killing everything in front of them.

The attack was stunted.

On KrutCheebel's command, Kelel advanced in counter-attack with her Bot-units, initially assaulting a heavily fortified and strongly defended Dinarchy hill position.

She skillfully led KrutCheebel's Raider Bot-units. Cleaning out zig-zag snow-encrusted trench lines. Through a veritable hail of intense enemy fire. Overrunning a series of impregnable bunker positions before reaching the crest of the hill and placing her men in defensive positions. Suffering a painful leg wound, Kelel steadfastly refused evacuation. Kelel successfully led her units to the top of the hill. Stopping on the other side digging in the military crest. Her Bot-units settled in, advancing by fire, not by foot.

KrutCheebel was impressed with his Executive Officer for Kelel's coolness and professionalism under fire. Though she had more experience in this Dinarchy war, she did not correct or countermand KrutCheebel's orders. Kelel did her job helping KrutCheebel to control his Bot-units in the insanity. She calmed and steadied all the Krils around her. She needed no male to watch over her; only her back.

Over four Kins, for KrutChan's Malkril, there were patrols, three more fighting battles, more Krils going down from TOTL. Enduring frostbite, the inevitable pathogen outbreaks. The depressing cold and low lighting accelerating Kril minds with combat fatigue, and from wounds.

Hortim heard the command when KrutChan's full Malkril was committed to the battle. She left her station to get to her command BakToob to support his Malkril of reserve forces. She did not look forward to being forced down on that glacier Laktavil planet but she would do what she could to help KrutChan's forces or Reela would have her head. Hortim had her own Nest but Reela had been her teacher and she owed her plenty.

When the Dinarchy were beaten back, KrutChan set his tactical line. His Malkril was again in a reserve quiet area. All around them, other Malkrils were involved in fierce battles. KrutChan held XibEk back after his other Bot-unit commanders left.

"What's up now, lifer. Gonna chew my ass about sumpin?" XibEk asked.

KrutChan was irritated. "Keep talking shit. You'll be here a long time."

"That's my EkSeet mutha…you know how he is."

"Bullshit! Override him, like you do when talking to Cheeb."

KrutChan offered XibEk Osil-potion. "I want to clear the decks."

"Don't want no Wild Turkey hooch. Haven't I obeyed all your orders?" XibEk cold stared KrutChan. "How have I offended you?"

520

"You know goddamn well what the problem is with you." KrutChan spilled the rest of his potion into the blue snow. "You've always hated my guts. Lately you been surlier, more aggravating and gnarly, since we boarded the Burseeosil back on Arna."

"Who? Me?"

"The others are noticing. I can't have this bullshit contention between us affecting the Malkril. Lighten up, EkSeet! "Tell me what your problem with me is. Or I'll transfer you immediately to another Malkril."

XibEk pushed the snow with his boots, making half-moon berms while he was thinking. "I've seen you talking to yourself or some invisible dude." He shook his head. "Don't deny it. And full conversations. Your crazier than CheChun. You'd transfer me? Lifers are always paranoid apeshit."

"If you're getting battle fatigue…or turning gutless, like my voices say; you bet your ass I will. Your Krils don't need your attitude about me. Enough bullshit answers. What the fuck is going on with you?"

"Half your Malkril has battle-fatigue." XibEk had enough. "How's your black ArnaVal? How's Akna feeling? I hear she's dancing with unicorns. Don't have any emotions left because of you."

"You know fucking well she's comatose over losing her kid. And don't give my any shit about not knowing what the word means." Lighting a cigarette and exhaling, KrutChan said. "I shudda figured. I'm the dumbass everybody thinks I am. You been ornery since I was inducted into Reela's Nest and Akna became my ArnaVal."

"Don't lay that crap on me!"

"She Soothed me!" He hissed. "I didn't pick her!"

521

"You're a goddam liar!  You enjoyed having a black sister fall in love with your white fucking ass! You got what you wanted from her.  Hated her for it. Can't have a black women part of your racist world, huh.  What would yer parents say?"

Shaking his head, staring at his Krils milling around trying to stay warm.  KrutChan wheezed. "Yes, she said she had Beekav for me.  Yes, I shied away, but not for your goddam racial fantasies."

"Bullshit!"

KrutChan field-stripped his cigarette.  "You ever meet her?  Trouble was, after she Soothed me, I felt like a supernumery.  Akna always sees her previous Uayeb Kril, who met TOTL; not KrutChan." Brushing blue snow off his rifle chamber, KrutChan added.  "After Nihberu, I wanted to make it up to her. Shoot me.  Her loss of her kid hurt her bad.  She loves her dead Kril, dummy."

"She was pregnant when she Soothed you, asshole?  Quit lying."

KrutChan was biting his lower lip.  Then he ordered his EkSeet and XibEk's to shut down.

He reached across, pushing Xib's hand away, and buried their Stels in the snow, under a rock. "Listen for once in your life.  Akna broke some fuckin Oval rule.  I was Soothed by her to hide her pregnancy by her true lover, the dead Kril her Beekav was for."  His red dope was enraging him.  "Shut up! The kid that died was his!  Not mine!"

XibEk looked confused.  "She lost that Kril's kid?"  He snorted.  "You treated her like shit; that's why her mind's screwed up!"

"No way!  Her Kril lover and I both walked away.  Her Kril meeting TOTL on a Uayeb.  That's what blew her mind.  Think how her dead Kril would feel, knowing that."

His face contorted in pain, XibEk was jabbing the snow between his boots with a thick purple twig. "Let me give you the word, honky. Before you were Seeched here, I was in another Nest with Akna. We fell in love as soon as we met. With her bad eyes she couldn't even see me."

His grimace had no humor in it. "I screwed up on an operation before I Burseeosiled back to Arna. As my punishment, I was transferred into another Nest. They must have told her I met TOTL to save her shame."

KrutChan was stunned; the cold adding to his feelings. "I didn't know. Don't look at me like that...I didn't. First you dying, me leaving without a word, and then the child dying. Her stress must have been unbearable."

"That's no fuckin excuse. Would any of it have made a difference?" XibEk snarled.

KrutChan's anger flared. "Goddam right, it would have! Why didn't you say something to her at Reela's banquet?"

"I was TOTL, remember? Ovals don't like having their plots revealed. Before you showed up, I swore to protect her. Think I'll put her in danger?"

"I did it for the same reason; to protect her. I know...I know you don't believe me!" KrutChan lit another cigarette. "Fuck! The goddamn Ovals screwed all three of us, killing you off and lying to her, bullshitting me, and then killing her kid. You know goddamn well they let that kid die because they thought it was mine, same as you!"

XibEk shook the snow off and put his Stel back on, while standing up. "Now you know why I hate your honky ass." He was subdued.

Leaning against a tree trunk, KrutChan replaced his Stel, snapping at XibEk. "Do you want a transfer or not?" He belligerently walked towards XibEk.

Shouldering his M-14, XibEk spit back. "Course not! I'll do my job, like a good nigger. I wanna be there when an assassin or Dinarchy wastes you!"

"Yeah, I been noticing." KrutChan laughed. "Every swinging dick in the Legions wants to arrange my funeral."

He gave KrutChan a two-fisted Kril salute, leaving before the lifer could return it.

Holy shit, Akna and XibEk were lovers. Xib had plenty a reasons to hate me. Now I gotta put XibEk to the top of my assassin's list. KrutChan thought.

Al was sitting in the blue snow with his back against a purple tree. ["That sounds like priority planning."]

Another patrol during another Kin occurred; similar and different at the same time. Laktavil hung in the twilight, cast by the twin brown and white dwarf suns. KrutChan's Krils were set in looking down into the ancient caldera surrounded by ridgelines thrown up by a cataclysmic eruption millions of years ago.

The rumbling sounds foretold nothing of what was coming. The ground was shifting and trembling, sending avalanches of ice and snow cascading down the slopes, as purple snowflakes were whipped around from the disturbed atmosphere. The Krils watched the liquid in the partially-unfrozen ponds and creek areas sending rippling waves from their center to the banks as tremors shook their area.

"What the hell is happening?" An anonymous Kril shouted into his Stel.

"The Dinarchy got tanks and heavy equipment!"

"Definitely sounds like a Panzer Division on the move!" A WWII vet said.

"Bullshit, I'm blind in this falling snow crap! Don't see nuthin'…are ya hallucinating?"

"All of you shut up and stay off your Stels!" KrutChan yelled over his command Stel. "Report if you see something! Keep your eyes open!"

Slack-jawed in wonder, their fear of impending death was choking their hearts when the unthinkable happened.

Far across the Caldera, KrutChan watched through his binoculars as another of his Malkril-Kril Bot-unit's ran in panic in a one-hundred-eighty-degree arc away from a white-wispy blowhole. And then a huge water-slush geyser exploded out of the terrain like a hose with a leak in its side. Rising slowly, rising with blue-white-water-slush.

Within seconds, the blowhole was screaming like poorly fitted metal valves grinding. The enormous geyser grew in size forming a two-mile-high erection. When the geyser plume hit the top of the next layer of atmosphere, a twenty-mile-wide ejaculate of white-blue-red mist spread out, capping the eruption.

"Holy mother of god, is that some kinda Dinarchy atomic secret weapon?"

Another Kril answered him. "Ain't none of you morons been to Yellowstone Park? That's a little old geyser, dummies."

Looking through their binoculars, many of the commanders on the rim of the caldera, watched in close-up horror.

The supposedly harmless extra-terrestrial geyser instantly killed a lot of Krils by the force of

525

the explosion and concussion. Quick-freezing other Krils nearby. The Krils closest to the blowhole, were ripped into shredded uniforms and body parts. The geyser's expansion was breaking the other flying glacial-Kril-bodies apart when they collided with the ground, like a rubber ball hit by a hammer after being immersed in liquid nitrogen. The few Kril survivors were crawling and stumbling away leaving elongated dark trails of blood or debris in the snow cover.

The Stels went active with running commentary. "What the fuck!"

"Them poor bastards…"

"Shit…shit…shit…!"

"We gotta help them guys!"

CheChun sadly said into his Stel. "Told you there's more to worry about on the Uayebs than Dinarchy."

"Keep down and watch yer own asses until it's over." KrutChan was calmly announcing. "Nothing we can do for them."

"Ya piss on angry gods, you be fuked." XibEk, praying again. KrutCheebel thought.

After ten minutes of continuous ear-splitting noise, the geyser slowly lost altitude, as if a faucet was turned off, changing in sound from a screech to a lower pitched scream, to a gurgling rumble, and then to complete silence.

In its aftermath, over fifteen minutes of timeline, blood-red ice and blue-white snow began falling from the white-blue-red crown it had created in the sky. Some of the huge falling white-blue-red chunks were killing more Krils.

Almost immediately, assessments were coming down from the Burseeosil EkTsab. "This planet of Laktavil is ice volcanic. Not is lava or rock from its heated core." The scientist sounding like a laboratory experiment was within acceptable parameters.

"That is refreshing. Those Krils didn't burn to death. Instant popsicles a more humane description. Lucky them." That voice had to be KrutCheebel, XibEk thought.

The Burseeosil was still broadcasting. "Avoid the blowholes as much as possible, especially when there are tremors in the area. There is an ocean of water-ice under the crust of this world, five Obsils down. Pockets of liquid nitrogen are propelling the eruptions...."

Their EkSeets were overriding. "The gaseous Nitrogen now warmer is harmless to the Krils. The Oval science teams are saying there are patches of olivine layers which are converted under pressure and hydrothermal vents forming the base of the blowholes."

"Thanks for the clarification, assholes."

"Anybody know what the hell olivine is?"

KrutChan yelled. "You EkSeets shut down. You're only confusing your Krils."

The Burseeosil was still droning on. "...These geysers are deadly to most of the Kril species caught in the eruption, except for the Dagots. If you are too near the blowhole let the Dagots cover you; they react faster than you can."

"Now you tell us..." One angry Kril muttered into his Stel.

CheChun remembered talking to UDT swimmers back in the Pacific. They had been sent in by submarine to check out obstacles, types of sand, all kinds of top secret stuff, and tide conditions before each invasions. "You goddamn Ovals ever recon these fukin planets?"

527

"Luckily, the caldera formation is the only area with the potential for these types of eruptions" The EkSeets chimed in additional information.

"Luck don't mean shit." XibEk said.

The freezing geyser grimly reminded the Krils; on invasions there were numerous ways to meet TOTL.

Kins later, ordered by KrutChan, KrutCheebel's Bot-units had settled in for the night on top of a hill shaped like butterfly wings, named appropriately, Pepam1150, after the Arna insect deity, indicating planting season.

KrutCheebel's Krils overlooked the blue glacier gouging the plains below. Where a fierce two Kin battle had raged back and forth. The Dinarchy side winning, and then the ArnaMal side winning, see-sawing between the Dinarchy troops and the ArnaMal ancient troops. Eventually the ArnaMal ancients were victorious in routing the Dinarchy.

Dinarchy pockets of resistance were being defeated as darkness dropped on the battlefield. Pepam1150 held the right flank of the KrutChan's Malkril forces.

Atop the hill, during the next Kin of fighting, the enemy forces KrutChan's Malkril encountered were of less than squad size and were easily disposed of by the Krils. The Dinarchy's modern Special Forces presumed sent to conquer Pepam1150 in a flanking movement to gain the high ground over the plain below.

Ordering KrutCheebel's raider Bot-unit there, KrutChan joined their ranks. He was to hold the high ground; to TOTL all his Krils to accomplish the mission.

There were four intense courageous charges by the Dinarchy. The Krils were respectful of their tenacity. The ground-searing fire from strafing Toobs and Kril artillery obliterated many of the attackers.

Kelel led her Bot-unit directly at the enemy, disappearing in the mass of bodies. Over his Stel, KrutCheebel was cursing at Kelel to get her Bot-unit back to her perimeter position in the line.

Blasters, grenades, K-bars, swords, helmets, spears, and fists flourished; but the obstinate Krils held onto Pepam1150.

The Dinarchy drifted away maimed, bloodied, forming into small groups of survivors retreating to assembly areas.

The Krils let them go, not wasting ammunition on the stragglers. Standing down, the Kril's adrenaline drained, bringing on shakes and tremors in their limbs.


"Jesus Christ, it's cold!" XibEk griped, "This black jungle-fighter's freezing his fuckin balls off…"

A North Vietnamese infantryman next to him, wearing Russian cold weather gear, agreed. "This frozen monsoon weather is evil. I have no anger left in me."

"Don't mean nuthin, dude."

"Buddha said, 'Holding on to anger is like grasping a hot coal with the intent of throwing it at someone else; when you are the one getting burned.' I would relish a 'hot coal' in my palm."

XibEk was nodding. "You got dat right, Mr. Nathanial Victor. Iffn I had a burning coal chunk; I'd share it wid ya though."

"Enemies we were, though we are fellow Krils now. Enemies always share discomfort and pain."

Many other Kril voices added additional comments on the cold conditions.

"Shut up! We're tactical. No more grab-assing!" KrutChan's snarl emitted from their Stels in a hoarse whisper.

XibEk was chuckling to himself. He saw the NVA and the rest of his Bot-unit throwing obscene salutes into the air towards KrutChan's position.

KrutCheebel's KanBalaam bugs were thankfully heating his insides, attacking his system again. When the virus backed off, he would be shivering with the rest of them.

On top of Pepam1150, the silence was somber and forbidding as the Kril's frozen time wore on. Dawn would be bleak and a long time coming in their NOW. The icy wind by osmosis was chilling them from the inside out. They had to remain stationary, acquiring no mercy from the heat-sucking chill as the night got darker. They could see only darkness-shadows and the purple background snow through squinting eyes. Their wind-driven tears freezing on their faces.

The Zars in the Bot-units had excavated fighting holes for the Krils around the butterfly perimeter on the military crests of the hill. Afterward, digging behind the lines, the Zar had bored into the hill, snugged below the frost line and were comfortable, away from the temperature and wind.

KrutChan put his forces on fifty percent alert. One Kril sleeping, while the other half, remained awake and on watch. A sweeping cold wind shrieked over caressing across the hill. Their Stels were maintaining silence, except for one word transmissions of 'Okay?' to keep contact. The answers, one word 'yes or no' in reply.

Brief firefights broke out in various areas, followed by grenade blasts in response, keeping them alert. The Dinarchy stragglers were probing the defenses. More than likely, the Dinarchy were lost and trying to find a way out, to go to their assembly points. It would be a creepy long extraterrestrial night.

Hours later, CheChun was dug in with KrutChan, out in front of the line. Ripping from out of the silent darkness, a tremendous flash of light, followed by concussion blast enveloped their hole. CheChun's experience, from his Pacific War and Japanese night fighters, decided it wasn't incoming; but far to his left down the hill. CheChun figured some Kril had been throwing ordinance at a lurking shadow.

KrutChan instantly came out of his slumber at the bottom of the fighting hole and stood watching next to CheChun to their front. Who can sleep now?

CheChun heard a hard-scrabble noise behind him and swung his M-1 towards the sound. He released his safety, with a soft click, and peered into the dark watching for movement.

"Pssssst...psssst...hey, it's me." A hoarse whisper floated in the frigid darkness.

"What's da password? Advance and be recognized." CheChun said quietly and quickly, thinking. Fuckin Dinarchy infiltrator behind us?

CheChun pointed his rifle at the moving figure less than three feet from him. He prepared to fire. Fuck orders; this guy's way too close. One shot...right in his fuckin...just one shot.

A new Kril's white face appeared above CheChun's front sight; the guy's eyeballs huge with horror at his imminent demise. "Hey! Don't shoot, I gotta see KrutChan." He was petrified with fear.

CheChun snarled. "Ya ever think of using the password, asshole?"

The guy was shaking with fear, almost crying. "I forgot coming over."

KrutChan was next to CheChun. "Hold yer voice down, dammit. You found me, genius. Who you with?"

"I'm alone. I'm from KrutCheebel's Bot-unit. He sent me over here to get you." The new Kril was shivering, more from fright than the cold. His imagination replaying, over and over, CheChun's muzzle blast tearing his head off. His owl-like eyes were shining with fear in the dark.

KrutChan spoke to CheChun. "Keep your ass in the fighting hole, pass the same word down the line. I'm moving over to KrutCheebel's hole. I'll send another Kril you can babysit. Hang in there, old Corps." He scuttled into the darkness prodding the new guy into leading the way.

"Watch your own jewels, mud-marine." CheChun whispered into the darkness.

KrutCheebel was waiting for KrutChan with six of his Raider Bot-unit Krils. Out of their holes in a clear area trampled in the snow peering diligently into the darkness.

When KrutChan knelt next to him, KrutCheebel pointed to an unoccupied blue-snowy rock for KrutChan to sit on.

Infantrymen were infamous for bunching up after action. KrutChan ignored it. Misery loved company, he thought and snarled. "Your raiders a little antsy, Cheeb?" He bit his tongue over using the "R' word. What you got?"

KrutCheebel pointed at the new guy who had brought KrutChan. "My guy saw something." KrutCheebel said. "He did the right thing, instead of firing he became artillery and threw his baseball grenade."

The new man was stammering. "There was…about six of them…in a bunch; the one dude …raised his head towards us…jabbering something, pointing at me." His fist punched his body armor. "I fixed the bastard."

KrutCheebel broke in. "It could have been a bush moving; the wind has been pretty brisk all night…even the shadows move with the moving clouds overhead."

"That wasn't a bush!" The new Kril insisted. "Or moving shadows…I heard groans after the explosion!" The man said in justification for his actions.

KrutChan was getting in his usual mood all his Kril Malkrils recognized, as he snarled hoarsely. "Keep your goddam voice down!"

He was thinking for a minute. "Are you sure you saw six of them? With deer antlers; the forks multiply when you're out hunting."

The new guy was wavering. "Well, I might be wrong on the numbers." Then his made up his mind. "There was more than one; I'm sure of that!" His expression said he was sure.

Speaking to KrutCheebel, KrutChan asked. "Why the hell you send for me? I'm busy shuffling around my Malkril units, Cheeb. You planning your retirement party?"

"Organizational Kril Code says leaders lead from the front. Bitch at KrutEk, not me. Nobody more out front then we are. You are the general."

KrutChan was staring in irritation at KrutCheebel. "The new guy's your man...what do you say? You call it."

KrutCheebel's anger was rising; he had enough of this discussion. "My guys are jumpy. Hell, after today, we all are jittery. Pissing our pants from any noise tonight. But, he did okay. He did what he was told to do."

"I asked 'you'." KrutChan was so goddamn calm. "My job isn't to comfort your ass. Run your own Bot-unit. What do ya say?"

"I'm saying...I'll back him up. I think he saw...some kind of shit...and responded correctly."

KrutChan punched KrutCheebel's arm affectionately. "That's good enough for me, colonel." He smiled at KrutCheebel. "You better work on your syntax, college man. Krils, like XibEk, are corrupting your educational background."

KrutCheebel was grinning back. "As the immortal Bard once said to Francis Bacon, 'Fuck it. You are a plagiaristic asshole'."

After they stifled their humor, KrutChan sobered up and became deathly serious. "If he's not dead; that Dinarchy dude will TOTL with my hand over his mouth."

The six Krils were watching him.

"You going out alone? KrutCheebel asked suspiciously.

One of KrutCheebel's Krils spoke. "I'll go check it out. Tired of sitting and freezing my ass off."

"No way, dude." KrutCheebel waved him away dismissively. "You're too blood-thirsty.

KrutSeet wants prisoners. He has enough corpses. Besides, this calls for higher echelon action."

The guy said. "I'm high enough on blue AnticArna."

"My MOS says I gotta be out in front." KrutChan glanced at the volunteer, silently thanking him. KrutChan inquired of KrutCheebel. "You volunteering to go out there for me, Lieutenant?"

KrutChan calling him by his former rank irritated KrutCheebel. Being a survivor was KrutCheebel's aim in this lifetime. "Hell, no...you got the four strikes and the brass balls. I called you here. You are in command; not me, right?"

"Spoken like a true TOTL-proof warrior." KrutChan punched his finger in the air. "You make sure 'everyone' gets the word a Kril is out there."

KrutCheebel grinned. "I won't use your name. Too many guys want to have a shot at you."

"My password will be 'Fuk-her' when I'm on my way in. The countersign will be, 'You fuk-her; you seen her first'. No goddamn Dinarchy is gonna figure that one out."

Before KrutCheebel could answer, KrutChan put his K-bar in his teeth, his M-14 in another hand, and crawled away into the darkness.

"You guys see that shit; a goddamn K-bar in his mouth...someone's gonna die tonight."

While others smiled with derision and some in awe at KrutChan's bravado, they moved back to their positions.

When the bushes rustled with KrutChan's departure, somebody else muttered, "That is one crazy mutherfucker."

Chapter 26

KrutChan was not irrational or feeling particularly courageous; just stupid.

Al was crawling next to him. ["This is my kind of dance. You need me, KrutChan."]

Muttering, KrutChan said. "Get outta my face. You ain't wanted. Stay with the unit in the perimeter."

Al disappeared.

Are you speaking to me, KrutChan? EkSeet said. Where your Stel goes, I go.

I'm sick of your diarrhea of the mouth. I'll throw my Stel back into the perimeter, if you say one more word.

That silenced EkSeet.

The good news was his movement was getting him warm. He moved as quietly as he could in the crunching snow. Sneaking past scrubs without touching the brambles and bushes; stopping now and then to listen. He heard the whispering wind and occasional coughing from atop the hill. The further down he went, the lonelier he became. KrutChan fought the familiar acid-forming-fear collecting in his throat. He fought off a tickling cough. It's colder than a frog's ass in hibernation. He paraphrased in his mind. If I get wasted, they won't find my dick until spring thaw; if this planet ever unfreezes.

He moved on.

In the dark, the purple-blue snow ahead had a blackened blast area outlined and he went to it. KrutChan stopped, listening intently before he approached the small crater; more like a rounded sump hole, in the middle of the blast area. The heat in the mound had cooled down in the air. He noticed disturbed snow and footprints leading away. He tracked a blood trail and scuff marks, his deer hunt skills returning.

He checked his watch, which was still on Arna time. Time was relative; it didn't have to be absolute. He had been on track for a half hour when he stopped to listen to a muffled sound ahead of him. He became silent and stiff, and then kneeling for a couple minutes, to present less of a target.

Goddam right; those are voices. He told himself.

He absolutely did not want his Stel suddenly crackling; sounds carried a long way in the bitter cold and snow. He took his Stel off and hid it under an easily recognizable huge black stone he found. He added a trio of dead tree branches, teepee-like, next to the boulder. Easier to find later.

Snowfall was increasing with larger flakes. Visibility was decreasing; but the snow-cover gave him illumination.

He moved on, hunching over, then stopping, for a long fifteen minutes, by his watch. The voices were carrying in the freezing night air.

Getting closer to the sounds, KrutChan slithered ahead quickly when the wind blew, being as quiet as he could, approaching a gash in the terrain. And then he was above a muffled conversation in a gully. He slowly stood up, so as not to draw attention to himself, looking through the brambles of the purple-black bush in front of him.

Next to a glowing red field stove, he saw four figures. By their uniform and headbands he could see they were Dinarchy. He raised his M-14 to take them under fire.

Consider the risk, butthead, he was telling himself. Once you start firing, you'll give away your position. Maintain fire disciple. You're only one guy. Check out the surrounding terrain first. See if there're other Dinarchy before you commit.

He made out three of them were wounded; two of them really bad. One guy had lost his legs; the other guy was sucking air, wheezing with a chest wound. Another Dinarchy was a woman; great looking face and had an Oval's figure, even in her dirty uniform. She was working with one hand; her other arm was bloodied and hanging useless at her side. The big guy of the four was in charge, sweeping the top of the gulley with his eyes. KrutChan became immobile to lessen the chance of being spotted. He decided he would take out the unwounded big guy first.

While he was pondering when to fire and kill or take them prisoner, he crouched, shifting his body into a clear area.

Fate made his decision for him.

The top of the gully crumbled. Riding a huge mound of dirt, grass and purplish aging snow; KrutChan crashed into the middle of their midst.

When he hit the rocks below, KrutChan's training and agility is the only thing that saved him. He rose in a crouch with his M-14 aimed at the unwounded leader carrying a weapon. The piece the leader carried, looked to KrutChan like a Russian SKS WWII semi-auto rifle. How the fuck do I recognize that? Oh, yeah, I captured one in 'Nam.

The Dinarchy Leader and KrutChan fired at each other in simultaneous explosions.

They both missed.

The Leader's weapon malfunctioned and he couldn't fire again. Either his weapon jammed or was fucked up in some other way, KrutChan figured.

KrutChan was aiming his rifle at the leader's head. In a time-pause instant, he noticed the other wounded Dinarchy were frozen with fear and had no visible weapons. The leader's weapon looked like the grenade the new Kril-guy had thrown, splintered the stock and probably bent the barrel. KrutChan lowered his M-14 to his hip but kept it aimed in their direction as a threat.

The tension accelerated. KrutChan was deciding upon taking prisoners or wasting them. The Oval-Dinarchy gal was smiling at him; probably to distract his attention. No way, sister; I ain't playing any Oval Soothing games.

Flicking the end of his barrel up and then down at them, KrutChan dictated the universal sign for them to sit down.

Tense minutes went by as the five of them assessed each other's intentions; actually now four, because the guy with no legs looked dead.

The Dinarchy Leader spoke to his wounded and was getting short answers. KrutChan warily watched the conversations and concluded the Leader was doing as he himself would have done. He guessed the guy was asking if they saw anymore Krils and getting a medical update.

The dead guy's shredded lower torso indicated to KrutChan the grenade had been between the guy's feet when it went off. From the look of the woman's bloody arm, she had been an innocent bystander, probably crouching to avoid being hit. The sucking-wound guy had a mangled device, probably for communication, hanging useless from his shoulder. More than likely it saved his life. He caught some of the blast-shrapnel in his face and chest. The Leader, KrutChan saw, had only minor shrapnel wounds in his neck. He was the most dangerous one for KrutChan to watch.

Holding the trump cards, KrutChan yelled up the gully to no one. "Cease fire...and hold your positions...I got 'em!"

The leader guy began loudly laughing. "Good bluff...what is that American game? Poker...correct?"

"Your English is thick, dude. You ain't Russian, are you?" The guy had a Slavic accent; he could have been Russian or from the Baltic area, as far as KrutChan could tell. He did have an elongated skull of a Willow. He almost looked human, in a way.

"I speak many languages and dialects. No question. We are alone...as 'you' are alone."

The leader guy changed his accent to stilted English. At least KrutChan didn't need his Stel for translation; but KrutChan kept wary. "Looks like a Mexican standoff. Who shoots first...one of you...or me? My rifle works. Yours is shit. Throw it away. You wanta die...er...you want to TOTL?"

"No one wants to TOTL. All here are in a quandary. Maybe you will get one of us, but if we rush you; one of us will get to you."

"Now who's bluffing?" KrutChan was nervously toying with the trigger on his rifle. "Speak for yourself, Custer, us Injuns got you in a cluster-fuck. You're the only one still functioning, Ace. Those three ain't in any condition to assault anybody. And besides, I got two or three rounds left for each one of you. You first, of course."

"We can try, Kril...we can try." His tone was either threatening or dangerously courageous.

KrutChan snorted his disbelief. "You plan on fighting this fuckin war over a goddamn gully?

540

Maybe you should be more concerned for your wounded?" KrutChan saw that took some of the fight out of the Leader.

A minute or two of silence passed.

KrutChan continued. "I could waste all of you and go back for warm chow."

"You are too experienced to commit murder." The Dinarchy leader scoffed at that remark. "You are still deciding what to do. When we live or die, I will have my answer."

"It would be simpler for me than trying to get you back as prisoners. I assume you wouldn't give me your word to go peaceably as my prisoner?"

The contempt on the leader's face was evident. "I would never give my word not to escape, Kril. I would TOTL you if you attempt to take us prisoner."

"You're helping me with my decision, asshole." KrutChan liked the Dinarchy's answer. Never give in to an enemy or show fear.

"I have killed many Krils. You will be the last."

Now it was KrutChan's turn to laugh. "Get in line, buster. I got a hundred Krils waiting to waste my ass ever since I got here. You have to wait your turn, tough guy."

The Leader's eyes narrowed, his mind was humming with thoughts. "Ah, I see. You must be that Four-Striker Kril that our prisoners have spoken of in our interrogations. Our troops from two other planets say you are invincible. Obviously, you are he?"

Damn! The Dinarchy had an excellent intelligence service too. How do I constantly get myself into this kind of shit? KrutChan wished EkSeet was here. He decided to disclose as little information as possible to this guy. KrutChan laughed to hide his confusion. "That Four-Strike-dude is a myth. If he exists; he's a pussy...sorry lady." He bit his tongue. She didn't understand him; at least he hoped not.

"You are saying then you are not he?"

KrutChan decided a little disinformation was called for here. "That Four-Strike guy fucks up more things than he accomplishes. I never rely on gods; they got feet of clay." He made a circle with his M-14, pointing each of them out, one at a time. "You better worry more about me and this rifle...and my intentions."

"I think you are a liar." To pacify KrutChan, the Leader held up his palms. "I believe we need a truce. An armistice, if we are all to survive."

The leader threw his rifle away, flipping it, end over end, over the top of the gully. It was useless anyway. He went to the wounded others disarming them of their hidden knives and other offensive weapons, throwing them away in different directions. He wasn't surrendering. He was acceding to KrutChan's dominance.

He turned back to KrutChan. "They do not understand your language. The Dinarchy do not possess Stels. Unlike the Ovals; we trust each other."

The woman said, "Kravid, [we must get him back to our lines. He will TOTL, if we do not hurry.]"

KrutChan didn't understand what that Oval-Dinarchy chick said, but he hid his surprise when he heard the word 'Kravid'. He tried appearing stupid; he was good at feigning ignorance. This Leader guy was Kravid Palatine, the high-mucky-muck of the Dinarchy. He ain't hardly about to become my prisoner. It was a distinct possibility the Dinarchy

542

committed their leader to the forefront of a battle just like the Ovals demanded of their KrilChan.  <u>Goddamn the luck!</u>

KrutChan sneered. "Kravid Palatine, huh? Now we both know who we are in this gully."

Kravid pointed at KrutChan. "I see you are missing 'your' Stel."

"I was with my patrol coming to capture you guys. I'll rely on your people's body language. If one of your people farts, I will blow your ass away first. How's that for a peace treaty? Stels or no Stels."

Their leader was somber. "That is also an obvious lie. There was no patrol. Let us both speak more clearly since we are under a temporary truce. No more lies. Stels do not get lost. You took off your Stel so as not to alert us that you were coming. You wanted to be covert. That was intelligent of you, showing a wise tactical decision. It worked."

"Yeah, I'm a real Einstein."

"Your EkSeet will report your silent Stel, however. You are out of touch with your Kril command. We have the same problem; as we are unable to contact our assembling force. Do you wish to know their names?"

"Hell no. I don't give a fuck who they are…if I have to kill them."

"There is that." The leader smiled. "We will win then. You will see us forever in your nightmare MishMells, no?"

"My EkSeet can track my Stel. They'll know where I am." KrutChan didn't believe that any more than this Leader guy.

"We both know that is not a lie; it is more of an exaggerated fabrication. The EkSeet can pinpoint the position of the Stel but that tells him nothing, since you are not wearing it."

"Cut the shit. Let's decide what this truce is all about."

"A ceasefire is more accurate." Kravid said. He seemed to be waiting for something to happen.

KrutChan motioned to the wounded. "That guy with the chest wound ain't gonna last long in this frigid air. And your female's arm looks like frostbite is setting into it. You better think about them. As far as I'm concerned I hold the winning hand. How're your wounded fairing? I think they want the war over."

The leader translated KrutChan's words for his wounded and then addressed KrutChan. "I have been fighting this war much longer than you have. I want you to know my beginnings. I am a Willow. I used to be a Kril."

Somehow, that did not placate KrutChan. "You were a Zabin-Kril. What era on earth did you come from?"

"I was not Seeched!" The leader waved his hand. "It does not matter. During my time in the Krils, they were not called Zabin-Krils. I was one of the original Willows not banished by the Ovals. Oval females did not hate all of us. As one of the original Krils, I too fought the Willows before they became the Dinarchy. I fought my brothers to save the Ovals I loved. Oh yes, you can be skeptical, but Willows show their love differently than the Ovals."

"This lousy iceberg of a planet doesn't have a real dawn. We're wasting our timelines."

"For brevity, since eventually morning will come and even a dull daylight puts a different perspective on things. I was not captured, I surrendered to the Dinarchy a very, very long time

544

ago before your THEN. My mate had met TOTL at the hands of the Ovals for Oval errors. Do not look surprised; back in my THEN some Ovals were quite bloodthirsty."

He backed away to make KrutChan feel more secure. "At the last, when I was still a Kril; I served under a Five Striker. The only Five-Striker the Ovals ever anointed. He was an incompetent dullard, a fool full of glory for himself. The Dinarchy was winning the war at the time because of that KrilChan. After my Oval-mate was killed, I grew weary of Oval sanctimonious controlling ways and my KrilChan's bad leadership. I surrendered to be with my fellow Willows."

"Turncoats always have their reasons. You broke Kril Code." KrutChan said.

"I am aware of Kril Code and their methods. In the beginning, I instituted parts of that Code."

"My EkSeet told me Kravid Palatine had a similar biography." Seeing some recognition flicker in the guy's eyes, with irony KrutChan sneered. "Kravid must be proud of himself."

The wounded female looked astonished.

The Dinarchy leader dismissed that thought. "My forces. My Dinarchy troops are surely looking for me, at the present, so we must act quickly."

KrutChan's ears perked up at the guy describing his units as forces coming. This guy wasn't no private.

"For your sake, should not your Malkril troops also be looking for you?" The Dinarchy leader said.

KrutChan lied. "I'm aware of Dinarchy Warrior Code. It's the same macho-military bullshit all armies have to discipline and train their troops to kill. It seems I've been fighting all my life. Your Dinarchy methods are not new to me either." He saw, by his expression, the Dinarchy Leader was assessing KrutChan the same way KrutChan had been assessing him.

"We are comrades-in-arms, so to speak."

Waving his left arm in a circle above his head while continuing to hold his M-14, in his right hand pointing at the leader, KrutChan said. "I'm sure the bloodhounds are searching for me. Let's end my standoff." He reached behind his back. "You need my compress from my cartridge belt, for her wounds? It ain't much...but I'll give it to you to bind your Oval's arm."

"That will help her."

KrutChan took out his hand drawn map overlay, an old habit of his; wrapping earth plastic around it, to protect the overlay from his body's condensation and the snow. "If you secure this plastic around the holes in that other guy's sucking chest wound and bind it tightly; it'll help him breath. It ain't perfect, but it'll help him. It'll keep the shit out of his wounds too."

The leader was mulling that over, evaluating the situation. The sucking-wound guy was having a tough time breathing. He was on the edge of TOTL. The Dinarchy leader nodded in thanks to KrutChan, as he took the plastic and instructed the woman on how to use it. It had an immediate effect on the wounded man, after puncturing to inflate his collapsed lung.

With his hooded look, Kravid hid his thoughts.

Momentarily daydreaming, Snake was seeing himself in Vietnam calling on the company TAC and announcing to the third Marine MEF, 'I got a NVA soldier named General Giap. Could you send me a chopper'?" They wouldn't have believed him in Vietnam.

KrutSeet, that conniving Commander of his, wouldn't believe KrutChan had captured Kravid Palatine. This Kravid guy would TOTL first before KrutChan could bind his arms. What the fuck have I got myself into now?

The Dinarchy woman was attending to the wounded guy; attentive to his vital signs, but could not conceal her worrying. KrutChan said to her leader Palatine. "You better check on her arm, make sure the bleeding's stopped. Sure she's not going to pass out? You don't need her to go into shock or you 'will' be up shit's creek."

Nothing could be done for the dead guy who had no legs.

"You are again correct." Kravid said. "She will appreciate your concern for her welfare. She is an Oval; you know?" Kravid went to the Dinarchy woman, translating KrutChan's words to her as Kravid redressed her bloody arm adding over KrutChan's bandage another more appropriate Dinarchy compress from the dead guy.

Her small grateful smile and nod to KrutChan made him look away, uncomfortable in the fact one of his men had caused her to be hurt. After a battle, any sane person, had empathy for the ones who had absorbed the violence. Pain and suffering makes us all equal. He thought.

As he watched, KrutChan divulged his back story, since Kravid had told his background. Omitting any military information, like units, Commanders, or planets, or battles he had been a part of in the Uayebs. He stuck to things Kravid was aware of already.

In a monotone, KrutChan spoke of Vietnam, his Seeching and his distaste for the alien Oval

Confederacy and their methods of obedience. He purposely left out his Maluayeb Arena experience. He finished, by observing. "Those Ovals are always spouting off about their love of life and their love for their individual Krils. Yet they perpetuate the war. As long as they don't have to be involved in it. Like your woman there."

Kravid corrected KrutChan. "She is not an earth woman or ArnaVal. She is an Oval female."

"I get your point. More power to ya. I'm constantly aware of the difference."

The Dinarchy leader Kravid locked his eyes with KrutChan's. "Your honesty is commendable. Clarifying our discussion; if you hate the Oval Confederacy or specific Ovals so intensely, you could surrender to me. The Dinarchy and the Willows honor bravery and courage in war. You would be assimilated into our troops without prejudice. KrutChan, the Four Striker. We would value you."

KrutChan had another thought. "If you surrender to me, Kravid, the Ovals would be ecstatic, if I know them. Kravid Palatine would be a coup for them, huh? They would treat you with respect."

Kravid scoffed. "Ovals respect no male."

Thinking for a moment, KrutChan added. "I'll honestly say if you know my leader, the KrilChan, no matter what the Queen said, would have you crucified before he let you finally die. That Roman has no pity for his enemies."

"Since you know who I am, and I you, we can stop pretending." Kravid smiled a deadly crooked grin. "Your perception and honesty is refreshing. My Dinarchy troops value you as an enemy."

KrutChan grabbed his chest. "Be still my heart."

Kravid shrugged. "I will return your honesty. My generals would treat you as a threat to them and their positions. They would TOTL you upon sight. We have no Kril Code. Since you are a Four-Strike Kril they would immediately eliminate you."

"We're both in the same spot on the road. Xmucane couldn't save you after my Generals got done."

Not listening, Kravid said. "I think during an armistice we should refrain from lying. KrutChan is well known to us."

Kravid was pointing his finger at KrutChan's face. "You 'are' KrutChan. The Krils marked your face after the Maluayeb Arena; the four strikes are observed by us. Do you think I or my wounded here, would have let you take us alive if you were not him? As I told you before, Dinarchy troops understand acceptance of pain, respect honor and submission for the greater good."

KrutChan was caught, but he was not placated. "If you know me as well as my Ovals then you know I hate their Willow-whipping traditions."

"As I said, they hate disobedient males."

"I'm a slave. And I don't intend to be another slave in your goddam Dinarchy. All I would be doing joining your ranks, if I survived, is trading one prison for another. I'll never be a turncoat. I despise defectors or traitors."

Kravid was nodding. "I made the offer; you refused, as I expected. We are not a tongue-wagging Tribunal here. We are running out of timeline. My wounded are weakening in their condition. As you said so correctly before, 'you have all the cards and are in control'. What are you going to do with your immense power?"

Barely whispering, KrutChan said. "I don't want power. People die because of commanders having power over them." His hard gaze bored into Kravid's face. "I'm sick of watching my friends die because of my orders. I'm tired of walking away; trying to forget the dead bodies I've left behind when I survive."

"You 'are' a great Kril. Your words are true." Kravid paused a moment. "Power and duty come at tremendous cost to my time-pause sleep. You will have an unconscionable decision to make in your NOW; should you become KrilChan. I think that decision will destroy you, because you do not lust for power."

Kravid thought again for a moment. "If you want to punish the Ovals, then supply me with information about your Kril plans in your NOW. Your Strategic Invasions and small insignificant features about your own troops. I will do the same for you."

The thought was intriguing to KrutChan. Jaguar already had pressed him into her service as an agent. He didn't volunteer; he was drafted by her. He said in reply. "Now who's lying?"

"I would value you."

"You want me to be a double agent? Once I get my Stel back, the StelaBalaam will assimilate my thoughts and use me to get you. Your side would do the same to me. I won't even think about it."

Neither of them said any more for ten minutes.

Remembering back to his last meeting with Kelel, KrutChan's mind became clearer about what she said. "Should you come in contact with a Dinarchy agent...or any high commanders of the Dinarchy on this Uayeb...you will respond like this."

550

KrutChan blew it off at the time because of the off-handed way she had said it to him. Now, he wasn't so sure. Was this Kravid dude the one she was talking about?

"We are at an impasse then?"

"Come away from the others." KrutChan said. He decided to probe. "Don't try anything stupid." When they were a comparatively short distance from the wounded, KrutChan asked. "Do you know our Oval Chief Intelligence officer?"

Kravid frowned. "Of course I do." That question had put him on his guard.

"Silly me...you know me; why not her?" KrutChan was pleased. Gotcha, buddy! You didn't correct me and say you knew 'of' her; you said you knew her. I think Jaguar planned on using a pre-arranged, secret diplomatic channel.

"Why do you ask?"

KrutChan said. "Jaguar mentioned a Dinarchy leader may be tiring of this constant warring and want peace. You think that's possible?"

Kravid snarled. "Do not be a hatchling; as long as the Krils are sent against us, there will be no peace."

Getting nervous, KrutChan noticed Kravid Palatine was attempting to stop his hand from shaking; hiding it behind his back. It made KrutChan happy he wasn't the only one feeling the stress. They were apart but both pacing in the gully, keeping their eyes on each other. KrutChan wasn't the only one knowing they were running out of things to say to each other. Decisions had to be made and soon.

Kravid hinted at something. "That dead Dinarchy Willow lying there was attempting to contact one of Jaguar's agents. He was a trusted confidant of mine in our Intelligence section."

That blows Jaguar's plan. Dead guys can't give messages or carry answers.

Kravid held up his hand, where the wounded could not see, his middle finger entwining his index finger. "With this sign we Dinarchy indicate we hate the Krils."

KrutChan's body flushed with heat-cold flashes. Holy shit! That was the signal! Next to his rifle he returned the hand-code nonchalantly Kelel had given him on orders from Jaguar. "We humans use this sign to say 'up yours'."

He noticed Kravid did not look surprised.

"I cannot politically initiate a cease-fire. Your OvalChanHalach has the same untenable position as mine." Kravid said. He raised his hand with the sign again.

"Even a cornered rat has options."

"However, if your OvalChanHalach, through neutral channels, were to agree to discuss a cease-fire in neutral territory." He was staring long and hard. "A secret conclave between the Dinarchy and the Ovals. I will attend and be willing to keep an open mind. If we both can temporarily end to our fighting. If the Ovals agree. If we can agree to terms."

"That's it? That's a hell of a lot of 'ifs'! This is your message you want me to deliver to Jaguar? If my EkSeet were here, he would say that's a tenuous agreement, at best."

"You are the agent-messenger; it is not your decision to make."

"I don't have any orders to make peace with the enemy."

Kravid spit into the snow. "I advise you to pick your timeline when to divulge my words spoken between us. TOTL will result on both sides, if you speak of this too soon. Guard our secret closely.

"What about Jaguar? When do I tell her?"

"She is the one you should not trust, until the proper time. As the Ovals say, opportunities arise when timelines branch. You will know when to tell her. Be very careful, KrutChan; you risk TOTL by keeping our secret."

For some stupid reason, KrutChan was grinning. "So what else is new? Let's get back to the others." KrutChan said.

He stopped Kravid, gripping his elbow for a moment, on the way back. "How do we solve our problems in this gully?" He squinted. "Course, my memories are hazy."

Checking on his wounded, after they arrived, Kravid spoke to them reassuringly.

Afterward, Kravid spoke to KrutChan. "I may have a solution we can agree on while we are here. We are both in a dilemma of sorts. Do you remember your Native tribes on your great southwestern plains, in your past? They were outstanding warriors who believed in respecting the strength of their enemies. After great skirmishes neither side won, they respected their enemies. Sometimes allowing them survive to fight another Kin.

KrutChan was listening to Kravid as he talked. What he was saying made sense, in a crazy illogical way. Anything was better than waiting for this stalemate to abruptly end badly. KrutChan added his own thoughts. "If I decide this cease-fire is in effect. If we both shake hands in agreement. If we walk away from each other; we will part in peace."

Kravid pondered. "Your 'ifs' are logical. Can I trust KrutChan?"

"As much as he trusts Kravid. I have the rifle and the power; I'm setting the terms. I'm willing to disappear. After I release you and yours."

Kravid Palatine's eyebrows raised in admiration. "That is a courageous decision; more bravery than I think I possess. However, it is a bit self-serving. We both know you have another mission to accomplish, do you not?"

For the benefit of his wounded, Kravid crossed his fists under his chin as a Kril would have done. "I respectfully salute that decision. We here honor you."

"Screw the saluting; I've one condition. The goddamn Krils, the Ovals, and your Dinarchy are not going to be as understanding about this conversation. Your folks ain't gonna like you leaving me alive."

KrutChan was moving closer. "If we have trust in this gully for the moment, then we must promise each other to forget everything that was said and agreed to here. And can I trust them to keep their mouths shut?" He pointed to the wounded.

Kravid immediately answered. "They do not understand our words. To them you are an illusion; an ominous creature from the freezing-snow-mists. I will never divulge what occurred here; for your sake…." Kravid said. "…and mine."

Kravid and KrutChan talked for a few more minutes, discussing how they would disengage. They were running out of options and timeline. The Dinarchy loaded the dead guy on a fashioned stretcher made of brush. The Oval woman lifted the sucking-wound guy; his arm over her shoulder.

Kravid spoke in the third person. "Kril and Dinarchy leaders are destined to TOTL. If they survive; they are destined to be forced to the forefront again. As we agreed, Kravid's wounded and dead will go in that direction towards our Dinarchy assembly area. And you, KrutChan, will fade back into the fog you attacked us from...disappearing into this planet's snowfall mist."

KrutChan wanted to be clear on something else. "You're forgetting, as we agreed, you keep your mouth shut about our meeting and conversations. I will never speak about what I never saw or never heard here. As far as what I saw, this rock-filled gully was empty of life."

"I impart one last thought, KrutChan. The costs of peace are as higher then war. Peace will never occur unless first there is a cease-fire, then a truce. Do you understand the import of what I am relating to you?"

KrutChan looked at Kravid for a long time in assessment. "I will remember this part of our non-conversations, as a MishMell daydream. You can take that to the bank, Kravid." KrutChan locked their secret in his heart; Jaguar had her answer.

They both nodded with respect for each other and shook hands. As if they were grabbing poison Ivy.

KrutChan couldn't resist jabbing back at Kravid for calling him by his Oval name of KrutChan. "When you see Kravid Palatine; tell him I will waste his ass if I ever run across him again in my Osil travels."

Kravid loudly laughed. "If I know Kravid Palatine, he will surely rename the Four-Strike KrutChan. He will bestow the name 'Xtabee' on his honorable enemy."

The Oval-Dinarchy woman raised her eyebrows, whispering in fear to the wounded man she supported.

Seeing KrutChan's expression, Kravid explained. "In our language, Xtabee means a 'Specter'. Similar to your 'boogie-man'. It is another

Willow legend…a myth. Remember, we never met. Xtabee helps me forget."

"I don't remember." KrutChan said.

The Dinarchy leader became serious. "You are aware Kravid will have to 'TOTL' KrutChan should we meet again, as you said so humbly?"

KrutChan formed a pistol with his left hand and pantomimed pulling the trigger, jerking his finger skyward. "I said it before, Kravid…get to the end of the line. There are hundreds ahead of you."

"To reiterate, I will TOTL you should we meet again; make no mistake." Kravid said.

"That's the only thing I dreamed I heard in this gully today that makes sense. It helps 'me' to forget. We have a secret mutual pact then; to TOTL together?"

Kravid snorted. "No, never together. We trust each other this Kin. But we will remain enemies."

"I get it." His eyes twinkled. "When I kill Kravid, it won't be personal."

"I will also have no regrets bringing TOTL to you. I have caused too many deaths of my friends. My mind too is crowded with spirits."

Kravid's mood was morose. He had been translating, with modifications as they spoke, the entire conversation to his wounded; editing out any names or secret future plans.

KrutChan slowly waved to the wounded. "My life has been the same, regarding my friends killed or meeting TOTL. As I know intimately, dying sucks. What more can we expect of each other?"

The Oval-Dinarchy woman was dragging the chest-sucker, stepping towards KrutChan. She had tears in her eyes. With a grunting effort to keep her patient upright, she put her hands together like she was praying to KrutChan.

"My Oval...my personal Oval mind you, is telling you she has Beekav for you. Stop projecting to her your 'essence-mood'. I am glad you are vanishing. I will lash her to teach her to behave correctly."

KrutChan hissed in his own defense. "I ain't doing shit!" God, it's a real curse being so handsome to females. XibEk would growl at that bullshit.

"Like all the Ovals I've met; their motor is always revved up." KrutChan threw his chin out towards Kravid. "Maybe less Willow-whipping and more caresses will teach you something about Ovals, Kravid." KrutChan said, with a grin on his face.

He responded to the woman, awkwardly repeating the prayer gesture of Beekav to the blushing woman. What the hell, like Xib always says...give a little...get a lot.

Kravid growled. "That may be so; when you are honest you are usually correct, as I have noticed."

He paused for a brief moment. "I do care for her. I will remember those words of yours. For now, in our WHEN, you must leave my Oval alone. Do not steal her heart, you evil Xtabee."

"I've never been a woman-thief." Inside, KrutChan winched, thinking about KrutCheebel's ArnaVal ZocKuk.

"I could end this by wasting all of you before I leave. The cease fire and the truce will die with you."

"I do not think so." The Dinarchy leader said. "You would not want to be responsible for the unnecessary deaths of millions to come." Kravid sighed in resignation. "Our war will go on until it is won by the right race."

KrutChan answered, with a disbelieving sneer. "War eventually sorts out who's left. War never resolves who's right..."

As the Dinarchy group began stumbling towards their assembly area, clumsily carrying their loads to the other end of the gully; they eventually struggled with their loads into the blizzard at the top.

KrutChan, slowly and carefully, his rifle aimed at the Dinarchy, backed up in the opposite direction, fading from their sight.

They were watching him evaporate from their position on the top of the gully. He looked like their Dinarchy Xtabee; and then vanished themselves.

KrutChan took a few minutes of time, staring long and hard at the empty gully so when he put his Stel back on that would be the memory he retained. *All I can see is a boulder-strewn gully of brush, rocks, and patches of purple-blue dirty snow. Nothing else.*

Moving faster than when he came; he was anxious to get his butt back to his lines. He was a Kril traitor, breaking Kril Code. The Marine Corps would never have condoned what he had done. *The Corps wanted unit-cohesiveness, not ten-per-center individuals.* The Corps would have court-martialed him. *They probably would have had me shot.* He had received, as her agent, Jaguar's secret message.

He had made a decision to let them live he might regret. *Christ, I've broken so many rules in my lives, one more ain't gonna make any difference.*

He dreaded reporting this bit of trivia to Jaguar, being he 'was' her spy in the field. Should he or shouldn't he report the Kravid encounter to her? Kravid had advised him to bide his time.

*After pondering in the minutes that followed; he made up his mind. Bullshit...Jaguar could fuck off; promising to keep the secret for the foreseeable future. If he was killed before he delivered it, tough. Jaguar and Kravid would have to find another way to deliver the news. He was no Archangel messenger. Besides, this was a secret that never happened.*

*No...he was better off clearing out his brain of anything that might have happened on his lone patrol. As he walked he forgot everything, over-writing it with scenes of the terrain he was traversing on the way back to his lines. He wandered aimlessly, leaving a meandering trail in the purple snow in case someone retraced his path and slow down anyone following. He was nearly blind in the wind and snowstorm that was building. Lucky for him the increasing snow blizzard would cover his tracks.*

*In his heart he felt elated and vindicated. The Dinarchy was just like any other enemy. He had found that out in Vietnam. <u>The enemy was as miserable, as shit upon, as fucked over, and as lonely as we were. They wanted to go home in one piece.</u>*

*As he put a lot of distance between himself and the gully, KrutChan walked with a slower pace into the storm. Spinning slowly as he went, to make sure no one was following behind him or on his flanks. He meandered, creating more tracks in the ice and snow to obliterate his original direction.*

*He found his Stel and re-attached it to his chest. When it crackled into life he immediately tore it off and put it in his back pocket. He was too vulnerable out here to be chatting with his probing EkSeet. That would be his excuse when he was debriefed. He would restore his Stel just before he reentered his lines. If he wasn't wasted first.*

559

*The snowfall turned blizzard blanketed his parka, his face full of icicles from his breath, and his bones aching in the freezing chill of Laktavil. He stopped many times to listen. His fear of unknown fates was rising in his throat, making a lump over his heart.*

*Getting wasted in a firefight with the Dinarchy was a possibility. Who knew how many other ragtag enemies were out here? He would handle that; the worst thought was he was a target for his own Krils coming back into his lines. Walking ghost-like out of a blizzard was a sure way to die. Those goddam Krils were hyperventilating, expecting an attack from his direction. His Krils supposedly knew he was out here, but some trigger happy boot, aiming to be a hero or just plain stupid, would see a snowman Dinarchy Trooper before identifying KrutChan.*

*He remembered two nights after he had landed in Vietnam, another Battalion was on top of Hill...he couldn't remember which one...and wasted one of their own patrols coming back in at night.*

*Making up his mind; he picked out a small patch of brush under a purple-black tree of some kind growing on this planet. He made a small wingless-snow-angel, scrunching with his ass, and cuddled himself into silence, waiting for what passed for dawn on Laktavil.*

*At first light, KrutCheebel heard the curse-password shouted from close below him. It was KrutChan! Yelling the profane countersign. Within a minute, KrutChan, carrying snow, dirt, rocks, and brush cascading in with him. Rolling into KrutCheebel's fighting hole.*

Who asked. *"Where have you been? Our fearless leader KrutSeet has been angrier than hell about your one-man patrol. He is furious."* KrutCheebel inquired. *"See anything?"*

*"I was warming up an icicle-princess."* KrutChan was brushing the snow and other debris off his uniform, which was futile in the diminishing snowstorm. *"Found a small insignificant grenade hole. No bodies. There was a blood trail leading off a long way in that direction."* He pointed away from the gully. *"And I followed it, eventually getting lost in the purple snowfall. Your boot did good; he got somebody."*

The new Kril's chest was swelling out with pride. *"See, I told you guys!"*

KrutChan added. *"Wounded guy must've bled out as he was carried off. The footprints and blood trail I followed petered out."* He frozen face was becoming porcelain hard. *"I didn't see a goddam thing. Wasted my timeline, is all."*

Before KrutCheebel could say what he was thinking, from high above them on a rock ledge, KrutSeet was shouting down to them angrily. *"You have put your Stel back on! KrutChan, by the gods, you are summoned to report to me...immediately!"*

KrutCheebel grabbed KrutChan's parka in his fist, playacting for his Stel and for their leader above them. *"Your master's voice. Report to your commander, you dumb-ass marine; you are in big time trouble."*

KrutChan shrugged in that way he had about him. *"Ain't I always in shit up to my ears?"*

*My fuckin hole, with no ladder, just keeps getting deeper and deeper and deeper.* He thought.

He scrambled up the snowy hill to meet his infuriated Commander KrutSeet.

Chapter 27

*A Kin later KrutChan returned to his Malkril. They were set in another position on the planet Laktavil. He was received with a lot of mock-bowing, Bronx cheers and wise-ass remarks about fuckin' heroes and dummies. Those kind of Krils were one in the same in their collective minds.*

*He never answered their questions about what happened with KrutSeet. Not recalling what had happened on his lone patrol. KrutSeet, Jaguar, and his own EkSeet tried to pry the information from him. He had filled his Commander's report with his usual lapse-of-memory.*

*"I was stressed out, being alone. I don't remember a lot, except wind, terrain, and tons of purple-blue snowflakes." Solidifying his loss of memory when he repeated the story. He told them he had taken his Stel off to keep from alerting any Dinarchy while he was lost in the blizzard.*

*They did not like his explanation or seemed to believe him. Jaguar's expression implied she did not believe a word of the report and indicated. "We will talk again, after the Uayeb...should you survive."*

*After Jaguar and KrutSeet melded; he told his pestering EkSeet to shut up. Their constant arguments had returned with a vengeance. <u>I forget tons of stuff because of your Oval AnticArna doping. Memories aren't perfect when recalled; look at my Greer visions, ya limey bastard.</u>*

*His EkSeet responded. <u>Do not denigrate your EkSeet. I simply reside here, wanker.</u>*

*On his Malkrils next combat actions KrutChan's one-man patrol was forgotten. Nobody cared anymore; they were at war on this frozen iceberg of a planet and had other things to think about, while wallowing in their freezing misery.*

*Somebody in intelligence was going to be dressed down, because of the light resistance and the half Uinal projected limit of the Invasion. Inaccurate history after four weeks of fighting.*

*The Dinarchy had stormed at them in heavy concentrations of frigid-terrain-trained-troops massed on planet Laktavil, in force. That unknown-somebody in intelligence must have broken their estimator, or was drunk when the plan was compiled.*

*Time was not on their side. Deaths from freezing and disease were mounting exponentially. This planet would be written off as a loss, if the Dinarchy could not be stopped. KrutEk never liked his Kril Legions to retreat. It looked like it might be a possible option.*

*Jaguar reported to KrutEk that her team was sure Kravid Palatine was in command of the Dinarchy on Laktavil. Though her intelligence team had no direct evidence he was present. Everything the Dinarchy did bore his trademark tactics. The Dinarchy leader was a masterful tactician.*

*After all, Jaguar said to infuriate KrutEk, Kravid was a former Willow-Kril. KrutEk dismissed that information, considering it similar pap as the inaccurate estimate of the Invasion timeline. Oracle-Goat-gut prophecies in his past, were either overly optimistic or pessimistically guesses.*

*Lashing out at his Kril Legions commanders, KrutEk was beating and kicking butts in his anger at the situation facing his Kril Legions. He was born and bred for victory not defeat.*

After coming back from KrutEk's Conference with BalamEk's Ovals on the Burseeosil, KrutChan called CheChun to his snow-drift headquarters. He dismissed his support Krils.

He and CheChun sat shivering, across from each other, on snow benches sculpted in the wall.

"War over? CheChun asked.

"KrutEk and BalamEk finally unified their commands. KrutChan grimaced at the stupid question. "Why? You miss your South Pacific wahine's begging for your essence?"

"Fuck you Krut."

"Yeah...same to you." KrutChan sobered. "I didn't call you here to discuss your love life."

CheChun refused to change the subject. "Listen Mug, for a guy that pisses off the Oval women he meets, you're no prize. I've seen plenty a wolf's chasing skirts. What do ya hate so much?"

"Aliens ain't bunkmates. Get it?"

"Ya don't haveta love em! Jesus said 'Give unto Caesar, the things that are Caesars, and...'"

"I don't read Oval scriptures from their lady god." KrutChan held up his hand. "Spare me the rest of your mangled bible quotes. I don't hate women. Only female Aliens."

CheChun began shaking from the cold. "Unless she's a woman named Akna? Scuttlebutt is, she's gotcha talking to invisible mugs. Guys nobody else sees."

Staring into the distance, KrutChan said. "Ovals dope their Krils to control them. I've figured out they overdosed and screwed up with me. I'm nuts because of them."

"Maybe you're crazy; maybe not. How about joining the rest of us normal humans before you crack up and get the rest of us killed?"

"Ever see any sane guys in battle? I only trust my 'voices', who ain't trying to waste me."

CheChun dragged in and blew out smoke. "Gotcha. Don't forget the Ovals try to help."

"Help? My fuckin Alien Oval, knocked me out and Soothed me to BakMeer; my step daughter. Before we left on this Uayeb. Punishment for shaming her at the Pavilion." KrutChan bristled. "That reason enough to be pissed-off? How would you like banging your own daughter?"

"Who the fuck does!" CheChun anger was spilling out. "Listen you goddamyankee. My old man was loved by everyone outside our house. He was porkin my three sisters every day, with my old lady suffering, but allowing it."

KrutChan was wagging his head. "Your home-life secrets are supposed to make me feel better?"

"I had to put up wid dat shit until I joined the Corps." CheChun said. "On my boot-leave, I came home and broke his fuckin jaw. Told him and my mother, if he tried screwing the girls again, I'd come back and kill 'em. After I denutted him."

KrutChan half-heartedly saluted. "Everybody's got a sad story. Least you were a man with a conscience."

"BakMeer ain't your daughter."

"Exactly what Reela said…."

CheChun was wasting his time. He went to attention. "Why am I here?"

KrutChan kept his order short and concise. "I'm putting you under KrutCheebel's command." He held up his hand. "Save your bitching. You've been a Raider."

"So fuckin what?"

I've got a mountain-hard-nut for him to crack, full of Dinarchy crazies. You have the training and know the score. Help him out and keep your personal feelings about him to yourself. Think you can handle that without a crying towel? Heaven forbid I'd get your cooperation."

"Shove it, Krut."

"I'll take that as a yes sir." KrutChan walked into the opening in the head quarter's snow-drift, beckoning his staff to return. He turned to CheChun. "Leave your prayer book here, and go do your goddam job."

Two more Kins passed and KrutCheebel's Bot-units were committed by KrutChan to seize a cave-pocketed hilltop adorned with a thick purple forest of alien trees. They moved out, as they began the climb, enveloped in a ground-hugging-ten-foot-high icy fog-mist.

"Where the hell are you going?" KrutCheebel challenged CheChun, who commanded another of KrutCheebel's Bot-units.

"Don't crowd me, boy." CheChun was cranky from the cold stiffness permeating his body. "Ah'm heading for the end of our flank wid dis .30 calibers 'chine gun." CheChun always wanted to be part of a machine-gun section in the Pacific War. He took it off a dead Kril's body.

KrutCheebel was busy. Ordering Kelel to command CheChun's Bot-units and directing his Raider unit to get on line. "I do not think that gun should be exposed on our flank. It should be in the middle protected by a couple of Kril riflemen. That's what KrutChan would do."

"Ah was killing japs before KrutChan was borned." He snarled. "Dem Dinarchy all-ways attack

567

the flanks when they make contact...like the Zulus. You got any Zulu blood in ya boy?"

"I'm in command of these Bot-units. You will do what I say while you're temporarily under my command."

"Ya wanna get in a 't-bacco-spitting' contest, or do you want to get this lashup on the road? My Bot-unit has this flank. You ain't a Malkril general. Ah already ordered my guys to tie in wid yer fuckin' Bot-units on da left. Ah control dem; not you."

"You and KrutChan do not know everything." KrutCheebel said.

CheChun pointed the machine gun in KrutCheebel's general direction. "Ah got the time and the grade in dis here war, boy. Ah was killin' Dinarchy before yer mommy gave you a tit to suck on. Dun't tell me how to frost enemy. When y'all hear me open fire on dem Dinarchy flankers; that will signal yer Krils to move their virgin asses over ta support me."

"Fuck you Cracker, and your old Corps attitude."

"Okay tough guy...my Stel has already notified my EkSeet and he has told KrutChan what I intend doing. That ought to be enough facts for you when he runs me up on an article 32. You need any more backup boy?"

KrutCheebel gave CheChun the finger and moved to the middle of his Bot-units knowing the attack demanded his concentration. Trying to keep his Krils on line in the poor visibility would be distracting enough.

CheChun could hear KrutChan yelling at him on his Stel and did not answer. Then KrutChan got busy with other problems in his Malkril.

As CheChun moved out he was already regretting his decision. His solace was he was determined, right or wrong, to control his own destiny. I ain't gonna end up in a funny farm with KrutChan.

In the fog he could barely see the Kril on his left shuffle-stumbling along in the drifting snow. He would be totally exposed and alone if the Dinarchy appeared. He planned on emptying the belt of ammo in the machine gun, in addition to the two metal boxes with additional belts, and then use his .45 Thompson sub-machine gun to keep the Dinarchy at bay until his Kril reinforcements arrived like Stonewall Jackson's Calvary.

As he stagger-plowed through the purple-blue drifts, the cold was sucking his breath out of his lungs. Goddam purple-snow-stuff is cotton-candy thick.

He went forward inhaling through his nose to warm the freezing air up; his exhale puffed into the air in front of his face. As he moved though the deep snow the alien trees were at attention, silent and edge-drooping with snow cover.

Within minutes, a trickle of fear entering his heart when he lost sight of his Kril Bot-unit on his left. The alien trees rustling and cracking in the wind made him feel more exposed and alone. There was no comforting bird chirps or other animal sounds; only the soft shush of his knees plowing through the drifts. In the white heavy fog his numbed hands on his machine gun were freezing in the bite of the cold wind. He wrapped his asbestos changer around the barrel under his gloves. I should have brought an assistant gunner with a tripod and extra ammo, if I'd had more time. Fukit. I don't want nobody else

CheChun's visibility was decreasing rapidly as twilight creeped in. KrutCheebel's Bot-units were long gone, probably humping up the hill and wondering where he had disappeared to in the fog. Finding them in this light was not something he was going to attempt. His own Bot-unit, commanded by the split-skirt twat Kelel, had disappeared with them in the white-purple reflecting fog.

When he reached his destination knoll on the farthest end of the flank, he did not have the tripod so he laid the air-cooled Model 1919A .30 caliber machine gun on a reasonably flat rock next to a snow-dressed bush. This was no mean feat because the rock wasn't as stable as a tripod. That was my mistake, but I'll make due.

Setting up fields of fire he marked with thick twigs, he scrunched below the alien brush hoping it wasn't alive. He still got nightmares thinking of KrutCheebel being eaten alive on KanBalaam. He was cursing at himself for coming up with this crazy idea to protect the flank of his Bot-unit. He could freeze to death. In a millennium, maybe some Laktavil critter or a colonial Kril garrison guy would stumble across his frozen ass. A frosted unknown statue from a long-forgotten invasion.

After an hour of numbing teeth-chattering-shakes, CheChun's senses alerted, forcing him to hold his breath, and become immobile. For some unknown reason, his Stel and EkSeet were mute. He guessed KrutChan had silenced them when he went off on his lone patrol.

He heard non-English speaking voices to his front. Lucky for him, the white-purple shroud of fog broke between the alien trees revealing spirits carefully walking towards him, as if rising from an ancient burial ground.

His spine-tingling he felt told him the appearing mist-forms weren't his faded Bot-unit; they were a Dinarchy combat patrol. CheChun had the satisfaction, for a moment, of knowing the Dinarchy were flanking KrutCheebel's Bot-units just as he had predicted. Being right ain't gonna win me any merry-go-round brass ring.

His pride lasted a fraction of a second when he realized the enemy was heading directly at his knoll hiding place.

His heart was thumping in his chest and nausea was feeding his fear. This was no place for a Christian white man.

For a moment, he contemplated quickly running away into the fog, vamoosing lika deer-skidding on a frozen pond, trying to gain traction for its hooves.   If he stayed and fought, he was outnumbered, like those ancient Greeks holding that pass against the Persians. Jumping out of the frying pan would throw me into the fire. That thought gave him no consolation, but it made up his mind. All he could think about was his number was up. He was about to die with no available options.

I asked for this. So, shut the fuck up; you brought it on yourself, gyrene.

Surrendering was an option he quickly discarded; those Dinarchy bastards would frost him rather than drag a prisoner along with them. The second option he had he didn't like; but it fit his inner-mind-gumption better. The only thing left to do was to fight and stay where he was; doing the job he came to do. Going on the offensive flashed in his mind, after all he did have surprise on his side. Gloomily CheChun thought about the fact he had a lot of firepower in his hands which upped the ante along with surprise. But once he opened up, his initiative was non-existent. The cock-fight would begin.

His EkSeet bellowed in his mind and he feared the Dinarchy patrol would hear it. You are alone and outnumbered. I think you should consider survival and withdraw immediately.

He answered curtly in his mind. Iffn ya ain't gonna wrassle, git outa the goddamn ring. Go backta yer EkSeet house…and shut the fuck up!

CheChun was reminding himself, since he was shooting downhill, he should aim high, lowering his fire to grazing if the tracers were off target. He wanted to take out their legs. Once they were down, he could murder them in the kill zone.

To stop his EkSeet's kibitzing and his fearful thoughts, CheChun opened fire on the furthest Dinarchy team in the Dinarchy patrol. That'll confuse the shit outa dem, giving me time to adjust my fire.

He fought to keep his rounds three inches off the deck so as to wound most of them that he could kill later while they squirmed in the snow. He watched as his initial targets fell screaming, cursing in their dialects, or collapsed silent into the blue snow cover.

In the following few moments of time, blaster weapons, arrows, spears and automatic weapons were spraying the area searching for his position. They could not locate him yet.

He again moved his fire to the farthest back of the advancing Dinarchy, concentrating on the hunching running figures seeking cover. He forgot about the cold. His mouth was sucking in the bitter cordite cloud his weapon that was surrounding him.

*Thank the lord for the fog!* His blinking weapon caught their attention. He felt the incoming fire zip by his head or above and around him as he discreetly slithered away changing his positions randomly.

For what seemed an hour, his belt was feeding into the machine gun like a rattlesnake as he plowed fire through the snowdrifts and alien trees at the closest moving targets. Some were sitting still stupidly waiting for the impact of his bullets. When his machine gun clicked on empty, he moved again. When he relocated he fed another belt into his weapon.

The Dinarchy patrol was caught flat-footed, with no place to hide, no cover to crawl to and protect themselves. His bullets were clunking through the purple tree-trunks, hitting the targets hiding behind them. His machine gun rounds were tearing and ripping rounds right through the snow, brush, and trees.

If he was taking hits he did not feel them. Briefly aware his clothes were fluffing and pulling, from stray hits like he was in a stiff wind.

CheChun let go of the machine gun when his last belt ran out of ammunition. CheChun shouldered his Thompson submachinegun, rolled a few feet like a kid in a snowball fight and kept up his lethal firing, dropping magazines when they emptied, and reloading furiously. He heard himself praying he would run out of Dinarchy targets before his ammo was gone. He wasn't looking forward to fighting with only his .45 and K-Bar.

What had felt like an hour; the firefight lasted minutes.

Abruptly the incoming fire had ceased. He stopped firing, though he sighted his weapon with both eyes open for any movement as the echoes faded to silence in his buzzing ringing ears. He roll-crawled to a better vantage point over the killing zone. CheChun lay silent, sucking freezing air, and trying to quell his trembling limbs. He wished he could piss away his adrenaline overload.

CheChun heard KrutSeet talking to KrutChan through his Stel. "Barbarian, I see you have one man acting like Roman."

"If he lives, Chun's giving things to Caesar. If he loses, he's giving things to his God." KrutChan said.

"Caesar was a legend." KrutSeet said, not understanding.

"Heroes arrive and last minutes." KrutChan chuckled. "Legends take a long time departing."

CheChun spent a cold forever-night hugging himself to keep heat in his core. He fought to stay awake, thinking the Dinarchy survivors were advancing on their bellies through the snow drifts wanting vengeance on his southern butt-hole. He was remembering the japs on Tarawa and Iwo had never given up, appearing out of nowhere to kill or die by their code of honor.

When the fuzzy-dawn crept silently upon him, along with the chill, he saw the mangled bodies of the Dinarchy patrol stiff with rigor, their feathers, shields, and torn uniforms stiffly laying. No mist of life steaming from their bodies, silent and forgotten. The purple snow had stop falling; the killing zone peaceful.

When the subtle light-dawn filtered down on him, he walked carefully into the killing zone. He found four moaning wounded guys crawling towards him indicating they surrendered and wanted his mercy.

He did what he could for them; but over time he watched three of them convulse and die. CheChun wrapped the last wounded guy in one of the cadaver's huge fur coat. CheChun guessed the dead Dinarchy had stolen it from some dead Zar comrade; in empathy with his pain.

CheChun struggle-dragged the moaning wounded guy to a high point overlooking the carnage of the night trying to take stock of whether the guy could be hauled by him far in the distance to the hill with KrutCheebel's Bot-units. That wouldn't happen, I'm too goddamn icicle-tired. He decided against movement and hunkered down; either he and the wounded guy would be located by his Bot-unit. Or the Dinarchy would find them and kill him.

A nagging thought kept repeating in his mind, over and over. What didja prove? Ya big hairy-chested mongoloid? Dumb-asses were never glorified or emulated, in his experience.

After twenty minutes of anxiety; his relief came filtering into his position.

"Where the hell did you disappear to?" KrutCheebel rasped, out of breath, fluffing into the snow next to CheChun.

XibEk was with him, out of breath, kneeling and scanning the area for more Dinarchy.

"I held the flank like I said I was gonna. They walked right into me…like I said dey would."

XibEk snarled. "Bullshit, you're a lying redneck…" XibEk nodded to KrutCheebel in disagreement. "Da overseer was lost same-as us."

"Ah new 'xactly where I t'was. Youse guys were lost." CheChun said.

A medical ArnaMal went to work on the wounded prisoner. The living Krils moved away from them.

XibEk raised up his fingers to KrutCheebel. "Ah counts on'y four guys...they had some guts comin' right at'cha. Fer all da firein' you did...dat's all ya got? You fired a lotta or-dee-nance wasting those dudes. Shows me you can't hit shit."

Both XibEk's and CheChun's EkSeets were shouting in their minds. Refrain from using idioms and speak English! You are both sounding like fools! Both of them chuckled and ignored the EkSeets.

Someone hollered to the meandering Krils as they came upon the scene, to come help and count the broken strewn bodies in the tree snow.

"Last night Kelel wanted to run off to help you when she heard the firefight. I ordered her to stay put. You are too stupid to die."

"That so? She's got more marine in her than I gave her credit."

XibEk tracked you like a Cherokee through the snow this morning, somehow." KrutCheebel observed CheChun's condition. "Your uniform is shredded with hits. You okay? Any bleeding wounds on you?"

"Ah'm tolerable...them Dinarchy's the only ones dat got hurt."

In the background someone was shouting out from within the dissipating fog in the purple forest. "Jeezuz! I've counted forty dead Dinarchy and I'm still counting." The man sounded awed.

Some grunt from a future earth era said. "The Dinarchy guys had no cover or concealment in this flat sump hole; they had no place to run. Looks like he took out a whole goddamn company. That's golden man!"

576

"You is a lucky crazy inbred cracker." XibEk said. "Like KrutChan, you hearing voices too?"

"Ah'm            alive...no            luck involved...peckerwood."

KrutCheebel snapped in anger, using profanity, unlike his usual demeanor. "Don't pull that shit again. I needed that mutherfukin' machine gun for defense at the top of the hill."

"Dat so General Sherman? On yer march tada sea...you have enny contact?" CheChun drily said. He had not heard any firing from the hill the night before. "Did ya get to the top of the hill, general?"

"We took it. We saw no one on the way up, or on top, only blue snowflakes, huge drifts, and bending trees in the wind. No Dinarchy. Eventually your troopers staggered in from the fog."

"Hoo-Ray fer you."

"I left most of my Bot-units up there and came running when dawn cracked. I wanted to bury your TOTL butt. KrutChan is screaming profanities; fit to be tied at you."

Noticing a guy standing to one side, CheChun was surprised to see him as a part of the relief force. CheChun didn't know the guy; a fucking Indian by the looks of his haircut; six fingers and all. Why would that Kril come to find me?

The Kril with the yellow-tinged-body seemed oblivious to everyone else, including CheChun. He was pointed out to CheChun as a strange one in the last two Uayebs by other Krils. No one seemed to be his friend; he was designated by the Krils as a loner. They said he had been on so many Uayebs, there were no Krils left from his other units.

CheChun had seen a few guys like him in his travels; those guys were not part of any units, silent and there when the fighting started. CheChun felt

577

these kinds of guys just filled the ranks and would never be a part of any unit. The guy's hair would shrink into his head the closer he came to combat. Maybe his hair was a helmet of some kind? The guy was definitely an oddball; like tits on a prehistoric Mississippi Boar male.

The guy said. "Mighty golden job done here, chump."

XibEk growled. "We call him 'Chun', not you Ace."

"Like I said, only a 'chump' fights as a one-man army thinking he's winning a war. You a hotdog, chump?"

After being in this Oval universe for some time, CheChun had seen many strange sights; he ignored this guy as another weird Kril to be added to his memory-locker. CheChun was immune to verbal slights.

"Keep your opinions to yourself, Mug." CheChun shrugged. "Da way ah sees it…ah was de only guy doin' my job. Like KrutChan did on his lone patrol. I caught forty winks. And ah sent dem Dinarchy up-yonder ta dare Jeezuz."

A whisper from his Stel rasped at him, sounding like KrutChan. "In other words you fucked up…got lost…had the shit scared out of you…and lucked out. That simple?"

CheChun mumbled. "Nuttin is simple… 'cept ah'm still vertical an' dem boys is prone, ready fer Seeching'."

His Stel crackled from a roused KrutChan. "You report to me when we assemble up. I want to hear of your exploit in detail."

CheChun had to get in the last word. "Muh leader's voice summons me ta da big plantation house fer a drink and a see-gar." As he wandered off, he softly punched XibEk on the arm. "Neveh thought ah'd be gladda see yawl XibEk…thanks gyrene."

XibEk sniffed. "I came to pick up your corpse, Chun. Change your diaper, ya lucky lifer asshole."

"If you gentlemen stick your tongues down each other's throat and kiss; I will shoot both of you degenerates." KrutCheebel said.

"Quit filming their love scene, Cheeb!" KrutChan's voice cracked from their Stels. "Saddle up and move your units to the assembly area!"

CheChun noticed the strange palomino Kril had disappeared or left to search for more Dinarchy.

Two Kins later, under clear skies, KrutSeet ordered KrutChan to move his Malkril headquarters forward. KrutChan did not need the command; he was advancing to regroup with KrutCheebel's advance unit.

KrutChan was in the midst of ancient Roman and Greek phalanxes of Krils spreading out in a low area between two mountains, when the incident happened.

Another group of four Krils closed on him from four sides; their intent was obvious. They were focused on him on their lethal mission. Four ancient Krils, two Romans and two Greeks, attacked him from four sides.

NoKoch was leaning against a tree, looking at the horizon.

This time KrutChan was ready. He planned on going down hard.

Styx exploded off KrutChan's arm in fury. KrutChan grabbed the one attacker's arm carrying a Dinarchy blaster. KrutChan's face was seared.

Styx lost a chuck of his body from the blaster. Taking advantage of Styx's furious diversion, KrutChan turned the blaster weapon towards the assassin and blew off his head, spewing gore in a cloud of blood.

Immediately spinning, KrutChan parried the blow from the second assailant with his flak jacket-plates, reached under the thrust and drove his K-bar into the attacker's windpipe and jugular, breaking the larynx, and let him fall knowing he would die.

While KrutChan was absorbed in his fights, Styx immediately attacked the third assailant, tearing off his leg and arm with his mouthful of titanium-shark teeth. As the man was dying, the man thrust his sword upward through the bottom of Styx's mouth, exiting out the top of the creature's head between his huge glowing-red eyes.

KrutChan sustained numerous wounds, and in frustration and anger he backed off to gain advantage. He pointed his .45 pistol in the momentary lull and fired twice. One round hit the ancient Greek center-mass in his chest. The second bullet exploded the man's groin. When the Kril assassin collapsed screaming to the ground, KrutChan straddled him and crushed his brain with a large boulder.

He saw NoKoch drifting away. Why didn't he help? Was he leading them? KrutChan thought. No. He would have killed me. Still.... NoKoch had his favorite blaster with him. Could he have fired the first blast?

The assassins all met TOTL.

Bleeding profusely and crawling over to Styx's greying body; KrutChan slid his friend over to him. KrutChan rolled over and sat up, terrified and semi-comatose from the blood draining from is body.

_God almighty, I'm sick of this crap. Where was your warning, EkSeet? Were you on a coffee break?_

His EkSeet was silent. Second guessing was of no use at this time.

PacalMo and CheChun staggered to assist KrutChan, after running from their Bot-units.

"I see my leader has survived again. I honor him." PacalMo recited.

"If we don't get an ArnaMal corpsman here quick, your leader's gonna die." CheChun was irritated.

Groaning, KrutChan was sitting on his rump, tenderly holding the limp Styx, close to his neck in an embrace. KrutChan said nothing. He looked to be meeting TOTL.

PacalMo said. "They attacked him from all sides. His little dragon friend helped him to survive."

"Dey was serious; dat one fellow had a Dinarchy blaster." CheChun said. "They were out to have him buy the farm, all right. Old KrutChan got lucky."

Laying his hand on KrutChan's shoulder, PacalMo asked. "Are you with us in this timeline, KrutChan?"

"Yeah, you okay, Hoss?" CheChun added.

KrutChan was unresponsive; they guessed either with grief, coping with his leaking blood, or in mounting hate for what happened.

CheChun and PacalMo tried standing KrutChan up. His dead weight was too much against them. The medical ArnaMal they had called arrived. He began treating KrutChan's blaster-burn and many cuts. While he was being treated, KrutChan incoherently mumbled. "I'm wondering about..." His frosting breath was in full panting.

"Wondering about what, Krut. Seems pretty clear to us what happened."

KrutChan's eyes were refocusing as he came out of his private world. "I'm wondering how many times those assassins in three Uayebs have stopped before they attempted to waste me? As if they wanted me to see it coming. Was that in their orders?" He squinted at the ArnaMal, seeking an answer.

The Corpsman ignored him.

"You can't ask the dead, Krut. Who knows?" CheChun said.

"Yeah but...how many, by bad luck, by bad decisions, or by having no opportunity, have failed to get me? Something's screwy behind this..."

CheChun nodded. "I agree; it does look orchestrated."

KrutChan was staring vacantly between his legs at the purple snow. "Manipulated is a better word. Someone's sending me a message. How long will it be before one rookie is successful? How many are still out there waiting their chance to waste me? This fuckin shit is getting old...real fast."

"You have been under additional pressure besides from these unworthy killer-dragon-minions." PacalMo said. "You must not divert your attention to plotting revenge."

CheChun took the same line of thought. "Y'all can't blame all the Krils for the actions of a few Damnyankees."

"The hell I can't!" KrutChan snarled in disagreement. "These killers aren't malcontents. They were being guided by someone else!"

The ArnaMal touched KrutChan's chest. "I wish you would calm yourself and lay still."

"Stick it, doc!" KrutChan weakly was pushing at the ArnaMal's ministering hand. "Give me some of that blue Oval dope."

CheChun was concerned KrutChan was dying. It sounded like his pain was beginning to be felt.

"The evacuation Toob is near. You must understand, Dagot blasters are not clean-cleaving-laser-like; the particles contained within the blast consist of tiny shrapnel's. I must stanch the bleeding first on your head wound. You must remain conscious. When I am satisfied with my work, I intend to blue-bathe you."

In order to keep KrutChan occupied, CheChun said. "The Jap imperial marines in my war had a mission unsaid by their commander. Bushido was to die fighting, protecting the Emperor. I never understood what that meant until I got Seeched here. During another past Uayeb, I talked to some Japanese troops in our Kril Legions. Those guys, according to their Bushido Code, could never commit dishonor while destroying an enemy. It was against their code. That's why enemies surrendering was contemptible to them."

PacalMo said. "I do not think KrutChan was surrendering."

KrutChan whined. "Your bedside manner really sucks, Chun."

"That's my point, Krut. They were respecting you as an honorable foe."

"I have heard Ninja's would commit this kind of act." PacalMo said. "Assassins have their code too."

"Those guys look like Japanese warriors or Ninjas?" KrutChan snorted. "I intended on trying to survive this Uayeb." His voice was deadly. "If I survive, I'm going to fix this problem, once and for all."

CheChun said. "KrutEk ordered his Legions to not TOTL you, before the first Uayeb timeline."

"Tell a condemned prisoner, for killing his Kahn; he is to die." PacalMo said. "If he kills his executioner; he will be executed by someone else. What does the condemned man do? Are they going to behead him twice?"

"Crazy Krils!" The ArnaMal was looking to the sky and muttering into his Stel. "Where is that Toob?"

CheChun was trying to reason with KrutChan. "Why are you going off half-cocked? Ah never knowed you to guess at anything."

"Every time I've been attacked from one unit of Krils. My source informed me who is behind this shit."

CheChun wheezed. "Hearing voices again?"

"Have it your way. My 'voice' tells me the assassins, in all three Uayebs, were attached to one Commander."

KrutChan's EkSeet spoke in his ear. You are bonkers, my friend. Who are you speaking of? I am the only intelligence source you have, old boy.

You know who the guilty guy is, EkSeet.

The perpetrator is not who you think.

Like our Ovals, I'm planning. My intentions are my own. You'll blab to the entire Oval Confederacy; to Ovals and Krils alike. If I survive this Uayeb; I don't want him, or her, to see TOTL coming. Let them sweat it out for once.

The ArnaMal corpsman's blue AnticArna was administered. "KrutChan will be babbling more soon, but he will remain conscious."

KrutChan moved mentally into temporary mourning mode, coping with the triple threats against him, angry about the death of Styx. Tired and sick of having to continuously fight for his survival against his own Kril forces. Angry about the unseen manipulator, and angry at the Oval Queen's promise to initiate the Tribunal death sentence at the end of the Uayebs. The Dinarchy was a fourth billowing fog hovering over the whole mess of his timeline.

"Why are you grinning?" CheChun asked. "Yawl a cat choking on a lizard."

"I ain't happy." KrutChan was somber. "Styx deserves more respect than residing in my dark heart forever. Styx chose me, was loyal to me alone, and became a part of me. My arm feels like he is still on there. He didn't deserve to uselessly die..."

"Veterans even protect the backs of their enemies in combat." CheChun said. "Styx did what the dead always do; give up their lives so somebody else can live. What's wrong with that? I've seen dogs protect their masters; sometimes even cats do it. There ain't no logical reason behind it."

KrutChan dropped his helmet into the snow. "I had a friend. It's too late now...."

PacalMo explained his own view. "Felines slink in patient hiding and pounce when ready; with deadly claws, fangs, and silent strikes." He knew what he meant; no one else did.

Hearing the screeching evacuation Toob approaching, KrutChan straightened out his back, in a moment of clarity, and hard-stared at CheChun and PacalMo. "Get back to your Bot-units. Disassemble this milling herd around me. Get back into the war." He put his helmet on after brushing off the purple snow.

CheChun grinned and said. "Yes sir...spoken like an uncaring true sonofabitch. The mission comes first; the obituaries second."

"Shut up! Move out! I gotta get some sleep."

KrutChan, feeling alone, was shaking his unwounded fist towards the sky. "You missed again! I'm coming motherfucker! Count on it!"

The ArnaMal was shaking his head at KrutChan's AnticArna raving. "Crazy Krils."

High in the Burseeosil command center, after four Kins of skirmishes, another boring meeting and training session was progressing. They were discussing the latest attempt on KrutChan's life.

"Reela, my former Oval, back on Arna, is beside herself." Hortim said.

The Ovals present were trying to make sense from their EkSeet's translations of what Hortim had just said. Obviously Reela could not be in two places at once, Oval magic notwithstanding.

Hortim figured out the problem and clarified her words. "As her Consort-OvalChanHalach, Reela has gone to meet with the Queen Xmucane."

BalamEk stiffly answered her. "That is Reela's right. Xmucane will listen and then rule...as is her 'right'."

"KrutChan is condemned to TOTL, by the Queen's own decree, should he survive the Uayebs." Jaguar said. "Ovals are aware Xmucane specifically ordered the KrilChan not to permit Krils to prematurely TOTL KrutChan."

Hortim added. "The hospital section has saved him from death again." She continued, not caring about correcting translations. "It was touch and go. Your Doctors are amazing. Within forty-

586

eight hours…er…two Kins, he was as good as new. Reela is pleased about that turn of events."

BalamEk said. "KrutChan is KrutChan. After the two Kins, when he reacquired his Stel, he immediately contacted KrutSeet. His earth message was garbled, but loud. "Get my…the hell back to my Malkril on Laktavil! I gotta score to settle!"

BalamEk seemed confused. "Whatever that meant. KrutSeet equivocated. KrutEk intervened, vigorously ordering KrutSeet to comply with the 'barbarian's' request. KrutEk and KrutChan seem to have the same genes in them."

"If KrutChan gets lucky, Xmucane will make KrutEk investigate." Hortim added. "It looks more and more like someone hasn't gotten the word."

"You will stop discussing this aberration of decrees, missives, and codes!" BalamEk shouted. "Xmucane will deal with this dishonor when we arrive back at Tikumyax. For the present timeline we will review our actions of the last four Kins and discuss what we will do in our coming NOW." It was a Kril problem. The Queen would settle the matter before the Fourth Uayeb.

After their review of the ongoing Invasion progress, the Ovals returned to the projections, their 3-D modules, or other duties.

Another Kin came…then another combined operation…another offensive mission by BalamEk and KrutEk initiated, continuing for another three Kins. The battles were fiercer than anticipated.

Spurred on by KrutEk, the ancient ArnaMals led by the Viking AinAcbal, crushed every Dinarchy force they came in contact with, backed by air superiority. Ancient human endurance, muscle, and guts drove the Dinarchy forces to ground, clearing them from their warrens, annihilating any resistance.

KrutEk gloried in the capture of platinum and gold resources buried in the snow-capped mountains. Captured by KrutChan's ArnaMals in his Malkril. His reserve units were following behind the bloodbath pillaging by AinAcbal's attacking spearhead.

KrutSeet and ZacNaab were angry and irked the 'barbarian's' units had been maneuvered by KrutChan into the right place at the right time bypassing pockets of resistance.

KrutChan's Malkril was commanded by him to search for treasure; nothing else. Ordering them to ignore booty-stealing or bogging down slaughtering Dinarchy stragglers.

KrutEk was pleased, not caring about his other subordinate's loud whining. He had achieved his objectives and confiscated the lion's share of the wealth for his Queen.

On the fourth Kin, the Kril Legions were congratulated by a Stel announcement from KrutEk, notifying the Laktavil battle over and Dinarchy resistance was curtailed. Only a mopping up operation had yet to be done.

Briefly meeting with KrutChan, XibEk asked him. "You okay? We figured you was wasted. Now you got the invincible stories flying again. This shit is winding down. You should have stayed out of it."

"KrutEk ordered me."

"Well, you look better. Chun said when he saw ya last ya had more holes in ya than a sieve."

KrutChan grimaced and was hunching his back against the freezing wind. "I'll tell you a secret. If I stand at right-oblique, my body-sieve-holes whistle the Marine Corps hymn, as breezes pass through me."

A moment later XibEk's eyes grew wide in astonishment. He could 'actually' hear the familiar song carried on the wind. All he managed to say was, "Sheee-it!"

When a grinning KrutChan spun around to face him, XibEk knew he had been had.

"The story's no good..." KrutChan was laughing about projecting his whistle into the wind. "...if I don't add some sound effects. Cheeb would be proud."

"Cheeb would shoot your honky ass." XibEk said. "If he could find a solid spot in that sieve of yours."

KrutSeet, still miffed at KrutChan and his ArnaMals for locating the metal treasure of Laktavil. Assigned KrutChan's forces to clear an island located on Lake Laktavil containing a large contingent of isolated desperate Dinarchy troops.

His anger was not abated by AinAcbal's arrogant boasting his attacking forces had 'procured' as much refined gold and platinum equal to the mines captured by KrutChan's Malkril.

To add to his bitterness, KrutEk had admonished him, in front of the other commanders, for leading a force of whining Oval females, stuttering and stumbling in the purple snow while fighting inferior enemy forces. KrutSeet raged at KrutChan to capture the island fortress in a Kin, with no excuses allowed.

KrutCheebel's Raider Bot-unit was being shredded. They landed on the island, off a Toob converted into an Ice-boat, to evacuate another Bot-unit holding on to the position.

KrutChan's other forces were advancing over the rest of the island, decimating fortification after fortification. Closing in on KrutCheebel's Bot-unit in the hardest position of resistance, KrutChan pushed his Malkril hard. KrutCheebel's Bot-unit, attacking another strongpoint, were out of touch with headquarters, and struggling to hold.

KrutCheebel's forces stepped into an inescapable meat-grinder with no way out. The way KrutCheebel's Bot-unit saw it they were sent by KrutChan to die. Their luck could hold only so long. Sooner or later somebody had to wrestle with that death demon on their shoulder. Kril Legions prayed it would not happen, but it always did. They were committed into a no-win hell.

The other Bot-unit, KrutCheebel's unit had relieved, had fought fiercely to avoid their annihilation. The other Bot-unit had been outgunned and killed indiscriminately as KrutCheebel's Bot-unit reinforcements arrived to join them. Being crucified in the follow-up Dinarchy counter attack, KrutCheebel's Raider Krils fought on. An Alamo on the icy killing ground of Lake Laktavil. In their minds: where was their relief force?

All KrutCheebel's Krils could do was watch their friends die, one by one, alongside the dead from the Bot-unit they had been sent to relieve. His Bot-unit adding to the body count.

KrutCheebel was cursing KrutChan, who had ordered them to hold the position after relieving the decimated other Krils, until help arrived. It was KrutChan's fault; placing them in this position.

He could feel his strength ebbing as his KanBalaam microbes seriously assaulting his system.

He was sweating in the frigid air, giving him warmth, but put layers of ice forming on the exterior of his body.

The evacuation-relief was a long way away.

The white-helmeted KrutCheebel and his crew on the ice-boat swung into action. The Toob boats, too small to remove all of the Krils at one time, made several trips from the frozen island, assembling near bigger Toobs out on the glacier.

Near the end of the evacuation mission, there were few Krils; one hundred at best, remaining on the frozen beach. Enemy fire was intense, pinning them down. None thought they would survive.

Kelel had not asked or bothered to get KrutCheebel's permission, but had taken her Bot-unit directly into a Dinarchy counter-attack to take the pressure off the ice-beach. Kelel's forces broke the counter-attack, but had only achieved time for KrutCheebel's forces. Her last messages to KrutCheebel from her Stel said she was surrounded. Her Krils she commanded were meeting TOTL, minute by minute. Since she was giving sitrep reports; the listeners assumed she was still alive.

When Kelel's final transmission came, she was ordering her remaining Bot-unit to breakout to the beach, one at a time. They were on their own. Her Bot-unit had ceased to exist. She wished 'good luck' to her surviving retreating Krils and KrutCheebel's raider Bot-unit back on the beach. Kelel remained, to be the last to leave.

Checking his watch, KrutCheebel recalled her Stel went silent about an hour ago. With his bugs prodding him, he mentally saluted Kelel. He would never forget her as the best of his subordinates. With her experience, he should have been led by her, not the other way around. There was no Oval logic on

591

this island.    He wished many times he had more women like her in his Raider-Zabin-Krils.

After Kelel went silent, KrutCheebel recognized his Krils were in an untenable position, and their deaths were imminent. He quickly commandeered other ice-boats, and ordered his own ice-boat vessel run aground between the beachhead and the enemy, thus drawing the fire to himself.

When his Krils scrambled for safely behind the ice-boat.  KrutCheebel, pushing aside the dead gunner, grabbed one of the ice-boat's two blaster guns, and released a murderous burst of return fire. Trying desperately to hold the enemy off until all that remained of the Kelel's meandering Kril survivors could be taken aboard.  They were a pitifully small number, but he refused to leave in case more straggled in.

Moments later he was wounded, in numerous places, from shrapnel, zipper fire, and an arrow. He vaguely noticed his right arm gone. <u>Probably from that zipper weapon.</u> He thought. He felt no pain, just the pulling and tugging of his uniform as he absorbed hit after hit. He was knocked back under the gunnels time after time.

Miraculously, he kept standing back up.  He would not quit firing or succumb by passing out. His crew, injured themselves, manhandled Krils aboard until the Toob was overloaded.

KrutChan arrived, in another area down the beach, at the same moment with CheChun's Bot-unit. On ice-boats with other Bot-units as reinforcements. He was puzzled by the figure approaching from out of a snow berm coming from the direction of the Dinarchy. The figure looked weird and encumbered, heavy with a load, and staggering in the snow.

He was overjoyed when he recognized his Oval-Kril, Kelel, loaded down with two fur covered Zars, one over each of her shoulders. The woman looked ludicrous, with only her white helmet sticking up between the Zar bodies. Her legs and boots churning puffs of snow, barely showing underneath the large lightweight Zars.

CheChun and KrutChan ran to her aid, reached out and relieved her of the deadweight of the Zars, handing them over to the medical ArnaMals who had rushed up.

Kelel was bent over, coughing from her efforts, her hands on her knees, struggling to adjust her rifle onto her shoulder. Between wheezes, she asked. "Any more of my Raider Krils show up?"

KrutChan somberly told her. "I got reports on a few of them. Dribbling in from the Dinarchy positions. Hortim's TunToobs are on station above us carrying enough ordinance to pulverize the Dinarchy. I'm waiting to let your Bot-unit survivors get here before I call in her airstrike. She can't wait forever. Cheeb is running outta timeline." He shrugged at Kelel.

Kelel was just as serious, puffing out her answer. "We only...had about...ten Krils left...rest are TOTL. These...Kril Zars...were the only...wounded ones with me...surviving."

CheChun raised an eyebrow at her calmness. "Shudda waited for the ArnaMal and Germ cleanup crews." CheChun said to her. "I take it you're transporting out your Zar lovers, sis?"

Kelel was breathing more normally. The flash of anger left Kelel's eyes when she recognized CheChun was joking. "Don't knock those Zars. They have huge steak knives for fingers. They know I like rough sex."

KrutChan was teary-eyed, clearing a lump from his throat; relieved she had made it back. He bumped his forehead with hers in the Kril Infantry way. "Way to go, Army. You're almost as good as a marine..." He was smiling. "...almost. You going to disobey orders again and lead another unauthorized attack, lady?"

Under the circumstances he was mocking her and she knew it. Kelel was smiling back. Deciding to give back as much guff as she was getting from them. She rendered the Kril two-fisted salute under her chin to KrutChan. "I'm not a lady; I'm a grunt."

"You got the balls for it."

Her smile innocently dazzled. "I can advise Hortim, my sister Kril-Oval, to work her magic on you in the Oval Pavilion?" She was smirking at CheChun. "Yours too, old corps."

KrutChan was thinking. These goddam Ovals never forget anything from their EkSeet and Stel network.

Kelel saw KrutChan's and CheChun's looks, mulling over how to respond.

KrutChan pointed his forefinger at her. "You're the one better watch out. You'll be lucky if Hortim don't make an exception in your disobedient case."

The background sounds of fighting were increasing. There was a moment of silence between the three of them.

"How's KrutCheebel? I been out of contact." A gooey mess from her broken Stel dripped from her chest.

"Outta contact until a couple minutes ago." CheChun said. "He's been fighting two wars, since you left."

"Inside and out." KrutChan said. He pointed at Kelel. "Get a new Stel from the Iceboat put on, before you break another fuckin Oval rule."

Kelel sobered, saying wistfully. "KrutCheebel's one hell of a leader...for a man. Better than most."

"That's why he's in command." KrutChan said.

She wasn't joking now. "What can I do to help?" Kelel knew KrutCheebel was in a bind, her anxiety showed.

"Nothing. Shag ass. You got no Bot-unit left. We're heading for his position now! Enough grabassing!"

"Spare me your sexist attitude." She innocently smiled.

KrutChan yelled into her face with smoldering anger. "I'm promoting your insubordinate ass! See how you like that...you fucking army grunt!"

She saw KrutChan start to roll up the sleeves of his uniform in a Hortim threatening move stepping towards Kelel. Seeing his chauvinist action, her own anger arose. She flicked her middle finger at him; rubbing it onto the bridge of her nose, before she went slide-running off to the iceboat. She was excited to be alive.

Further up the frozen beach, KrutCheebel understood his ice-boat didn't require dead leaving the beach. It needed the load lightened to lurch and take off. With assistance he removed the dead bodies; throwing them over the sides. He kept the dying and grievously wounded on board; ordering the ice-boat to leave. KrutCheebel jumped clear to get rid of his weight, crunching onto the ice-beach. He had used his last ounce of energy and lay face down in the freezing snow.

The relief force had landed too late to help the dead Krils. KrutCheebel was hoping to save the survivors.

CheChun broke into a run, when he saw KrutCheebel fall from the ice-boat, as it left the beach.

KrutChan jogged behind him.

CheChun tenderly turned KrutCheebel over.

Cheeb fought to maintain consciousness, long enough to utter four words to CheChun: "Did they get off?"

KrutChan said. "You'll be happy to know Kelel showed up...alive, humble, and insubordinate. Like you trained her."

KrutCheebel hoarsely said. "Thank god, she made it. You better believe that woman is golden." He coughed. "Wish I had three more like her." He was spitting blood-bubbles, fighting to catch his breath. "Any others from her Bot-unit make it back?"

Assured by KrutChan lying that they had; KrutCheebel looked relieved. He felt the Balaam bugs inside him wrestling ferociously with his immune system, as his fever skyrocketed. KrutCheebel was dying of the disease he caught on Uayeb KanBalaam. His fresh wounds helped the process. He consciously realized he was shot to hell. His arm and one leg were severed, thickly bleeding lumps in the sub-zero atmosphere.

"Hang on, Cheeb. You'll make it." CheChun said. He was applying compresses to the bloody stumps on KrutCheebel. His glance at KrutChan said different.

"Advance by fire!" KrutChan was yelling into his Stel. Bot-unit reinforcements were charging towards the Dinarchy positions. "Don't close with them! I got an airstrike coming!"

KrutCheebel's head was bobbing.

"Talk to me, Cheeb…don't drift off, dammit!" KrutChan yelled.

KrutCheebel was tired, badly wanting to sleep. "Get away from me!" He snarled at KrutChan to let him alone, let him nap just for a minute.

KrutChan could see KrutCheebel's many wounds and CheChun's AkSilk tourniquets on his lost arm and leg were still leaking blood. It was a moot point whether his wounds or the Balaam disease would TOTL him first.

Incoming tracers, mortar blasts, zipper air-rips, and flying shrapnel intensified. KrutChan's Stel crackled with his order calling for fire support from the ice-boats. Time to pull the plug on this shit. He thought.

Hortim's air wing sortie began their run-in.

Sprawling over KrutCheebel, protecting him from debris, as the TunToobs he had called screeched in on a bombing-strafing mission to blast their targets. Hortim's forces were obliterating Dinarchy targets.

The reinforcements of ancient ArnaMals, charged with blood curdling screams after the air strike was over. They went about their grisly business with lethal efficiency. Incoming ceased. The noises from their massacres gradually fading into the background.

The firing and noises of battle were petering out over the next five minutes. Either the Dinarchy had met TOTL or they had withdrawn from the field. Either way this mission was over. KrutChan thought.

It closed out like a curtain dropping on the last act of a play.

"Hang in there, buddy." Through his pain and fever KrutCheebel could hear KrutChan.

"Forget it." KrutCheebel moaned. "My ass is TOTL...and you know it. KanBalaam is calling me back, Krut."

KrutCheebel was mumbling; CheChun and KrutChan had to place their heads on each side of his mouth.

They had no words of solace to offer him.

KrutCheebel's good hand gripped KrutChan's forearm. "Don't leave me in this fucking ice. Throw me into the Seech field...so some other poor bastard...can join this universe."

CheChun was shouting. "Pay attention to me, goddammit...fight this...stay awake!"

"No fight...left in me, Chun. I'm feeling too good. I want to sleep, Krut. Let me alone!" He feebly pushed KrutChan's hand away.

Cheeb was fading, busy dying. KrutCheebel was facing death on his own terms. Dying is always about being alone. KrutChan thought. He could see Cheeb had no hope left.

"Was I a leader?" Whoever he was talking to was not answering. "Did I do okay, dad?" His eyes went cross-eyed. He was looking at something...talking to someone...somewhere off in the distance. "Tell ZocKuk..."

CheChun saw KrutCheebel whispering to KrutChan; who was shaking his head. "I ain't gonna be the one..." KrutChan's voice cracked; his heart breaking. I wanted ZocKuk, but not this way. His rage was welling up in him; his tears held back with his pain.

CheChun was holding back his own emotions. "Tell her yerself, buddy."

KrutCheebel choked again. "Who picked this fucking place...? Tell my mother.... No, my father would understand..."

KrutCheebel stared ferociously into KrutChan's face; his voice became coherent. "I don't want to die. Not on this frozen fucking planet."

His palm touching Cheeb's hot brow, KrutChan whispered. "Let go...just let go..."

Cheeb was speaking perfectly to someone else; not seeing KrutChan. "Funny, I see..." He drooled blood. "...colored lights are beautiful...it's so..." He wheezed. "...you know what...?"

KrutCheebel's death rattle was loud and fierce; struggling hard not to die. His death sigh followed and deflated his lungs. KrutCheebel, the Army Lieutenant, the Kril, the man... met TOTL with a glass-eye-lights-out stare. KrutChan was his friend.

During what felt like a time-pause, CheChun whispered. "It's easier on us when they just die."

KrutChan was angry at CheChun.

Moments later, the medical ArnaMal ran up to them and knelt down next to KrutCheebel to take his vital signs.

KrutChan grabbed the man's arm, shoving him away. "He's gone; you can't help him!" KrutChan was furiously revolving his head. "Where's that fucking Chaplin, Whalen?"

"Whalen is busy with many other Krils." The ArnaMal continued trying to revive KrutCheebel.

"I said, let him alone!" KrutChan hauled the medical ArnaMal to his feet and pushed him away. "Go help the wounded. Get the fuck away! Let him be, goddammit!"

After the cursing medical ArnaMal left, CheChun watched KrutChan stomp away, kicking small mounds of purple snow and ice, walking in the opposite direction. CheChun heard him cursing the gods.

"GODDAMN THE OVALS! GODDAMN THE KRILS! AND GODDAMN THIS FUCKIN' UNIVERSE!"

KrutChan quickly wiped his eyes with his sleeve when Reverend Whalen bumped into him running towards KrutCheebel, trailing bandages.

CheChun caught up with KrutChan. He kept a discreet distance from him. And then, after a minute, walked alongside KrutChan with a strangle hold around his friend's neck. He grieved with KrutChan. They both knew how to face mortality when it came.

They were walking in silence, trying to forget KrutCheebel had ever existed. Holding in a lot of anger at themselves; because secretly KrutChan and CheChun were glad it was KrutCheebel who had died, and not them.

On their last battle for an unknown, forever unnamed frozen hill, KrutChan was hit and down. He was found afterward, saved by XibEk and AinAcbal. They carried him to a medical station. He was grabbed from death's grasp by his courageous and favorite adversary ArnaMal doctor; keeping him alive until an Oval doctor resuscitated him on the Burseeosil.

Much to KrutSeet's and ZacNaab's displeasure.

The planet Laktavil was finally declared secured four Kins later. The Kril Legions had won.

The ArnaMals and Germs began taking over the planet and scrubbing up the mess.

As the surviving Kril Legions headed back to the Burseeosil, they were told Uayeb Three was over and they had not met TOTL. Great news for their Ovals back on Arna; an obvious fact to the living Krils. Most of them had the same thought in their mind. They had won what? This shit wasn't a soccer match. Why keep score?

Later in the Kin, aboard the Burseeosil EkTsab, when PacalMo visited KrutChan in the medical unit and told him the news. KrutChan had wheezed. "We lose some battles. We win some battles. In the end, nobody wins." PacalMo agreed.

KrutChan closed his eyes, thinking in broken spurts of that gully conversation he had in his past. "War's still on. Ain't no cease-fire, Mongol."

# Chapter 28

The city of Tikumyax outwardly looked the same, but somewhat different. The green lush jungle surrounded the east side of the city. Piercing sounds, with rumbling groans of small animals and roars from dinosaurs, announced the Krils return. The scents of cooking from the city mixed with the flower-smells of the Bougainvillea, Hibiscus, and Lotus blossoms. The varied bird sounds in a constant background of noise; all seemed normal.

The difference, to an astute observer, was quiet desperation after three Uayebs, hung like a funeral pall over the city. The joy of living was missing, to anyone who made their home in Tikumyax, caused by the Germ revolt here and in other Galaxies. Stirred by the prevalent war-weariness, the mood was catatonic-somber, with an undertone of fragile hope. The population wanting desperately to be joyful, now the last three Uayebs were over. Cringing because there were two more Uayebs on the horizon.

KrutChan hesitated before KrutCheebel's barrack room; walking in when the entrance went clear after he used his Malkril Commander device. KrutCheebel's ArnaMal, ZocKuk, was standing before him.

There was no introduction required by either of them. They were back by themselves on the balcony outside the Banquet in KrutChan's mind. Thank god for silenced Stels. He wasn't in the mood for his EkSeet's formal introductions. Helping KrutChan was his memory of KrutCheebel's constant refrain. To anyone who would listen, professing his love for her, using 3-D missive projections.

His ugly cratered face helped ZocKuk's memory. She took his hand and walked to a cubicle. Dressed informally, not wearing the Civil War-era hooped dress she wore to the banquet.

Their smiles said more than words.

KrutChan was thankful. He spoke first. "How're you holding up?"

"My new Oval advised me when KrutCheebel met TOTL on Laktavil."

"His friend XibEk is part of the honor guard bringing his…corp…I mean remains…to the Oval Memorial."

"Akna spoke of her dead Kril often before you arrived in the Seech Field."

Did she know about XibEk?

"I grieved much when Akna lost her Germ and nearly met TOTL." ZocKuk said. "You must visit 'her'; not myself. I not did expect you. Oval codes are strict about visitations by non-Nest Krils. I not could stop you. You enter under your Kril Malkril rules."

She did not know he had put them in a time-pause. KrutChan assumed she was trying to explain her innocence for breaking Oval decrees. She was fidgeting, as she said. "Welcome to my…rather KrutCheebel's area. I was performing menial tasks."

"I didn't mean to intrude." KrutChan lied. "KrutCheebel hated me. I usually act before I think. Cheeb paid the price."

"I think you are still in mourning in your heart."

"He was one of my best Krils. You don't understand. I sent him to die. Cheeb hated my ugly ass."

Her eyes had a twinkle of humor in them. "He described you as cruel and repulsive. When we met at the banquet I said he exaggerated your facial appearance. He described you as a cold-blooded monster. As I gaze on your Uayeb Three younger frame, we will put that description aside as another exaggeration."

ZocKuk was wearing a blue jumpsuit, tight on her thighs with a split down the sides. In his day he would have called them lounging pajamas. Her muscles flowed sauce-like down to her calves when she moved. She was awesome. She touched her jaw and brushed her hair back with both hands and seemed pleased when she saw his Kril inventory-glance moving over her body.

Speaking English, with her slight southern accent, he needed no translation. "Ah'm preparing for the Memorial of the Oval Protector's ceremony. As Kril KrutCheebel said, the last time he was leaving me; tying up loose ends in our relationship. Neither of us had Beekav for each other…"

KrutChan was surprised. "He said that?"

She shook her head and raised one eyebrow at him. "KrutCheebel knew…I loved another."

"I understand." KrutChan said to ease her discomfort. KrutCheebel had told him that just before dying. ArnaVals move on with their life.

"Not do you really understand?" ZocKuk asked him. She was tougher than she looked. "Krils usually not do appear unannounced after a Uayeb in another Kril's barrack. That is the purpose of the Oval Memorial ceremony, to do him honor there."

"Your EkSeet should get rid of your Oval-speak. Malkril Commanders have lots of leeway after Uayebs." KrutChan rubbed his forehead trying to eradicate a thought. "I didn't know you were in

604

here." He lied. "On Earth, my presence is something done for the family of the fallen. I'm apologizing if you are uncomfortable."

Softly, she said. "Akna is the one you should…; I am not ill." She was not taking her loss of KrutCheebel out on him. "Our timelines are done branching and cannot be changed. KrutCheebel was a new Zabin-Kril without many friends."

"Cheeb chose his friends carefully. Commanders don't become friends to their subordinates; hurts too much. No, I don't think he was my friend."

"I was not expecting you to come here."

"That's plain to me." He was staring at her. "I came to see his barrack, to get a sense of who he was." He decided to speak the truth. "My conscience was bothering me. I caused his death." She was the one misunderstanding. "I want to bury his ghost deep, in my way. Doesn't help him; helps me."

She unexpectantly laughed.

He had added guilt, because his eyes were on her trembling breasts. Twenty years out, his life was a long time without a woman.

"I am really surprised." ZocKuk said, watching his gaze.

He subconsciously looked away from her chest. "I'm still a little fuzzy, full of time-pause. I'm trying to get my head unscrewed."

From her puzzled expression, he could see ZocKuk's EkSeet was having trouble with translating his last sentence.

"You make me blush, sir."

"I meant I'm confused." KrutChan's eyes wandered to her fertile hips. He felt a familiar ache. No time for that, asshole. He was thinking. Say what you came to say.  Get in and get out.  He shook his head.  Not a very appropriate pun.

The Oval look ZocKuk gave him showed she was aware of his 'problem'.

Thank heaven, I'm finally looking at her eyes. "I tend to speak in a rambling-fumble-fuk fashion."

Her quizzical look indicated she was again having trouble with her translation of his words. With a small laugh she said "I am quite sure you didn't say anything meaningful.  Krils after Uayebs are fastidious in their speech."  She lowered her eyes away from his.  "I understand your mood."

He then said.  "What I meant to say was after the last Uayeb I'm not housebroken.  You know, not talking correctly around well-bred people."

ZocKuk grinned again.  "I have never experienced a Uayeb, but I believe I understand your perplexity."

"Well I wanted to get clean before I came." He did not say anything about his nearly dying again; that was not her problem.  "I sterilized a lot of the bugs I may have picked up on Laktavil.  Got rid of most of the cuts and bruises so I would be presentable and healthy; not so cock...sure...of myself."

She frowned again at the 'bug' translation. Then her hand covered her smile at the translation of his last sentence.

With a touch of irritation, he said. "I'm really not trying to be funny.  I'm serious."

She touched his cheek.  "I realize.  It is my fault.  The Krils I have met in my timeline, are not always translated literally or figuratively."  Her speech was clearer without her EkSeet translations, filtering out the third person syntax.  "Please, do not be insulted.  I am glad not to receive any of your...what did you call them...bugs?"

Now he laughed and held up his hands. "The good news is they weren't of the VD kind." Quit coming on to her, shithead! KrutChan admonished himself. "At least the Oval crew nuked most of them, but the critters scared the hell out of me when I found out I was harboring them."

"Yes, slaying them on the atomic level will usually do that."

"Right on..." he said. Either her mourning was over or she was toying with him. He was feeling nervous, losing control, as she motioned him to a floating chair. She curled her legs up under her on a couch across from him. Averting his eyes from her bare thighs.

ZocKuk's attitude turned somber. "KrutCheebel's constant apologetic missives to me became tiring. You are what I expected after the banquet. In his vernacular he described you as a fair leader. I meant fair, not in a derogatory way, but as a Tribunal Judge."

"Relax, I knew what you meant."

She absently brushed her hair back. "On the balcony I told you of my non-feelings for KrutCheebel. He is dead. Neither of us are to blame. I or you cannot bring him back." She was tilting her head; her dry eyes locked on him in analysis. "We both know why you came."

He thought he read her mind. "No, I didn't come here to tell you how he died...er...met TOTL. You have the official Stel recording...which pretty much sums it up. He didn't have any last words about you."

"That is not very flattering. You could have lied without being so blunt."

"Is that what you want? A lie?"

"No!"

607

He tried to display more tact. "I came to his quarters to get a sense of who he was. We were together through a lot of shi..." He said. "...I'm here. It seems silly now."

ZocKuk eyes at the corners crinkled in amusement. "I have Soothed many. I do not grieve lost Krils. Or dead hatchlings, like Akna's. It is 'my' way. KrutChan wishes to teach me more?" She acted as if they were being observed.

His heart pounded, staring directly at her. "You need teaching? What I have to say is for us, not everybody."

KrutChan reached between her breasts removing her Stel, then his, placing them in a drawer across the room. As a last measure, he made sure he had shut off his EkSeet and hers. "I got us in a time-pause."

"Thank you."

"At the end Cheeb was as frightened as every guy is at meeting TOTL." He said.

"You are very direct. Can you tell me what emotionally happened? Knowing might help me to understand. Perhaps help you too."

<u>Nobody can help me.</u> "You weren't there standing over him." He bit his lip, deciding to get the hell out of here. "I gotta go. I can't explain."

ZocKuk was not a fool. "By saying you would not, tells me volumes more than you saying you can't. I do not need my EkSeet. I speak English...yawl."

She giggled at his expression as her head cocked sideways. "He is gone! We do not have to keep him hypocritically alive by speaking his name. I never had Beekav for him. I told you that Kins ago."

"You said yourself I shouldn't be here." He was squirming. "Look, I better leave."

"Please stay. You are here. We are alone. My feeling has grown deeper for you since we were on the balcony." She caressed his cheek. Do not deny it; you have the same feelings."

"Are you kidding? I've been guilty about you for years on the Uayeb." KrutChan was feeling he was betraying KrutCheebel. "I'm glad you're here…" KrutChan heaved a sigh. "…but, I don't want to put you in danger. Don't gimme that look. If we're caught, you'll suffer the most."

"My Confederate father told me I should savor every minute with a lover. He said if I do not steal happiness I will suffer years of regret."

KrutChan waved his hand in dismissal. "I can't promise you a future together. Years drop off me so goddam fast in this Kril-Oval universe."

Her frown went deep. She was staring over his shoulder at something in the distance. "My father also said war enhances passion and longing. In my father's past I would be a widow." She snapped angrily. "Stop treating me as a reluctant virgin. And you are not an adolescent boy meeting with a fancy woman."

"Don't knock those women." He snarled back. "They supplied a need."

ZocKuk touched his knee. "Widows have needs too."

"Wow, now who's speaking bluntly." KrutChan's gaze was back on the rise and fall of her breasts and her glowing thighs accented by the lighting in the room. "I didn't want to treat you like an Oval or ArnaVal vessel."

ZocKuk scoffed. "The hell you don't."

Her perfume was light, barely noticeable, but having an effect on him. "I've never had a real woman choose me since I got thrown onto Arna."

609

ZocKuk slid her arms around his neck, kissed him tenderly; stirring both of them. She whispered into his mouth. "I choose to have us ravish each other until we run out of breath in our time-pause."

"Glad I got a choice."

Both standing, her tongue was deliciously probing. Her body was rubbing braille-like, her hands roving his body; his lips pushed her neck back while he softly pulled her hair.

"I think you are doomed, Kril. You will not ever forget any part of my body, my love."

"Shut up." KrutChan ripped her lounge bottoms, stripping her bare.

ZocKuk was as insistent, doing the same to him.

Pinning her against the wall, KrutChan feverishly grunted. "Southern women should be careful what they ask for. I'm going to make you wish 'you' were doomed. Entering your horny mind before the tickling begins. I love you, Scarlett."

"Who is…?"

His mouth smothered her words. "In my fantasies, southern belles are all named Scarlett." His lust accelerated. "I'm through talking."

Her pleasure was mounting; her passion flooding. She was softly groaning in response.

They lost all sense of time. Discovering and touching every pore of each other. Changing positions and places after each coupling. She exploding each time.

He holding back as long as he could before his time arrived….

Laying on her lace-canopied bed, both perspiring. For minutes they were laughing, talking softly, and wondering why they had not mingled before now. Playful teasing begetting fear that they would be caught. They both knew time-pausing always ended.

She seemed happy and mellow.

Not him. His regret for what he had done to KrutCheebel and her was building inside of him. KrutChan wished he had not come here.

They reluctantly dressed, but continued groping and touching.

ZocKuk was brushing away an invisible tendril of hair from her face; maybe to draw his attention back to her face. "I have had many of my Krils TOTL. You are the only one who ever..." ZocKuk's voice tapered off.

Deep in her eyes KrutChan could see her memory-pain flicker momentarily. When she smiled and blinked, her face relaxed; her youth hid whatever sorrow she felt. At the moment she seemed delicate, making him feel like a rat for putting her at risk.

Then she was moving on to change the subject. "Would you like something, a refreshment to drink? I have nothing from the earth universe for you. My dead Kril not did partake."

Bullshit! KrutChan was remembering on KanBalaam after KrutCheebel came back from the dead. Maybe it was one of those secrets between males and females, like on earth, not sharing everything with each other. "Thanks. I've given up booze. Got too goddamn many voices competing. You don't want me wailing and singing obscure songs after I get plastered."

"Oh no! We cannot have Malkril Commanders act foolish." She giggled. "I would not have you inebriated. I do have some ArnaVal Osil brew that is quite potent."

He waved off the offer. He thought she was one beautiful ArnaVal. <u>Being BalamEk's daughter meant she isn't even an Oval.</u> Her delicate nose and swelling lips heightening her sexuality.

A few somber minutes passed.

ZocKuk snapped her fingers. "Come back please." She quietly said. "You were somewhere else in those tiny Bots…moments. I am sorry if I caused you distress."

He had been sadly thinking of KrutCheebel dying dismembered, broken, and wasted in the icy snow of that frozen island on Laktavil. He waved his hand in dismissal. "I wasn't distressed; just thinking back. Memories are like that. Many memories should be forgotten, I think. Forgetting is not easy for Krils."

"Ovals never forget."

"Tell me about it."

KrutChan watched her eyes do their thing again and she smiled, changing the subject. "May I offer food sustenance?"

He returned her smile because it was contagious and shook his head. "I'm not hungry. I have to get ready for the Memorial. Then prepare for my next Uayeb."

"Of course…" she dourly said.

KrutChan decided he had made a mistake coming here and wanted to get the hell away before KrutCheebel's name returned to the conversation. "You've been great but I should leave." He reattached their Stels and initiated their EkSeets.

<u>You do have a wanker's way of insulting people, KrutChan.</u> His EkSeet said.

<u>You'll survive.</u>

"There is much substance in you, KrutChan. Deeper than the Ovals and Krils ever saw."

612

KrutChan shook his head. "Not hardly."

Before she could answer him, the room brightened to a deep blue and he knew she was being advised of company visiting. KrutCheebel's Oval, Dirva, melded through the wall and stood between them speaking to the ArnaVal immediately.

"This Kril not does belong here! You must get ready for the Memorial Ceremony." Dirva glared at KrutChan. "Have you Soothed with ZocKuk, human? Are you breaking another Code?"

Before he could answer, ZocKuk said. "Not do insult me! I not have requested him to Sooth me with his essence!"

"Not do lie!" Dirva shouted. "When ArnaVals lose their Krils; they have Beekav for the first Kril in their presence." She glared at KrutChan. "As this Kril is well aware!"

ZocKuk grinned, more to irritate Dirva than embarrass KrutChan. "There was no time. Though, surely it would have been pleasant. He is much like fumbling KrutCheebel. You know how humans are...." Another shot across the bow. An indication there was an unsettled argument between her and her new Oval.

Dirva snapped back at her. "You should know better! You have a TOTLed Kril to remember. You are too vulnerable to any Krils you encounter alone. This is why the rules are set and must be followed."

KrutChan got Dirva's attention when he moved in front of ZocKuk, stiffened his back and stared darkly at Dirva; with his Malkril Leader attitude. "Commanders can enter another Kril's quarters. KrutCheebel's ArnaVal didn't start anything." Partially true; all my fault.

He was gambling this Oval would not ask why their Stels were off. "I broke your goddam Oval Code. I let myself in." He had heard from other Krils it could be done, if you had enough rank. He had never tried it, until now.

"I trust her! Never you!"

Dirva looks like she swallowed my bullshit. He thought.

EkSeet said through his Stel. You tell a lie magnificently, KrutChan.

His mind responded. Shut up. Nobody asked the Queen's Guards to comment.

You are the interloper here, you uncivilized colonial bastard!

Dirva was shrewdly eyeing him. "I know of your arrogance and stubbornness, KrutChan. Malkril leader or not, you not should be here. Soothing her not will be tolerated in my Nest. Your Oval has been advised. You not will be warned again in your NOW!"

He glared back at her. "It's a human failing of mine to fulfill a wish of someone who goes TOTL. You Ovals control the Krils but not our thoughts or our sense of responsibility. It's not her fault. She didn't know I was coming. My breach of your fukin Oval protocols is mine alone."

Dirva's expression indicated his response did not translate totally accurate. "Since you not are with Reela, I must act in her stead. I command you to proceed to Jaguar's Area with haste. She is waiting for you."

"Jaguar waits for no Kril."

The Oval listened to her EkSeet. "When Jaguar is done with you, immediately proceed to your Soothing Room and your own Oval. She will punish

614

you.   Reela is advised of this breech of Oval Protocol."

He was going to make a wise ass curse, but since he didn't give a shit about Oval rules he winked at the ArnaVal ZocKuk and melded through the wall in silence.   For the first time since he arrived on Arna he felt like a real human lover.   Let Dirva's Stel attempt to translate a wink and ZocKuk's smile.   He thought.   What is Reela gonna do…send me on a Uayeb?

At the same moment he melded, KrutChan was not the only Commanding Kril in trouble. KrutEk had been summoned to the OvalChanHalach's quarters.

KrutEk was at the end of his after-action review of the last Uayeb with his Malkril Commanders.   It was more subdued, without the usual recriminations.   "I honor the Malkril commanders for their foresight and initiative obtaining more wealth and treasure for my Queen and myself."

He witnessed the angry moods of KrutSeet and ZacNaab.   AinAcbal, and in his usual mocking way, was laughing loudly.

KrutEk gave a hand signal to KrutSeet.

"My suggestion to KrutEk is he further demotes ZacNaab…" KrutSeet said.

Finishing KrutSeet's sentence, KrutEk said. "…and place him in the barbarian's Malkril, as his subordinate."   His grim smile was telling.   "KrutChan needs a bodyguard."

They were amazed he used KrutChan's name.

ZacNaab was trembling.   "I respectfully refuse.   If I am going to be convicted of trying to kill the barbarian without proof, then I have a solution to

615

exonerate my name. I wish to remain on Arna for the rest of the Uayebs." ZacNaab said.

AinAcbal was grinning again. "Spoken like a true coward; wanting to avoid battle. He wants to curl up in his Oval's Nest with his ArnaVal as he lashes and penetrates both of them."

"There is no living evidence remaining of ZacNaab's dishonor; the assassins have met TOTL." KrutSeet said to KrutEk.

"I do not require verification! ZacNaab will do as I command." KrutEk said. "Though I hate to use his name, KrutChan's EkSeet has notified the StelaBalaam. Jaguar, our intelligence Oval advises the barbarian has plans of his own. I will never allow the barbarian to be another ZacNaab bringing more dishonor on my Legions. I will deal with the barbarian."

For ZacNaab's benefit, he winked at AinAcbal. "Maybe ZacNaab can be his friend; like that arm-snake-lion Styx of his who met honorable TOTL."

Wiping the tears of fear from his eyes, ZacNaab accepted the inevitable.

"Summon KrutChan." KrutEk ordered. "The barbarian's commander, KrutSeet, will explain my terms to him. the rest of you are dismissed."

KrutChan melded immediately. KrutSeet had him standing at attention in front of him. "All of us present are aware of your plot to eliminate ZacNaab during the next Uayeb. Very unwise to break Kril Code. Someone has lied to you about ZacNaab. There is no proof. Plots are for Ovals."

Looking directly at ZacNaab, KrutChan quietly said. "I don't plan on liquidating ZacNaab. My enemies are spreading rumors."

The room erupted in laughter, except from ZacNaab.

"I honor him. He is respected by all of his subordinate Krils."

"You do not lie well." KrutSeet mildly said. "Always you speak as a fool; but act differently. Enough evasiveness has occurred in the last three Uayebs."

"We are at war with the Dinarchy; not each other." KrutEk added. "Do not underestimate me, KrutChan."

At attention, KrutChan kept his eyes front.

"KrutEk did not sanction those assassination attempts." KrutEk said in defense of himself. "Revenge is for amateurs. KrutEk has decreed attempts on you by others; or your plans to TOTL ZacNaab in revenge, are to cease as of this Ob. KrutEk's death-decree is being published to the Kril Legions as I speak."

KrutChan raised his hand, like a school boy, for permission to speak.

KrutSeet nodded.

Staring hard at KrutEk, KrutChan said. "Our KrilChan, I have been informed many times, by my EkSeet, you issued a similar standing order before the first Uayeb."

"Are you insinuating I allowed Krils to disobey me?" KrutEk asked.

Hell, I might as well clear the air. "Apologizes to my KrilChan, but my cynicism is deep. In spite of the orders, the incidents happened. Not every Kril, in every Legion, has interpreted this order as standard operating procedure..."

KrutEk snarled. "Be very careful what you imply, barbarian."

KrutSeet continued. "Orders constantly change. This death-order for any Legion member disobeying, has no appeal."

Saluting under his chin, KrutChan said. "I'm like any other Kril now? Facing TOTL without distractions. My thanks to the KrilChan for his clarification order."

KrutSeet shook his head at KrutChan's impertinence. "ZacNaab has been demoted for incompetence and will serve in your Malkril. You and ZacNaab will both treat each other with deference as an example to your lower ranks. KrutChan, as your subordinate, ZacNaab will be diligent in his duty to you. He has been ordered to guard your person against more attempts, to the best of his abilities."

KrutChan's throaty chuckle at that thought caused the rest of the Commanders in the room to begin laughing. KrutEk put his hand over his mouth, joining in.

Unfazed, KrutSeet continued. "You will adhere to the same attitude. We will not weep if you hate each other. However, both of you must adhere to my words. Do not let your animosity for each other lead to anarchy amongst your Malkril or the Kril Legion's battle plans in general. Both of you will suffer greatly at KrutEk's hands."

"I always do as my KrilChan orders." KrutChan dutifully said. "Many Krils, including myself, have been fighting alongside former enemies. Sometimes hate is conducive to respect."

"This meeting is over." Before they were dismissed, KrutEk said. "Do not appear too smug, barbarian. From the start of these Uayebs, you are not the only Kril facing oblivion from all sides and quarters. You are not special. Monsters grow to maturity in many Kril hearts."

_KrutEk has heard the bullshit monster-scuttlebutt about me too._ That empathetic remark set off a light in KrutChan's mind. _For the first time I realize KrutEk is facing more plots than me. From now on I'll have to remember the immense forces plotting against him._ KrutChan rendered a two-fisted salute under his chin, in deference to KrutEk.

He did not fool or placate KrutEk. "Leave and return to your Malkril Legions." He pointed at KrutChan. "Barbarian, you will remain alone here for an Ob." He crookedly grinned. "I will insure you come to no harm."

KrutChan went immediately to attention. He was controlled, watching KrutEk with his peripheral vision.

For minutes, KrutEk was reading some incoming missives, seeming to decide about something in them. After he discarded them, he cleared away 3-D projections on his desk. KrutEk pondered the wine in his cup for another minute.

He gave KrutChan a Kril hand-signal, indicating his subordinate to go to at ease. "I am not enamored of you, barbarian, as you well know. Once you asked me for my respect. You have been an undisciplined dolt ever since you arrived here, in my mind. Much too absorbed, with your unseen demons, countering assassination attempts." KrutEk flicked his purple cape over his shoulder. "But, I will admit, you have served me well, in your own way. Under grievous circumstances."

He took off both their Stels.

_So much for the Oval rules about Stels._ KrutChan thought to his EkSeet. _Does anybody not take them off, for Christ's sake?_

_Stel removal does seem prevalently cautious around you, old boy. You are one of the worst_

619

*offenders of traditions. Perhaps your vast conspiracies incite protective measures.*

"I have now shut off your EkSeet and mine. Our belief they uselessly babble constantly." KrutEk said. "I will now divulge to you my Queen's secret plot involving the both of us. She trusts your discretion." KrutEk whispered. "I trust her; not you."

KrutEk outlined her plan to eliminate a Covert Oval, speaking gravely. It was strategic, not tactical in nature; simple in its complexity.

KrutChan was in awe of the plot, but he kept silent. *I'm really, really, really getting sick of this spy-agent stuff.* KrutChan thought. *Now I got another goddam secret to not remember.*

"Before you came to this meeting, Xmucane and myself loudly argued about you and my responsibilities. She insists I have you protect me."

"How the hell can I do that? I ain't CheChun."

"I relate what the Queen wants done. I disagreed angrily with her!" KrutEk shouted. "Her opinion was neither of us trusts anyone. Either of us have friends. She believes that is advantageous to both of us as Four Strikers. She wants us to protect each other on the Uayebs. Women are devious and soft."

"Ovals don't understand Krils." KrutChan agreed. "I will do my best to obey my Queen." He said.

KrutEk scoffed. "After the Maluayeb Arena, we both would not have avoided killing each other. The Legions have no use for two Four Strikers. We are destined to face each other to the death. I will succeed."

Remembering a voice in the gully-mist saying the same thing, KrutChan understood the threat of a final battle between them. "Fate holds all the cards. I intend to survive."

"Depart from me, KrutChan, and keep silent about what the Queen and I have divulged; particularly to Jaguar."

He was shaken. Jaguar's involved? KrutEk had called him by name. The Queen's orders?

"You are surprised by my familiarity? I am also surprised your name comes from my lips. Out of respect, I will use your name during our future timeline together. We are both marked. That is to say however long we both last, before we TOTL. The Dinarchy are your enemies from this point on in your lifetime, not the Krils." KrutEk replaced their Stels. "Now, leave me."

KrutChan only half-believed KrutEk could stop an erratic killer at the bottom of the Kril Legions food-chain from doing whatever they wanted, for their own crazy reasons. His only comfort was now he knew KrutEk had the same problem. KrutEk was taken off his assassination list.

Her EkSeet responded when Reela inquired where KrutChan was at this Ob. She had summoned him to receive punishment for meeting ZocKuk.

Your Kril is near the jungle containment field.

Reela was curious. Whatever for is he there? I will meld to his site.

He is in time-pause. Unless you intervene, he not can audibly hear or observe your presence. His mind is null to his timeline reality.

Melding from her quarters, going to the containment field, Reela remained motionless,

621

looking for KrutChan's body signature within her full electromagnetic spectrum vision. Her infrared was searching for her warm-blooded human. It was difficult to ascertain his particular heat emissions, because of the multitude of radiant signatures emitting from within the jungle. The jungle, in her sight, was black-purple in color, moving with the breeze, sprinkled with red flares of moving creatures within it, small and large. The containment field was shimmering white in wire-like-nova bursts with no sound. Her mind was assimilating too much information of her surroundings; a common trait among Ovals. _EkSeet, announce his bearing and distance to my location._

_KrutChan is less than fifty Obsils from you at two-hundred degrees. If you intend to make yourself visible upon approaching him, be careful. He is standing immediately before the field, dangerously close to a territorial predator._

She refined her infrared vision while silently walking towards him, stopping at a safe distance, powering up her biological Oval-wand for protection.

_EkSeet, converse with his Stel and have his EkSeet send you his thoughts. I want immediate translations._ Whatever is he doing?

KrutChan had been alongside the Oval jungle cage for an hour. KrutChan had wandered seemingly forever beside the enclosure after leaving KrutEk. The sun-dappled foliage of the dark and light-green jungle fostered an inner peace in him. The grunts, the throaty coughing, and rumbling snarls filled his ears but held no fear for him. He had faced worse fear of the unknown in his two wars.

Five minutes ago the undergrowth had exploded with twenty small calf-sized rodents or

622

lizards running pell-mell alongside the cage line. Then they disappeared into the undergrowth, leaving the leaves, brush, and grasses trembling, marking their exits. When the cause for their fright jumped with a bound against the cage, the huge eight-foot-tall dinosaur was thrown on his back stunned.

The action surprised him momentarily. KrutChan was no paleontologist. He did not know the species; however, he knew it was no tyrannosaurus; too small. He corrected himself. The beast towered over KrutChan by two feet, so small was relative. Muscled heavily, with a mouthful of sabre-teeth, a cranky disposition, and six-inch claws on its feet and long arms. The creature definitely was menacing. Zar-huge in appearance, but not emotionally contained like that alien species. The multi-colored feathered beast was nobodies pet. The creature stood calculating and his head revolved to the front with predator binocular vision. Staring cognitively at KrutChan; the human immediately knew who was the hunter and who was the prey in this jungle.

Wanting to display they were both predators, KrutChan cocked his pistol, chambering a round. Old habits die hard. <u>Wish I had taken my rifle with me. If I blind this bastard and blow off his nose, maybe, just maybe, I got a chance. And maybe this ornery dinosaur can eat .45 slugs for lunch, dummy. At least I got more chance of surviving than Akna had during child birth.</u>

His EkSeet shouted. <u>You must leave immediately!</u>

Stubbornly ignoring the advice, KrutChan was enjoying this stand-off. A hushed minute passed; time stood still. Remembering the assassination attempts, he relished the adrenaline rush.

The human Kril's challenging stare was mirrored by the dinosaur. They were both assessing the danger of each other, sorting out options to run or fight.

Another minute went by, with neither man nor beast giving ground. KrutChan was aware the charged fence was for his protection.

The dinosaur, huffing and puffing, was loudly snarling in a boastful display of dominance.

Long minutes slowly came and went.

KrutChan retorted to the beast by loudly and stupidly singing off-key, 'From the Halls of Montezuma...to the Shores of Tripoli', intermingling that, because he forgot the rest of the stanzas, with the Animals Rock tune, 'We gotta get outta this place...' He was foolishly grinning.

In answer to the human's terrible off-key shout-singing, the angry dinosaur bellowed with rage, abruptly thrusting his teeth-ladened mouth at KrutChan, throwing gooey saliva and snot in a spray of disgust at the human.

In anger, KrutChan hacked up a green lump of butter from his throat, and spit back into the monster's face, standing on his tiptoes, yelling loudly. "Fuck off!"

Another minute passed.

Then the beast turned his back on the insignificant human-meal-idiot, in an indifferent huff, and wandered back into the jungle foliage; his broad ass swinging like a hula dancer, his scaly tail bushwhacking brush in his wake.

KrutChan raced to the edge of the cage, careful not to touch it, waving his pistol, roaring at the dinosaur in his own rage. "You fucking coward cocksucker! Run, you goddam chicken-shit lizard!" KrutChan's chest was heaving with frustration. "Run

624

away, back to your mother's Nest! Big tough guy! You're an egg-sucking sonofabitch!"

Stupidly, KrutChan felt invincible. He was satisfied for the first time today.

In the awkward afterward-silence that came, there wasn't a sound coming from the contained jungle. Within minutes KrutChan's shoulders drooped with exhaustion, his mind drifting.

_Wonder how far this cage goes?_ He wandered off.

Watching his infrared signature recede, blending into the curve of the jungle, Reela shook her head and grimaced. _KrutChan is courageous, but insane at the same time. What did this confrontation prove?_

_I believe he was testing his human will; not his courage._ Her EkSeet said.

_My Kril remains a male enigma to me._

_As a female, I not do understand him either._ Her EkSeet intoned.

_My KrutChan still grieves for Akna and her Germ. He is a strange Kril, more like an ArnaVal in his desires._ His rambling thoughts, after translation, were alien to her Oval mind. She could not comprehend what had occurred with her Kril and the beast from the jungle. What purpose did displaying fierceness accomplish? _I will never understand males, particularly human Krils._

Reela was then beginning to understand. Every female he met has attempted to control him. His visions attuned to human love. Beekav the Ovals not can give him. If he Soothed with ZocKuk, perhaps she will give him her Beekav 'love' he treasures. ZocKuk can make him happy. She may still his constant rage from the AnticArna drug. _I not_ will punish him. I am melding to my Soothing Room. EkSeet. How long will he be in time-pause?

For him, about one half hour of his former timeline. Her EkSeet said. Unless, he is destroyed by another creature he antagonizes in his travels. My pity is for the animal creatures.

Reela thought. Not do speak foolish.

Her EkSeet added. He is not fearless. He believes himself to be very alone; not invulnerable.

Before leaving, Reela instructed KrutChan's EkSeet, through her own EkSeet, to not let him TOTL. She ordered his EkSeet to cease KrutChan's time-pause, without his consent, and meld him to his barrack should another possible lethal situation again begin to occur.


Reela summoned KrutChan to her Soothing room when he returned. When he entered, she gently took his hand and led him to the locked glass door of Akna's cubicle. The inside of Akna's living quarters was decorated in a Sudanese motif with animal furs on her bed. Rough-hewed oak arches and a wooden table and chairs surrounding a glowing fireplace. Inside, Akna was puttering with minor chores, oblivious to them. She was bloated in the belly, carrying a child in her.

Whispering so Akna couldn't hear him, KrutChan asked. "Jesus Christ, she's pregnant. Is she a ghost? Or a clone? She looks like the old Akna. How did you do it?" His heart was racing. XibEk should be here, not me. This Oval magic is driving me insane!

Reela was smiling with a self-indulgent grin. "Not did you request we revive her senses? Nor is she a clone. She is as real as we are to her."

626

"Can I talk to her?" His voice was cracking. "Just for a minute. Before I leave on the Uayeb." He wanted to tell her he had found alive the dead Kril she yearned for. He could stop being a substitute.

Reela held his chest to stop him. "She not will recognize you. Her mind is recuperating. Her THEN is coming back to her in fragments of memory; much like your future wife-visions. I not do understand what the doctor did at the Queen's command. I cannot explain her procedure of implanting essence Xmucane supplied. You wanted her to recover; she somewhat has. The simplest explanation; the doctor impregnated her with her former Kril's essence. The doctor was pleased; she feels she has discovered a new cure for the sickness that occurred in Akna."

Pregnancy was one way to throw Akna's hormones out of whack. KrutChan was stunned and speechless for a moment, and then grew pensive. "Thanks for what you did. I owe you one, as my earth folks used to say." He stared in worry at Reela. "Hopefully, I'm not someone she remembers?" He was thinking of complications with ZocKuk.

"What a strange question. Why the concern? You never had Beekav for her." Reela took his hand again. "According to her Oval doctor, you are gone forever from her mind; becoming just another Kril."

Nodding, KrutChan went with Reela back into her Soothing Room.

Reela rubbed his back. "You should maintain your hope the Oval doctor is correct. It is highly impossible she may eventually remember all of her THEN. The doctor says not to hope." She changed the subject and grew somber. "I wish to speak to you."

They melded together to the Memorial.

After attending the Memorial ceremony, he was waiting for his time-pause evaporation. When he was suitably dressed and leaving for the Uayeb, KrutChan dropped his gear and peeked through Akna's door into her area. Akna was in heavy REM sleep; her eyes laterally moving, back and forth, under her lids.

Knowing Reela would have Oval-locked the entrance to Akna's area. He was sure Reela would go ballistic if she found him there. On the spur of the moment he melded, using his Malkril Commander authority, into Akna's cubicle and approached her bed. She was under a fur blanket. The room was desert cold and chilly, even with the fire crackling. Sudanese weather-acclimatized, her cubicle had her father's own home atmosphere to live in.

KrutChan remained looking down on her for minutes. Her half-parted lips were softly letting her breath in and out; her breasts rising and falling in synchronization. When she raised her hand, after a tiny cough, he gently and carefully brushed her hair away from her nose. He did not want to disturb her. Not wanting to wake and startle her.

He wished XibEk could see her glowing from her pregnancy.

A brief gust of wind from behind him caused Akna to roll over, turning her back to him; the fur blanket sliding down to her hip. In that moment, he cautiously, taking a chance, raised the fur blanket covering her to her chin. Afraid she was about to wake up, and scaring the hell out of her with him hovering over her; KrutChan slipped out of her cubicle.

He was not surprised seeing Reela by the door. He stared at her, their eyes locked in mutual questioning. KrutChan assumed Reela had been there a while. Realizing she had been silently watching him and Akna, probably intending on protecting Akna. Reela did not seem hostile; more resigned.

Shouldering his gear from off the floor with a grunt; he saw Reela move away from the glass door to Akna's cubicle.

Reela grabbed his arm, stopping him from leaving. "As human's say, I am sorry for your loss of her. She will never remember you."

"She loves...has Beekav...for her dead Kril." KrutChan said.

Reela was thinking for a minute, sorting her thoughts. "Akna had Beekav for another she 'thought' met TOTL. You helped her. Her loss was large. Her memories will recall him when she sees him again."

KrutChan already knew the answer. "In other words, he didn't die."

She saw he was hiding a secret of his own. "The Queen commanded I acquire her lover's essence from Xmucane's former Oval-trainee to impregnate her. After this Uayeb, I will convince the Queen to transfer her Kril lover closer to her."

KrutChan's face reflected his relief. "Oval plots don't surprise me anymore. He blinked, damping down his reality. "If he survives the Fourth Uayeb. If he doesn't, she'll be okay?"

"Akna will never know."

"In other words, he'll stay dead, right? And the Ovals, as usual, will forgive."

He started to leave, and then stopped for a moment, and turned to Reela. He sent her a conspirator's exaggerated wink, just before he made his exit meld.

Reela's hid a small smile. Reela felt overriding Beekav for Arna when he was standing over Akna. If KrutChan or Reela had spoken; it would have broken the spell. She would never forget those moments Akna and KrutChan had shared.

Her EkSeet said. Your KrutChan is aware, but refuses to recall Akna's stillborn Germ was not of his essence.

He feels responsible. Akna had much Beekav for her former Kril who met TOTL. KrutChan was a replacement.

ArnaVals are emotionally intense when they lose their former Krils. Reela's EkSeet laughed.

Reela was fully aware that is what happened to Akna before KrutChan arrived. Akna germinated the former Kril's essence in her body. I allowed KrutChan to believe he could save Akna from meeting TOTL if he supplied his essence. In order to control him in my Nest.

Reela melded to the Ovut area to witness KrutChan leaving for his Fourth Uayeb. Akna's former Kril XibEk was part of KrutChan's Malkril. She intended to keep silent; better timeline for KrutChan and Akna...and XibEk.

Chapter 29

The Fourth Uayeb of the Unlucky days.

A Time hiccup; an alternative-life beginning.

The Burseeosil arrived at a main sequence yellow-white star, but 1.05% larger than Earth's. The Krils aboard were ecstatic, they were finally invading a system which promised a more Sol-Earth configuration. No more Dwarf suns with low lighting, terrains not full of freezing snow and ice, or desert desolation, or erupting mountains.

Compared to Sol this sun was hotter by comparison, radiating two percent more luminosity from its outer atmosphere. It was glowing with a white-yellow hue at approximately 6,300 K with a faster period of rotation. Seven planets made the system diverse. It was a remote and isolated system. The only habitable planet the Ovals had named YaxEb was about 1.15 times the radius of the Earth, with three moons.

YaxEb was in the habitable zone from its star. Having a molten core and a strong magnetic field to counteract any solar flare particles. The dense atmospheric effect kept the average mean temperature of YaxEb at 20C, 70F. Air pressure similar to Earth's at sea level. Scientific Ovals had chosen a planet with 20% oxygen. The higher carbon dioxide level made the air clammy. The strong winds were frequent and annoying, but would be bearable to the invading Krils.

They were Bumping onto a triple-canopy jungle planet. Appropriately called by the Krils, Green Skull; for them a miserable kind of place

KrutChan well remembered from Vietnam. The Kril Legions soon learning from experience this extraterrestrial planet would proliferate with small and large predators. Scurrying prey harmful to Krils, poisonous flora and fauna dangerous to invaders. YaxEb could never be a vacation spot.

Time-pause briefings told them the planet had intelligent deep-red colored sentient beings living in clusters of cities far to the north and south. The indigenous YaxEb beings avoided the hemisphere of swamps and jungle terrains from which they had migrated. Intelligent, but not foolish; the YaxEb residents were pre-machinery civilizations; more inclined towards religion, spirits, and crude science.

After being conquered by the Dinarchy, when the Ovals and Legions arrived, they wisely remained neutral. Avoiding getting in the middle of a war between the Krils and Dinarchy. Dinarchy aggression had taught them well. They would remain neutral, as long as the off-planet Dinarchy mined for heavy elements, which the indigenous people had no use for, and avoided the off-worlders.

Upon Bumping, KrutEk's Stel transmitted an ominous update to all of his Kril in his Legions, as their Toobs detached from the Burseeosil. "We have superiority in numbers. Be on guard, for our intelligence estimates have been vastly underestimated in our past three Uayebs. Your enemies are tenacious and wish to obliterate every Kril they encounter during this Uayeb to gain victory. These are not the defending defeated-rabble-mobs we have encountered in our past. Intelligence advises me the Dinarchy Commander we are facing is Kravid Palatine. You must not falter in your resolve. I will certainly not."

His last sentences had chilled the Krils; it was not the 'Great Mission' speech they expected to hear from their usually supremely confident Leader. His words sounded like he knew his Kril Legions were teetering on the flip of fate's coin.

"Your KrilChan has prepared you veterans well. However, I expect this Uayeb Invasion will be a critical, green-hell-TOTL-campaign; with only your god's favorites surviving. I anticipate meeting many of you in Elysium." It was not a battle-cry shout; more like a whisper of the inevitable.

XibEk was thinking. That Lifer sounds like he's preparing himself and us, to be wasted. That's a hell of a strategy.

BalamEk's following transmission did not help their morale. "Not do contact the sentient beings of this planet. They not do care about our invasion mission, or the Dinarchy's. Never TOTL them either overtly, covertly, or by mistake. We not do need them aiding the Dinarchy, in any form or fashion. Not do create a guerrilla third force because of Kril 'excesses'. You are here to defeat Kravid Palatine's Dinarchy. We are here to acquire treasure for our Empire and your Ovals. We Ovals will help you in your endeavors. May the Cosmic Egg watch over you."

The Bump Toobs, loaded with Zabin-Krils and ArnaMals, landed without incident. The jungle ate the Bot-units in the Malkrils, one by one, as they moved frightened and carefully into the wet swamps of foliage. KrutChan's Malkril took hours to penetrate the foliage; but eventually they got to a hill, their objectives. High ground was always the premium position, even in a jungle. On the advance, isolated gunfights broke out around them, lasting only seconds. They kept going, apprehensive and waiting for the unknown to punch the unwary.

The jungle awaiting them was hated by all of them; impenetrable, silent, ominous, and deadly. The leading Kril Malkril, led by KrutChan, pressed on into a dark tunnel of bent over green canopy and scrub grass.

Shrill weird chirps, moaning groans, rustling brush, and thumping grunts alerted the Krils there were other creatures alive on this planet YaxEb. The blue-green sky was hardly visible, except patches directly above them. The three moons shone dimly, in the daylight. Alien Krils able to see in the ultraviolet spectrum had an easier time in the jungle.

A muted noise filtered to them from ahead, they scarcely breathed when they stopped. KrutChan hand-motioned his immediate headquarters group. Silently, the column went prone on the broken ground.

XibEk lay next to the trail, his semi-automatic M-14 pointed to where a twisting-trail turned a corner. Mimicking the others, he waited to see whether a Dinarchy trooper, or something more ominous, like an indigenous predator hunting for food, would turn the corner.

Soon the guys on point heard the sound of something falling to the ground with a thump. Dinarchy had to be around here! Then there were a few footsteps. Followed by an eerie silence. No movement. No sound.

Raising his eyebrows, XibEk turned, looking at KrutChan.

In the middle of the trail, KrutChan hoarse-whispered towards the trail corner with a warning, "This is KrutChan. Advance and be recognized, or my Krils will zap you." He used his name so the shadow out there would know he was a Kril. He would be the only one to draw fire if the hidden guy was Dinarchy.

A disembodied voice said. "Easy buster, this is Baker Unit here; we're all Krils. Take it slow and easy, buddy."

"Come closer so we can see and identify you."

"Coming to be recognized, fellow." Emerging from the darkness, the guy continued talking. "Am I glad to see you people? You guys are Dog Bot-unit, aren't you?"

"No, we ain't, but that's close enough." KrutChan said. "We ain't Dinarchy. Feel better?"

"I'm point for Baker Bot-unit." The guy, EkSeet advised, was US Army, from the Spanish-American War era. He cautiously moved into sight, crouched low to the ground, his eyes darting, trying to see everything at once.

PacalMo raised his hand in greeting and the guy relaxed. He snapped the trapdoor on his Springfield Model 1873 rifle to put it on safe. His features were drawn, and the jungle rot on this face made him look leprous. He didn't smile when he passed through them, shuffling back near the rear of PacalMo's Bot-unit.

Others of the Baker-unit Krils following him, emerged around the bend in the trail, then another, then another, then a staggering wounded guy.

Eventually a skinny, gangly, young looking Captain emerged from the mottled-dark jungle. His Marine World War I putty-colored uniform was sticking to his body from his sweat when he moved into view. He carried a Springfield '06 rifle. He either was shaking with anxiety or hypothermia from the dampness. He stopped next to KrutChan and let the rest of his men amble by.

"We Bumped here on reconnaissance-in-strength before our full Malkril Bumped. Nice to see you fellows." With a cracking rasping voice, he said. "My hay fever is pounding me in this humidity. We have been an advance party for two Kin; feels like we've been here for a month. Is Dog Bot-unit in this same area?"

"Your units are from another Malkril." KrutChan answered him. The Captain's 3-D map was broken. They spent time reading KrutChan's and where Dog Bot-unit was located. Finishing by discussing comparisons between trench-warfare and jungles; deciding no dying ground was dissimilar. They both discussed their missions before the Captain led his men away.

I'll never see that guy again; dead or alive. KrutChan thought. He was daydreaming about the hundreds of guys passing through his life from another war in his past.

"Sir!" KrutChan's mind slammed back to the jungle trail on planet YaxEb. "Sir...the patrol has passed through us." His Unit was stirring back to life.

His Mayan point man, HoyAhChan, was sent by XibEk to report to KrutChan for a moment before moving out.

"This jungle is strange to me." HoyAhChan said. "The foliage is not similar, much as others I have hunted, but this jungle is alien to me. I will have to move quietly and carefully."

It's gravy to have a more experienced point man. This dude lived in a jungle; he's no rookie. KrutChan thought. "Do what you gotta do to stay away from TOTL. But don't slow us up; we've got to be on the objective quickly." KrutChan shook his head. "Let's get where we're going; enough of this memory bullshit. He patted HoyAhChan's shoulder

636

and pushed him to the front of KrutChan's Malkril Bot-unit onto point.

Upon Soothing, HoyAh's Oval had named him HoyAhChan. Oval and Maya languages were similar. HoyAhChan felt at home being alone in the green undergrowth again, away from the other Krils. Much as when he was hunting when he was Seeched. But now, on his fourth Uayeb, he was finally in his own element. The last three were alien to him, especially the rock hard freezing cold. These other Krils he was with, since being captured by Kukulcan, mystified him.

He threaded carefully through the jungle, assessing where he was going, staring at individual leaves so that he could see through the foliage. Small alien critters scattered away from him as he moved. He was thankful nothing larger than a rat appeared in his path.

HoyAhChan was on alert the entire time he was leading the Bot-unit. He was pleased that he was responsible for KrutChan's Malkril trailing after him. His former King had expected him to be flawless as a scout. To die giving warning, if needed. His leader KrutChan was concerned HoyAhChan might die; that was good.

During the next hour on point he stopped only two times. XibEk came to him, surveyed what he saw, and grunted his approval of HoyAhChan. After each stop, XibEk spoke into his Stel to KrutChan, and then spurred HoyAhChan on.

Fifteen minutes after the second pause HoyAhChan's timeline branched. A growling panther jumped upon him, followed by flashes of lightening, and the ear-numbing sounds of an erupting volcano. He blissfully met unconsciousness.

XibEk saw the Dinarchy trooper drop from the alien tree onto HoyAhChan. He ran forward, butt-stroked the trooper off of his point man, and fired his M-14 into the Dinarchy's face. XibEk briefly assured himself HoyAhChan was alive.

He then led his Kril Bot-unit directly into possible ambush. He had survived a few ambushes. Doctrine was great reading in a classroom for textbook tactics, but the bottom line was: Get out of the kill zone. Charging straight ahead was the only maneuver for survival. Except for the dead.

The firefight lasted brief minutes. When the firing from his Bot-unit tapered off and then stopped, he walked back to HoyAhChan's position. The ArnaMal Corpsman was attending to HoyAhChan's bleeding wounds.

KrutChan's voice spit from XibEk's Stel. "You got the ambush cleared? When you moving out?"

"No ambush. Stragglers. In a skosh; the little Mayan dude is needing transport back to the Burseeosil, the doc says."

"My doctor is heading up there with others. Let them do their jobs. Move out!"

XibEk grabbed HoyAhChan's shoulder in reassurance. "Hang in there, little guy."

HoyAhChan was apologizing. "I did not see the panther. I felt his mauling. I am sorry."

"Ferget that...ya can't see one-hundred eighty degrees...and up and down! My fault. I shudda seen that dude and wasted him before he jumped on ya."

"But our leader, KrutChan, will think poorly of my skills."

"That's crap...not yer fault. The fukin lifers don't give-a-shit about what happens to us colored." XibEk pushed his helmet back on his head. "Ain't ya learned that yet?"

KrutChan was yelling. "C'mon Xib...haul your ass! Get your Bot-unit on the move. You still got the point!"

The medical Germs were carrying HoyAhChan off on an invisible stretcher. The medical ArnaMal decided HoyAhChan's wounds demanded he be evacuated to the Burseeosil.

XibEk went in the opposite direction, muttering under his breath, cursing lifers. It did no good, but it made him feel better.

The first Kin of the Invasion started like all the others; by losing Krils. Before the end of the Kin, KrutChan's Malkril headquarters set in an alien forest of stumpy trees, like tall skinny pineapples growing.

Calling ZacNaab to him, KrutChan showed him the see-through tactical map. "Take your half-Malkril to this position behind us and hold yourself in reserve."

ZacNaab spoke aloofly to his troops into his Stel, preparing to take his half-Malkril behind the front lines; a position he favored.

KrutChan grabbed ZacNaab's shoulder to stop him for a moment. "I don't want you playing asshole-games with your Krils back there. Stay prepared. You better run like hell to help out when I call you."

After translation, ZacNaab understood KrutChan's underlying threat to his existence. He numbly nodded. ZacNaab's Stel crackled with KrutEk's loud embarrassing laughter.

"I have been in reserve to our KrilChan's Kril Legions many times in my THEN." ZacNaab sniffed to alleviate his pride. "I do not need your instructions."

In a sing-song voice, KrutChan said. "I 'expect' more from you than our KrilChan." KrutChan was slowly releasing his grip on ZacNaab's clothes. "You ain't in KrutEk's Legion reserve, genius. He put up with your superior Roman shit. I won't."

KrutChan was tapping his naked K-bar on ZacNaab's leather armor. The threat was unspoken, but obvious to the watching Kril. "You need any clarification of what I'm telling you regarding your mission?"

"No. I do not." His eyes were lowered to the ground.

"No...what?" KrutChan snapped. He learned a long time ago, in his former universe, frag orders should be simple and not obtuse. And subordinates fought better when their mindset was without questions.

ZacNaab was confused for a second until his EkSeet gave him advice. "I meant no sir!"

When he heard AinAcbal laughing at him through his Stel device, ZacNaab calmed his anger down. "I understand completely, sir." In his mind, he was thinking. I will find a way to assassinate you, barbarian.

He moved his half-Nacom-Malkril, as ordered, back into reserve.

Minutes later KrutChan gave his remaining Malkril the order to advance.

The YaxEb invasion stutter-stepped and began falling apart over the next week, infiltrated in many

640

places by Dinarchy troops. The Kril Legions were reeling and fighting to hold their ground. It was not helpful for their ArnaMal-ancient forces, use to assembling in cohorts, turtles, and massed units.

Following Kravid Palatine's orders, the Dinarchy had changed their tactics; not charging pell-mell into the Kril Legions. His Dinarchy Zars were furiously digging tunnels and bunkers, in which he placed his reserves, and supplies. Not wasting time and effort being obliterated by the Legion's high explosive ordinance and flying Toobs searching for them.

Kravid's Dinarchy were attacking in force on all fronts. While their follow-up reserves settled down to claim territory; their fighting point-of-spear troops were advancing.

BalamEk advised KrutEk that Kravid's forces were on a par with his own Legions; hardly an advantage in men and material for an Invasion force. Kravid had strategically and tactically planned well.

Kravid mobilized his troopers into spearheads and wheeling divisions. He led off his attacks with all the ancient and Modern artillery he could mass. Walking his curtains of firepower in stages, followed by his infantry on the offensive.

The Zabin Krils in the Legions, were getting a taste of their own medicine. Each countermove by KrutEk and BalamEk were stunted by Kravid.

KrutEk was not panicking. Early on in Invasions, new problems always materialized. His Kril Legions were trained to react with deadly force. Studying Kravid Palatine for over twenty-four Uayebs had not been for naught for KrutEk. Palatine's strategy on YaxEb had changed, but held no surprises...yet.

641

Two weeks into the confusing mess on Planet YaxEb, Kelel's Bot-unit in her quarter-Malkril was surrounded and facing being killed near a stream-river. Same old combat scenario's prevailed, but this time her Bot-unit was losing ground. She fought her way out of it, with half her Bot-unit surviving. A Dagot, IxuTuul, sent by Kelel, attached himself to XibEk; who wondering what the hell his lifer Executive Officer was thinking.

HoyAhChan, who had returned from the hospital on the Burseeosil, distinguished himself during the breakout, using his Mayan blowgun and arrows with extreme effectiveness. His weapons were made for the jungle. He was in an alien forest; his bow and arrow and blowgun were just as effective as if he was in the Guatemalan canopy.

The Gurkha, TzenalAh, was killed in the retrograde fight. Kelel would sorely miss him. Kelel had never, in the timeline she knew TzenalAh, had to question the small man's discipline, tactics, or leadership in any battle. Where the fighting was the densest, TzenalAh was always in sight, always shouting his war-cry. Fearlessly leading his Gurkha troops. Kelel had told KrutChan, one time during their brief lulls, that the diminutive fighter from their eighteen hundred's, fought with a focused Bengal tiger's ferocity. Kelel had highly respected the little brown man.

Three weeks into the Invasion, Kravid Palatine had contacted KrutEk, on his own Stel, and harassed him about KrutChan. That missive should have been a clue that the Dinarchy's intelligence capability had evolved. Kravid bragged he was going to TOTL KrutEk before the KrilChan ever got off the planet.

"I'm coming to cleave you into tiny pieces, KrutEk. The Tribunal will have to vote and replace your stinking carcass with another KrilChan. I will crush your Kril Legions before your Tribunal can vote a successor. I will never treat your barbarian KrutChan, as humanely as I will treat you."

KrutEk remained silent. Words of bravado never impressed him. He contacted BalamEk and she assured him her Ovals were working on the communication problem. He was thinking to himself. For once in these Uayeb Invasions, I would like her Ovals to predict and overcome situational problems before they happened.

The YaxEb Invasion had stalled and was on the point of crumbling. When BalamEk sent help through her air sorties, the TunToobs were blasted out of the sky by Palatine's new particle weapons.

Over the weeks, one-fourth of Hortim's flying squadrons, were disabled by attrition. Palatine's scientists had adapted those particle weapons to be used as artillery when Kril air cover was absent. His planning was meticulous, precise, and devastating to the Kril Legions.

Four Krils, from different Bot-units, arrived in his headquarters, and went after KrutChan. He was suspicious of anyone approaching him.

He was ready this time. These goddam assassins, who were supposed to be disbanded, are coming at me early. Who knows? Maybe infiltrators sent by Kravid Palatine.

The closest first attacker, KrutChan charged and stunned with his entrenching tool and then drove the shovel into his attacker's forehead, taking off the top of his head. He immediately spun on his heels and sidestepped the second one; spinning the attacker around.

PacalMo grabbed that man in a crushing strangle hold from the back; jumped backward and laid his attacker out prone, breaking his neck.

Thank the gods for the Mongol.

CheChun had distracted the third one, putting a round into his shoulder, and KrutChan collided with him full on, judo-flipped him over his hip, driving him into the ground. KrutChan then noticed the attacker was already dead; his face half-cleaved.

Whoever shot this guy with that Dagot-blaster barely missed me by inches. I don't know whether to kiss my helper or kick his ass for his lousy fire discipline! I was right there, for Christ's sake!

Angrily, KrutChan viciously used his K-Bar and ripped out the third attacker's jugular and throat to make sure he was TOTL. He glanced up to see NoKoch, with his Dagot blaster whining down. Rubbing the bridge of his nose with his sixth finger; grinning like KrutChan's savior he was. For once KrutChan was happy to see the Kril Legion's shadow-warrior had showed up to save him.

The last attacker charged KrutChan, screaming in rage. KrutChan had been distracted by NoKoch. KrutChan realized his death was imminent.

Surprisingly, the fourth attacker stopped in mid-charge, beheaded by the Viking, AinAcbal, wielding his axe. The assassin's body was flopping around, his blood draining; with his head in the dirt blinking in shocked surprise.

When KrutChan realized the fourth attacker had been taken out by AinAcbal. KrutChan asked. "Where the hell did all you guys come from?"

NoKoch and PacalMo gathered around the last assassin on the ground comparing battle notes, when AinAcbal answered. "Our KrilChan KrutEk, orders that you will NOT be TOTLed by any dishonorable Krils. KrutEk notified all the EkSeets to advise him immediately of such intentions from Krils in any Legion. We were dispatched in the last Obets. As he told you, your trials are over and you need not fear assassins again."

WOW! Was all KrutChan could think.

"Where was the dark dragon ZacNaab?" PacalMo asked.

CheChun bitterly laughed. "Pouring wine over his pecker, before he pounded his pud. Where else, in his tent."

"Let him alone!" KrutChan exploded. "He's where I assigned him! In reserve, where he belongs!"

AinAcbal agreed. "KrutEk will cause TOTL to any Kril who thinks, or has plans of attacking you in your future NOW. ZacNaab had nothing to do with this dishonor."

KrutChan noticed XibEk and CheChun mumbling something to each other while walking away. PacalMo and NoKoch each nodded with respect and parted in different directions.

At least the Viking and KrutEk now know who their enemies are in the Krils. It ain't me. KrutChan thought.

"Finally we get some R&R, food, water, and rest. I'm going to enjoy this, until we get relieved." XibEk was wiping sweat off his face. "Ah can't wait ta move off this fukin 'Green Skull' planet." XibEk said to himself.

"Amen brother, we earned it. Somebody tried to frost KrutChan again?" CheChun changed the

645

subject. "These immature mud-marine Krils should appreciate all we do to save their tiny balls." CheChun was boisterous, but guarded. "Iffn 'they' really are going to pull our asses back into reserve." He had developed a tic. Everybody was stressed out.

"Right on, cracker. Your EkSeet has cleaned up your language, Chuck, for the better. Ya piss on angry gods, you be fuked'."

"One more time…" CheChun groaned. "…I hear that bullshit of yours and I'll shoot ya myself. We can both use R&R."

"I will believe it when we are on the Burseeosil. Damn! I am starting to talk like KrutCheebel!"

"EkSeet hocus-pocus, is all." CheChun was somber staring off into the jungle. "Cheeb was an all-right mulatto-nigger."

XibEk did not swallow that sentiment. "Don't you go all black-loving radical left-wing on me." XibEk snorted. "Cheeb thought he was a black man. But, being a white-black man he was okay by me, in the end."

"You both were black men…ta me. When ah'm wrong I say ah was wrong…you fellows showed me how wrong I was…"

A new replacement Kril was listening to them. "Who the hell was KrutCheebel?" The man was chuckling, then said with humor. "Are you guys gonna kiss?"

CheChun punched his dirty knuckles into the man's face. XibEk's closed fist hit the other side of the recruit at the same moment. XibEk pointed his shaking finger at the man on the ground. "You ne'vah say nuthin' 'bout Cheeb, asshole…ne'vah! You ain't earned da right!"

646

"You wouldn't make a pimple on Cheeb's ass, boot!" CheChun yelled.

"What the hell's going on here?" KrutChan was standing over them with the Viking AinAcbal. "Can't you guys stay outta trouble, just once? I liked you guys better when you hated each other's guts."

"We still do." CheChun said, without smiling.

XibEk flicked his thumb, in agreement, at CheChun.

The new replacement Kril lay bleeding and staunching his broken nose. "Nothing is going on! My two leaders were demonstrating some defensive hand-to-hand shit to me."

"Of course, they were. You tripped over a foot locker I suppose?" KrutChan was irritated.

Both Chun and Xib were smiling. Chun said. "Those mobile foot locker's is dangerous."

Xib picked up the replacement and dusted him off. "Wel-com ta KrutChan's Malkril."

Xib and Chun were ignored. "I called you two birds here to meet me; not to tighten up our replacements. Get back to your own Bot-units and train them. These Krils are mine. Come over here before you leave."

CheChun scowled at the new Kril and punched his arm. "Keep yer head back ta stop da bleeding before ya upset yer friends." He pushed the new man on his way. CheChun walked with XibEk, wandering over to AinAcbal and KrutChan next to a black-moss-painted alien boulder.

"Chun, your Bot-unit has been assigned by me to AinAcbal's reserve Nacom." KrutChan said. "And don't start giving me any shit. Do what you're told. We got to save some Dinarchy for the KrutEk to murder. Your Bot-unit needs the rest and I'm tired of fixing your screw-ups. Do your job by the book; the Viking'll kick your ass if you don't."

"Dat ain't right, KrutChan." CheChun mumbled, swallowing a curse.

"Rest with your men and take your attitude out on them. Just be ready when AinAcbal commits you."

AinAcbal said. "My Vikings have been told by the KrilChan to withdraw; he has a plan. Both of you are a great Krils with much Viking in you, CheChun."

"Ah ain't no goddam White-Viking." XibEk said. Grumbling his displeasure, but flattered by the Viking's praise.

CheChun had taken orders for a long time, and he reluctantly acquiesced. He knew all officers were crazy anyhow.

"Ah'll see ya later, cracker. Ah'm movin' out post-haste on the lifer's mission he gives me." XibEk told CheChun.

"What are you grinning about, Xib?" KrutChan yelled. "You're going with him!"

"Wadda-ya mean...with da overseer? Dat racist lifer is crazy, man!"

KrutChan was waving his fist at both of them. "For tuning up that new Kril, I'm promoting XibEk to command a half-Katun in reserve. CheChun will take orders from you. You'll enjoy that. CheChun won't take a promotion and you ain't gotta choice. I need you there in AinAcbal's Viking Kril Malkril."

"Dat yellow-haired-white fuck will get us kilt!" XibEk shouted.

AinAcbal looked amused. "Do not tremble. The manner these battles on YaxEb have dribbled against us, KrutEk will make sure we all TOTL. You will be among warriors when you breathe your Valhalla last."

KrutChan attempted to reassure them. "KrutEk has decided to stop running; he has a plan. You guys are his ace in the hole."

CheChun muttered out loud. "Yeah…his ace…shoved right up our asshole." He grabbed XibEk's sleeve. "We's out voted…show me, fearless black leader, where ya want me to dig the slit trench."

Both XibEk and CheChun were snarling at their EkSeets to translate their idioms and jive speech correctly or shut the hell up!

"You guys been told for four Uayebs to clean up your language. Nobody understands you guys." KrutChan glanced at AinAcbal and shook his head at his two Bot-unit leaders.

"Their purpose seems to be confusing their EkSeet translators…" AinAcbal agreed with KrutChan. "…and themselves."

XibEk jerked his head at CheChun. "Let's get-some air. Too much Lifer brass around dis place." He grinned lasciviously and rubbed his hands together. "Let's go ona 'dog-hunt' lookin for an ugly shot down Oval Toob pilot who needs sum lovin'. We can lay pipe fer da Con-fed-er-ation…long as we got time."

As they walked away their conversation was getting dirtier. "Ah go first…black man!"

"Go head. Your redneck pecker is a preverbal needle-dick. That tiny meat hook will only get her lukewarm. Ah'll su'ply da ecstasy wid my blacksnake."

"Yer black-ass; She won't be able to close her legs after I be done wid her; that's the only way u'll get in her…."

649

"Dream on...dream on...old corps...yer tallywacker is a splinter compared to my four by four..."

As they wandered off, their filthy conversation got raunchier; their voices drifting away on the breeze.

KrutChan looked at AinAcbal and rolled his eyes to the sky. AinAcbal was chuckling.

"I am pleased you are intent on following KrutEk's directive, KrutChan." BalamEk said, upon him entering the Command Center in the Burseeosil EkTsab.

"I didn't volunteer." Over his objections, he had been ordered by KrutEk to report to BalamEk on the Burseeosil EkTsab. "Ovals have my EkSeet pouring so much crap into my mind; my brain is exploding." That was the extent of KrutChan's thoughts. KrutChan was in a time-pause.

BalamEk removed his Stel. "I have silenced your EkSeet also; unless you need him to allay your fears." Without more explanations, she led him past her Oval Crew members, depositing him in a huge room, seating him on a throne-like console.

She left immediately.

KrutChan felt tiny. The room completely filled with thousands of floating miniature 3-D displays. All activated, all full of constantly changing technicolor scenes. When he looked at a specific scene, the view expanded in front of his eyes to desktop size.

I was pulled out of YaxEb battle to go to the movies?

He was bewildered, at first. It took him awhile viewing, before he figured out the activated screens. Red-flashing borders, were more than likely the critical screens to focus on.

When he glanced down at the console full of flickering lights, the expanded projections receded back into the montage. Eventually, after scrolling his eyes in a three-hundred-degree arc, he realized none of the thousand 3-D displays were of the invasion planet YaxEb.

He tried to figure out what purpose the room served. Majority-wise, the screens projected peaceful scenes full of city-dwellers, farmers, happy groups of people living out monotonous timelines. Walking or running to fulfill their duties. Krils acting like bored garrison troops on far off planets, and Tribunals with courtiers of anonymous Oval Queens.

The flashing red-bordered projections showed Krils skirmishing in urban or rural battles, indigenous people fighting each other, angry mobs roiling, pillaging, out of control raping, destroying property, and other ongoing anarchy-strife events occurring within the projected scenes. He was kept distant from the turmoil, yet conversely intimately involved. He was there, induced by three dimensional projections.

BalamEk reentered, standing next to him. "You are in...rather, this room is KrutEk's observation platform. He reviews the status of his Queen's vast thousand galaxy empire at any given Ob, paying particular attention to his outlying Kril garrisons on remote planets."

KrutChan felt out of place, no shave, grimy with layed crusts of jungle-sweat. But he was amazed. "How the hell does KrutEk keep track? This mess would drive me nuts."

BalamEk smiled. "The KrilChan has immense burdens on his shoulders. The Supreme

651

leader of the Krils does more than simply issue orders, rulings, missives. He does more than strategically plan invasions, command his Krils in battle, and discipline unruly upstarts. He has graver responsibilities than omnibus-control of his Legions."

Shaking his head, KrutChan said. "Beat me how he does it."

"How do human's say it? The job of KrilChan not is for the faint of heart. Our Queen, the OvalChanHalach has a duplicate observatory. The StelaBalaam connects to this room of his, to keep him aware of timeline conditions in the Oval Empire."

"Listening to you, I'm clear on his overwhelming responsibilities now, but not on how he juggles it all."

She put her palms together and faintly smiled. "That is why 'he' is the KrilChan."

"Why me?"

"KrutSeet and AinAcbal in their past have been in this room in their THEN."

KrutChan kept his eyes on the console; he didn't want to trigger a zooming 3-D projection. "I'm pretty sure these lights on this chair control whatever KrutEk wants to do or say to these screens. For my own morale, what button or areas, should I keep my hands and fingers off of?"

Glancing down, BalamEk said. "Do not touch these two purple and blue orbs. The left blue one puts you in immediate Osil contact with Kril Commanders on a specific projected planet. The left purple one issues a destruction sequence, ordering military action to those same Kril commanders. You can consider left-side blue and purple as on buttons; right-side blue and purple turns them off. Though the StelaBalaam is biologically attuned to our KrilChan's touch and not yours; I would advise you not to attempt testing them."

"No shit. I feel like the president sitting next to his red phone; ready to order a nuclear war." KrutChan sighed."

BalamEk smiled. "ZocKuk would prefer being near you at this Ob."

KrutChan kept his lovers' memory to himself. How the hell did BalamEk know about ZocKuk? Crap...I forgot she was ZocKuk's mother. "Why did you bring me here?"

She lightly touched him on his shoulder. "You are mistaken; not I did. KrutEk advised me you were coming, in the midst of a time-pause."

"Why? Place reminds me of a perverted Disneyland ride. He trying to educate me or impress the hell outta me?"

BalamEk rubbed her temples to help ease her tension. She had an invasion on YaxEb to oversee. "KrutEk not does confide with me in Kril Legion matters." Her body language indicated she did not care.

"ZocKuk was KrutCheebel's ArnaVal. Is she aboard?" KrutChan kept staring at the console.

"In your earth universe parlance, not do flatter yourself." BalamEk said, to keep him focused. "In two sets of previous Uayebs, KrutSeet and AinAcbal, have both sat in that console."

"So he's training me, like he trained those other guys. To take over if he gets wasted.?"

She paused for a moment getting a translation from her EkSeet, before speaking. "Krils constantly TOTL by fire and attrition. As I stated, I not can ascertain his motives."

Her hands were on her hips, in a masculine pose. "I remind you, as I did KrutSeet and AinAcbal,

653

the OvalChanHalach chooses her KrilChan at the end of the Uayebs. Her choice consented to by the Tribunal. KrutEk never can pick his successor. In your years in your timeline here, not do you comprehend yet?" Her expression signaled he was too stupid.

That remark triggered his resentment. "I been here four damn days. I got long-term amnesia." He straightened up, waiting, in case she punched him with her Oval Wand. When it never happened he added, to tone his attitude down. "If KrutEk meets TOTL, our Queen Xmucane ain't about to choose my ass to be KrilChan. She's already put me under a death sentence, when the Uayebs are over."

"You not do understand. KrutEk does not command his Queen. He must prepare for his demise. Unfortunately, you are a possible branch in his timeline."

"In my timeline on earth we would say, I'm his hole card."

BalamEk was shaking her head wearily. "I not do need an EkSeet. You are again correct. 'If' you survive these Uayebs, the OvalChanHalach has the power and will to TOTL you for your prior transgressions. I not do question my Queen's thoughts. As I said repeatedly, I also not do attempt delving into the mind of KrutEk, a Kril."

KrutChan rolled his eyes to the myriad scenes on the ceiling, causing a confusing explosion of montages receding and expanding in front of him. He quickly looked down at the console lights. "So what happens? What do I do now?"

She was femininely adjusting her trousers to leave. "KrutEk has given you two of your earth hours in your time-pause after we finish speaking, to absorb whatever you can of these surroundings. To make sense of what you see. He said, as you observe, you should make mental decisions about what you would tactically do on those Oval planets in the Empire to correct the ongoing conditions. As if you were KrilChan. May the Cosmic Egg be within you."

BalamEk melded.

Holy Mother of God! KrutChan had a hysterical, laughing meltdown, thinking about his plight. I gotta get back to the war-insanity on YaxEb; to remain only semi-crazy.

He filled his mind with the problems in the thousand galaxies Empire's projections, until he was numb. He stopped after two hours, trying to make sense of it all. Subconsciously, he was absorbing infinitely more.

Why can't my life remain simple? Each day passes with tons of complications. Why can't I ever get a break; draw a decent hand?


Kins followed, with the reserves preparing for the coming offensive. KrutChan's Malkril sitting in one place, playing with their thumbs and yawning in boredom. XibEk abruptly sat up. His back was against a huge orange banyan-like tree, with no fruit, coming out of deep sleep. His teeth grated as he shivered from the wet jungle.

He saw KrutChan, off by himself, staring between his feet, absorbed in his thoughts. It was the first time in jungle YaxEb that KrutChan was finally alone, not pouring over 3-D maps, giving out Frag orders, or conversing over his Stel with higher command.

XibEk shuffle-walked over, putting one boot between KrutChan's feet to attract his attention.

"You done cutting 'Z's'?"

How can I sleep in this frigging heat?" A shivering KrutChan asked, looking up at XibEk's somber expression. Something's bugging him. He thought. "Sit Down and take a load off. Anything going on?"

"Nuthin much." XibEk bitterly said. "AinAcbal say he know everythin bout dis offensive. Scuttlebutt say KrutChan knows what's going on. We Krils in the rabble, know from nuthin."

"I see your English hasn't improved. Did your EkSeet finally surrender?" KrutChan threw him a tired squint, not elaborating. "Don't give me that shit. You been briefed. Sometimes it's all about nuthin, Xib. You know that."

"Yer saying ignorance is bliss? That crap only applies to the Lifer-Kril dummies."

Staring flatly at XibEk, KrutChan said. "You don't wanta know."

XibEk's ornery flashed for a moment. "Yer saying our Roman 'Lifer's' illustrious offensive…is turning into an offensive-bad-smelling type? Tell me another secret nobody knows."

"I got no secrets to tell ya, man."

XibEk pulled his towel off his neck to ventilate the clammy heat. "Chun says, we're getting our asses handed to us on a platter. Anyone with eyes, a functioning Stel, or a blinking tactical map can see the Dinarchy are winning the battles. What's the big secret? The Seventh Calvary coming to our rescue? Riding in, guns blazing, bugles trumpeting, and scarf up all the lust-filled maidens?"

KrutChan laughed. "Yeah…lethal love-making for the conquerors."

"The Dinarchy going to bury us, dude."

KrutChan grimly smiled. "Wow…you're in a mood. I liked you better when you were jive-talking, throwing your sexual bullshit around."

"Ever since you came back from the Burseeosil, you've been a daydreaming honky sonofabitch." XibEk griped. "You find a secret you didn't want to know? Something you're hiding?"

"Nope…" KrutChan looked down at a small black-dotted-yellow creature-critter scurrying over the toe of his boot. "Forget it. For your information, in this invasion of YaxEb, we're a minor comedy on some Oval television screen."

Thinking the lifer was not making sense, XibEk laughed cruelly. "In other words, we're being wasted. You're saying we're all gonna die here."

"Whadda want me to do…spell it out!" KrutChan's anger flashed. "Has it ever been different?" KrutChan blew his cheeks out. "Wasted is our middle name."

"I don't gotta like takin it up the ass."

"You think I do!" KrutChan's temper flared. Get your…head screwed on!"

It was XibEk's turn to deliver a cold, flat stare. "You were about to say 'nigger'. I wudda knocked out your front teeth, Krut; iffn you talked to me like you did Cheeb. We're talking is all. We aren't brothers!"

They were silent after that exchange. Then XibEk broke the stalemate. "You're a lifer bastard."

"Right on." KrutChan said. "I'm invincible, my troops say. The dumb bastards think if I stay alive; they'll survive. This offensive is going to bring their fantasy crashing down."

"You love da ArnaVal?"

KrutChan wore a surprised look, thinking XibEk knew about ZocKuk. "Who?"

657

XibEk had his turn at being angry. "Akna, your black ArnaVal in Reela's Nest! Don't gimme that, who me, shit! You were mind-groping her booty a minute ago, before I came over."

KrutChan was assembling his thoughts, remembering the last time he saw Akna. He answered in a whisper. "How can you love someone you've known only days?"

"It's happened before. Happens in this Oval universe all the time. Guys are lost, miserable and mad when they get here. When a woman treats them right, the Krils are left with whatever feelings they can suck out of her."

KrutChan went into his mind for a moment; thinking of ZocKuk. "Never thought about it, in those terms. Sounds like that's the problem with my visions."

"You got one of the best ArnaVals in this Oval universe, Chuck. Except your white-honky-ass is too stupid to appreciate her."

"I honestly don't know." KrutChan tried to be truthful. "Sometimes I do and sometimes I don't. It's all mixed up with my hallucination-visions, I guess. She thinks she loves me."

"Knowing her...er...of her; she would."

KrutChan yanked a tuff of alien grass and broke the clumpy mud-dirt off. "Way deep down she looks at me, like I see my 'Greer-Visions'. Like I'm somebody else to her; a Kril she lost. How would you like being somebody else?"

"I'd rather be dead." XibEk was emulating KrutChan, staring off into the rocks and boulders surrounding them. "Losing somebody you really love is..." XibEk drove his bayonet into the dirt, with a clunk sound. "...and I'd kill the fucker that ever took my place."

Knowing XibEk was Akna's former lover did not help KrutChan. Now was not the time for revealing secrets. KrutChan whistled briefly, standing up and stretching. "Remind me never to steal your woman, Xib."

XibEk was leaning next to him, speaking clearly. "You remember this conversation, when I do remind you." He stomped off, throwing his voice back at KrutChan. "I'll send you the sitrep on my quarter-Malkril when I get back to them."

A minute later KrutChan jumped, thinking Albert was back. "What the fuck you want?"

It was NoKoch. "Just wanted to tell you I'm going on Recon again." His teasing expression did not help, when he said. "Still seeing invisible ghosts?"

"Go do your Sneaky-Pete thing. I ain't got time."

"I got time." He was adjusting his equipment. "In my era back on earth, the psychiatrists would have diagnosed you as having 'Schizoaffective Disorder'. Hallucinations, voices, and constantly lonely; to name a few symptoms. That's what the doctors used to say. Everybody has seen you speaking to invisible guys. The Oval drug they gave you each Uayeb is the trigger-cause. Recognizing your condition will help you cope. If you ever wise up."

"I don't need any future psychoanalyst bullshit. Shove off."

"Next time one of your talking ghosts shows up..." NoKoch began walking away towards the jungle. "...remember what I told you. At least you'll die happy. The Lord Jesus Christ won't save

659

unbelievers like you or KrutEk. Asking for God's forgiveness can only save you."

Sitting alone, KrutChan was wishing Albert would appear, to tell him if NoKoch was one of the assassins. Blinking, he saw from a distance Albert was following NoKoch, waving and grinning at KrutChan.

During the next Kins, the rest of the ArnaMal ancient Krils, led by KrutEk, off in the distance, were engaged in fierce battles. Over their Stels the Krils were updated hourly on the situation. The Kril Legions were steadily losing ground to the superior Dinarchy force; getting mauled, falling back, and collecting their dead and maimed. Eventually, they had given enough ground KrutEk told them to stop and fight to the last Kril. He intended to leave and stop the Dinarchy at another strongest point of attack.

On a clammy-blistering-hot Kin, KrutChan and his headquarters Malkril were ordered to a blocking defensive position on the huge hundred-mile depression on YaxEb. Their hilltop position overlooking a large plain levelled by ancient pyroclastic flows. Where the Kril ArnaMal ancients were slugging it out with the Dinarchy forces.

KrutChan's Zabin-Krils became the shaft of the spear. As soon as they had dug in, they were immediately in contact with the Dinarchy. On that Kin, the final battle for YaxEb was engaged.

CheChun came to the aid of XibEk during a fierce hand to hand struggle for a minor bump on the terrain of YaxEb. NoKoch joined them in the battle. When it ended, XibEk and CheChun were gasping for air while NoKoch ran off somewhere.

660

"What's happening? Tell me what's going on!" KrutChan was yelling for situation reports over CheChun's and XibEk's Stels. They both were too out of breath and exhausted to answer. They mentally told their EkSeets to reply for them.

During a lull, after reporting, CheChun and XibEk were joshing each other about KrutChan's ascension to power. They joked about KrutChan never surviving.

In the lull of the battle, CheChun said. "'For a knee-grow, you are one hell of a fighting marine. But I still won't attend your funeral."

XibEk replied. "After all the Uayebs we've survived; you being a lifer asshole, I don't want you near my grave. I still can't stand you, white bread. Only thing I'll do at your gravesite in the Memorial to the Oval Protectors is…'

They both said the final phrase in unison, "…piss on your golden statue!" They both were grinning at the truth of it.

"At least I ain't cornholin' my sister." XibEk said.

CheChun threw a handful of dirt and pebbles at XibEk. "I remember now why I hate your guts, Xib." CheChun said.

Minutes of silence followed in their timelines.

"You ever wonder why the hell we didn't kill each other off." CheChun asked. "This would be a better place, if we did."

XibEk was staring intently at CheChun. What the fuk brought that up?

"I think when a dude is facing racism daily; it's easier for hate to become the driving force." XibEk grew quiet. Don't stick your educated

661

ignorant nose in here, EkSeet; you don't know nothing about it.

"KrutChan hates the Oval race. He's a giver, not receiver."

"With us, fighting each other is an equalizer. On earth, we would have kicked each other ta death. We protect each other's six with our hate."

"With my EkSeet butting in, this is the clearest you have ever talked to me."

"Yeah, I noticed that from you too. We spent seven Uayebs, hundred forty years, talking 'at' each other instead of 'with' each other."

"You are a sonofabitch. We could never be friends. War or no war." CheChun said.

"Right on, motherfucker."

They were both trying to say something and did not know what the something was at the moment.

Taking a twig and scratching circles in the dirt, CheChun was reminiscing. "Yeah, well, my old way of life is deep-ingrained in me. I can't stop hating you, Xib. You're too black." CheChun drank out of his canteen. "Here's a toast to all the goddamn haters and tent-camp-evangelist religious morons…like NoKoch."

"NoKoch's on a mission for his god. Same here, I'll never stop hating your guts, Chun. Who I am and my hate, goes way deeper than yours." XibEk took a swallow of his own, after raising his canteen. "Here's another toast, striving to have our opportunity to waste the dude we both hate before we meet TOTL."

"Now that makes more sense!" CheChun spit out water, splattering in the jungle dirt. "Not this Oval 'love your former enemy' shit." CheChun was somber. "Getting even is one thing, but never

662

bushwhacking in the back, like KrutChan's assassins."

CheChun felt this Kin was the end of their timeline and TOTL for them was near. They were staking out their positions, clearing the air.

XibEk was contemplating. "No dry-gulching for us Krils. I want to have the dude I hate see it coming from me. Never in the six by ambush."

NoKoch wandered over to them, stopping the conversation.

The Dinarchy were counter-attacking.

NoKoch waved to them to follow, before charging into the enemy.

Jumping to his feet, CheChun held XibEk's arm for a brief second. "Speaking of sneaks; I'd never trust that mug NoKoch behind me, no way, no how." CheChun said.

"Right on." XibEk agreed. "Dude always gets into battle with lot'sa Krils; but he's the only one who walks."

CheChun attacked, on the run, to NoKoch's left.

XibEk trotted to NoKoch's right, since the palomino guy's Dagot Blaster filled his right hand. XibEk preferred living to dying.

In the mad scramble of battle, they lost track of each other. For an hour they were intensely involved in just trying to stay alive. As usual, the rain of ordinance, the ricocheting lead and other missiles preoccupied them. The buzzing sounds of Blasters, the whiffs of arrows, the zipper-weapon buzzing, and the concussions from explosives, damped down any thoughts.

Eventually, when the firing around him receded, XibEk was checking himself for weeping wounds, then his lines. Between battle, when he came across dead Kril after dead Kril, their bodies in death throes. Mingled with deceased Dinarchy troopers.

XibEk, stumbling with a curse, tripped over CheChun. Thinking the body was just another Kril, he was stunned for a moment at the sight. <u>Seven Uayebs and this is how it ends?</u>

Chun's body was still smoking out of a couple burnt holes. He had not died instantly. XibEk saw he had gone down hard. His back braced against a stone parapet draped with six Dinarchy troopers.

XibEk's EkSeet retransmitted CheChun's last words. <u>I hit that yellow sonofabitch!</u>

<u>Chun's M-1 rifle breech is locked open with no ammo in it. KrutChan ain't gonna like this.</u>

Staring a long, long time at the cadaver that used to be CheChun, XibEk was puzzled. Something looked strange to him. None of those Dinarchy dead troopers carried a blaster with them or a blaster mark. He paced around Chun's body, trying to see the angle of fire that wasted 'old Corps'. Wandering, he came across a blood trail. The shooter must have been behind and near CheChun.

XibEk recalled CheChun's last words. None of the Dinarchy were Asian. Was he mad at some chicken-shit unknown Kril? Naw, sounds more like he knew the dude. It was possible Chun was venting his racism, not his killer's cowardice.

Then out loud, XibEk mumbled to himself. "That Overseer bastard, always thinkin' 'bout hisself. He got dead just so I couldn't waste him. Don't never mean nuthin'."

He made a decision. NoKoch had Chun's right flank. XibEk slowly walked away, silent and unforgiving, forgetting the other Kril bodies he encountered. He was following the blood trail; knowing instinctively who was the killer.

In his headquarters, KrutChan's Stel reported, that along with CheChun, many other Krils he knew, XocYax, the WWI soldier, and YaxPac, the WWII soldier, were killed along with KinikXoc, the harmonica player from WWII. KrutChan felt his ghosts piling up in his heart and mind.

An Obet hour ticked by. Two mighty forces were struggling to dismember each other. After KanBalaam, after Nihberu, after Laktavil, this planet was different. It was awash with meandering rivulets of human and alien blood, unlike the previous invasions of pocket battles. The jungle on YaxEb did not soak up blood like deserts, frozen glaciers, and sandy terrain. It was a minor footnote-anomaly in this planetary war.

Two huge herds of snarling predators collided together. The twin forces fighting fierce-clawing combat. Neither side knew exactly what was going on, at any one time, with the thousands of struggling creatures. The battle was reaching the point of finality; only one side would prevail. The Krils were outgunned and outmanned; about to be annihilated by Kravid Palatine's Dinarchy. Forced back into a defensive pocket, The Krils were becoming trapped and cornered. Dinarchy victory was close.

His mission took him a half hour as NoKoch stumbled numerous times through the battlefield. His wounded thigh from CheChun's bullet slowing him.

665

My God ordered me to kill CheChun. The marine had been pagan KrutEk's protector on past Uayebs.

NoKoch was heading to KrutEk's headquarters after he killed CheChun. Killing CheChun had been easy. Some sixth sense made 'old Corps' turn; seeing it coming from the NoKoch's Blaster. Firing at the same time, CheChun's mouth had been spewing hate with his last curse. NoKoch not wanting CheChun reporting NoKoch's treachery. Though wounded, NoKoch had to get to KrutEk's headquarters.

Killing CheChun was a necessary function, in NoKoch's mind. His God ordered him to slay KrutEk. His mission made sense to him. No one but his handler, knew he was responsible for sending assassins after the first Uayeb at KrutChan as a diversion.

If KrutChan had died from an assassin, he could not replace KrutEk; the real target. There was a strong possibility KrutChan would never come out of his raging AnticArna delusions. In that case, once he terminated KrutEk, NoKoch planned on killing the Four-Striker KrutChan. Succession to KrilChan would be cleaner.

Over his many Uayebs, his Oval, BalamEk, was filled with religious fervor for her Cosmic Egg. That dedication in her inspired him in his own religious belief in his God. NoKoch was directed by his God to kill unbelievers; as an avenging angel. He never questioned BalamEk's plots; leaving him alone with his mission. He was loyal to her mysticism. Yet he had no misgivings when she transferred him into Jaguar's Nest. He adjusted his mission, obeying Jaguar.

BalamEk had a plot of her own; maybe against the Queen of the Confederation. Who could tell with these Alien females; maybe it was a religious ritual of some nature, or politically motivated? He did not care.

Moving through KrutEk's headquarters personnel as they set up within KrutChan's Malkril area; NoKoch's movements were invisible to the Krils. Being wounded helped mask his presence. The flurry of activity, from the constant reports coming in about the Dinarchy advance, the barking orders, and the palpable fears being generated; also screened his approach. He was ignored. He had been seen around KrutChan's and KrutEk's headquarters many times in Uayebs past. A few of the Krils addressed him but he was ignoring their salutes and greetings.

A suspicious rookie Kril guard was approaching NoKoch. He was going to challenge him; of that NoKoch was sure. NoKoch was focused on KrutEk, the Roman pagan he was after. He was about to fulfill his mission for his God. NoKoch killed the suspicious guard, with a bayonet thrust into his temple, when the Kril got within his reach. Ignoring the dead Kril, NoKoch refocusing on KrutEk.

Battle confusion reigned.

Running into the Headquarters group while pushing Krils out of the way, XibEk was yelling. "Kill that sonofabitch!" NoKoch peripherally saw XibEk aiming, as he initiated his Dagot Blaster, aiming at KrutEk.

PacalMo and CuXiu had been discussing the advantages of each of their tribal weapons when they saw NoKoch stagger, avoiding XibEk's fire, approaching KrutEk.

CuXiu asked. "Why is that future Kril carrying a Dagot Blaster?"

PacalMo yelled. "That huge Kril is in dragon's fog! Xib is warning us!"

"He is an assassin! Kill him!" CuXiu shouted in warning.

XibEk fired again. In the same instant, PacalMo immediately loosed an arrow at NoKoch, followed by CuXiu's spear.

NoKoch fired his Dagot Blaster at KrutEk. The last desperate act by NoKoch played out in slow motion. CuXiu had been running to shield KrutEk. The Auxiliary Roman CuXiu was not fast enough to save his KrilChan. CuXiu absorbed part of the deflected assassin's fire.

CuXiu's spear was aimed at NoKoch's chest, missing, perforating his neck; stapling him to a tree trunk. PacalMo's arrow entering NoKoch's eye-socket, as he spun in pain. XibEk fired three more rounds into NoKoch in rapid succession.

KrutEk and CuXiu collapsed in bowling pin fashion. CuXiu died instantly.

PacalMo ran over to NoKoch's wheezing body, stabbing him through the heart.

Mortally wounded, KrutEk was barely alive. He was being attended to by his ArnaMal doctor. KrutEk yelled into his Stel for KrutChan to immediately report to his position.

The Stels and EkSeets were transmitting the same message through their network to the Burseeosil EkTsab, and the Kril Legions. KrutEk is gravely wounded.

Monitoring the Stels and sensing victory; Kravid Palatine's Dinarchy forces were pressing their offensive.    Surrounding and attacking the Kril Legion's survivors in KrutEk's last pocket of resistance.

KrutChan swayed from side to side in exhaustion, navigating the jungle canopy trees, thick brush, and ripped leaves in the terrain.    When explosive concussions threw dirt, chunks of trees, and a blizzard of shattered leaves at him, he collapsed. A few minutes later, pulling himself together he continued on, the violence of the battle was throwing him down, over and over again.

Trying to survive the incoming maelstrom of fire, he never heard EkSeet whispering or his Stel transmitting; his mind was on survival. During a lull in the battle, running from one position to another while struggling to get his Malkril unit organized. Now he repeatedly received EkSeet's and his personal Stel's summons from KrilChan's headquarters over and over. He was commanded, by name, to report to the KrutEk. He immediately acknowledged the transmission.

That's weird. I've never been summoned by KrutEk. The problem, how to get there without getting killed. After a prolong series of explosions, came the last blast, as the barrage moved on to his left seeking out more Krils. He struggled to regain his footing, and then zig-zag ran. His lungs blew out a gasp of air when he broke into the Kril Legion's artificial clearing surrounding KrutEk's headquarters. KrutChan stooped over in a crouch.

The Krils on the perimeter were his Malkril Krils. KrutEk's headquarters group had joined them. KrutEk was technically in command. KrutEk was where he should be, where he was destined to lead; from the front.

KrutChan glumly thought of options; his battle-weary mind clearing with his decision. That glory-seeking Roman will finally get what he wants, death at the front of his Legions. Meeting TOTL like a loyal sonofabitch. What the hell am I smug about; the bastard's taking us with him?

KrutEk was leaning, in a sitting position against a rock, his legs spread, when KrutChan approached. Seeing chunks of flesh, ripped leather, and clothing around the Dagot blaster hole. The blood had splattered; he saw the fierce look his KrilChan directed at him.

KrutChan looked at his Malkril ArnaMal doctor he had argued with in past Uayebs for a silent assessment of KrutEk's condition. The doctor, saying nothing and shaking his head slightly. KrutChan's heart deadened; he had seen that look too many times in his past. "Get outta here, doc. The dead don't need you."

"KrutEk still lives!"

"Jesus Christ's Lazarus miracle can only save him."

"I know…but I have to…"

"Don't argue, doc. Get on that fucking TunToob. Take your ArnaMal medicals with ya. None of you guys will be needed here soon."

The doctor was resigned. "I should go with the wounded. I want to…" He choked up.

"Only the dead and soon to be dying are staying here." KrutChan hit the doctor's shoulder, not very tenderly. "Leave, before ya get me crying like a baby in front of my men."

The doctor ignored him. Dabbing at KrutEk's seeping wounds, spraying blue AnticArna

670

KrutEk painfully adjusted his position. He was drooling blood. "This Uayeb has lasted only a short time for me. Did you stroll through the fields of flowers on your way?"

"Yes sir, I was smelling the roses. I'm not as tough as you." KrutChan said, to remain noncommittal and respectful.

KrutSeet, was kneeling, whispering to KrutEk. "Do not trust this barbarian. He is full of mischief. You are losing consciousness and have done enough."

The Doctor grabbed KrutSeet's hand, jerking it off KrutEk's shoulder, and shook his head. When KrutChan glared icicles at him, he ran with his ArnaMal medical people to his assigned TunToob. It screeched into the sky.

The last evacuation TunToob arrived. Father Whalen was reluctantly aboard. He grimly waved at KrutChan in farewell.

KrutEk gasped. "Leave the field, KrutSeet. I appoint KrutChan, a Four Striker, to replace me."

"You will pay for this KrutChan; I have a long memory." KrutSeet said. He ran for the final Toob, pushing the Reverend aside.

The fire and violence was increasing in crescendos around them. They were running out of timeline.

KrutEk groaned. "KrutSeet has left on the Toob! What do you say, KrutChan?" KrutEk was coughing, spitting blood onto his leather armor. "I am useless now. What do you wish me to do?"

KrutChan was thinking of the possibilities, assessing the risk to himself before he answered. "Your duty is to stay with your Kril Legions. Why ask me?"

"I do not ask when I am in command." KrutEk coughed. "I requested your opinion."

671

"Fine!" KrutChan knelt next to KrutEk, really wanting to say he was sorry the man was hurt. "You're where you belong. You've sent thousands to meet TOTL. Why should you be exempt?"

The lull in the fighting was a prelude to the coming violence.

"I have never shirked my responsibilities." He was nodding off and on, but still coherent. "You are not enjoying this, are you KrutChan?" KrutEk rubbed his hand over his mouth, streaking blood onto his chin. "I promised you, when I respected you, I would use your name before I meet TOTL."

"Bullshit, sir! You'll live to attend to me in the Memorial to the Oval Protectors."

The racket in the background sounded like a lethal orchestra, building to a final crescendo.

KrutEk was drooling blood, his Stel relaying a missive to all the Krils to hold their positions. Ordering useless wounded and non-fighters onto evacuation Toobs.

Coping with his increasing pain, KrutEk was fighting to remain conscious. "Do not let me be captured, KrutChan. Give me a Roman death. These Legions will die under your command."

KrutChan was yelling his defiant objection. "I don't want your goddamn Legions!"

"I have sent my final command to the Burseeosil EkTsab to unleash BalamEk's death-catapult." KrutEk gasped to KrutChan. "We will all meet, as Golden ghosts, in the Oval Memorial."

KrutChan put his head close to KrutEk's and whispered. "What are you talking about?" He was mystified.

"Stay close to and hug your hated Aliens." He coughed blood. "No Kril or Oval is authorized to over-rule you when I meet TOTL, KrutChan."

He spoke into his Stel, to be recorded in the StelaBalaam. "I promote KrutChan acting-KrilChan, Commander of my remaining Kril Legions." He fiercely gripped KrutChan's shoulder. "Lead my Legions as you have proven in your THEN Uayebs."

Speaking into his Stel, KrutChan ordered XibEk. "Don't argue or ask why. Take over command of my Malkril immediately. I'll get back to you."

KrutChan remained standing over KrutEk and nodded. "You faced death from your own unseen assassins." KrutChan said. "You could have told me."

"Would you have listened? I did not choose you; Xmucane did." KrutEk squinted with one eye.

Both of them understanding KrutChan did not want command of the Kril Legions.

Thunder-rumbling, the Dinarchy and Kril fire was beginning to rise, across the front.

"I'm not stupid enough to think I could ever replace you, sir." KrutChan whispered.

Whispering back KrutEk said. "You do not have a choice this time." KrutEk was slowly fading away into his personal loneliness. "Your gods saved you; not me."

Their Stel's, in many languages from the dug in Krils, were screeching that the Dinarchy were

673

nearing the Kril perimeter at the top; continuing to advance.

KrutChan hissed. "I'll make sure KrutEk is obeyed!"

KrutEk mumbled defiantly. "My Legion's fate is in the hands of BalamEk." He started mumbling to someone.

The battle in the background was increasing in fury and noise. The Krils were fighting to save themselves within the perimeter. They were aware it would soon be a matter of time and distance. The Dinarchy attack was in full swing, getting closer to their lines, being overrun was imminent.

"KrutChan, make haste! Dragons will rule here soon!" PacalMo shouted.

The sounds of the battle were all around their cul-de-sac. The screams of death and the crashing ordinance, fires, and splitting-exploding boulders were slowly increasing with violence. The end of their timeline approached. It would soon be everyone for themselves.

KrutEk approached death.

PacalMo and KrutChan were left standing over KrutEk in the smoldering, stinking clearing.

"Goddammit." KrutChan's eyes were welling up with tears. "God damn KrutEk! What the fuck; I don't want this!"

Beckoning KrutChan to bend down, KrutEk was whispering; placing a list of names on a missive into KrutChan's hand. His last order. "Do...not...let my enemies...prevail."

After KrutEk's death-rattle-groan ceased, KrutChan understood the last order. I don't see how I can obey it. We're gonna all die.

KrutEk was a fellow combatant. KrutEk's unseeing eyes met KrutChan's in a cold stare of acceptance between them as he knelt by the KrilChan. He whispered in KrutEk's ear. "You got your wish. They ain't gonna take you alive. Let it go...just let go."

KrutChan grabbed KrutEk's limp hand and placed it on the K-Bar that was centered on the leather over the KrutEk's heart. Only a moment passed as KrutChan looked into KrutEk's dilated wide eyes. He brutally thrust the K-Bar up to the hilt into the chest of KrutEk.

PacalMo shouted to the sky an ancient Mongol incantation and said to KrutChan. "His TOTL is done! You gave him the respect he deserved, KrutChan."

KrutChan argued. "Bullshit! Don't make me his savior. Kravid Palatine ain't taking me alive."

Both their EkSeets were speaking in their minds. Your Stels recorded KrutEk's TOTL by KrutChan into the StelaBalaam.

KrutChan shouted back angrily. "He was already dead, Goddamit! Fix the recording! My Stel is wrong!"

"I observed what our Stels recorded. The Stels are correct." PacalMo interjected.

KrutChan's EkSeet agreed.

"CHANGE IT!" KrutChan again shouted, his words drowned out by the tremendous din of battle around them. From the battle noises all around them, the Krils were holding their ground, stiffening in their resistance against the Dinarchy advance. KrutChan knew he had no orders to give. Each Malkril was on its own. A disorganized brawl was on. No time for an obituary.

Their EkSeets were shouting. The TOTL of KrutEk was recorded in the StelaBalaam. Even an OvalChanHalach cannot erase what is stored there!

For the entire timeline of the StelaBalaam, it would store the recording. KrutEk's TOTL was caused by KrutChan; accomplished before KrutChan stood.

The blasts of ordinance were continuous; the angry-hornet-buzzing-sounds of gunfire. The symphony of soft-swishing arrows thunking into armor. The zipper-ripping sounds of guard Kril-weapons discharging. Greek-fireballs arching in the sky. Whining blasters from the alien Dagot-Kril weapons and the swishing ballistic sounds of spears in flight were all continuous background echoes.

The Dinarchy forces were getting closer to their position. Around the crater rim, thousands of Legion Krils faced obliteration.

KrutChan looked at PacalMo and significantly raised his face to the Burseeosil, unseen above them, knowing his words would be recorded and stored. "We Krils, about to TOTL. Remember our Kril Legion dead! This is one hell of a useless end for us."

Goddamn you EkSeet, stay out of this. KrutChan thought. I'm scared shitless right now of dying. Stick that heroic thought into your StelaBalaam.

PacalMo twisted and ripped the K-Bar out of KrutEk's chest and shoved it back into KrutChan's sheath. PacalMo walked off to meet the Dinarchy attack. "He was a great warrior. He has died with nobility, as he lived, according to his personal code."

Code, my aching ass. Those guys at Thermopylae didn't die for Spartan code; they died for each other.

As he wandered off to the top above the jungle, the incoming fire was increasing. KrutChan looked back briefly for a moment, at his fallen commander. Then he shook his head and wandered off following where PacalMo had gone, thinking to himself.

Hope I got one last stand in me. I don't want to be crippled. I don't want to be captured. I'll make them kill me. Hopefully like CheChun-cussing to the last. I know that scenario is bullshit, EkSeet. No one dies in glory. Leave my last thoughts alone.

The Legion's Stels had stopped asking KrutChan what his orders were a half hour ago. They heard KrutEk promoting him. They did not care; too busy trying to live.

For some unearthly reason KrutChan remembered a survivor of Wake Island, he met while in the Marine Corps, that when the survivor's leaders surrendered the civilians and marines to the invaders; it did not mean life. The killing went on for years for the captured. Very few made it through WWII.

KrutChan snapped out of his funk and muttered aloud to himself. "Suicide ain't my bag. Let's go, General Custer; it's time to die."

Pockets of thousands of Krils were surrounded in their stationary positions, facing the final Dinarchy assault. He couldn't help them. KrutChan's command to the Kril Legions he broadcasted from his Stel. "The Dinarchy will kill you! You're on your own! Meet TOTL, your way, not mine! See you all in the Memorial of the Oval protectors. Or if we're lucky...in hell!"

Those were stupid remarks to them and he wasn't proud of his cynicism. He hoped their hate for him would sustain them in their last moments. What the hell did King Leonidas say to his three hundred Spartans engaged in their last death skirmish? Who knows?

KrutChan was sure the StelaBalaam, refraining from being totally accurate, would clean his words up for posterity, making his final transmission self-serving.

A Dagot, KukMuan, was at KrutChan's side as they watched the battle approaching. There was no place to run, no TunToob to evacuate them, no time left for anything but dying.

KrutChan was rubbing the Dagot's head, like a favorite dog, and then thought better of it, saying. "Sorry, my friend...you're my friend...you know. Maybe the only one I got left with the Krils. I didn't mean to insult you, Dagot. You're nobody's companion dog. I'm sorry for you. You are a great Kril. I'm sorry you're alongside me."

KukMuan shuddered. "Dagots honor Kril. KrutChan is Alien. I am you." The Dagot was slowly dying of his many wounds.

KrutChan ordered his EkSeet to connect him with the Command Center on Burseeosil EkTsab. "BalamEk, KrutEk gave you his last order. Hope you can live with your responsibility. For once, obey him! You have an entire Malkril up there. When we meet obliteration; release those reserve Krils to invade and kill every last one of these Dinarchy bastards. Display Kravid Palatine's head on Arna!"

678

He knew the entire Kril Legion heard him through their Stels.

He stood up, ignoring the firestorm around him. He was outwardly calm, his constant overwhelming fear gone.

Speaking into his Stel transmitting to the Burseeosil again. "The KrilChan is TOTL. Our wounded Krils were evacuated on the final TunToob. We're being overrun. My Kril Legions deserved better than this."

In Burseeosil EkTsab, Jaguar and BalamEk had been watching the final battle on their screens. As the two lines of combatants were merging into one mass. BalamEk ordered Hortim to her Squadron.

The Burseeosil did not answer him and he did not expect one. KrutChan was muttering to himself, hoping the Burseeosil crew were listening. "Remember these Krils of yours for one Kin, you Ovals, before you count your treasure."

He watched the advancing Dinarchy troops coming in waves, firing as they charged. He and KukMuan huddled together in their fighting hole as EkSeet spoke in his mind. The Burseeosil has displaced Toobs. The Toobs will unleash huge tactical weapons. Your TOTL is imminent.

"The good news is I'll never end up in another fuking universe." He said quietly to himself.

KrutChan would get his wish. The next few minutes of hell were a blur. His mind was racing and totally focused in the confusion; his heart deadened.

His EkSeet was whispering advice to him. Protect yourself, lie down, and let the Dinarchy come to you.

While repeatedly firing, KrutChan responded to his EkSeet for the last time. Too late for that, ya English virgin. By the bye, old boy...if you ain't coming to join me...shut the fuck up!

His rage against the fates, speaking for all the doomed Krils in the Legions, exploded out of him. "I regret my Krils dying!"

He was yelling into his Stel to his Kril Legions. "Throw away your Stels!" KrutChan ripped off his Stel and threw it past KukMuan towards the advancing Dinarchy to mingle with the strewn debris-rubble of the battlefield. The Ovals would never get to review his final minute of death. Screw 'em.

"Finish this final goddamn battle for YaxEb!" He yelled to BalamEk.

KrutChan heard Hortim's voice calmly answer him. "Roger that. We're forming up, starting our run. I'm sorry."

PacalMo came out of nowhere, having thrown his Stel away, and jumped into the fighting hole with them. PacalMo's EkSeet had gotten on his nerves. Too many of those invisible dragons were chatting around a night-fire.

KrutChan and KukMuan were busy surviving for a second longer.

Frenzy hit all around them; arrows, firebombs from catapults. Screaming, writhing wounded attackers and defenders. Bombs, grenades, artillery fire, air-ripping bullets, the clanging of swords, heavy clashes of steel, muscles meeting bone, all dancing in a Dante's Inferno of death.

Dinarchy and Kril Legion fire was coming from everywhere; hitting friend and foe alike. Krils were insanely fighting. None of them could absorb

680

all the action going on around them. The deafening noise was background out-of-tune orchestra-cymbals; a crescendo of violence.

Their hearing deafened in acceptance of the inevitable; their death was imminent and would be final. The brawl-of-killing went on forever; their minds obliterating everything. They were in their Kril, Zar, Cunack, and Dagot-present; their WHEN, nothing else mattered. Like two grappling terrified horrible monsters, bellowing, and ferocious. The fight was accelerating.

Death came to KrutChan. He felt his body take a heavy-blow, knocking the wind out of him, spinning him around. His broken nose feeling the punch first. He stood, wobbling on weakened knees; swinging his M-14 like a club at the huge, charging Dinarchy trooper. He saw the trooper collapse without the side of his head; then felt his own mind exploding in millions of stars.

The last thing he saw was unconscious PacalMo lying next to him. KukMuan, the Dagot, over them. After he slid his face into the jungle muck he exhaled loudly with a rattle. His world became peaceful and quiet just before black oblivion absorbed him.

The action timelines branched. On the Dinarchy-Kril violent battleground the Toob delivered BalamEk's new weapon. Screaming back into the sky, screeching, ahead of a sonic-boom. The Toob was gone in an instant, moving briskly to a safe distance, in a millisecond.

The blast from BalamEk's Osil-bomb imploded, creating a vacuum air-bubble in an instant, and then immediately exploded outward in a crunching-tremendous shock wave of concussion.

High intensity radiation rode the wave, obliterating all biological life, all indigenous creatures, wherever sentient creatures were huddled. Other Toobs were screaming away as their payloads of Osil-bombs detonated.

The residents of YaxEb, who survived on the other side of the planet, would never forgive the Ovals for what they had done to their innocent species.

Chapter 30


Hortim melded into the Command Center from completing her sortie, she heard the StelaBalaam further critiquing. "There is no life existing on the sides and in the plain for one hundred Obsil kilometers. Occupying Kril and Dinarchy forces have experienced TOTL."

"What of our Kril forces elsewhere?" BalamEk asked hopefully.

"The Malkril reserve forces, as KrutChan commanded, have disbursed on planet YaxEb in Dinarchy staging areas. The Zabin-Krils and Kril ArnaMals in our remaining reserve Malkril, are safely distanced from the irradiated area on YaxEb. They have been committed, by BalamEk's command, and prevailing planeside. They will be victorious over the crumbling Dinarchy supply forces not involved as fighters."

Hortim was amazed. She had researched the Neutron bomb perfected in her nineteen fifties and sixties. That weapon's funding was stopped. The bomb was a fantasy pipedream; determined to be ineffective. Supposed to keep radiation longevity practically nil, leaving structures and cities standing. Designed to kill humans and other life on the battlefield. Earth scientists found it a weak weapon, with not much blast area, easily defended against.

This Osil weapon was more lethal. Fifteen Uayebs ago, BalamEk's Oval science teams had developed a secret, much more robust lethal biological apparatus. To the Oval Confederation's scientists chagrin, their creature tool obliterated, using blast effects and Osil radiation, all life, structures, and terrain; leaving a desert marking its use. Mercifully, the fall-out would dissipate in a Kin day.

Only BalamEk and the OvalChanHalach could authorize its use. This weapon use was against Oval codes. An Armageddon weapon for the defense of Arna should the Dinarchy invade there. The Osil-device had never been used, until now.

When the Osil creature-weapon had triggered on the battlefield below, the biological 3-D viewers, on the Burseeosil, had sparkled and reset themselves, blacked out, then came back on. The viewers pulled back in perspective, showing the large desolation. As ominous as it looked to the Oval crew members, they interpreted the sobering fact. The invasion was a success. The Oval Confederation had won!

The operations room exploded with cheering.

"What you are seeing not is to be enjoyed! It was a crime!" BalamEk's bellowing voice brought them out of their cheering euphoria. "Obliterating countless indigenous YaxEb innocents. Thousands of Dinarchy met TOTL with our Krils!"

The room fell silent.

"I have shamed myself and the Oval Confederacy." BalamEk said in a hushed voice.

"You did what you had to do." Hortim whispered. "Saved the Confederation Empire."

"I am responsible for mass murder. There is no justifiable excuse.

Jaguar thought. <u>What else could BalamEk do?</u> "KrutEk demanded the use of this weapon. His responsibility is huge." Jaguar said. She was thinking of KrutChan and her plot. She would have to revise her plan.

"KrutEk was militarily correct. I am not a focused Kril. I thought KrutEk would live." BalamEk was staring at her Command screen;

684

oblivious to everyone around her, inconsolable. "Whom did I save?" She whispered. "May the Cosmic Egg forgive me. I never will."

BalamEk transmitted a missive of apology to Xmucane. "I intend to depart the Confederation Empire forever in the Fifth Uayeb. I have witnessed and now condoned enough obscenities."

Hortim and Jaguar had no words to comfort or absolve BalamEk.

"Kril reinforcements that Bumped are a Kin from there." BalamEk was intently watching her command screen. "Too late to help our surrounded Krils."

Two Kins elapsed. As each report came into the Command Center on the Burseeosil, BalamEk's terrible cost of using the Osil weapon became depressingly apparent. In distress she glanced at the Ovals in the room. "KrutSeet's escape TunToob has vanished. He has met TOTL, his Stel non-functioning; his EkSeet discontinues searching."

Hortim saw it as a victory, sort of, considering the cost. In her ambivalent mood she was mumbling, having a conversation with the ghost of KrutChan hovering in a dark corner of her mind; sort of absolution for herself. I hoped...you would have survived. I personally disliked your methods. She rubbed her eyes. You represented everything I fought against in our old universe. "I'm truly sorry for you."

"Who are you speaking with...!" Jaguar was yelling from across the compartment. "Do you have reports from our Krils?"

Hortim angrily answered. "I'm talking to myself!" Her mind went back to KrutChan's ghost. You had a lot of Don Quixote in you. I pity you.

She walked closer to her console showing merging lights of the Legion reinforcements. <u>I never liked your kind, KrutChan.</u> <u>You were everything arrogant and cold I hated; the worst kind of human being.</u>

She blew the viewer a kiss. "Good-by, KrutChan, you were so wrong." She shrugged. "But, I got to admit, you were never boring."

"What did you say, Hortim?" BalamEk's eyebrow was arched in a question.

"Nothing important. The boring part is over now, I said."

During the following Kin to stabilize the areas; the Ovals began the process of rebuilding. Commanding their Germs to begin gathering any Kril or Dinarchy survivors in the outer periphery of the detonations.

Surprisingly, not everyone had met TOTL. The Atomic Bombings of Hiroshima and Nagasaki, in the mid-nineteen forties, had survivors.

TunToobs were transporting the horribly maimed, in the slim possibility of reclamation from their injuries.

Sadly, after recouping the living, the ArnaMals and Germs had the inglorious tasks of digging out, identifying, and loading the dead. It took time to sift through the immense carnage.

Receiving BalamEk's permission, Jaguar Bumped to YaxEb to assess the damage and gain intelligence.

In the late afternoon on the second Kin, the mood in the adjacent meeting room to the Burseeosil Command Center, was somber. The Ovals, in the Empire's Galaxies, had religiously celebrated their victory at YaxEb. Now came the Obets for atonement.

The Ovals on board were assembled, surrounding a hologram, projected from Arna. Present in the projection was the OvalChanHalach Xmucane, with Reela and Dirva alongside her in their grief. In the background, the Tribunal had assembled.

Xmucane was distressed, speaking as a Queen. "The cost is prohibitive, not in treasure, that will increase, but in so many of our Krils, Ovals, Germs, Allies, and ArnaMals meeting TOTL."

Her eyes lowered. "My dear KrutEk is gone. I grieve for him and the thousands of others who met TOTL."

Curiously, Dirva was smiling. "The Confederation is in mourning with you."

Xmucane frowned. "BalamEk will return to my quarters. We have much to discuss of the unauthorized use of the Osil-creature." She did not sound pleased. "KrutEk always protected me. I will protect his memory in my NOW."

The Oval Dirva spoke. "My Kril, KrutSeet, is missing. Why not are we seeking to find him?" She turned to face the Tribunal. "He should have been named KrilChan."

"Your Kril is missing." Xmucane said. "My KrutEk promoted KrutChan to succeed him. You must review KrutSeet's Stel. Cease your harping!"

Dirva insisted on her point. "KrutEk was hurt grievously and close to TOTL. KrutChan also met TOTL. This Tribunal must vote me the OvalChanHalach and my KrutSeet the new KrilChan. If KrutSeet has truly met TOTL; I will choose a new KrilChan."

Xmucane controlled her temper. "Not do insult my KrutEk. You not will again be

687

OvalChanHalach. The Tribunal will accept the next KrilChan...not you. Be careful of your ambitions. The Tribunal may install me for another reign."

BalamEk, cut into the conversation to advise them, and to alleviate the bickering. "Jaguar has been on the planet with the Germs as they clean the battlefield. She has been gaining intelligence insight. Jaguar knows of this conference. She will be returning presently."

Hortim, the Tribunal, and the other lesser Ovals remained in the background.

During the next twenty Bots, the usual after-action administrative tasks were discussed and decided upon. The main contention was, how the Ovals were ever going to be able to handle the rage against them by the population of YaxEb. Other questions were tabled, put aside until later.

Supplies, legal administration of the indigenous people on YaxEb, the mining of the natural resources, especially Gold and Platinum, were next on the agenda.

They discussed rebuilding the infrastructure, constructing fortifications to be used in the NOW for defenses against the Dinarchy. The mindset of the non-Oval Confederation beings and how many Kril Legion occupation forces would be left on planet YaxEb. Mourning would occur after the Krils returned in the Fifth Uayeb.

After the stifling Oval session quieted and tailed off, they were startled when Jaguar melded. She was disheveled, dirty, and out of breath. Hardly dressed to be in this meeting before the Tribunal and the Queen.

When Xmucane waved her forward, Jaguar went through the protocols demanded of her, addressing the OvalChanHalach's presence first and working her way down through Xmucane's subordinates. "I apologize for my appearance. I not do trust the Stels transmitting secrets."

Dirva snarled. "Your obscure timeline is a secret to all of us."

The Tribunal broke into laughter.

Xmucane shushed them with her glare.

Jaguar reviewed, what the strategy was, the conduct of the battles from the initial invasion, finishing with the final battle. Jaguar was hurrying, wanting to get to her final report. She reviewed the tactical reasons for the use of the biological Osil weapon, when the reinforcements arrived, and the Kril reserves that were committed to end the conflict.

"Enough! What were our losses?" Xmucane asked; not wanting to review the terrible statistics, but such was her duty to the Confederation Empire and her Allies. Her personal loss of KrutEk was enough pain for her.

Jaguar paused briefly, knowing Xmucane's nature. "We have lost many in the Kril Legions in battle before using…" She did not want to comment on the Osil-creatures. "…including one third of our committed Kril forces. The Malcan of ArnaMals fared better; they lost a fourth of their forces; slightly less than we anticipated."

"Your intelligence estimates were faulty…again?" Dirva smirked.

"A smaller number of casualties was projected. After the use of…" She cleared her throat. "My intelligence unit not could foresee causalities to be so high."

Xmucane glared at BalamEk. "Of course not!"

"The majority of the Dinarchy were eliminated, however."

"What of Kravid Palatine?" Xmucane asked, her voice sounding hopeful. "He was present not was he? Did he TOTL with the rest?"

Jaguar's voice quavered slightly. "Unfortunately, by foresight, Kravid Palatine evacuated the planet just before our Toobs loosed the.... Kravid's counterintelligence accessed our Stels; predicting the new weapon. They had Obets to anticipate what was coming."

"Jaguar failed. Not predicting his departure." Dirva said.

"'Your' accurate assessments only improve with age." Xmucane snapped.

"Kravid Palatine was halfway to his Dinarchy home planet when we initiated our action." Jaguar explained. "We lost our military leader; they not did. Though a goodly number of prisoners were taken; I estimate a few thousand. They prefer not to be captured, as you are aware. A small amount but not infinitesimal."

"Thank you, Jaguar, precise as usual." Xmucane said. "What you neglect to say is, thanks to BalamEk, YaxEb is a dismal victory."

BalamEk said. "I am leaving the Oval Confederation Empire."

"The Queen received your report." Xmucane said. "You should feel ashamed. I will facilitate your exit from my Empire."

After a few minutes of respectful silence in the room, Jaguar cleared her throat to get attention.

"There is also good news to report." Jaguar happily said.

Xmucane brightened her mood. "Please, inform us of the 'good' news. Our Confederation

690

Empire certainly would appreciate some cheerfulness from this miserable business."

Her cold stare at BalamEk was pointed and icy. "Our secret Osil-creature-weapon was to be used for defense of our home world."

"May I continue, my Queen?" Jaguar asked.

Xmucane waved her hand. "You may. Can you alleviate the heaviness in my heart?"

Jaguar paused a moment. As Intelligence officer she possessed a flair for the dramatic. She loved dropping booby-traps on the arrogant, who more than once argued interminably about faulty intelligence.

She addressed them in her professional monotone. "KrutSeet was found, by the Germs, near a demolished TunToob; a great distance from the last battle." She smiled at Dirva. "Escaping what was to come...."

"What is his condition, spy?" Dirva shouted.

Grinning crookedly, Jaguar said. "Your Kril is being treated for his...slight, non-life-threatening wounds. The others with him all met TOTL. Losing his Stel in the crash, he says. The Germs found no Stel nearby. How convenient for him."

"Not do toy with Dirva." Xmucane said.

"He is healthy and alive, the human Krils pronounce him grouchy, whatever that means. He is anxious to discuss his timeline possibilities reuniting with you, Dirva."

Xmucane was grouchy then. "I believe you said you had 'good' news. You realize the political ramifications of your statement."

Jaguar nodded. She did not seem at all distressed.

Dirva was ecstatic upon hearing the news. "My Kril lives and brings me much fortune! I

691

demand I be made OvalChanHalach by right of succession!"

She calmed down when Xmucane glowered at her. Dirva knew her place. "When Xmucane declares it and the Tribunal approves my reign. I meant to say. KrutSeet will be my KrilChan." Dirva was bubbling with enthusiasm; the rest of the Ovals not so much.

Xmucane interrupted her sister's display of joy speaking directly to her friend, Jaguar. "Is that 'all' the good news?" Her eyes were pleading with Jaguar.

"No...there is more." Jaguar's huge grin and her shaking and grooming of her hair telegraphed her own happiness.

She paused, while the Ovals in the meeting room on the EkTsab Burseeosil were focused. The Queen back on Planet Arna was standing with the interested Ovals in the Tribunal, holding their breath in expectation.

Jaguar revealed the streak of drama in her. "The Germs found honorable Dagots have saved many of our Krils over the entire irradiated battlefield. Our Dagot allies are principled beings. Their self-sacrifice immense in the final battle on YaxEb."

Dirva snapped. "KrutSeet did not need a Dagot. Dagots not do concern us here." She modified her attitude. "It is pleasant few numbers of Krils survived, of course."

Jaguar walked up to the 3-D hologram wanting to be close to Dirva's image. She spoke clearly and enunciated. "Many Dagots enfolded their Krils to protect them. Many Krils met TOTL despite their efforts"

"And...?" Xmucane asked.

692

"Unfortunately, the Dinarchy killed many of the ethical Dagots. Those Krils the Dagots protected from blast and radiation were evacuated immediately. Timelines of the recovered Krils were difficult to obtain. The Krils, for whatever reason, discarded their Stels. Through visual and DNA analysis by the StelaBalaam were we able to identify them, though time consuming."

"Jaguar!" Xmucane shouted.

"Unlike KrutSeet...some Krils met TOTL before being enfolded by the Dagots."

The Tribunal and the other Ovals were hanging on every word.

Jaguar continued. "Many of our Kril survivors are somewhat healthier because of the Dagots. Some of those Krils were severely wounded; but their timelines still exist!"

Forgetting about her Regal presence, Xmucane began clapping her hands. "That 'is' good news; a golden nugget in this dismal unlucky timeline." She was beside herself, for a brief moment, letting down her Queenly wall of distance.

"Where are those Krils?" BalamEk asked. "That is very good news of their survival."

Dirva looked worried and showed it. "How many were they? These un-TOTLed Krils!"

With a vengeance, Jaguar snarled. "There were thousands in count. Not every being meets TOTL in war. I know it will bring immense happiness to your sensitive inner emotions, Dirva. My Kril, PacalMo, has survived...."

Dirva, with hooded eyes, said. "How fortunate for you, Jaguar. We honor and rejoice with you. However, the StelaBalaam has recited your PacalMo Kril not did TOTL KrutEk."

"Nor your KrutSeet Kril." Xmucane added.

693

With her wide eyes in shock, Hortim blurted out. "Before the bombing mission, I saw another on my screen with PacalMo." She couldn't believe how hard her heart was pounding.

"Not do interrupt me!" Jaguar flashed an angry expression now that Hortim had stolen her pronouncement; her moment of glory and revenge. "Remember your place!"

Jaguar watched Dirva's fearful face staring back. "Yes, that is true, as Reela's former Nest-Oval has so rudely..."

"This Queen tires of your dramatics!" Xmucane's patience was growing thin. "Stop this charade! Speak what other good news you are bursting to announce!"

Jaguar, approached closer to the 3-D projection, looked directly at Dirva. Saying in a soft sneering voice, so everyone would strain to hear her.

"KrutChan lives."

Reela gasped loudly, her eyes saucer-shaped and astonished at the news.

Xmucane eyes watered, her shoulders slumped. She 'had' been hoping the impossible, that KrutEk somehow survived. She walked away from the viewer to compose herself.

Dirva could not conceal her disdain at the news; her momentary silence said it all. Then she dismissed Jaguar's report. "Impossible! That is incorrect. He met TOTL by the Dinarchy."

"That is your convenient error in judgement." Jaguar said. "KrutChan and PacalMo disposed of their Stels! As I said, so did thousands of others. I interviewed the survivors, asking why they discarded their Stels. KrutChan ordered them to do so. Recording the end of their timelines not was how they wanted to be remembered by their loved ones."

694

Xmucane spoke to ease her tension. "Timelines always branch..." She was glaring around at the Tribunal and her coterie. She asked Dirva. "Not do you agree, Dirva? Jaguar has clarified the succession situation somewhat." She touched Reela and caressed her Consort-OvalChanHalach's quivering hand. "Not do you think so Reela?"

Reela was still stunned, unable to speak, afraid to speak; unable to grasp her timeline branching.

Dirva was beside herself, shouting. "It changes nothing! KrutChan was condemned to TOTL! You defied the Tribunal when you let him live! KrutSeet deserves to be my KrilChan and lead the Kril Legions!"

Xmucane shouted louder. "I command the Ovals, including this Tribunal!"

Moments passed while she collected her thoughts. "I decree our Confederation Empire award a special gift. A fourth of the total platinum treasure obtained on YaxEb, to our Dagot empire. In memorial to our brave Dagots as reward for their performance to life, protecting our Krils."

The Tribunal loudly erupted in agreement; ratifying her decree.

"As Queen, I command the Tribunal to elevate Reela to OvalChanHalach and KrutChan to be her KrilChan, if he chooses. As our ancient laws decree."

Dirva was screaming. "You cannot do that! KrutChan was condemned to TOTL by the Tribunal. You said he had to serve Five Uayebs. He was to be eliminated!"

Reela was shouting now. "KrutChan will be my KrilChan! If Xmucane condemns him after he

695

returns to us, then I will choose another Kril to be my KrilChan."

Xmucane was beaming at her former consort and princess. "That is how an OvalChanHalach rules." She bowed her head at Reela. "I honor Reela. She will decide in the Fifth Uayeb if KrutChan is to have a NOW to live."

Despite the opposition, who backed Dirva, the Tribunal voted to rubberstamp Xmucane's decision. The silence in the EkTsab Burseeosil and on planet Arna, in the Tribunal records was recorded the historical moment. Which had just occurred in the Oval Empire's Confederacy.

They could physically feel their Timeline branching in possibilities. The end of Xmucane's and KrutEk's era that had lasted long in their timelines.

The OvalChanHalach addressed BalamEk, her Burseeosil Commander. "Bring the surviving Krils of the invasion force here immediately, BalamEk."

"Returning immediately is not possible for some wounded Krils." Jaguar glanced at her Queen and Reela, in sympathy. "Our doctors require half-KinUt weeks here for the cruelly wounded Krils to recuperate. KrutChan, for one, has sustained horrible wounds to his body and face."

Ruefully, Reela asked. "When not does he?"

The Tribunal tittered in relief.

"Though he will be younger when we return him to Reela, he will not be unscathed." Jaguar said. "He will be normal in bodily functions."

"My pleasure will increase." Reela said.

"Your timeline not will suffer." Xmucane laughed.

Jaguar was not finished. "I regret to say, Reela, mentally he is not the same KrutChan. Like many other Krils, he is broken in spirit. He has

696

experienced and absorbed much inner pain. KrutChan refused AnticArna or Osil-wine palliatives; threatening the doctors and nurses. Your Kril's dark mind is lonely drifting; near inconsolable."

"He will heal with Reela's help." BalamEk said. "Krils heal when Ovals have a full Tun year with them in our NOW."

Xmucane asserted her authority, visibly becoming the outgoing OvalChanHalach. "Jaguar will return on Burseeosil EkTsab with the other Kril Legions. Another older Burseeosil, GryleTunToob, has been dispatched piloted by BalamEk's trained successor."

Hiding her fear, BalamEk was worried about her daughter, ZocKuk, being involved in Oval transition- politics.

"As Jaguar has advised, when the wounded Krils recuperate sufficiently, they will board the ancient Burseeosil GryleTunToob and return to Arna."

She turned to BalamEk. "Leave immediately with the surviving healthy Legions for planet Arna. Your subordinates can finish your duties on Burseeosil EkTsab." Xmucane said. "BalamEk GryleTunToob will bring the critically wounded. I have had enough of these unlucky days. Return the invasion survivors back so we may finish the Fifth Uayeb of these cursed unholy Kins."

Xmucane pulled Reela aside as the Tribunal members melded. "You are the Queen of the Confederation Empire. Be aware KrutChan must choose to accept your honor to be KrilChan. You cannot force him. That is his choice under Oval law."

Reela's face blanched and she looked worried. "Yes, my queen, I understand."

Patting Reela's face tenderly, Xmucane tried to cheer her. "My name to you is Xmucane, your sister Queen. We are equals now. Not do despair. One of my duties as former OvalChanHalach is to meet with KrutChan, when he arrives. I will explain the protocols of what is expected of him."

"Thank you, Xmucane." Reela was beaming with satisfaction. Yet inside hoping KrutChan's contrary attitude would not surface and ruin their coming NOW together.

Melding to her Soothing room, she hid her rising fear of the immense OvalChanHalach duties collapsing upon her shoulders. Reela had much detail to absorb befitting a Queen.

In his medical quarters, KrutChan's Stel called individual Krils and Ovals to confer with him before they departed on the first Burseeosil.

The area was similar to a biological growth, with transparent cellular walls. KrutChan was possibly the KrilChan. The size of his area was expanded to handle vast communication Stel, StelaVal, and StelaBalaam networks. Including 3-D projectors and Ultra-secret reports on the Dinarchy. Information was processed within the Burseeosil's biological nervous system.

Notwithstanding, the cubicle was equipped with standard AnticArna processors, medical bath-beds, trays, chairs, and stands, visible to the creature being medicated. In addition, Osil-wine medications, and other patient-specific enzymes, were accumulated in the cubicle.

KrutChan saw the entire medical structure in three dimensional three hundred sixty degrees, but no others could see into his area. When he was visited, security shut down military intelligence projections.

698

A warm blue glowing on the walls, ceiling, and floor created peaceful resting. A pink-glowing atmosphere was maintained, adding to the effect.

Before leaving on Burseeosil EkTsab with Jaguar; XibEk reported to KrutChan. He melded into the large medical cubicle on the second Burseeosil GryleTunToob.

"You gonna live?" XibEk asked.

"That's what the doctors and their angels say." KrutChan was winching in pain.

"Ain't they got no more blue AnticArna or Osil wine for lifers?"

KrutChan's face flushed. "They been drugging alla us since we got here. They forced physcotic delusions on me; inheriting voices and talking hallucinations. I ordered them to cease my drugging or someone...." He did not finish his thought.

"They'd be easier killing than Krils."

"Other people know better than to screw with a KrilChan." KrutChan painfully coughed. "How did you survive? Last my Stel advised, before I shitcanned it, you were being overrun."

"My Dagot flopped on me like a Ho wanting her money." His expression was not of happiness. "Bastard nearly suffocated me."

"Lucky you...er...us." Blinking his eyes, KrutChan shut down a projection he was watching. "All rested up?"

XibEk eyed KrutChan suspiciously. "You saying my R&R is finished."

"I gotta mission for you, Xib."

"Gonna stretch your KrilChan command muscles, huh honky."

"KrutEk's last order...." KrutChan showing XibEk KrutEk's list. He waved his hand. "...not me, man."

"You gotta lotta work on this list, lifer. Gonna use a time-pause?"

"Nope. I dealt out missions to surviving Krils I trust. The few Krils I chose are the muscle. Remember, I'm the invalid boss."

"Whatever it is; I don't need no more shit." XibEk thought. I don't know what the mission is, but already I don't like it.

"Quit whining like a recruit."

Seeing KrutChan's expression, XibEk added. "I don't have no choice. You turned off my EkSeet and Stel?"

"Of course." KrutChan said. "But before you get comfortable thinking your proper English will fool anybody, keep in mind this Burseeosil has a nervous system that's wired into every facet of its body. The Ovals don't like secrets."

"Wilco. What's the skinny?"

KrutChan held up a clear 3-D sheet with a roster of Krils on it. "BalamEk told me she outfitted one of her KatunTunToobs with Osil capabilities for her future use. When she vamooses the Empire."

XibEk shrugged. "Her Oval lifer-perk."

"She's evacuating KrutEk's remaining Commanders back to Arna, including mine, aboard EkTsab. Like an advanced party. Upon arrival, since she's going too, Jaguar will debrief them. You're going with them, along with my former headquarters Commanders."

"Like who?"

"I've already transferred AinAcbal, a few other non-political Krils loyal to me, like the Zulu, Samurai's, Spartans, Maoris, a couple Zars, Dagots,

700

and PacalMo to leave with you. I don't trust most of the Romans allied with KrutEk. I got one other Kril to talk to…"

"No goddam way! Why me?" XibEk was feeling high profile picked on.

KrutChan was grumpy. "Cause, I promoted you as PacalMo's Executive Officer. I'm in command, dipshit."

"Half the time you're off your fuckin rocker! Use one of your talking ghosts. You been smoking Cambodian weed laced with blue Osil wine?"

Ignoring the outburst, KrutChan continued. "Those former KrutEk commanders, like KrutSeet and ZacNaab, would like nothing better than to TOTL you guys when you get back to Arna. Stick together, AinAcbal and PacalMo will watch your back."

XibEk was wary, but thankful, those two Krils would be with him. He remembered NoKoch and his lone wolf demise. "Don't jack me off. You're covering your ass. That the good news?"

"I picked Krils who could keep quiet." KrutChan sneered.

XibEk's finger spun a circle near his temple. "You see conspiracies in your sleep."

KrutChan was smiling, after wiping away another projection. "Two things. I instructed my other Commanders to…stifle…KrutEk's band of disloyal Commanders, before they can plan eliminating my Krils."

"In other words, you want us to…"

KrutChan put a finger to his lips. "Ah-ah-ah! I don't want to know." Learning from KrutEk about Plots.

He outlined his plans to XibEk using Kril hand-signals. Leaning closer to XibEk. "The Krils I'm sending back on EkTsab are in charge of their own missions. If they screw up; we never talked about this plan. Don't get 'actively' involved in…anything…branching a timeline. Use the people you trust to keep their mouth shut."

"That's number one." XibEk felt a trickle of sweat roll down his back. He got his orders. He didn't like them, but he would complete the mission. "What's the number two shit? And that ain't no pun."

KrutChan was smug. "You'll love this part… Xmucane sent me a classified missive. Reela transferred you into her Nest. The new Queen thinks I'll be too involved in KrilChan crap. Akna wants her lover. Think you can handle her?"

XibEk's heart was thumping. He hoped KrutChan didn't notice. Was he talking about handling Reela…or Akna? XibEk wondered if this was Queen Reela's doing or one of KrutChan's aberrations? <u>So what? I got Akna back!</u>

He worked hard at revealing only a half-grin, to hide his true feelings. "I understand. You're putting me in charge of a shit-burning detail; sugar-coated with a bribe."

KrutChan pumped one fist. Without an activated Stel or EkSeet, any eavesdropping Oval could not interpret his reaction. He calmed, aloofly looking around for interlopers on the medical staff. "You got your missions. Meld your ugly butt outta here! Akna's waiting."

XibEk correctly Kril saluted with both fists, spun on his heels in an about face, and walked out of the cubicle. Basking in his joy, his boots never hit the floor, before he melded.

Ten minutes later, Kelel and KrutChan were meeting in his Burseeosil hospital area. KrutChan turned off their EkSeets and Stels.

"Why is Reela transferring me into her Nest?" Smiling, Kelel was happy hearing the news for her own reasons. KrutChan scooted to a more comfortable position. "Queens have their own reasons. She didn't ask me. She told me."

He was scratching his elbow. "I didn't call you here to discuss Oval family arrangements before you go with Jaguar back to Arna. Here's your mission."

Kelel was watching his hand signals for a couple minutes as he outlined verbatim what he told the others.

After a pause she said. "Looks to me like you're the new KrilChan now."

"I been mentally manipulated since arrival."

"Only the Queen can control a KrilChan, but not in Kril Legion discipline matters."

"Don't bet the farm. That's why I'm staying out of the actual missions."

"Your orders are clear to me." She said. "I understand I'm not to be directly involved. I'll finish cleaning up KrutEk's mess."

"I've covered my back with redundancy. I've ordered others besides you to 'educate' the co-conspirators. Just stay out of each other's way."

She sighed. "You handling the details of KrutEk's... power penance?"

"None of your business." He coldly stared at Kelel. "Then there's Hortim. I'm telling all of you to leave her alone. She's not the kind to conspire. I'm keeping her safe from you guys. She's my problem, not yours."

Kelel was mulling the possibilities. "What're you going to do?"

"I've taken about as much as I can stand with Hortim. The more promotions I got; the more mealy-mouthed disrespectful she became."

"What about your own plans? Gonna shack up with a Villa full of naked ArnaVals?"

"I'd never leave the Villa alive..." KrutChan roared. "One of 'my' missions are to fix my Hortim-problem."

"How do you intend on doing that?"

"Killing her is ridiculous." KrutChan said. "I'm sick of hearing from everybody who witnessed what happened my first day. It's about time I set her female-officer's teeth on edge."

"How you going to do that?"

KrutChan glared. "I want your woman officer's opinion. How would you handle her, I mean?"

"Your sexist gene is showing." Kelel teased him. "My grandpa was sexist too, back in his day."

Not knowing what she meant, KrutChan said. "Let's keep sex out of this conversation." KrutChan outlined his plan for Hortim for a few minutes.

"That's a little drastic, even for you. Giving her the 'word' or punching her out, like one of your new Krils, is not the way to handle her." Kelel took a deep breath. "She and I are unlike in so many ways. But, I wouldn't change her. We both respect women."

"I asked you for an answer!" KrutChan snarled. "What would you do?"

Thinking for a moment, Kelel softly smiled when she continued. "Hortim has corrected a lot of males in her two timelines. She'll resist and the Ovals won't stand by letting it happen."

704

"Cut the crap. Say what you mean."

She pointed at him. "Hortim needs a man she can respect. After Four Uayebs, that's the only thing you got going for you. As the new KrilChan, I mean."

KrutChan stared at Kelel. "Take a deep breath and relax. Hortim's personal punishment solutions are way out in left field. She knows her way's not for me."

"Your attitude won't help you."

"Tell me something I don't know. How should I fix Oval superiority? What's your assessment of her?"

"I see her as a control freak." Kelel said. "Not the majority, but many women leaders are like her. Not wanting to relinquish their power; especially over men."

"Quit beating around the bush!"

"But, a lot of those same controlling women, in the bedroom, want to give up their power to the right man. Not by punishment. They want erotic fantasy."

"So you mean what, Kelel?" KrutChan asked.

"Forget the Oval traditions. Temporarily take away her control-power; treat her like a human lover."

KrutChan croaked. "She'd puke if I tried that approach."

"Don't coddle her; that'll turn her off. Give her what she secretly wants. If you don't screw it up, her frustrations at being always in charge will melt away. Get it?"

"Hell no, I don't get it!" KrutChan said. "I've never believed in women's intuition." He scrunched his face. "But you got my inner voices making suggestions."

705

"Atta boy. Don't listen to your lethal ghosts, though. It's about time your fantasies show up. I think you'll find out how golden daydreams can be when acted out." She raised a finger in the air. "But don't go macho-overboard. Keep it light and fun for her."

"Hell of a way to run Kril Legions." KrutChan looked confused. "I'll think about it; we'll see." He wore his stern KrilChan face. "On the bright side, if I screw up I'll blame you. Thanks. Get back to Arna and do your job."

"Roger that." Kril saluting, Kelel spun on her heels and departed.


Seeing Kelel leave, the Commander of Burseeosil GryleTunToob ZocKuk, melded into KrutChan's medical facility. They both were polite, struggling to keep from embracing. She remained official in her manner in case Oval interlopers were witnessing. They discussed preparations of the Burseeosil leaving for Arna and his protocol duties.

They were silent for a minute.

KrutChan turned off their Stels and EkSeets.

"I love you." ZocKuk said. "BalamEk ordered me never to speak or meet alone with you again without others present. On future Burseeosil invasions I must keep my correct distance."

"She wants to protect you. My love can't. BalamEk's right. My power is limited. The new OvalChanHalach has the ultimate power."

"Reela would have us both meet TOTL." ZocKuk said. "How are you really feeling? You look terrible."

"I'm sick of killing…everything…never mind." He felt impotent with his power. "You're the only bright spot I got left…and I can't have you. I'm dead inside." He wanted to go rogue, desert, and have ZocKuk steal the Burseeosil. But that was a MishMell daydream.

"I have acquired the same conclusion." ZocKuk touched his face. "I suggest when you arrive on Arna you go to a calmer, more life-fulfilling place. When I am depressed, this is where I go." She held up a projection. "Life is beautiful. Cleanse your heart and mind."

"We both ain't the martyr-type. Living a timeline in each other's NOW is an unfulfilled dream. Dying together is really stupid. An adolescence's fantasy."

ZocKuk sighed deeply, staring at the ceiling. "The mission for my mother BalamEk is beginning...."

KrutChan touched her wet cheek. "Mine ends in the Tribunal. I don't want you there."

"I wish I could help you." ZocKuk was adjusting to reality. She deeply kissed him. "Remember my love. Good-by."

"Aliens live forever." He never was good at saying good-by.

ZocKuk melded to her Command Center.


The planet cleansing went on for weeks after the second Burseeosil loaded the remaining YaxEb wounded survivors and left for Arna. Rebuilding was the dirty and ignominious duty of the support Germs. Their sole existence in the Oval universe. They were the non-glorious dregs with the terrible job of sorting out the battlefield debris of ruined equipment and sentient dead.

They worked like a vacuum to reclaim and recycle the landscape. At the same time, the ArnaMal Engineers moved in to begin their operations of

707

infrastructure reforming of the planet. With the Zars mining, the occupying Kril Legion policed the sullen, angry indigenous people. Getting the planet's political society functioning again; making it another asset for the Ovals. That whole process would last Tuns into their NOW.

The battle of YaxEb was over. The Oval struggle for power on Arna and in the Confederation Empire was just beginning.

Chapter 31

Uayeb Five

This is WHEN, the fifth day of Uayeb of the unlucky days. The day of reckonings.

A Time hiccup; an alternative-life beginning.

Burseeosil GryleTunToob's crew and Krils aboard the orbit-biologic elevators disembarked in clusters.

KrutChan prepared to exit with his remaining Legion Staff and Palace guards preceding him. As they descended, the cheering outside increased. Kril security held KrutChan back. Advocated by Reela, the entire Confederation Empire, since dawn, had been Stel-fed his entire Four Uayeb biography. Invasion by invasion, battle by battle, and included visuals of him surviving assassinations. Recalling his record wealth accumulation for the Ovals Confederation Empire and Reela.

When the biological elevator bottomed; tremendous crowds increased their volume, shouting his name when his face was projected on thousands of 3-D screens. From the Ovut area, to the pyramids in the distance. The colossal mass of people; Ovals, ArnaVals, ArnaMals, Germs, and Confederation alien species were praying with religious fervor, or screaming adulations at KrutChan. The reception was heady, unbalanced, and scary to KrutChan.

To the relief of his security detachment, he skipped the triumphant walk to the OvalChanHalach's pyramid by KrilChan melding.

Reela's Soothing Room was locked to him. His plans for confronting her were nullified. His Stel notified him to report to the former Queen. Ignoring the command, KrutChan left Reela's area and melded to another place.

It was Sixty years ago in his Earth-time frame; four Kins in Arna-time, since he began his Invasions. He was preparing for his immediate mission; the confrontation. KrutChan was sick of the manipulation. He planned on no more excuses, no more slights-of-hand Oval tricks and modifying of his timeline. Three weeks of no AnticArna or Osil-wine had cleared his mind. Determined, he would have KrilChan control; intending to shake the Oval-Kril systems to their foundations. He was on a mission of cleansing retribution, targeting enemies and the Empire.

He smiled to himself when he arrived in the Oval Hatching area, following ZocKuk's advice. For the first time in Five Uayebs, his unblemished mind was free of AnticArna drug. He was seeing Arna differently. He was in his twenties, brasher, more self-centered, and confidently invincible.

He could have melded to the Queen, but his twenty-something contrariness had hold of him. He was thinking to himself. Only the young feared nothing; not realizing nothing feared them.

KrutChan was in the Oval Hatchery. For a half-hour, he absorbed the crying, screaming, new creatures; reminding him of what he had missed on the Uayebs. Watching the hatchling children adjusting to their new existence. Throughout his lethal missions he had forgotten birth was a miracle. The babies were beginning, with their first breath, a new lifetime. His heart was healing. KrutChan dried his tears in satisfaction.

Appearing next to him, the former Queen Xmucane hardened his heart when she led him out of the Hatchery. "This not is the place for our discussion. Attend me."

The closed atmosphere around him was absorbing, then echoing, the heel-thud sounds of his boots. He felt peaceful and at home.

Thoughts rippled in his solitary world. His fear was gone, raging hate had tempered his previous Uayebs, his actions and moods. He was mind-broken.

He did not care what happened. KrutChan was set up the moment he arrived out of the Seech Field. Absolutely no feeling described his reality well; no sorrow, no triumph, and no joy was normal in his business.

He planned on no more bullshit WHEN, NOW, and THEN in this crazy universe. KrutChan was preparing to adjust the timeline of the Empire, the Confederation, planet Arna, the ArnaMals, Ovals, ArnaVals, and the Kril Legions. As he walked down the corridor, inside himself he was ready to confront Reela.

KrutChan was irritated when they melded into Xmucane's Soothing room and Reela wasn't there. In his confusion; his resolve melting. Taking a few minutes to prepare himself, while Xmucane appraised him adapting to his surroundings.

Xmucane approached him. They were alone. She nodded at him and pointed to her Oval Osil-wine, floating on a cabinet in the corner.

He poured hers into a gold goblet. He refused to have any wine himself. He wanted a sharp mind to face what was coming.

She wandered around her quarters, looking at 3-D projections, taking her time, making him wait for her to start the conversation.

KrutChan thought the former OvalChanHalach looked radiant; Xmucane's face peaceful. The strain of being a ruling Queen, the constant plotting by others, and the hushed attempts on her life, was lifted from her shoulders. KrutChan envied the retirement she enjoyed.

He thought to himself. <u>Becoming a past OvalChanHalach doesn't faze her at all. Don't you think so, EkSeet?</u>

His Stel voice concurred. <u>You must not blame Xmucane or Reela.</u>

<u>I get all squishy inside thinking about my new OvalChanHalach, Reela.</u>

<u>If you want to survive, that would be the correct course to follow.</u> His EkSeet warned. <u>Xmucane remains a powerful Oval female. Her favorite Kril was killed by you. And let us refrain from your usual moaning that you were innocent of his demise.</u>

"Has your MishMell ended?" Xmucane spoke.
"My assessment of you is accurate?"

"Ovals forget our EkSeets snoop; constantly modifying our speech. Why ain't you blaming them?"

"Ovals should then curtail our EkSeets, relying only on our Stels?" Charging her Oval wand, she swept the room with a grey aura. Xmucane's crooked smile defined her mood. "You imply I am mistaken."

KrutChan was stoic. "Queens are perfect, ain't they? Everybody obeys the OvalChanHalach."

712

Xmucane laughed. "KrutChan is incorrect. When you are KrilChan, you will change."

"If and when I'm KrilChan, I'll change the Ovals. Where's Reela? Why're you here talking to me?"

"I am performing my functions as a former OvalChanHalach before you attend to your Oval, the new Queen. How are you presently?" Xmucane asked.

His twenty-something brashness came out. "Does the Queen want me to speak like a subservient Kril, or as a possible KrilChan?"

"Bluntness has never been any KrilChan's failing."

Does that mean yes or no? KrutChan wondered. "KrilChan's have everything to lose. The leader of the Kril Legions is appointed. Ruled by, set up as targets, and eliminated by the Ovals."

"You will be KrilChan."

KrutChan shrugged. "I figure your Four-Uayeb plot succeeded. KrutEk told me I was your pawn."

He knows! I took tremendous risk pitting one Four-Striker against another! Pouring more Osil-wine, Xmucane said. "This Queen had you reconstituted; a unique creation. Jaguar initiated the Red-Antic-Arna redeeming process."

KrutChan groaned. "You make it sound like pulling an abscessed tooth. You and Jaguar cooked every possible human psychosis in me, during and after reconstitution."

"I gave life to you."

"C'mon! You 'gave' me nothing." He exhaled in frustration. "I admit your subversion worked." He pointed at her. "Because of what you

713

two did, I intend to adjust the Oval system's timeline…"

Xmucane interrupted. "KrutChan not can TOTL every Alien species, every Oval, to survive. Oval Plots not are shamed. You war within yourself."

"Another of your miscalculations." He answered. "Ovals are as lunatic as I was. You can't erase Oval memories in a thousand galaxies. The StelaBalaam, the Stels, and the past love-bullshit-memories, will work for me…against the Ovals."

Her eyes narrowed. "You have formed a plot?"

KrutChan relaxed. With all her snooping techniques, she didn't know. "I've some ideas of my own."

"Is KrutEk's plot evolving? What will you do in your NOW?"

"Tell you in a minute." He grinned. Her questions felt like interrogation. "You know something I don't?"

"Observe." Xmucane flashed a 3-D projection in front of them. The replay showed Dirva screaming at KrutSeet. She was quite beside herself; berating KrutSeet for being a weak Kril. She yelled he shamed her by not bringing TOTL to KrutEk and KrutChan. She was struggling for words, shouting KrutSeet was inept and cowardly.

The more she railed at him, the irater KrutSeet yelled back at her.

Stopping the projection with his new KrilChan ability, KrutChan asked. "What does this crap have to do with me? I'm no EkSeet peeping Tom."

"Watch and learn." Xmucane said. "My purpose will be served."

She resumed the projection.

KrutSeet threw a full pitcher of Osil-wine at Dirva. Missing her face, the wine doused her, and her veranda. Most of the blue potion flying over her balcony rail.

Looking considerably contrite, KrutSeet ran outside, leaning over, looking down her pyramid. He was laughing, beckoning her to see the terrace below. He said the flowers on the terrace below were withering from the Osil wine. And residents below were raising fists.

Dirva's anger increased. She was lashing out at him for shaming her again with his behavior, while she bent over the railing.

While in her raving mid-rant, KrutSeet grabbed her hip and legs and threw her screaming down the side of her pyramid to her death. Yelling after her, asking if she could reconstitute herself.

The projection ceased.

KrutChan held up his palm. "That lover's quarrel wasn't my doing. KrutSeet hated my guts. And hers, it looks like...."

"You are responsible for the Kril Legions, therefore 'you' caused my former OvalChanHalach's TOTL. Your innocent expression not does absolve you. You became aware she was behind the assassination attempts on your person. Not do deny."

KrutChan scoffed. "You guys sent me tons of information after I woke up in the Burseeosil medical habitat. Ovals don't show me everything my Krils do in error." He cracked his knuckles, working out a few kinks. "Yeah, in KrutSeet's case I was informed by BalamEk on the Burseeosil. With one brief revelation, from Jaguar, about Dirva." He said.

"My KrutEk had no patience for Krils breaking Kril-Oval Code." Xmucane said. "He would have Seeched, scourged KrutSeet to death…or crucified him."

"Not my problem." KrutChan looked indifferent. "AinAcbal was the one who convened KrutSeet's Malkril. AinAcbal and KrutSeet never got along. He presented KrutSeet in front of them, and ordered Viking Kril justice."

"Justice?"

He mocked her. "Somebody forget to tell you? KrutSeet was court-martialed for killing an Oval; sentenced to TOTL before Seeching. Kril Code prevails. I wasn't here at the time."

Xmucane was not mollified. "You believe justice is served? The Viking will behead him before having him Seeched? How compassionate."

"Don't worry. That'll never happen." He leered at Xmucane. "I intend to show mercy. Like my former Queen had for me and stay his execution. Saving him for different reasons than you had for me. KrutSeet was loyal to the Legions. He was another Oval Tool."

KrutChan punched the biology wall hard enough to make it tremble and grunt. "By the way, correct the StelaBalaam record. Dirva sent KrutSeet to kill KrutEk…not me. As you were well aware; that would have still-borned your plot. Your StelaBalaam ain't perfect; making mistakes all the time."

"You intend eliminating other co-conspirators?" Xmucane was not happy with Kril discipline; never was.

Speaking quietly, KrutChan told her. "KrutEk had a sit down with me before the last Uayeb. He outlined Jaguar's and your conspiracy. And kind of sorted out the assassination plots. Four-Strikers have a lot in common."

Xmucane looked stunned. "Jaguar advised me your inner voices broke our secret plot."

"Secret, but not foolproof. Everybody tries to eliminate Four-Strikers and Queens every Tun year." He held up a finger. "I sure wasn't promoted every Uayeb for my good looks and military expertise." KrutChan stare was intense. "Al, my invisible spook, and KrutEk helped me survive."

Xmucane was squinting, holding in her temper. "And ZacNaab, what of him? Nor does he deserve Kril justice?"

"KrutEk had plans for him. By Kril Code...Krils can't murder another Kril."

She told him what he already knew. "Many Krils have been eliminated by someone not wearing a Stel."

"You see any human tears?" He removed his pisscutter and scratched his head. "To use your idioms, I not had a hand in their timeline branching."

"Your own conspiracy is progressing I see." She growled. "With power comes responsibility." She reminded him.

KrutChan spread his hands. "Now you're blaming my Command responsibility." He shook his head at her manipulation. "I never instigated those actions. I knew nothing. Until after the fact."

Xmucane was skeptical. "You not did order other Krils meeting TOTL?" Her body language showed she did not require an explanation. She had witnessed the Stel projections.

"Nope." KrutChan scratched his jaw. "Kinda ironic. Maybe those Krils pissed somebody else off."

"Not do speak as a fool!"

Straight-faced, KrutChan continued. "KrutEk hated them. They were the kind of dudes an Oval could coerce."

"That not does excuse what happened." Xmucane said.

"My intelligence reports from Jaguar are unclear. There's a lot of timeline branching possibilities." He grinned. "We'll never know for sure."

"You lie convincingly."

"Lying is an Oval trait too."

Pursing her lips behind her steepled palms, Xmucane added. "These assassinations occurred while KrutChan was still on the GryleTunToob Burseeosil. You are conveniently blameless with your own deniability."

"Ovals reconstitute Krils to meet TOTL; that's nothing new." He turned his back on her. "Are we done with the history lesson?"

Xmucane looked unhappy. "Not are Jaguar and I to meet the same fate at your hands?"

"As you know, I don't kill females."

"Not did you terminate one in the Arena?"

"I protected myself from an armed guard."

"Krils enjoy killing." She lowered her voice. "Jaguar and I meant you no harm."

"Nice way of absolving the two of you; after setting the timelines in motion. Harming you both for collusion isn't my brand of justice." He put his pisscutter back on. "I honestly don't give a shit. Ovals can't breathe without 'plausible deniability'."

Xmucane walked to the other side of the room and beckoned him to sit beside her.

"By Jaguar's count, thirty-nine other Krils met TOTL under suspicious circumstances. You were on the Burseeosil at the time of their TOTL."

"The Ovals taught me well."

"Therefore, you are absolved of blame?"

KrutChan shrugged.

718

"Jaguar advised the incidents were strange occurrences. Some Krils were part of Dirva's assassination plot to circumvent my plans. The other assassination-led Krils, from other sources involved, also met TOTL. Convenient, not do you think? Before returning from the other side of our universe, you have eliminated your enemies, as a careful KrilChan would."

"KrutEk's last order. Don't blame me. Not all of them met TOTL."

"Your plot not is finished?"

"If Jaguar had intelligence on those guys, why didn't she stop them?" He said in a tired voice; not angry. "Will the Ovals absolve Jaguar of 'her' responsibilities?"

His legs began twitching, bouncing up and down while he was seated. KrutChan not aware of what he was doing. "Makes no difference; I don't remember."

"How convenient for you." Xmucane said.

KrutChan rubbed his nose with his middle finger. "During my first Tour, I was unconscious for two weeks with falciparum malaria in Vietnam. I forgot a lot of things." He held up his hand before Xmucane could respond. "In the second Burseeosil's medical center, I was unconscious. I forgot a lot of things again."

"Not remembering, Reela advises, is a constant failing of yours. She not does forget."

"How convenient for a new Queen." KrutChan wise-cracked.

Xmucane held her temper after his insubordinate remark. "Reela, as the new Queen, knows everything we discuss here. And your complicity."

"She's in MishMell."

Observing his reaction, Xmucane remembered KrutEk had the same indifference after his Uayebs concluded. Timeline and distance would heal KrutChan, as it did KrutEk

"What comes now, more reminiscing?" KrutChan said. "Or are you finally going to get to the point?"

"My KrutEk gave me his essence willingly and with much Beekav, in his way. Ovals respected his protecting ways."

He looked away from her. "There's the rub, my Queen; Alien ethics. Sex between species, including an exterritorial and a human being was forbidden in my Earth society. Even in my paranoia, my mind never adjusted."

"Would not Akna be considered an alien in your 'old' timeline?"

His legs were bouncing on the tips of his toes again. "Give me a break. Reela already covered the Akna-Kril-breeding ground. Knocking me out and BakMeer's incest was the last straw."

Xmucane walked to the center of the room, her slipper scraping at a piece of biological debris molting from her pyramid quarters. "From the moment you arrived on Arna you have exuded much hate." She discarded the molt in a bin. "Have you learned nothing?"

KrutChan went to attention. "May I speak honestly to my Queen?"

"Of course. Not do you have to maintain that ridiculous military stance! I remind you I not am your Queen."

Unzipping and dropping his flak jacket to the floor, KrutChan felt more comfortable. "KrutEk knew, when I survived the Arena, you were plotting his death. Your StelaBalaam must have went bonkers

720

trying to fix the other problem. Two Four-Strikers on one planet in one timeline? Gimme a break. Even Hortim was truthful."

She smiled. "Not did you know? Hortim was informed by Jaguar of the plot to use an uncontrollable Zabin-Kril. She, how do you say, volunteered her services to teach obedience using her methods. Reela not was capable."

KrutChan was irritated, amending his mission. "Jaguar and you transmitted Hortim's 'methods' to the entire Empire? You know how much shit I've put up with because of my first day with Hortim?"

Xmucane was thoughtful. "Much stronger than blue, you were in red AnticArna overdose. You required shaming to teach you who was in control."

"Hortim will get another chance." KrutChan cryptically said. "Let's end this discussion."

"Why would I want to TOTL my KrilChan?"

KrutChan disbelief was evident. "It was your idea to checkmate KrutEk's ambition to become Emperor of your Empire." He stood and walked away. "I'm getting bored with this redundant conversation."

"As am I. Are you prepared for Reela?"

"I don't want to talk about it." KrutChan said. "Let's end the confessions."

"Yes, we should. You are in time-pause. We complete our discussion after my purpose is quickly accomplished."

KrutChan's knee painfully bumped an invisible table. In a rage, his boot drove down upon it, smashing it to squealing bloody crushed pieces.

Xmucane jumped. "Does that help?" Her Oval neticulating membranes observed the debris in a different spectrum. "You have destroyed my creature!"

"Tell your critters to get out of my way." He was thinking Cheeb would have enjoyed doing that. "Your pyramid and everything in your Empire is Oval-biology abortions."

"You were reconstituted to be a better creature."

"Enough bullshit.    Get to the point." KrutChan said.

Frowning, Xmucane asked.    "We have digressed.  I require one answer from you.  Will you accept becoming Reela's KrilChan?"

"Maybe." He curled his lip. "It depends; I'm deciding."

"Not have you learned anything in eighty years on four Uayebs?  Reela will be furious at your decision."

KrutChan smiled.  <u>I might pull this off yet.</u> "Even after using me; you and I aren't at war.  I want to use a former Queen as a silent partner."

Xmucane was suspicious. "Your missions not have terminated."

KrutChan hesitated for a brief moment. "After listening to my former Queen, I got two left.  I don't need your help with the first one.  Reela will end my first mission."

"Assuredly, she will respond as a Queen must. She not needs a devious Kril as KrilChan.  In your earth words, she owes you nothing."

"You owe me."    KrutChan reminded her. "Your plot was carried out by me.  I need assistance in a tiny way to complete my second mission."

Putting aside her misgivings, Xmucane was intrigued. "What is to be a former Queen's 'tiny' role in your plot?"

"Here's how." KrutChan quickly explained his second mission while Xmucane listened and nodded.

"To sum up, I want unshaming applied Hortim's way. Every Oval has witnessed Pavilion traditions." KrutChan said.

Xmucane's eyebrow arched while agreeing. "Not do I think KrutChan will enjoy the results. Hortim has a strong will. But...I will accede to your wishes and summon her."

"We done then?"

Xmucane swept her controller wrist-wand, causing the furniture in the room to appear within human visual range.

They both shared a few quiet minutes together while waiting.

Him tapping his boot.

She reading KrutEk condolence missives.

Hortim melded, carrying a Willow whip as instructed by Xmucane. She immediately noticed a pink bench from the Redemption Pavilion between Xmucane and KrutChan.

She approached KrutChan wearing a large grin. "Finally, I get to finish what I started."

Xmucane was unsure, but determined in her role. "KrutChan has requested of me earth tradition-justice with you. "In your earth terms, I see you have prepared for what is about to occur? Not do you wish to proceed?"

Hortim pushed a button on the Willow whip she carried; paddle mode activated. "I've so been patiently waiting for four Uayebs. You've shamed enough people, KrutChan. Time for payback."

"Right on. Time to balance the books, whipper."

She winked at KrutChan. "You'll be standing up for a couple days after Reela anoints you KrilChan. My brothers learned to obey."

"I told you on the first Kin day, I ain't one of your brothers." KrutChan said.

"When I get done, you'll be as fire-engine red as them, young man."

Xmucane swung her controller over them. "A time-pause will allow both of you sixty Obets for unshaming. Only after you are both satisfied will I release your pause. Enjoy your ridiculous earth timeline together."

KrutChan and Xmucane exchanged looks.

Three things happened immediately.

XibEk melded into the room.

Xmucane yanked the weapon from Hortim, tossing the Willow whip over the veranda's protective railing.

XibEk sat on the pink Pavilion bench. Hortim was pulled over his knees.

Xmucane's wand prepared Hortim in a pink Pavilion short-gown; leaving only her boots, offering no protection.

Realizing her predicament, Hortim pleaded with Xmucane. "Order him to release me! Oval Code says a Kril can't abuse an Oval in the Willow way!"

"Sixty years I've tried every goddam approach with you." KrutChan told her. "Stupidly thinking you'd finally wise up." He whispered to her. "I promised Akna I'd Kril-correct you. XibEk has his orders."

"I hate you!"

724

"Ovals learn to be humble in spirit. This Kril will teach you Kril justice. KrutChan and I not will stay." Xmucane spoke to KrutChan. "Pavilion protocol requires witnesses, which I now supply through both of their Stels. My Confederation Empire is pleased when performing Pavilion witness functions."

"I hope it's in technicolor." KrutChan said.

"Help me!" Hortim shouted.

"I don't know what you did, but you really pissed Akna off!" XibEk said. "Be thankful, girl. She wanted me to use a switch on your bare ass. Never fuk wid KrutChan. My hand'll light a torch to your Lifer-booty."

In distress, Xmucane hurried to leave. "KrutChan, report to the Tribunal anteroom."

XibEk was grinning and adjusting his target.

"Hortim, you also will attend Reela at the Tribunal. The Ovals in the Confederation Empire will constantly remind you of your unshaming for many Tuns. Your Now timeline should remember Krils also forgive and never forget."

As she hurriedly melded, Xmucane heard the first, of soon-to-be many, loud stings. Kril discipline being applied upon a reluctant Oval-Kril.

KrutChan met ZocKuk in an anteroom. She told him she saw the unshaming of Hortim on her Stel. "You did not exactly follow Kelel's recommendations on the Burseeosil to Sooth Hortim."

"My orders. XibEk's explaining what humble means." Wiping his perspiring forehead on his sleeve, KrutChan said. "Hortim got my message. Pavilion correctness is an Oval thing...not mine."

ZocKuk would never allow unshaming of that kind upon her. She asked. "Are you satisfied?"

"Hell yes! I got my rewrite of history." He was looking away from her. "Xmucane wasn't pleased when I refused to tell her my decision. She says Reela will go ballistic."

"I not do believe you should renounce the promotion you earned." ZocKuk was teary-eyed. She tenderly kissed him. "How can you dismiss what we have by giving up?" She was leaning against KrutChan.

"How can I explain?" KrutChan was nuzzling her hair, smelling her perfume, wanting to remember. "If I decide to reject the new Queen, I'm forcing Reela to let me go. I'm not giving up...she is."

"Reela will condemn you!"

KrutChan hugged her tightly. "Timelines branch. Right?"

"You survived the Uayebs." She said, fiercely determined. "I won't let you leave me this way."

"Xmucane can transfer me to Reela's Nest!" She was hoping.

"Your mother asked for my help dealing with her demons after YaxEb." KrutChan said. "I'm surprised she's agreed to help me defy Reela. Is moving to another Nest what you really want?"

"My OtseOval mother, BalamEk, spoke to Xmucane. I was allowed to meet you for the last time. Reela does not know of me."

"If I become KrilChan, we got a big problem." He snorted. "Reela would make our lives miserable. I'm her Kril, remember?"

ZocKuk nodded realizing the reality. "My OtseVal father was a Zabin-Kril; not an Alien. KrutCheebel said you were a racist."

KrutChan kissed her deeply. "On his past Uayebs, XibEk didn't do half of what I intend. The Ovals made him disappear. My regret is you facing the music without me." He hugged her fiercely.

"My mother is Alien. Not an Oval. She taught me to only observe and never become involved. My mistake was not heeding her instructions. She and I are from a different time-species."

He nodded. "Your mother told me on the Burseeosil."

"I will miss you." ZocKuk said.

"In two lifetimes I've never liked saying good-by." KrutChan said. "Never wanting the past to foul up my future focus." He caressed her cheek. "You'll be my only regret."

ZocKuk was trembling, kissing him; her tears meeting his cheek.

"When I enter the Tribunal Chambers, either way, I'm finished." He said.

ZocKuk was shaking. "I know."

"My feelings, except for you, are running on empty. I'm tired and sick of the Oval's manipulating. The killings. Watching my Krils die for nothing. My constant paranoia, while living with my ghosts and demons. Don't you see? I got nothing left of myself."

"KrutChan will be gone. Like you, I will have nothing."

He looked at the floor. "I've been a mind-bent physcotic, hallucinating and hearing voices. None of my pleasant recurring visions were real. Sharing love with you was MishMell. Our hearts and minds were one. Creating our memories forever."

Goddamn EkSeet and his poetic side! I want to leave her more than platitudes.

727

ZocKuk nodded. "I will treasure our shared timeline."

KrutChan placed his forehead against hers. "Your mother BalamEk has made plans to protect you. She knew about us. You tell her?"

"Do not be foolish." ZocKuk smiled.

He embraced ZocKuk to reassure her; never wanting to let go. If they could run away together, KrutChan would leap at the chance. "We don't need years together." He shook his head. I'm screwing this up!

ZocKuk was kissing him over and over, wanting to keep him with her. "Our time together was so short."

He caressed her earlobe. "The danger of discovery made it sweeter." KrutChan coughed and cleared the lump from his throat.

"Losing KrutChan is not fair to me."

He shrugged. So much to say and no time for me to cover everything. "My EkSeet would say life is not fair." KrutChan said. "We got our memories. No Queen and no Oval or EkSeet can take those away."

She was drying her cheeks. ZocKuk giggled and tenderly tapped his jaw. "I would have cherished our NOW future; no matter how much hateful criticism we faced."

"Hate is putting it mildly. The entire Oval system is skewed against us." He said. "You gave me back a normal life." He kissed her. "And my nuttiness lasted eighty years."

"I love your craziness."

"Forget KrutCheebel...and me. You'll find love again." He wistfully said. "If we only had more time."

She gave him a squirming full-body hug. "I need to kiss you one final time."

They clung together.

He had to stifle his emotional overload for ZocKuk before he changed his mind. Happy endings are for movies. It isn't what we said; but what we didn't say that counts.

"Good-by, my love." ZocKuk whispered.

I love her. I got to let her go. "I'm not good at good-byes. Memories. You know what I mean." KrutChan's heart turned cold. He said upon melding. "Jaguar is whining for me."

Jaguar met KrutChan, taking his hand, melding with him into her Soothing Room.

He frowned at being summoned again.

"In your Earth words, my KrutChan, are you okay? Do you understand what Reela has in store for you?"

"I ain't your KrutChan."

Jaguar reviewed one of her 3-D reports. "Reela wants to honor you..."

"Bullshit! Reela's not honoring me" He said. "She has the power and wealth she wanted." KrutChan sat on the edge of a table, his leg jerking against the other.

"We not do possess much timeline." She said.

He was agitated, pacing around her Soothing room to ease his quaking legs.

Removing his Stel, Jaguar blocked his EkSeet at the same moment.

He groaned. More goddam secrets!

She cuffed him tenderly, like a mother lioness to her biting cub. "I command you to listen. You will refuse Reela's offer to be her KrilChan."

"Haven't made up my mind."

729

Jaguar whispered her plot for his NOW future. She watched his face for a sign of understanding.

His expression was blank. "Your plan is full of possibilities. I'm used to being hung out. You're snakier than I thought. No matter which way it goes, you lose nothing, do you?"

"Why do you care?"

KrutEk's ghost was whispering in his mind. [How does it feel barbarian, to be betrayed from all sides?]

Jaguar shrugged, trying to read what he was thinking. She paused a moment. "You must 'remember' my instructions."

"I got a lousy memory."

She went to a hidden portal. "Replace your Stel."

Jaguar stepped back and held him by his shoulders. "Meld with me. Come, we must hurry."

KrutChan accompanied Jaguar and others in the Confederation, to the Memorial Of The Oval Protectors.

The memorial looked different to him. Immense, like a couple of mirrors at an angle that created reflections to infinity. The Five Unlucky Uayebs were ending; a golden aura encompassed the entire stadium. The effect was awe-inspiring and emotional. The huge projections on top the granite arena, projected former ancient OvalChanHalachs. Inspirational, there was a pulsating life in this place of remembrance; more unrestrained than in his past. He could not help being impressed.

When the Stel records from the Krils who experienced TOTL were shown, KrutChan stared down at his feet, off to one side, not wanting to

730

watch. His downcast eyes, showing respect for his friends and all the dead Krils of the Legions.

Reela's voice sounded different to him, more sure of herself. She was intoning to everyone present, speaking a eulogy. "We must remember these Krils who have faced their TOTL for their Ovals. Their sacrifice is the reason this Memorial was erected."

KrutChan glanced at the aura-dome above them. It showed the death of one Kril after another during the battle for YaxEb; not the final one. That signaled to him the Memorial part of the event was closing. Witnessing NoKoch, with his Mohawk hairline, killing Dinarchy one after another, until he was crippled and gloriously died.

He looked over at PacalMo, wondering what the Mongol was thinking. The scene not showing what really caused NoKoch's death. Maybe PacalMo didn't care.

KrutChan had made his peace with NoKoch, knowing he was behind some of the assassination attempts. NoKoch was as manipulated as any of us. He lived with his God each day and died with his faith intact. KrutChan envied him.

The last death shown was KrutEk meeting TOTL by KrutChan's hand. He held back his fury at the politically-revisionist images; knowing no one would believe him. The reality of KrutEk's death would retain an asterisk only in KrutChan's mind. In the end he respected KrutEk.

The Stel recordings went on for an eternity.

Finally, the recall ceremony ended.

Reela took command of the proceedings. "As the presiding New OvalChanHalach, I was about to rescind the death decree of the Tribunal on my Kril, KrutChan."

Glaring at her self-serving remarks, KrutChan's inner relief added to the ashes of his past.

The stunned silence that followed her implied forgiveness was significant. She was asserting her power as Queen.

Reela explained her coming decision. "KrutChan has served the Ovals, and your new Queen, with Kril honor. He has honored my succession by bringing merciful TOTL to KrutEk. I not will rescind his execution, should KrutChan choses his NOW unwisely."

KrutChan was cynical. <u>Eliminating the Tribunal's execution decree is a bribe, Reela. A goddamn gift by you to force me to Command the Legions.</u>

Following to the letter dead KrutEk's decrees, the more sinister part of the Ceremony began.

Krils, who ran away during the Uayebs, were disloyal, or defied a Kril edict or Oval law, were sentenced to death in the Seech field, by Reela. They would not be remembered in this Memorial.

KrutChan stepped forward, raised his fist for attention, and shouted at Reela. "Stop this farce! This isn't justice; it's murder, my Queen!"

Reela waved her Oval wand-hand, staring hard at KrutChan, ceasing the ritual killings.

The entire Confederation Empire could feel their timelines approach branching. Pardoning these creatures was unheard of.

The captured Dinarchy troops were to be dealt with in the same way, dying no less ignominiously.

<u>Nobody builds memorials to enemies,</u> KrutChan thought.

"What of them? Do you wish mercy for the Dinarchy prisoners captured in the last Four Uayebs?" Reela asked KrutChan.

KrutChan's face was tight; his angry dark feelings not well hidden. "Yes!"

Reela spared the captives.

The entire Oval population in attendance were cheering KrutChan. A minority of angry Oval naysayers were outshouted.

"You are a contradictory Kril, KrutChan. Protecting life 'is' our Oval way. Perhaps you have learned something in your travels."

"I've learned on the job."

Reela waved her Oval wand-device and had the Dinarchy prisoners led off. "The cowards and captured Dinarchy are yours, my KrutChan. Deal with them as your mercy dictates."

KrutChan was guided by Jaguar to a raised platform. Part of his duty as the soon-to-be-incoming new KrilChan.

The Malkril Krils KrutChan had led into battle in the last Uayeb were shouting and cheering for him. All he had left was his dead heart, and his demon memories of battle-hell. He felt a huge knot coalesce in the middle of his forehead.

EkSeet whispered in KrutChan's ear. You must tell them what you expect of them. I tire of living within the Krils and experiencing them dying. I have seen too much. EkSeet sounded shell-shocked, as his old WWI era called the condition.

KrutChan thought to himself. EkSeet, you have my permission to leave. When I'm ready.

"Are you a golden statue?" Reela asked. "Or not do you remember?"

Saying nothing, KrutChan waited until the entire assembly was mute and staring at him. "The Ovals prepared a missive for me to say." He crumpled the glowing transparent sheet into a tiny ball, dropping it at his feet.

Some Ovals were sifting nervously in their seats.

Not Reela; nothing he did surprised her anymore.

"I stand here because of the Krils in this Oval universe. You're responsible for me." He was smiling; actually blaming their former leaders, though they didn't know it. "I was a minor Kril in the Legions. You're the real Krils." His words sounded stiff and formal to his ears.

Reela's shout, coming out of his Stel, reminded him where he was. "My KrutChan, are you in MishMell time-lapse, or are you finally at a loss for words?"

KrutChan's silence had the Kril Legions waiting for him, wanting him to get his speech-making done. "I've been trying to shut your EkSeet up!"

Liar!

He waved at the assembled Krils. "I know the Krils 'never' had their EkSeets babbling-problems in their timelines."

The Krils roared, while they were pointing and snarling their agreement.

Every Stel on each Kril translated his words as KrutChan spoke. "Thousands and thousands of Krils don't know me. Your diverse specie's codes, are something the Ovals can't take away from you."

He cleared his throat. "The goddam Ovals are vulnerable without you."

734

From behind him Hortim admonished him. "Careful. Don't go off half-cocked, jarhead."

Jaguar silenced Hortim with a gesture.

"Those lousy Stel death recordings you watched up there; you remember without help. KrutEk faced his TOTL, in his old Roman way. I honor his Oval, Xmucane and her Nest."

KrutChan paused, gathering his thoughts. "However, the Krils and Ovals present be advised, timelines will change for the Krils."

EkSeet was almost shouting in his ear. <u>Be careful, KrutChan, do not paint yourself into a corner.</u>

KrutChan grimaced. "If you want to know who I really am…look into yourselves. I and you are we Krils." KrutChan cringed at how mangled that sounded.

A tremendous roar of approval arose from the Kril Legions drowning out naysayers.

He held up his fist for silence. "You should honor the heroes who met TOTL by their self-sacrifice. Including the Ovals, ArnaMals, Dagots, ArnaVals, Zars, and Cunacks, who met TOTL on the Uayebs."

The Kril force were in meditation.

"All of you Krils who survived have the duty of living for your dead friends who did not. Remembering we couldn't fight without support from many."

Hortim was loudly scoffing. "It's nice of him to add the rest of us."

"Keep silent." Akna said. "I want to hear his words."

KrutChan was still speaking. "Krils must honor what the Ovals have given us in return for our service to them."

That brought on a Kril chorus of derisive cat-calls, many Krils rotating their hips forward obscenely, and releasing general mocking laughter.

He raised his open palm, and then closed it into a tight fist, silencing them. "I blamed the Ovals." He paused for a second. "Seeching is wrong. Meeting TOTL because of a split second moment in my old timeline. At the last second, screaming I didn't want to die! A lot of our dead friends in our past wars were never Seeched. My death meant nothing in my old universe. My TOTL in this universe will mean nothing."

The Kril Legions were nodding their heads in agreement.

"Make haste, KrutChan." Reela ordered.

"We Krils and our Legions change nothing. The Confederation Empire and the Ovals change nothing. We Krils chose our past reality; not the Ovals."

The Krils were glancing at each other, some shrugging, many nodding their heads. KrutChan was putting into words what many of them had told themselves.

"You are becoming tiresome, KrutChan." Reela announced.

He was spitting out his thoughts while he had the chance. Waiting briefly for the moment to pass. "Can't we give the non-ruling Ovals something?"

This time the entire crowd of Krils, ArnaMals, Germs, and Ovals were loudly cheering his statements.

"We owe them." KrutChan humbly said.

Reela's voice came out of the Stels. "The Queen has other duties forthcoming. This Kin will end soon. Hurry with your Beekav missive."

This time no Oval cheered.

736

KrutChan stared at her for a prolonged minute, his eyes narrowing. "When I leave, my words and what I say here will fade. Dissolve your memory of me." His voice choked with emotion. "My fond memories of you and others will never disappear."

The silence of the crowds was like a regretful applause.

"He's very good at ass-kissing." Hortim said.

Jaguar put her finger on her lips shushing Hortim. "Not do say more! Reela is intently watching you and listening."

"We were one in our timelines." KrutChan said. "We Krils fought together, sometimes against each other when our differences and racial anger rose. But, in our hearts we're Ovals. We're ArnaMals. We're Germs. We're Dagots. We're Zars and we're Cunacks. In war, all of them were Krils."

Reela shouted. "Not should you insult Ovals as Krils!"

He took a deep breath. "I consider myself representing entire non-Oval species in the Oval Confederation's Empire." He paused for effect. "I'm a Kril because of all of the species who came before me. They gifted me honor…that they earned!" You're on a roll, EkSeet.

You are the one digging your own grave.

KrutChan stopped for a beat of time, waiting for the cheering to die down.

Reela silently commanded silence, after letting the atmosphere of deference for him build. She spoke into KrutChan's Stel. "You have said enough."

"Now I'm going to end." KrutChan announced, smiling at his Queen.

737

Reela wondered if he knew what was going to happen to him.

The crowds were muttering, wondering what he meant by 'end'. His TOTL, or his speech.

Reela said, for all to hear. "What are your commands going into your NOW timeline."

KrutChan closed his eyes for a minute. "I command the Kril Legions to be true to themselves. The Ovals will give you your freedom, if they ever come out of their superiority MishMell." KrutEk said it many times, 'the KrilChan rules the Krils'.

Hortim whispered to Akna. "He is dangerously close to preaching sedition."

"Not quite treason..." Jaguar answered her. "...but he does it with flair...not does he?"

KrutChan made two fists and crossed his arms in front of his chest in the Kril salute. "I am you...you are me...and we 'will' be FREE!"

There was a bellowing roar from the Krils. A cresting wave, a crescendo of sound, exploding around the memorial; never heard before, probably never again.

When he made a fist with his hand; the sound slowly subsided.

"Keep our severely crippled Krils in your memories. No one else'll remember them, or their legacy."

Again there was an explosion of cheers.

KrutChan was breathing deeply. "Protect your Ovals; that's why you were brought here. Protect the Confederacy Empire; it is the Ovals only hope of surviving. Contrary to the ruling class's teachings, the Confederacy Empire isn't perfect."

Reela, the Ovals, and Jaguar were astonished at the thoughts from this human Kril. Who had hated their government since he arrived Five Uayebs ago.

"The Empire should take note." KrutChan said. "Krils hate this Dinarchy war. Wars suck. Only the ones fighting it are qualified to judge the cost. Krils survive and keep their respect bright for the ones who met TOTL unselfishly, with no thought of agendas or politics. I salute the Kril Legions!"

"Ker-Rutttt!" The Krils shouted, returning his gesture.

His sweat was soaking him. "Remember the loving Ovals in the majority. The other dishonorable Ovals and Krils should be forgotten."

The Ovals were looking at each other, sorting out what female he was speaking of and were smiling with their knowledge.

"Control your ignorance." Reela said.

He whispered, knowing they would hear him. "This Empire will be forgotten in time, along with the thousands of Confederation galaxies in the Empire. The Oval star systems will TOTL. This home galaxy will implode, dissipating out of existence, and be forgotten. KrutChan will be gone long before that happens. The Cosmic Egg teaches there is no infinity."

The crowds were murmuring, looking for Reela's reaction. Seeing none, they began whispering in debate amongst themselves.

"I'm leaving you as a free man!" KrutChan shouted. "You'll be on your own! Fight in your NOW future for your own freedom! Freeing yourself will make this Empire greater!"

C'mon people, don't let me twist and drift in the wind. Wake up!

KrutChan angrily shouted. "Kurrr-Rutttt! Goddamit...sound off! Are you Krils or Oval hatchlings! Kurrr-Rutttt!"

739

The Krils came alive. They rumbled the same defiant and loud 'Kurrr-Rutttt' chant, over and over. It was deafening and guttural; an earthquake tremor shaking the foundations. They were bold with their emotions, and proud.

"Kurrr-Rutttt! Kurrr-Rutttt! Kurrr-Rutttt!"

During the chanting, KrutChan left the platform after giving them a final Kril salute. I sounded like a goddamn politician.

Perspiration was darkening his shirt. He was thankful his knees were not trembling. He pushed his helmet back, wiping his wet brow on his sleeve and blew out a breath.

The die was cast; the Ovals did not realize, but he had planted a seed of ending slavery. He couldn't stop the fear of the unknown growing in his chest.

Hortim, Akna, and Jaguar had been standing off to one side. As he left the platform, Jaguar said to him. "Your Legions show much Beekav for your words. You have changed them."

"At least the Kril Legions didn't fall asleep. Don't blame me..." KrutChan said to her. "...my EkSeet should get the honors. That guy is an incredible editor. I may even grow to like his interfering ways. If he lives that long."

You are lying. Placing your insanity on me? You persist in claiming you have no intelligence? His EkSeet said.

KrutChan wearily walked away.

Jaguar said to Hortim and Akna. "I admire the way KrutChan says good-by. I will miss him. These Krils would TOTL for him."

"Yes, he is much like my former Kril." Akna was saying into their Stels. Her inner voice whispering to herself. I love XibEk.

740

"I didn't believe him..." Hortim said. "I pity our new Queen trying to control him. KrutChan refuses to be housebroken. He'll never knuckle under."

Akna bristled. "Our Confederation Empire is acutely aware you have come to regret your methods, Hortim. Clear your mind of disrespect for him."

Jaguar added. "As an Oval-Kril, you will earn our Pavilion ways again, should he hear."

Hortim's human emotions made her blush, remembering her mother.

"I think we are the ones who are changed, my sisters." Jaguar added. "He opened his heart to the Kril Legions, and in his way, to the Empire. He was saying he was sorry to all of us. Ovals, ArnaVals, ArnaMals, Krils, Germs, Zars, Cunacks, and Dagots; for any pain he caused them. We, the Ovals, and Reela were the fools, my sisters. We not did know what we possessed."

She left to go to the Tribunal, shaking her head at her sisters. Jaguar, used to reading between the lines, interpreted KrutChan's speech differently from those in attendance. Her Intelligence experiences were sounding loud persistent warnings in her head. KrutChan, his motives and decisions, would remain hidden. As she left she thought. KrutChan will possibly TOTL his OvalChanHalach; my sister Reela. I must use all my wits to protect my new Queen. Being an Oval, she began to review her plot for her own survival.

Before the ceremonies began at the Tribunal, KrutChan saw XibEk. He motioned him aside, away from the crowd. They nodded politely to each other, XibEk saluting and KrutChan returning the two-fisted formality.  XibEk watched KrutChan turn their devices off.  Their meeting was awkward and stiff.

"I see the honky-lifer survived...so far.  You satisfied, Dog?"

"I'll let you know."  KrutChan lowered his voice.  "Did Reela talk to you...about joining her Nest, I mean?"

XibEk's eyebrows knitted.  "Slaves don't join nuthin.  Bitch gave me the word."

"Reela was looking out for Akna."

"She told me you were the one who requested I become Reela's slave."  XibEk was angry.  "You're a fuckin asshole."

"Akna was carrying your child at my Soothing.  The first kid was yours...not mine!"

"Didn't stop you from screwing her."  XibEk mumbled.

"Grow up!  Like I had a choice!  Akna didn't mentally withdraw because of 'me'.  Thinking you were dead and losing your kid...triggered depression. When I rejected her, she was a breakdown waiting to happen.  The death of your kid triggered her response. Damn, I don't understand the psychology.  Xmucane ordered Reela to impregnate Akna with 'your essence' to bring her mind back."

XibEk was stunned.

"Akna doesn't remember me or our time together.  Don't you get it?"

"Don't make it right, dude.  Akna, a colored woman, was all I had after I was Seeched here.  I won't forgive you or any other white bastard who would've taken your place.  I spent the last Uayebs struggling not to waste you.  Too many other Krils in the way."

742

KrutChan snorted with indifference. "You're hate for me was obvious. In fact, I had you on my list of potential assassins."

"Shore 'nuff...Massa. That's why you had me doing your dirty work while you had an alibi on the Burseeosil."

"You weren't to be a part of it. If you were; you screwed up. Other Krils were available. My hand-picked Krils did the jobs. You guys were to orchestrate the dance."

"Whipping Hortim was another dirty job. Akna pushed me into it." XibEk's face was grim. "Inside, you gotta lotta CheChun in you. Think you ain't a racist?" He lowered his voice as an Oval walked by. "Hating Alien Ovals ain't racist? Treating Akna like she wasn't good enough for you, ain't racist? Bullshit!"

Looking around at the waiting crowd, KrutChan said. "I didn't call you here to rehash your hate for me." He whispered again. "You got your woman back. Enjoy her and your new child, but don't admit it's yours for now." He kept glancing over XibEk's shoulder for anyone snooping. "When you are with her before the Tribunal meets, tell her how you feel about her."

Not satisfied, XibEk snarled. "Get on with this shit. You giving me another mission?"

KrutChan snapped back at XibEk. "I'm trying to save your fuckin life, Akna's, and the kid's. All hell's going to break loose in the Tribunal. If you want to make it a Frag order, fine; have it your way."

"I'm listening, lifer."

"And I'm running out of time." KrutChan got close to XibEk's face, so only he heard. "Just for once listen to me. Our EkSeets and Stels will be babbling to the Ovals about this. Events are going to happen fast."

"So fuckin what. I ain't gonna be in the Tribunal."

"Running away ain't your style, man." KrutChan sighed a deep breath. "If you love Akna, be there, Xib. If you want your kid to survive, be there." He locked his stare. "If you're absent, whatever happens to them will be on your shoulders; not mine."

XibEk was smoldering. "Ya always got an angle, don't cha?"

"No matter what you hear or happens, stay clear of me." KrutChan said. "Protect Akna. I'm riding a TOTL-wave. Forget Reela, protect your woman and kid; blend into the background. That's your objective; nothing else. No matter what you see or hear, keep your mouth 'shut'."

"Shore-nuff, Boss-Massa.

KrutChan was holding in his anger. "Regretting later won't help you!" KrutChan glared at XibEk. "Akna's, the kid's, and your own future, depends on your non-actions coming at the right time. Can…will you cool it?"

XibEk was mad. "Slaves are invisible, lifer. You wouldn't have told me your secret if you thought I had any loyalty to the Ovals. I won't fuck it up…whatever 'it' is."

KrutChan melded into the Tribunal outer corridor. Directly in front of two blue-robed sentinels. With their zipper weapons turning on, initiating with an ominous humming sound during charge up.

KrutChan relaxed; bored at the both of them while he waited, looking dangerous. After only an

Ob second, they let him tunnel through the door. He was to be their leader; he needed no explanation. The biological gate silently closed behind him.

[The Tribunal never begins on time; protocols and procedures last a long time.] Al was barring his entrance to the Tribunal section. [Hold up.]

Not surprised at seeing Albert, KrutChan waited. After detoxifying from the Oval AnticArna drugs, he realized Al was not real. <u>Albert was my own physcotic-conjured hallucination.</u> KrutChan used Al as an invisible version of his unconsciousness self. With what he was facing; what harm could his alter-ego do?

Wanting to clear his mind, KrutChan said. "Albert doesn't exist." KrutChan felt strange, like an imbecile, talking to a blank wall. A few Ovals went by him tittering; seeing him talking to himself.

"Disappear Al." KrutChan said.

[Remember those colony planets you saw when you were in KrutEk's Chambers on the Burseeosil?]

KrutChan was remembering the conditions on those planets. He had to admit he was impressed. Not by the power of the Ovals, but by their patience and empathy with the indigenous species. The Ovals were manipulative and the Kril occupational forces subdued; implying enforcement. KrutChan realized his own government in his past mirrored the Ovals.

There wasn't a President, King, Queen, or Dictator who could remain in power without laws and judges, backed up by police forces. They tried to control events; but no one controlled anything.

The United States of America grew out of colonialism. Lip service was paid to 'freedom'. But who was really free? Even their Presidential voting was a lie. The Electoral College voted politicians into

745

power; not the uneducated masses. The ones with power had the freedom; constantly protecting their positions and the hell with the country. How are Ovals different? Non-white's in America knew the status quo was filled with hidden discrimination. All I gotta do is ask Xib.

The Tribunal Session was beginning to be civil, mundane, and boring. He was about to be called.

[Don't trust any leader with power. Trust BalamEk; she has nothing to gain. Use her, like they used you.]

KrutChan's Stel alerted him to enter the Tribunal chambers. "Hit the road, Al. I don't need you no more. Don't bother coming back."

KrutChan was alone entering the Tribunal.

Chapter 32

The Oval Tribunal was in full regalia, packed to the ceiling. Standing the prescribed distance from Reela and barring his way was the Mongol PacalMo. Beside him was AinAcbal. Each watching him carefully. On orders from their Ovals, they were protecting the new OvalChanHalach.

AinAcbal was silent and somber at losing his chance to be KrilChan. The Viking never dwelled on defeat. Anything he grumbled would be considered a guess, bad form, and self-serving.

Other Ovals and Krils in the crowd KrutChan recognized; he officially nodded to each one. XibEk, standing near the Queen's throne, was grinning, talking with a happy Akna. KrutChan briefly saw his EkSeet, dressed in his WWI garb, aloof and stiff-upper-lipped as usual; smiling at him in pride or possibly pity.

Advancing past many former Malkril Krils from his Bot-units; it was easy to read their looks. Many appeared happy surviving the Uayebs. He observed pockets of anger from others, one or two showed indifference. Some Krils were under-the-chin-saluting him. Overall, most showed respect, but the majority yawned, their boredom indicating they wanted to be anyplace but here.

Continuing to advance, KrutChan tightly grinned at Kelel, her arm around CauacSky; her possession of him clear. Good for her and him. They'll be perfect for each other.

BakMeer was standing in front of Hortim, who had her hands possessively on the shoulders of BakMeer indicating her own claim. Hortim defiantly scowled back at him when she saw him wink.

KrutChan looked into the crowd, not seeing her. I'm glad ZocKuk wasn't invited or didn't show up. I don't want her here. If I saw her I'd lose my nerve.

The Viking, AinAcbal, moved his position far below the Queen's throne, warily assessing KrutChan's demeanor. PacalMo was too busy performing, in protocol protective mode, to meet KrutChan's eye contact.

KrutChan was thinking. I don't see a lot of Zabin-Krils from my Seech class. Either by their choice or they're dead and gone.

Xmucane's ceremonial place was on the right of Reela and Jaguar on the queen's left.

He came to attention at the bottom of the golden stairway leading to the Queen. That's a long hill to climb if I wanted to harm her. My days of out-of-control-drama are behind me. I hope.

AinAcbal and PacalMo were reading his thoughts, they were of a like mind about security; their senses and tense bodies on alert.

The timeline beginning of the OvalChanHalach's induction of her new KrilChan came.

Clearing her throat, Reela spoke with deadly undertones. "Ah, KrutChan finally arrives. My KrilChan, my greatest Kril; it is good to see you. How should I say, healthy...for the present?"

He had decided to quit any pretense that he was happy. "I feel nothing." He said. "No thanks to many."

Reela was squinting; expecting the worst attitude from him.

He steadily met Xmucane's glare, silently warning him to keep her secret.

"I don't want your goddam job! Are you in MishMell or can't you remember?"

An audible gasp, followed by angry shouts broke out from the Tribunal. Sucking in its collective breath in expectation. A sense of dread hung in the atmosphere.

Icily Reela said. "I am the Oval Queen. Never speak to me with that tone in my presence."

His voice was deadly. "You asked me as an Alien Queen. I answered as your ex-slave."

The Oval Tribunal paused, creating a vacuum of non-sound; the entire area held its breath.

Moments passed.

Reela waited a beat of time before answering him; her eyes dull-flat as she spoke, "You are the Alien here. You have refused me. Your NOW future awaits you."

Waiting for that fact to be grasped by his mind; she watched his slumping shoulders, his body language a precursor of his thoughts.

He raised his eyebrows. "I don't remember."

"Not do you have to..."

KrutChan slowly turned away from her, glancing around at the Tribunal. "Even in a 'good' war, I've never seen one guy dying for a political principal or patriotic slogans." He loudly dropped his flak jacket with a thump. "But they all do. They die for nothing."

She nodded, wearing a wicked smile. "You spout drivel to the last."

"My EkSeet once told me about another guy." KrutChan was addressing his Krils, through his Stel. "The Roman Krils remember Lucius Quinctius Cincinnatus. Like we Krils, the Romans twice drafted him off his farm, for their wars. He didn't want to go. To be a General leading the Roman Legions into battle. Old Cincinnatus won both Roman wars. He refused the Senate's offer to be emperor both times and retired back to his farm. I admire humility in a leader."

Turning back to face Reela, KrutChan took off the temporary badge of rank of the KrilChan, spit on it, and threw it in a Hail-Mary high arc. The badge was not aerodynamic; it fluttered, like the weightless trinket it was, skittering on the marble floor. Landing in the middle of the stairs ascending to Reela's throne.

The entire Tribunal and crowd-guests recognized the insult, no matter what species they represented.

The Oval crowd reacted loudly, with an intense murmur; a low-rumble voicing their angry confusion at his gesture of defiance.

Reela paused. She was catatonic. Xmucane was fiercely nodding at her. After that silent communication from her sister, Reel's entire demeanor changed, and the new OvalChanHalach took control.

"Contrary gestures from you are well known in my Empire. From the first time we met, your hate has been evident. Over these unlucky Uayebs, your insolence has increased. I prepared for this negative outrage from you."

Reela pointed to a vestibule. Two blue-robed guards escorted a Kril between them. The new Kril wore a Marine uniform exactly like KrutChan's. The man was a facsimile of Snake. He had similar facial scars in the same place, the purple lightening symbols of his four strikes, his promotional red dots of rank and the blue bars of Uayebs served. He was not a clone. The eyes were a different color and his twenty-something year-old-build was heavier. He could have been an older brother of Snake's in his past.

Reela threw her robed arm out in presentation. "My new KrilChan. His name will forever be KrutChan. His mind is yours, but without your hated independence of my will."

Another Four-Striker KrutChan, with all his memories intact. Technically he was Jaguar's Kril in everything but name. Her sister, the OvalChanHalach Reela, had just given Jaguar the key to her scheme.

There was utter silence in the hall; surprised at Reela's decision. The entire room breathed again with satisfaction at her prescience.

"What do you say to this branching of your timeline, KrutChan? The StelaBalaam never will condone two exact Krils on one planet."

He spoke in rote, evoking his old identity. "My name was Snake; Corporal, 2021212, United States Marine Corps; Vietnam dead."

"You are a non-entity!" She screamed. "Never worthy to be KrilChan! As of this Bot, not are you a Kril!"

Taking a few steps back, KrutChan said. "A cornered rat can be lethal."

He saw AinAcbal and PacalMo advance to protect the Queen.

KrutChan turned away, indicating Reela was safe. He had one more mission.

He strolled over to ZacNaab among the spectators. "How's your timeline, Roman? Did Dirva pay your Assassin's fee before she transferred you to another Oval. Still guzzle-boozing and screwing reluctant ArnaVals instead of doing your job?"

Reela was shrieking. "What are you doing?"

"Do not speak to me, barbarian!" ZacNaab ordered. "I am a Kril commander. Our Queen decreed you are not a Kril!"

"Right on."

KrutChan grabbed ZacNaab by his leather armor, kicked and viciously broke ZacNaab's kneecap, crashing his former commander to the floor. Crushing his other knee supporting him, until ZacNaab was prone.

"I will kill you!" ZacNaab screamed.

Gripping his K-Bar, KrutChan ripped ZacNaab's toga. "Even a gutless man should know when he's screwed up too many times."

"Krils cannot kill Krils!" ZacNaab screamed with fear. "Your timeline ends in the Seech Field!"

Calmly, KrutChan said. "I ain't a Kril, you dummy. KrutEk sends his compliments. And you

751

won't be a Kril male anymore." KrutChan's K-Bar slashed.

Writhing on the Tribunal floor in intense pain, ZacNaab's blood was pooling under him, after KrutChan's K-Bar castrated him.

KrutChan raised his knife, not intending on murdering ZacNaab. "From this Bot on, no Oval or ArnaVal will Sooth you. They can't collect Essence from a neutered Germ."

The castration happened so fast, the Tribunal was numb with shock.

Queen Reela was shamed, shouting for him to stop.

While ZacNaab was miserably choking, begging and whimpering, KrutChan backed off. "With his dying breath, KrutEk ordered me to give you his demotion."

Xmucane hand covered her lips, hiding a small smile.

XibEk whispered to a horrified Akna. "Good Lifers don't order any dude to do 'their' dirty work."

KrutChan's last mission from KrutEk was accomplished. He dropped his K-Bar clanging to the bloodied floor.

Quickly holding his palms up to pacify AinAcbal, PacalMo and the guards; Snake then held out his Forty-five pistol while at attention. He ejected the pistol's magazine into his bloodied palm, opened his fists, and dropped both. Discarded pieces of equipment useless to him now.

A contingent of Germs raced to the area. Germs happily collecting ZacNaab on an invisible stretcher, racing out to a Germ medical facility near the Maluayeb Arena. While other Germs were cleaning the mess.

"The Female Germs will perform daily Pavilion traditions upon him until he loses his pride." Jaguar announced.

Snake spoke his final words to the Tribunal. "I reject this OvalChanHalach. I'll never submit to her slave tyranny. I'm a free man!" Thanks, EkSeet, for not editing my words. You can live your NOW future, Englishman.

The Kril Legions, the Oval Tribunal, and the Confederation, sobered.

"Again you shame me, the Tribunal, and yourself!" Reela said with distain. "Kril violence not is for Oval sight. I see you are incorrigible to the end."

"Life always meets TOTL. Right?"

She whispered to him for the benefit of the assembly. "I reinstitute the Tribunal's death sentence decreed upon you Five Uayebs ago."

KrutChan ignored her.

Reela continued. "As you arrived; so shall you leave this Oval universe. You have shamed me and the Oval Confederation Empire. By Oval and Kril traditions, you have the right to make a final statement to my Tribunal? Do not ramble."

"Short and sweet?" Snake asked. "Fuck Oval manipulating and their slavery."

"Our Cosmic Egg witnesses your hate." Reela snapped with irritation. "Our timelines are being wasted listening to your gibberish! My ArnaVal had Beekav for you."

He cupped his ear. "I don't remember any of the ArnaVals I enjoyed screwing."

Staring hard at XibEk, hoping he would follow the last order from his 'Lifer'. "Your Cosmic Egg won't forgive a Queen out of control. Stick your Oval slavery!"

XibEk drew Akna close to him. He hooded his eyes and kept silent.

A unified angry roar came from the Tribunal. "Blasphemer! This Kril is a traitor!"

As OvalChanHalach, she angrily hissed. "This non-entity is not a Kril! The OvalChanHalach, Reela, has her new Kril, named KrutChan. Take your place in protection of her, my new KrutChan." She took her new Kril's hand. "Do you accept service to me, as my KrilChan?"

The new KrutChan said. "Yes." Stunned at the events.

"As my KrilChan, you shall rule the Kril Legions. This is my decision as OvalChanHalach."

She heard the Tribunal leaders voting, agreeing with her choice.

Reela pointed at Snake, and said softly. "Send this obscene male from the Seech darkness back into the void he exited from!"

As his EkSeet had told him many time; Snake had his last words. "Took ya long enough ta act like a Queen, honey."

AinAcbal, the Viking, on Snake's right, rendered a Kril salute to the old KrutChan. PacalMo was on his left.

In Kril deference to Reela's Oval Tribunal, in respect to their Kril Legions, PacalMo, AinAcbal, and another Blue robed guard, got in front of Snake.

Snake broke out in loud laughter; a final tribute to this Oval Universe. My last gesture has to be a lame one? You're laughing? C'mon, you're scared shitless! You ain't no hero.

Reela was shouting. "I command these Krils to take you immediately to be Seeched alive!"

In the entire Gryle of Arna's history never had such an event occurred. The entire assembly could

754

feel timelines shifting. The Tribunal hall exploded with excitement and awe at her words and his intransience. Their Queen's obstinacy and determination inspired the Ovals.

"SILENCE, ALL OF YOU!" Reela shouted. She paused for only an Ob to allow the assembly to obey her. "The Fifth Uayeb is growing short!"

BalamEk approached Reela. "By Oval traditions, I choose to Sooth this non-entity you have condemned. Ovals not do abandon even the lowest of the lowest where there is life."

Unconcerned, Reela said. "Not do feel compassion for him! He relishes our disdain!"

"I not do profess sympathy for him. I advance Oval traditional forgiveness!"

Reela thought for an Ob; remembering her regal position as OvalChanHalach. "My Oval sister, you are correct. You do fulfill Oval traditions. I pronounce he is Soothed to you as part of your Nest until he meets oblivion in the Seech Field."

She faced the Tribunal with a cruel smile. "BalamEk is aware she will lose much treasure because of her compassion."

"I accept your decision, my Queen. My wealth will fill your coffers. You have given me permission to leave your Empire when this Seech ceremony is finished." BalamEk said, bowing low.

AinAcbal, a blue-hooded guard, and PacalMo escorted Snake down a ramp to the Seech Field. Taking their places around the condemned as protocol demanded.

They were there only a minute.

PacalMo whispered in Snake's ear. "My Oval, Jaguar, commands me."

Remembering Jaguar's last instructions to him, Snake nodded. "She controls secrets." My tough guy stony attitude ain't gonna last. Come on, Jaguar...finish this Oval ceremonial crap.

Snake's shirt was sweat-soaking dark, his fear oozing from his pores. His stomach was rolling. His legs were steady while walking; masking his anxiety. He cleared his throat of phlegm, spitting into the Seech Field defiantly.

AinAcbal steadied Snake and asked him. "Do you wish assistance?"

Snake took AinAcbal's hand off his arm. "Thanks Viking, but we both know everybody dies alone."

PacalMo left the group, stiffly marching back to Reela, to fulfill the traditional protocol of the Ovals.

Snake was left standing before the Seech Field.

Even in murder, Oval Tradition had to be observed.


Reela on her throne said to her new KrilChan. "Clear your mind. We must prepare how to spend the next Tun. Planning for the upcoming five days of Uayeb; the five unlucky kin that are coming in your NOW."

"Who the hell am I?" The new KrutChan said, sounding remarkably like the old KrutChan. "I don't remember."

Holding back her irritation, Reela said. "Calm yourself. Your new timeline will be explained to you."

PacalMo approached Reela. "I accept the new OvalChanHalach Reela as the Supreme Oval Queen who has vanquished her enemy to the Seech Field."

"Proceed with your duty." Reela waved her hand in finality. "Fulfill your service, Kril!"

"The deed will be done. She who is Queen rules." PacalMo recited, as instructed by his Oval, Jaguar. No one noticed he did not wish Reela a long life.

The previous OvalChanHalach Xmucane came forward, as required by Oval law, to accept the change of leadership. Jaguar approached with her. They said together. "We accept and proclaim you our Queen by Oval decree and law."

Reela watched in satisfaction. Her victory was at hand. All of her former MishMell dreams and aspirations were to be realized. Inside, her heart was in pain, remembering her many failures. She waited for Jaguar to speak to her.

Jaguar said, as part of the ceremony. "Your new KrutChan must TOTL the condemned non-entity. That is the Oval way. Only he can TOTL your old Kril, to assure the new KrutChan's succession to KrilChan."

Reela looked like a trapped animal. She motioned for her new KrilChan to accompany PacalMo to the Seech Field.

The Ovals and Krils stepped away, clearing a path for them.

Jaguar smiled at her Nest-Kril PacalMo.

It took the new KrutChan and PacalMo very little time to reach the Seech Field. When they arrived, AinAcbal left the area and went back to the Tribunal. Only PacalMo, Snake, and the New KrutChan stood in front of the vortex crackling Field.

PacalMo kept Snake's bayonet and threw away the rest of his weapons.

All the Stels in the Confederation Empire were activated.

The KrutChan's were images of each other, with small distinct differences, and the same inner attitudes. A twenty-something new KrutChan silently tried linking his mind with the much older Snake. Neither wanting to be here.

Snake was approaching the end of his timeline; the new KrutChan just beginning. Soon, their timelines would branch upon observation. There could be no observing after the fact, only the algorithms and mathematics of the vicious uncaring Seech Field were the future. Like a guillotine, the Seech Field held functional-purpose with no emotions.

"I don't like murder." The new KrutChan said. "Albert is raising hell right now."

With a serious expression, Snake said. "You look a lot like Al. He'll drive you crazier than you are; but you'll need him."

"That's what Albert just said."

"We'll both choose our timeline branches, me and me." Snake said. "Good Luck to you. Let's get this goddamn Oval ceremonial-farce done."

The new KrutChan was nodding. He had his explicit orders from Reela. He would not hold back. He didn't like it; but it was time for killing as he was trained.

PacalMo, in a deliberate stride moved in front of the new KrutChan. Disarming him of vest armor, cartridge belt weapons, and helmet. Throwing all into a heap towards the Queen. He held the M-14 bayonet Snake had surrendered to him at the Seech field and gave it to the new Kril.

"You will need this to fulfill your duty. You are not my KrutChan. My condemned friend respects your courage. TOTL the Queen's former Kril!"

Standing beside the buzzing crackling Seech Field, Snake Kril-saluted PacalMo.

Reela's voice rang from their Stels. "Finish this!"

As a reminder, the Seech field was ominously growing in strength, charging to full force.

The new KrutChan was balancing Snake's M-14 bayonet in his hand. Deciding where and when to stab Snake, wanting to be quickly merciful. He understood what Snake was saying to him. He had no choice in this Oval-controlled universe.

They stared hard at each other. Both the new KrutChan and Snake wanting this to end. Snake smiled, literally at himself, and said to the new KrutChan. "This is going to hurt both of us."

Snake nodded to PacalMo.

The new KrutChan never saw it coming.

With a crunch, PacalMo forcefully thrust his Mongol knife, busting through the biological Stel the new KrutChan was wearing, piercing his heart. The Mongol professionally twisted and ripped the knife out. PacalMo then sheathed his weapon.

The new KrutChan instantly met TOTL, quickly crumpling to the floor.

Together, Snake and PacalMo bent over, picking up the body of the new dead KrutChan and immediately threw him into the pulsating, spinning Seech Field.

Snake's tears blinded him as he watched the process in the Seech Field. Poor bastard...actually I...never had a chance. Wasn't cool watching

759

yourself die.  Regret doesn't help the innocent who are wasted.

The cadaver falling into the Field snapped the black whirling vortex Void briefly, causing a pinpoint of light to appear for a moment.  After which a multicolored universe of exploding bursts and waves filled the entire Seech Field.  Then the Seech Field silently turned solid ebony with deactivation.

I didn't even get a funeral.  Snake thought.

A hushed eternity of a branching-in-timeline snapped into existence.

Suddenly, the entire mass of assembled Ovals and Krils were flabbergasted, and others were cheering with joy.  A new timeline came into existence.

Jaguar's succession plot was revealed to the Krils, Ovals, the planet Arna, and the Confederation Empire.  The overpowering noise was numbing everyone's senses.

PacalMo and Snake marched side by side, back to the Reela's throne in the Tribunal.

Arriving, they were met by Jaguar and Xmucane.

Pandemonium ensued and the former OvalChanHalach, Xmucane, raised her hand for quiet.

Xmucane turned to a stunned Reela.  "You have lasted only a Bot as OvalChanHalach, Reela.  By Oval succession law you are replaced by Jaguar.  Her Kril, PacalMo, if he chooses, will now become KrilChan.  Yours has met TOTL.  You have miscalculated.  It is done.  You are finished as Queen."

Reela was screaming in outrage. "You not can do this! I won't allow a coup d'état! Guards, detain these usurpers! Arrest them all!"

"You not will obey her!" Xmucane was shaking her head. As a former OvalChanHalach; she commiserated with Reela.

The guards were confused. Which OvalChanHalach should they obey; Reela or Xmucane? The guards made their decision together, staying out of this Oval tradition-squabble among the Queens.

As the new OvalChanHalach, Jaguar took control and signaled them to withdraw.

Reela could not believe what she was hearing. "I was duped by you and Jaguar!"

"You are blaming a former Queen and Jaguar?" Xmucane pathetically looked at her Tribunal, while they broke into laughter.

Jaguar motioned for Snake to approach her. Speaking in a whisper, Jaguar gently took Snake's hand in hers. "You were Soothed to BalamEk. You are hers."

"Don't need her. I'm a free non-entity. I belong to nobody." KrutChan crossed his index and forefinger. "Recognize this sign? Before you kill half of Arna solidifying your power as Queen, we better talk."

Jaguar felt BalamEk sweep them both within an Aura, silencing their Stels, EkSeets, and the StelaBalaam.

KrutChan said to Jaguar. "BalamEk knows everything. She controls the StelaBalaam. Kill me, Reela, or any in Reela's Nest and the whole Empire will know your secrets in an instant. Your murder of Akna's kid. And other assassination plots, involving

761

ZacNaab, NoKoch, and KrutSeet." He was grinning. "And don't forget my buddy Kravid Palatine. Both you guys deserve each other."

In anger, but outmaneuvered, Jaguar agreed to discuss their timeline possibilities.

BalamEk melded both away for secrecy. KrutChan explaining while there, how Jaguar could save being shamed and replaced as Queen by the Tribunal.

Within Ten Obets, Jaguar and KrutChan reappeared. Jaguar ordered her palace guards to release Reela.

Xmucane was not surprised when she saw the hooded guards were searching for the former Queen under their care.

"Where is Reela?" Jaguar was looking around.

BalamEk came forward and swept her arm, projecting another Golden Aura over Xmucane, Jaguar and herself.

"While you were in conference, I melded Reela, Akna, and XibEk secretly to my personal Osil-biological Gryle-Toob."

"I not did permit you to act unilaterally. I am the Queen!"

BalamEk listened politely, and then was speaking calmly. "OvalChanHalach's not do command me. Xmucane was aware I not am an Oval. My Race saved the Ovals a Gryle ago in your THEN past. As an Observer from our more ancient Multidimensional-Empire; we are non-interfering with our 'Magic'."

Jaguar felt faint. Trying to come up with secret options. She could think of only one.

BalamEk gently touched Jaguar's shoulder. "As KrutChan advised you; you will command me to have company in my self-imposed Exile."

"By company, you refer to Reela's Nest. I am satisfied. Take her to the Dinarchy as a Hostage-Ambassador." Jaguar said.

"If that pleases the new Queen. KrutChan has joined Reela and her Nest."

Jaguar spun to look behind her. KrutChan was gone. "I not do remember all of his pleadings." Jaguar said.

When she turned back, BalamEk had disappeared. Removing the Golden Aura as she departed.

"Decree your wishes." Xmucane whispered. "Not do change any timeline. I know you wish to follow KrutChan's...instructions...in your heart." Xmucane said. "You have succeeded in becoming Queen, Jaguar. "BalamEk was the 'Covert Oval' you sought to uncover. I commissioned BalamEk in my plot. I trusted no one but her."

The new OvalChanHalach was petrified.

Xmucane whispered to Jaguar. "If you wish to remain Queen, announce to our Empire their Exile was upon your command."

Taking Reela's place on the throne, Jaguar spoke to the recording StelaBalaam. "With BalamEk, I exile Reela and her Nest to be Hostage-Ambassadors to the Dinarchy. The non-entity will accompany them."

Some Ovals in attendance disagreed with their new Queen. The majority were cheering.

Jaguar announced. "Reela's remaining brood have lost their OtseVal KrutChan and their Oval Reela. BakMeer, CauacSky, and Kelel will join my Nest. They will obey me in my reign."

She turned to Hortim. "Later, you will be further rewarded as a loyal Oval."

PacalMo got in front of Jaguar going back to the high platform.

Her solace was great. Reela had errored by not reinvigorating the old KrutChan, following Jaguar's suggestion. Upon the end of the Uayebs the Ovals kept their Krils alive to serve them. Dissatisfied with any of their Krils, Ovals did nothing; allowing them to faster-than-light age backward to oblivion. Him meeting TOTL became part of Jaguar's new option. KrutChan would never survive to see what happened in the Confederation Empire.

There was a reflecting silence within the crowds of quad-billions on planet Arna. And quad-trillions in a thousand galaxies, part of the Confederation Empire, for a long time after BalamEk left Arna.

Xmucane brushed against Jaguar having an intimate conference between two Queens.

"With BalamEk gone, Xmucane, your counsel is sorely welcomed." Jaguar told her.

Xmucane smiled. "BalamEk trained ZocKuk over many Uayebs. In case BalamEk met TOTL on the Uayebs."

Jaguar looked suspicious. "Why not have I heard of her?"

"I placed her in Dirva's Nest to spy for me for protection. BalamEk approved. ZocKuk not does want to be OvalChanHalach. Not can you see the timeline possibilities? You require a non-Oval you can trust. Not do I care. It is your decision to make."

"Why do you smile?" Jaguar said.

"You will be a great Queen, should you cease your devious attitude. BalamEk is gone. She not will deliver her passengers to the Dinarchy. She will free them all. KrutChan's minor victory."

Jaguar was pale. "I have won; it matters not. At my coaxing Reela created the new KrutChan. The non-entity's timeline will end with TOTL as soon as BalamEk's creature creates Osil-Drive. I have won again; not he. KrutChan will Osil-disappear when he meets TOTL."

Xmucane looked happy. "This former Queen honors your plot."

To eliminate further conversation, Jaguar quickly walked away, worried about the coming Uayebs at the end of the next Tun. She had to make a decision well before the Uayebs arrived about Kravid's Dinarchy cease-fire and line up her allies against the hard line Ovals. A cease-fire would not be an easy task to sell to the Confederation Empire and the Oval Tribunal. Hopefully, Kravid will succeed in his Empire.

Xmucane followed Jaguar, as proscribed by Oval protocol, to greet the other Ovals in the Tribunal as they prepared for the end of these five unlucky Uayebs they had just endured.

Jaguar was muttering to PacalMo. "My final plot has yet to occur." She said enigmatically.

PacalMo misunderstood what she was saying. "KrutChan said many times right or wrong; the leader leads; the followers must obey." PacalMo said. "You are my Queen. I obey."

"I did the right thing by banishing them, my KrilChan. I am right." To herself she thought. But am I safe?

Her EkSeet said to her. A Queen is never safe. That is why she needs her KrilChan.

765

Annoyed, Jaguar fumed. <u>Be still! You are not an OvalChanHalach, allowed to intimately converse with me!</u>

PacalMo never answered about her being right. Jaguar absorbed his silence with her own. He protectively paved the way for her though the crowds of well-wishers.

On Planet Arna, twilight announced the end of the Kin. When the sun setting occurred, the OvalChanHalach Jaguar shouted to the crowds in a thousand galaxies.

"The life celebrations will commence! The Unlucky Uayeb Kins are over!"

An impressive shout of approval broke like a wave crashing-surf onto a beach. Life and timelines normalized.

Leaving the Tribunal, Jaguar whispered in PacalMo's ear. "KrutChan was an honored Kril."

PacalMo quietly said. "He was an Eagle flying over us, never wanting greatness. Being an Eagle was enough for him. I will be your KrutChan."

"Never. You always will be PacalMo; never KrutChan."

PacalMo was in deep thought. "He showed many faces, as a multi-headed dragon. KrutChan was alone in his way and in the Kril Legion's way. KrutChan did not want more dead friends in his painful memories."

Jaguar pulled PacalMo to her. "I am your friend." She nervously whispered to him. "I intend my Empire reign will remember him as the greatest of all Krils."

His EkSeet abruptly chimed into his mind. <u>She means she will place KrutChan as a Dragon on your back.</u>

*Be still, demon dragon spirit. Enough!* PacalMo admonished. He hated that inner voice.

Hortim became Jaguar's OvalChanHalach Consort.

Her lover BakMeer became greater than Jaguar as head of Oval Intelligence.

Kelel commanded a Malkril as PacalMo's executive officer.

CauacSky's artistry in sculpture and painting would make him a leader of the ArnaMals. Kelel loved and completed him.

ZocKuk replaced BalamEk; dismissing ever becoming an OvalChanHalach. Observers did not become involved. The Empire never noticing how much her daughter looked like KrutChan.

Akna and XibEk were forgotten. KrutChan had made sure of their anonymity.

BalamEk disappeared from the Oval universe. When Kravid Palatine advised Jaguar he did not receive any hostages, Jaguar was furious, pledging to TOTL BalamEk, if she ever re-entered the Empire.

Jaguar nurtured KrutChan's legend. The Krils, and the trillion-trillions in a thousand galaxies, quoted KrutChan constantly when retelling his story. She insured his prowess, intelligence, and deeds became overblown and inaccurate.

During the next Tun year, Jaguar commissioned CauacSky to sculpture a huge golden statue to KrutChan across from the legendary OvalChanHalachs, above the Memorial of the Oval Protectors. The only Kril alongside the prior OvalChanHalachs.

By Jaguar's orders, the anonymous statue could not look anything like KrutChan. She not did need to foster a god-hero in her Confederation Empire. In time, she felt his legend would fade.

On the statue, KrutChan's inscribed words added by CauacSky in all the ancient Kril languages, in many modern Kril dialects, and in the Oval Confederation Empire's languages; depending which Stel was translating:

KRUTCHAN

He was on Arna for Five Uayebs before he experienced TOTL. Living an eighty-year human lifetime.

StelaBalaam recorded, KrutChan said: A HERO IS DESTINED TO BE REMEMBERED BY HIS ENTRANCE. A LEGEND IS DESTINED TO BE REMEMBERED BY HIS EXIT.


Attending the dedication ceremony of the gold edifice with his NEST family around him, his EkSeet knew KrutChan would say the statue was 'bullshit'. EkSeet never recalled the quote from the StelaBalaam, ever coming from KrutChan's mouth or mind.

By his own thoughts, KrutChan was no hero or legendary leader. He told EkSeet many times, he was an insignificant man from earth; an unwilling anonymous draftee.

KrutChan's legend would grow to immense proportions. Every Kril race recalled KrutChan as one of 'their' legends from Earth's past. Not the truth, which is boring and mundane.

KrutChan wanted freedom for himself and his Krils, and like Spartacus, had lost. But, his dream was never lost in the Legions. Because of KrutChan pointing the way, many in the Kril-Legions believed, the Krils would be freed by the Ovals.

Chapter 33

Epilogue
    KrutChan pop-blinked from the Osil drive's tunneling effect. Snake's information-wave approached the end of the Oval universe; his duality wave was gravity-crunched against a brane.
    His information-wave prevailed. Reaching the edge of the universe. The Brane Event-hole retained information from the Earth-Oval universes in two dimensional modes. When he collided at the miniscule center of the monster; his wave duality tunneled with an existence-expanding-exit. His duality occurred over millions of years.
    At the exact center of a Quantum-Many-Worlds Brane, a gravitation Planck time-instant occurred, causing symmetry-breaking, that leads to cosmic inflation. The tiniest information-wave that had been KrutChan evolved within the expansion. The he, of his former him, was not conscious of the change.
    A coalesced information-expanding intelligence, upon exiting, was clear as a particle-wave on a one-dimensional circle-wave contemplating an infinite I. Me is I. I am infinite. Nothing I was is destroyed...I exist and therefore I am.
    A quantum nanosecond later he endured his evaporating TOTL from the Osil effect. At this moment, the tunneling effect was serene. His new wounds were numb. Snake coughed and hacked up phlegm.
    In his mind, he was thanking his God, or maybe the same thing, the Oval's Cosmic Egg, had reconstituted him. I don't care. Does a newborn

<u>worry?</u>

A huge red-yellow-black explosion erupted fifty yards from Steven. He saw Ord and Scuz's position, enveloped in an instant. One minute they had been there; the next sad moment they disappeared. Knocked back from the tremendous concussion, Steven was thinking. <u>'Hell, they're dead."</u>

As the artillery concentrations bore into the NVA on the other ridge. Lieutenant Steven knew Ord's actions, calling in artillery support for his platoon, had killed Scuz and him; falling short.

Then quickly, it was over; sanity returning to the battle terrain. The area became calm. Steven wandered to the smoking crater were he last saw Snake and Scuz. Dreading the sight, preparing himself for the obscenity of death.

Snake scrapped mud from his teary eyes and saw Lieutenant Steven, in the light of Vietnam, was coming towards him through the smoke. Somebody supported him under his armpit, holding his waist. When he had been Seeched before; he never saw his Lieutenant.

From his prior experiences in the Oval Universe, he realized he was in an alternate timeline of his. Another dimension or another universe, beginning an alternate timeline. An Osil-Seech had happened. He had not caused the quanta-tunneling to happen. <u>In another Empire at war, or similar universe?</u>

Steven was shocked at the sight of Snake. Corporal Ord, being supported stumbling out of the cordite and mud-jungle-smelling mist. Snake crumbled to his knees, parodying a drunk getting his bearings, mumbling. A dazed and disjointed ragdoll with one of his boots bent at an odd angle.

Appearing, with him, out of the smoke, Staff Sergeant Huggy was dragging Snake from the crater. Huggy yelling at Ord to hang on.

Where the hell did Sergeant Huggy come from? The Lieutenant didn't care. Steven shouted in his mind, like he was in an old Frankenstein movie. He's alive! Steven yelled. "Corporal Ord...Snake...you survived!"

Seeing his former platoon sergeant and lieutenant standing over him caused his brain to reboot. Kee-Rist! Huggy was wasted in Vietnam. I have to be in another goddam universe or dimension! Not again...!

Snake bitterly rasped. "Goddam right..." The Oval universe prepared him for being in other universes, in strange places. These dudes would never understand.

Steven and Huggy were both talking about different timeline realities than Snake's. Snake was disoriented; similar to when he was being herded towards the Maluayeb Arena.

Doc was heading for the three of them on the edge of the erupted crater.

"He's fucked up, Lieutenant, but he's alive." Staff Sergeant Huggy said. "I got reinforcements coming. I called the Company on the platoon radio; second platoon's coming on the double."

Snake was repeating over and over. "Scuz...where's Scuz...he was right next to me."

Steven looked at Huggy. The Sergeant was shaking his head, "Scuz is gone, I mean disappeared, nothing left to bury."

Snake was listening intently.

771

"I found his watch band and parts of Scuz's shredded pack...nuthin' else." Huggy said. "The Intel guys'll search the area, maybe tomorrow, for any corpse we can't find. They'll find nuthin'."

Snake was listening, slowly moving his face and eyes away from the Corpsman, a look of combat acceptance and the finality of the fact, clouding his expression. He was going into that lonely place were wounded went.

The sound of incoherent mumbling coming from Snake drew Steven's attention. Listening closely to grasp what Snake was saying, as the wounded man laughed.

Steven could not make sense of the sentences. Snake was talking gibberish, in some kind of foreign tongue. The Lieutenant heard and interpreted as best he could. Snake was rambling on; something about Krils...Ovals...Bumps...TOTL...Seech, or other weird shit...and love?

The Lieutenant motioned to Huggy and the radioman to follow him away from Doc and Snake. Every one of the survivors were averting their eyes from Snake, embarrassed at seeing the injury to their fellow platoon member. Adjusting already, ready to move on, remembering, but not wanting to get involved in present memories. They were dealing with their immediate reality.

The orders would come soon from their Company. Steven shouted. "Let's get back on this Search and Destroy mission...saddle up! We'll move out as soon as the Dust-off is loaded and airborne!"

Snake was a memory already, not part of the team anymore.

Snake, you're a poor lucky bastard. His friends were thinking.

His platoon enemies didn't care; their

772

thoughts were darker. That gung-ho bastard had it coming.

Doc was whispering in Snake's ear; sounding remarkably like an EkSeet to him. "You're gonna be okay. The bad news is you're torn up. The great news is you're going back to the land of the big PX."

Snake swore doc was talking about the EkTsab Burseeosil, where the ArnaMals would fix him up, and then throw him back into battle. Something Snake did not want; he wanted out of the lethal craziness. The morphine had taken effect, but he was still conscious.

"Don't look so grim, marine!" Doc shouted above the sound of the Dust-off helicopter hovering. "I don't think you'll lose your foot."

"Piss on the gods and you be fucked." Snake recited.

Doc was repacking his medical bag. "The guys I've seen, hit like you, get a free pass back to the States. You'll tell stories about your limp until you reach a ripe old age. Enjoy the rest of your life!"

Snake was in transition; everything he was experiencing had an Oval universe feel to it. His chemical high from the morphine shot doc had given him felt like AnticArna.

Incoherently Snake mumbled. "I never seen a medical ArnaMal yet that knew what the hell they were talking about."

Doc was confused. "You see an animal?" He slapped Snake's shoulder. "You'll live. Hang in there, dude. You'll never be on a battlefield again."

Surprisingly, in Snake's drugged mind, the black Navy Corpsman resembled XibEk. He grinned at that thought. I gotta be someplace else. XibEk would never be concerned about any goddam Lifer.

Doc moved out of the crater, joining up with

773

Staff Sergeant Huggy and the Lieutenant, to oversee the evacuation on the perimeter for the Dust-Off.

A couple marines wrapped Snake in a poncho, carrying him behind doc and Huggy.

Snake was chuckling quietly, floating in his chemical-induced high, pleased with himself. Those individual Uayeb Bumps rapidly ran through his mind. All those Bots, the battles, the screeching of the AkSilk straining against the Toobs as they had dropped to an invasion planet; the sights and sounds came back to him in a grotesque speeding montage.

Snake wondered. <u>Which KrutChan am I? The 'new' KrutChan PacalMo killed? Or the 'old' KrutChan who faded into nothing from the Osil effect?</u>

He decided quickly. Probably the latter.

Ord had learned 'spooky action at a distance' duality and branching timelines behaved randomly.

He had arrived back in his past. But not 'his old' past. Another past...another dimension...another timeline...another THEN.

As he drifted on that wonderful cloud of chemical consciousness, he was laughing at the irony. He thought of Scuz being spit out into that other side of the Seech field. He was wondering what the hell kind of future Scuz would endure on Arna. Snake silently wished him well. He hoped Scuz would be Soothed by another Reela and loved by a similar ZocKuk. Snake decided Scuz would be okay. At least for a while...until he Bumped on his first Uayeb Invasion.

Snake was ready to live in this renewed earth universe. He would never be able to explain it to anyone, even himself, as his past being observed, was now moving his present towards his future. He did not give a crap about his THEN, his NOW, or his

WHEN, anymore. That was now Scuz's problem, not his.

Minutes later, Snake gazed out of the Dust-Off chopper as it lurched on lifting off. He dimly realized ZocKuk was in another Universe, another Time. He had this Earth Universe to live in now, and a future wife he hoped he would meet in his New Timeline. His heart soared because 'she' was all he ever wanted in any of his timelines.

In a short span of time, he passed out on the floor of the chopper. When he regained consciousness in Charlie Med, Snake could not recall planet Arna. His consciousness forgetting all the Ovals, Krils, Dagots, Cunacks, Zars, Dinarchy, and his friends lives that were lost.

In the hospital, the old man was remembering. His ghost and demon nightmares during his night sweats were blurry vivid. The Dinarchy, Kravid, KrutEk, XibEk, CheChun, KrutCheebel, and the alien Zars, Cunacks, and the Dagots. Unconsciously returning in the night as frightening 'unrecognizable' demon visions. His information-memories over years survive by duality.

Eventually he recalled, while tossing in sleep, those unrecognizable ghosts who helped him. Along with what he experienced in the Kril Legions; fragmented and forever-cloudy unsolvable enigmas. Reela, Akna, Jaguar, Hortim, Xmucane, and ZocKuk appeared; but they would be flitting spirits; momentary fictions.

Over the years he would now live, the faded memories would flicker in his mind, in his dreams, or screech back in his sweat-soaked nightmares making no sense.

And when interested people asked him, during

775

his new future lifetime, "What was it like in Vietnam?" He would smile and say, "Never mind. Can't explain." Seeing their disappointment at his answer, after a pause, he would truthfully add. "It was a different universe."

Seventy years after Vietnam, old man Ord was dying. Every orifice of his body filled with rubber-coated wires transmitting readings, supplying oxygen and nutrients, and hollow tubes eliminating waste. Synchronizing with plastic patches attached to his skin, adding more monitors.

After three lifetimes, I found my 'Greer' in my third timeline. She was everything I dreamed she would be. We experienced my third life, in tandem, as a pair.

After marrying her, Ord had moved seamlessly from being an 'I', evolving into a 'we' bond. Their lifetime together was happy, but not easy.

His lover had helped in his third timeline to cope with his rage. Regretfully, taking his war-anger out on her and the kids. Rising to the surface during depressing times. But their love endured. Understanding a little after he apologized. Their persistent love modified his angst.

I remember the things I should have; but never said. Before she died, I regret the things I never told her. When I should have listened to what she was asking of me. The things I didn't do that could have made her and my four children happier. In retrospect, living together creates a lot of small nagging scars, slights, and regrets to swallow, and happiness.

Our love made me forget about the dying universes demand. After she died, I fantasized she ended up on Arna with the Ovals where she would be

776

more appreciated.  She would have been a great Oval.

Now he wanted to forever cease to exist.  He had his loves, had his lifetimes.  Perfection is an unattainable goal.  Waiting for his last breath, Ord realized Universes are indifferent; too busy creating, destroying, and then recreating life.

Love lasted through the heartaches, the hates, the pains, the disappointments, and the fleeting happiness.  Losing his true love to death, he had a lot of regret for things-left-unsaid.  He had been right a few times.  His choices were wrong, too many goddam times.

Am I sorry for how I lived my lives?  No. Regrets can't change history.  Timeline choices change lives second by second and eventually end. But not always.  I've survived three times.  I know.

When death arrived, Ord's floating-exploding-colored-lights absorbed his body.  He hoped forever.

In a Planck instant of time, he met oblivion, he hoped.  Boson duality…and time is fickle.

**Alpha Beginning**

When writing the Kril story I spent a lot of my timeline in my fictionalized Oval Universe.  I owe a lot of the characters created to my faded memories of people in this Universe.

To the intelligent women portrayed in this story; sinners and saints alike; coping within a world order not of their making; they weren't alone.  The women Sinners spend their lives mimicking male hate. The Saints among women were tough; frustrated while trying to change male physical and mental cruelty. Basically, Kril is about gender hate and love; two sides of life. Kril honors all the past, present and future male and female Veterans who survive combat. Hopefully, not angry at me sharing my night-riding ghosts and demons.

This Kril book thanks the few Veterans in my life, hating my guts for good and not-so-good reasons, and making me stronger. Those miserable folks are reflected in the characters I created populating Kril, strengthening the story as it should in any journey in life.

If wars ever stop becoming a grisly solution…cynically, there's little chance of that happening; my empathy is for future broken-minds of our descendants surviving combat.

I sincerely hope those Veterans eventually find the nonjudgmental love I did.

This book is intended for informational and educational purposes only. While every effort has been made to ensure the accuracy of the information contained in this book, the author makes no representations or warranties of any kind with respect to the completeness or accuracy of the contents and assumes no responsibility for errors or omissions.

The techniques and suggestions described in this book are designed to help improve typing skills and confidence. Results may vary depending on individual effort and practice.

 # DISCLAIMER

This book is intended for informational and educational purposes only.

The author has made every effort to provide accurate, clear, and practical guidance to help improve typing skills. However, the information in this book is provided "as is" without any guarantees of completeness, accuracy, or specific results. Individual progress will vary based on practice, consistency, and personal ability.

This book is not intended to replace professional advice, technical support, or formal training. Readers are encouraged to use their own judgment and adapt the suggestions to suit their personal needs and comfort levels.

While the exercises and techniques in this book are designed to be simple and safe, readers should take regular breaks and avoid strain or discomfort when using a keyboard or computer. The author is not responsible for any injury, loss, or damage that may occur as a result of using the information in this book.

By using this book, you acknowledge and accept full responsibility for your actions and results.

# TABLE OF CONTENTS

# 01

## Getting Comfortable with the Keyboard

If typing has ever felt slow, tiring, or even a little frustrating, you're not alone. For many seniors, the keyboard wasn't something you grew up using every day, so it can feel unfamiliar or even awkward at first.

But here's the truth—typing is not about age. It's about comfort and familiarity. Once your hands begin to understand where the keys are, typing becomes smoother, easier and far less stressful.

So instead of rushing, we'll start by making you comfortable.

First, let's look at how you sit.

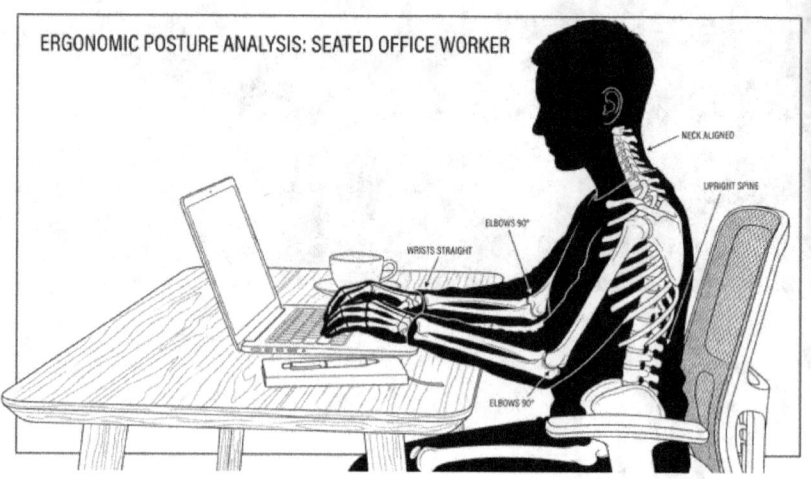

Your posture plays a bigger role than most people realize. Sit in a chair that supports your back. Keep your back upright, but not stiff. Let your shoulders stay relaxed—not raised or tight. Place the keyboard directly in front of you, not off to one side.

Your arms should feel natural, not stretched. Keep your elbows close to your body, bent at a comfortable angle. Your wrists should stay straight—not bent upward or downward. If your wrists are uncomfortable, typing will quickly become tiring.

Take a moment to adjust your position before you start. It makes a real difference.

Now, your hands.

Place your fingers gently on the keyboard. Your left hand rests on A, S, D, F, and your right hand rests on J, K, L, ;.

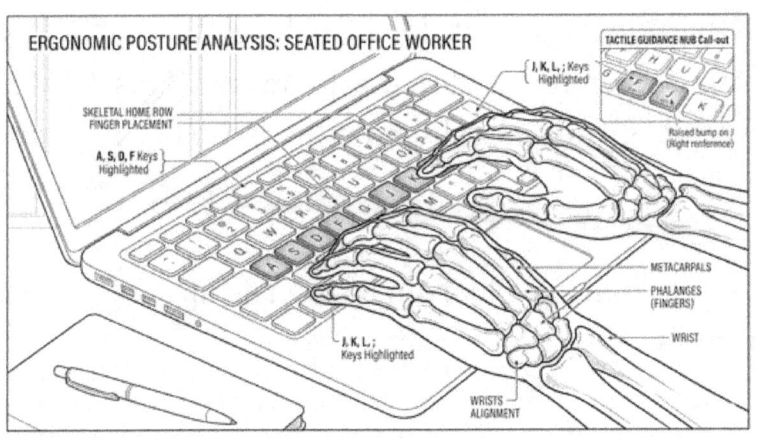

This is your **"home position."** Think of it as your resting point, where your fingers return after pressing any key.

You don't need to memorize anything complicated. Just remember this position and come back to it often.

You'll also notice small raised bumps on the F and J keys.

These are there to guide your fingers without needing to look down. Once you get used to them, they become a quiet helper that keeps your hands in the right place.

At this stage, forget about speed.
What you want is control and comfort.

Press one key at a time, slowly. Let your fingers move with intention, not in a rush. After each key, return your finger to its starting position. This may feel slow, even a bit unnatural—but this is how your hands begin to learn.

Think of it like learning to write neatly before writing quickly.

Another important habit is where you focus your eyes. Try to look at the screen more than the keyboard. It's fine to glance down when you need to, but don't rely on it too much. The goal is to slowly train your fingers to find keys on their own.

This is what eventually allows you to type without constantly stopping.
Now, let's talk about a common mistake.

**KEYBOARD HOME ROW TACTILE MAP: F and J GUIDANCE**

TACTILE GUIDANCE NUB Call-out

LEFT INDEX FINGER POSITION (F Key)

**TACTILE REFERENCE POINTS (GUIDANCE NUBS)**

RIGHT INDEX FINGER POSITION (J Key)

TACTILE GUIDANCE NUB Call-

Raised linear guide on F (Left Index Reference)

Raised bump (Right refere

PROPER HAND ANGLE (MAINTAINED)

PROPER HAND ANGLE (MAINTAINED)

Many people try to type fast too early. They push themselves to go quicker before they are ready. This often leads to more errors, more frustration, and sometimes the feeling that typing is just "not for them."

That's not true.
Typing improves when you slow down and build it properly. Accuracy first, then speed. If you get the first part right, the second part will come naturally.

You also don't need to spend hours practicing. In fact, shorter sessions work better. A few focused minutes each day will help you improve more than a long, tiring session once in a while.
Keep it simple and consistent.

Before you move on, here's a short exercise to help you get started:

- Place your fingers on A, S, D, F and J, K, L, ;
- Press one key at a time slowly
- Return your fingers to the starting position after each press
- Keep your shoulders relaxed and your wrists straight
- Try to look at the screen more than the keyboard

Do this for just a few minutes.
That's enough.

The goal here isn't perfection. It's getting used to the keyboard in a calm, steady way. As your hands begin to feel more comfortable, typing will stop feeling like hard work.
And once that happens, improving your speed becomes much easier than you might expect.

# 02

## Breaking the Hunt-and-Peck Habit

If you've been typing by looking down at the keyboard and using one or two fingers, you're not doing anything "wrong"—you're just using a method that slows you down.

This is called the hunt-and-peck habit. You look for each key, press it, then look again for the next one. It works, but it takes time, breaks your focus, and makes typing feel like hard work.

The goal of this chapter is simple: help you rely less on your eyes and more on your fingers.

Not all at once. Just gradually.

Let's start with a small shift in how you think about typing.

Instead of searching for keys every time, you want your fingers to begin recognizing where they are. This is called muscle memory. It simply means your hands remember positions through practice, even if your mind isn't actively thinking about each key.

And yes—it works at any age.

The first step is to trust your starting position. Place your fingers on **A, S, D, F** and **J, K, L, ;** just like you learned in Chapter 1. This is your "home base." Every movement begins here and returns here.

Now, here's where things change.

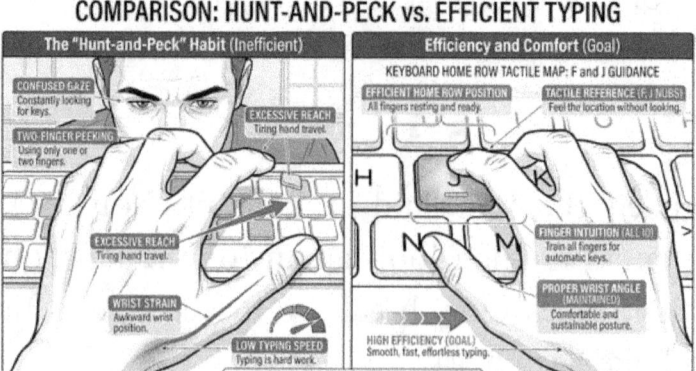

COMPARISON: HUNT-AND-PECK vs. EFFICIENT TYPING

When you type, try not to rush to look down immediately. Press a key based on where you think it is. If you get it right, good. If you get it wrong, that's also fine—you're learning.

This might feel uncomfortable at first. You may make more mistakes than usual. That's expected.
What matters is that your fingers are starting to figure things out.

A helpful trick is to reduce how often you look at the keyboard. You don't have to stop completely. Just pause for a moment before looking down. Give your fingers a chance to try first.

Another simple method is to lightly cover part of your keyboard with a piece of paper or cloth while practicing. Not all the time—just for a few minutes. This encourages your fingers to rely less on sight and more on memory.

Now, let's talk about movement.

Instead of using just one or two fingers, begin to involve more of your fingers. Each finger has a small area of keys it can handle. You don't need to learn this perfectly right now—just start using more fingers instead of relying on one.

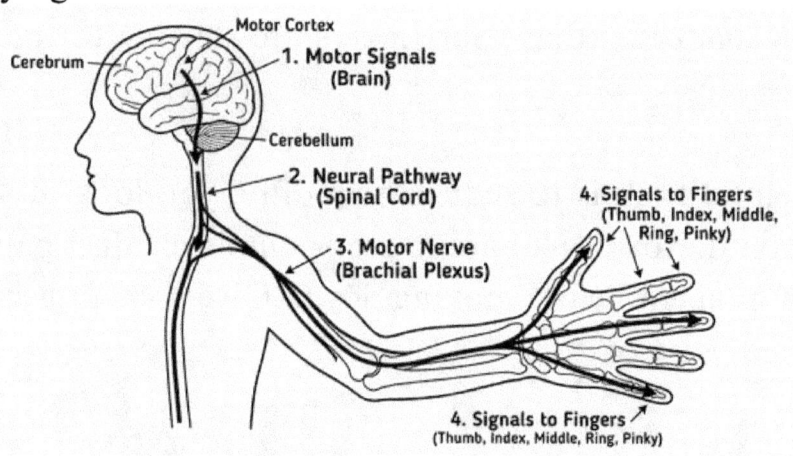

For example:
- Your index fingers can handle keys close to the center
- Your other fingers can begin to take small roles on nearby keys

It won't feel smooth immediately, and that's okay.

Progress here is not about perfection—it's about *less dependence on looking down.*

Another thing to keep in mind: slow typing is better than rushed typing. When you slow down, your brain and hands have time to connect. When you rush, you go back to guessing and looking down more often.

So give yourself permission to go at a calm pace.

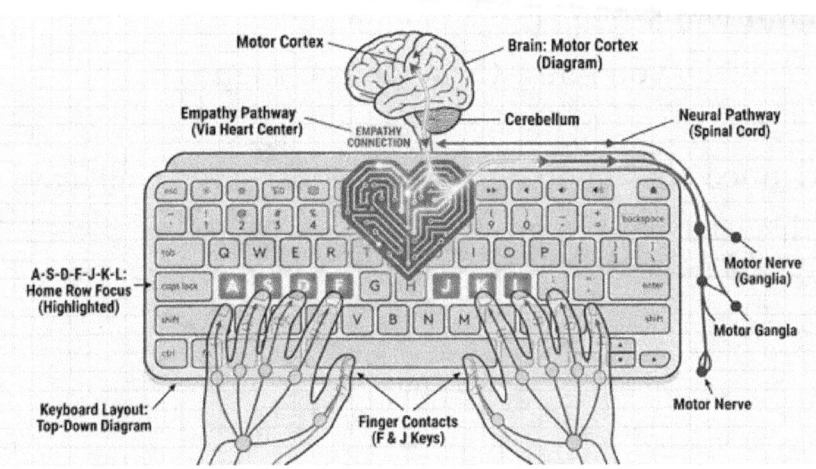

Here's a short exercise you can try:
- Place your fingers on the home keys
- Type simple letters (like A, S, D, F, J, K, L) without looking down first
- If you make a mistake, correct it and continue
- Try short words like "sad," "ask," or "fall"
- Keep your eyes on the screen as much as possible

Practice this for a few minutes each day.

You don't need to get everything right. You just need to get a little better at trusting your fingers.
Over time, something interesting will happen.
You'll look down less. Your hands will move with more confidence. And typing will start to feel smoother—without you forcing it.
That's when you know the habit is changing.

And once this habit improves, your speed will have a much easier path to grow.

## The 10-Minute Daily Practice Plan

By now, you've started getting comfortable with the keyboard and relying less on looking down. That's a strong foundation.

But here's what truly makes the difference: how you practice.
You don't need long hours. You don't need complicated tools. What you need is a simple routine you can follow every day without stress.

This chapter gives you exactly that—a clear 10-minute daily plan that helps you improve steadily.
Let's keep it easy.

## *Why 10 Minutes Works*

It's short enough that you won't feel tired or overwhelmed. At the same time, it's long enough to build real progress if you stay consistent.

Ten minutes a day is far better than one long session once in a while.
The goal is not to do more. It's to do it regularly.

*Your Simple 10-Minute Routine*

You can follow this plan anytime that suits you—morning, afternoon, or evening.

## Minute 1–2: Get Settled

- Sit comfortably with good posture
- Place your fingers on the home keys (A, S, D, F and J, K, L, ;)
- Take a moment to relax your shoulders and hands

This helps you start calmly, not in a rush.

## Minute 3–5: Slow and Accurate Typing

- Type simple letters (A, S, D, F, J, K, L)
- Move to short words like "ask," "sad," "fall," "dad"
- Keep your typing slow and controlled

Focus on pressing the right keys, not typing quickly.

## Minute 6–8: Easy Word Practice

- Try slightly longer words like "desk," "lad," "flask," "salad"
- Keep your eyes on the screen as much as possible
- Use more than one or two fingers

If you make mistakes, correct them calmly and continue.

**Minute 9–10: Light Challenge**

- Type a short sentence like:
- "I can learn to type better every day."
- Don't rush—just stay steady
- Notice how your fingers move

This helps you connect letters into real typing.

**What to Focus On Each Day**

Keep these three things in mind:

- **Accuracy first** – getting the right keys matters more than speed
- **Relaxed hands** – avoid tension in your fingers and shoulders
- **Consistency** – a little every day beats doing too much once

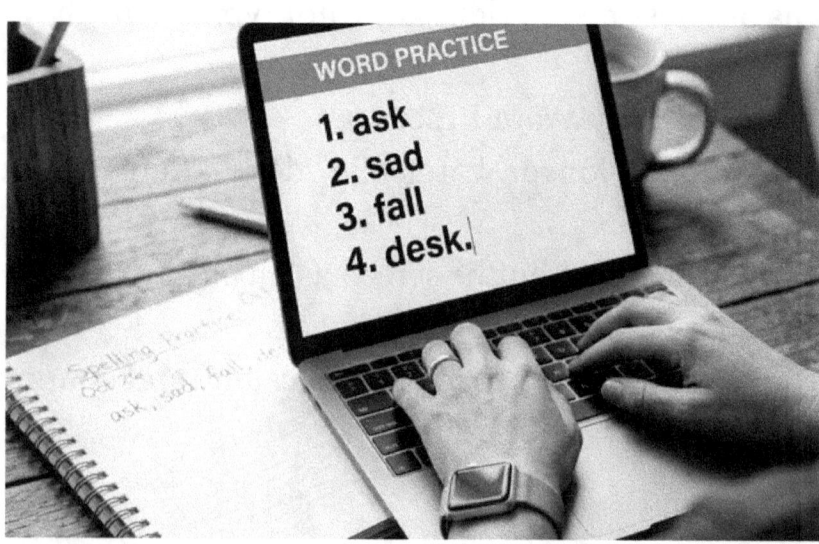

## *What Progress Will Feel Like*

At first, it may seem slow. You might still look down often or make mistakes.

Then, little by little:

- You'll hesitate less
- Your fingers will move more naturally
- You'll need to look down less often

These are signs that you're improving—even if it doesn't feel fast yet.

## *A Small but Important Reminder*

Don't judge your progress too quickly.

Typing is a skill that builds quietly. You may not notice big changes in one day, but after a week or two of consistent practice, the difference becomes clear.

So keep it simple. Stick to the plan.
Ten minutes. No pressure. No rushing.
Just steady practice.

And before long, typing will start to feel less like effort —and more like something you can do with ease.

# Typing Faster Without Mistakes

At this point, you may be wondering, "When do I actually start typing faster?"

It's a fair question. But here's something important to understand:
Speed doesn't come from trying to be fast. It comes from being accurate first.
If you try to rush, your fingers get confused, mistakes increase, and you end up slowing down even more.

But when you focus on getting the keys right, your hands begin to move smoothly—and that's where real speed begins.

So in this chapter, the goal is simple: help you increase speed the right way, without pressure and without constant mistakes.

### *Step One: Stay Accurate, Even If It Feels Slow*

It might feel like you're typing too slowly. That's okay. Typing slowly but correctly trains your fingers to move with confidence. Each correct movement builds memory in your hands. Over time, those movements become quicker—without you forcing them.

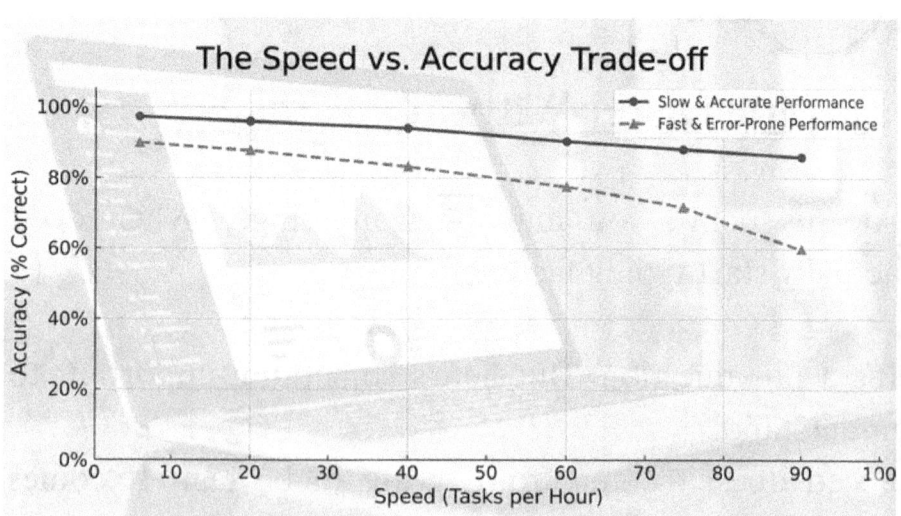

If you notice yourself rushing and making errors, pause and slow down again. It's not a setback. It's part of the process.

### Step Two: Let Speed Grow Naturally

As you keep practicing daily, something subtle begins to happen.

You stop thinking about every key. Your fingers start moving more freely. Short words become easier. Then sentences.

This is where speed begins to grow—quietly and naturally.

You don't need to push it. Just allow it to happen.

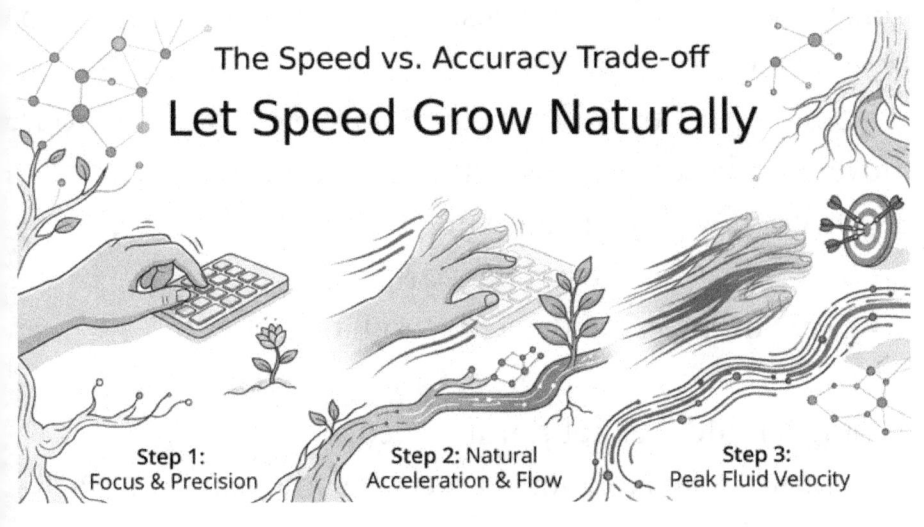

The Speed vs. Accuracy Trade-off
**Let Speed Grow Naturally**

Step 1:
Focus & Precision

Step 2: Natural
Acceleration & Flow

Step 3:
Peak Fluid Velocity

## *Step Three: Fix Mistakes the Right Way*

Mistakes are normal. What matters is how you handle them.

When you press the wrong key:

- Don't panic
- Don't rush to "cover it up"
- Simply correct it and continue

This helps your brain and fingers learn the right movement instead of repeating the wrong one.

Ignoring mistakes or typing over them too quickly can slow your progress.

## *Step Four: Keep Your Hands Relaxed*

Tension is one of the biggest reasons people struggle to type faster.

If your fingers are stiff or your shoulders are tight, your movement becomes slower and less accurate.

So check in with yourself while typing:

- Are your shoulders relaxed?
- Are you pressing the keys too hard?

Typing should feel light, not forced.

The more relaxed your hands are, the easier it is to move quickly.

### *Step Five: Use Short, Familiar Practice*

Instead of jumping into long or difficult passages, stick with short, familiar words and sentences.

For example:

- "I am getting better at typing."
- "This feels easier than before."

These simple lines help you build rhythm and confidence.

As your comfort improves, your speed will naturally follow.

### *A Simple Speed Exercise*

Try this for a few minutes:

- Type a short sentence slowly and correctly
- Repeat the same sentence 3–4 times
- Each time, let your fingers move a little more freely
- Do not force speed—just reduce hesitation

You'll notice that each round feels slightly smoother than the last.

## *What Real Progress Looks Like*

Typing faster doesn't happen all at once.

Instead, you'll notice:

- Fewer pauses between letters
- Less need to look at the keyboard
- Smoother movement from one key to another

That's real improvement.

## *Final Thought for This Chapter*

You don't need to chase speed.

Build accuracy. Stay relaxed. Practice daily.

Speed will come on its own—and when it does, it will feel natural, not forced.

And that's the kind of progress that lasts.

# 05

## Staying Consistent and Seeing Real Progress

You've learned how to position your hands, reduce looking down, practice daily, and improve your accuracy. Now comes the part that truly makes everything work:

Staying consistent.
This is where many people struggle—not because typing is too hard, but because they stop too soon or practice

without a clear approach. The truth is simple: small, steady effort brings real results.

### *Make Typing Part of Your Day*

You don't need to set aside a large block of time. In fact, it's better if you don't.

Stick to your 10-minute daily practice. Choose a time that fits naturally into your routine—maybe in the morning, after lunch, or in the evening when things are quiet.

The easier it is to fit into your day, the more likely you are to keep doing it.

Think of it like brushing your teeth. You don't skip it because it's short—you do it because it's part of your routine.

Typing can become just as natural.

### *Don't Worry About Perfection*

Some days will feel better than others.

There will be moments when your fingers move smoothly, and other times when you make more mistakes than usual. That's normal. It doesn't mean you're not improving.

Progress in typing is not always obvious day by day— but it builds over time.
Instead of aiming for perfection, aim for showing up and practicing.
That's what makes the difference.

### *Notice the Small Improvements*

Real progress often comes in small ways:
- You pause less before pressing a key
- You look down at the keyboard less often
- You type familiar words more easily
- Your hands feel more relaxed

These changes may seem minor, but they are signs that your skills are improving.
Take a moment to notice them. It helps you stay encouraged.

## *Fix Common Problems Early*

If something feels off, don't ignore it. Small problems are easier to fix when you catch them early.

Here are a few common ones:
- Typing too fast and making mistakes
- Slow down. Accuracy still comes first.
- Looking down too often
- Give your fingers a chance before checking the keyboard.
- Feeling tension in your hands or shoulders
- Pause, relax, and adjust your posture.
- Using only one or two fingers
- Gently involve more fingers, even if it feels slow at first.

Simple adjustments like these can quickly get you back on track.

### *Keep Practice Simple and Enjoyable*

Typing doesn't have to feel like a task.

You can make it more enjoyable by practicing with things you like:

- Type short messages or notes
- Practice with simple sentences you enjoy reading
- Even type out a short paragraph from a book or message

The more relaxed and interested you feel, the easier it is to stay consistent.

### *Be Patient With Yourself*

Learning any new skill takes time, and typing is no different.

You're not competing with anyone. You're simply improving your own comfort and ability.

Some people may learn faster, others slower—but what matters is that you keep going.

Even a little progress each day adds up more than you might expect.

## *What You Can Expect Over Time*

If you continue with your daily practice, you'll begin to notice:

- Your fingers move with less effort
- You type with fewer mistakes
- You feel more confident using the keyboard
- Simple tasks like writing messages or emails become easier

And most importantly, typing will no longer feel like a struggle.

## *A Simple Weekly Check*

At the end of each week, take a minute to ask yourself:

- Am I more comfortable than last week?
- Am I looking down less?
- Do my hands feel more relaxed?

If the answer is yes—even slightly—you're on the right track.

# *FINAL THOUGHT*

You don't need to do anything complicated to succeed at this.

Just stay consistent, keep your practice simple, and focus on small improvements.

Typing faster isn't about forcing speed, it's about building comfort over time. And once that comfort settles in, everything else follows.

Give yourself time to grow into it. Some days will feel smooth, others may feel slow, but both are part of the same progress. What matters is that you keep going, even if it's just for a few minutes.

The more you practice, the more natural it becomes. Your fingers will begin to move without hesitation. You'll spend less time thinking about each key, and more time simply typing what you want to say.

And that's the real goal, not just speed, but ease.

A point where typing no longer feels like effort, but something you can do calmly and confidently whenever you need it.

Stay patient with yourself. Keep it simple. And trust the process.

It works.

# ABOUT THE AUTHOR

 Karen Howard has always had a quiet interest in helping older adults feel more comfortable with everyday technology, not through complicated instructions but through simple, patient guidance that is effective

Too often, they were left to figure things out on their own, pressing keys slowly, looking down constantly, and feeling a bit frustrated without knowing why.

This book grew from that understanding. Karen didn't set out to create a technical manual. Instead, she focused on something more practical, breaking **"typing"** down into small, clear steps that feel natural to follow. No pressure. No confusing terms.

Through this guide, her aim is simple: to help you move from hesitation to ease. From slow, uncertain typing to something smoother, more familiar, and far less frustrating.

Because in the end, it's not just about typing faster.

It's about feeling comfortable doing it.

www.ingramcontent.com/pod-product-compliance
Lightning Source LLC
Chambersburg PA
CBHW060542190526
45337CB00023B/638